REHABILITATION MEDICINE

Third edition

REHABILITATION
MEDICINE

HOWARD A. RUSK, M.D.

Professor and Chairman of the Department of Rehabilitation Medicine,
New York University Medical Center, New York, N. Y.

With 36 collaborators

With the editorial assistance of

EUGENE J. TAYLOR, A.M.

With 231 figures and 24 charts

THE C. V. MOSBY COMPANY

Saint Louis   1971

COLLABORATORS

Present and past colleagues of Howard A. Rusk, M.D., in the Department of Rehabilitation Medicine, New York University Medical Center

Edward H. Bergofsky, M.D.

Ernst Bergmann, M.D.

Harold Brandaleone, M.D.

Sophia Chiotelis, M.A., O.T.R.

Ellen Coffman, R.N.

Donald A. Covalt, M.D.

Michael M. Dacso, M.D.

J. Herbert Dietz, M.D.

Leonard Diller, Ph.D.

Daniel J. Feldman, M.D.

Saul H. Fisher, M.D.

Menard Gertler, M.D.

Joseph Goodgold, M.D.

Leon Greenspan, M.D.

Albert Haas, M.D.

Jack Hofkosh, M.A., R.P.T.

Nancy Kester, M.D.

Edith Buchwald Lawton, M.A.,
R.P.T.

Mathew H. M. Lee, M.D.

Edward W. Lowman, M.D.

Hans R. Lehneis, B.S., C.P.O.

Morton Marks, M.D.

Martin E. McCavitt, Ed.D.

Eugene Moskowitz, M.D.

Joseph Novey, M.S.W., Ed.D.

Ann M. Powers, M.S.W.

Allen S. Russek, L.R.C.P.S. (Edin.)

Martha Taylor Sarno, M.A.

Meyer S. Siegel, M.A.

James F. Sullivan, M.D.

Samuel S. Sverdlik, M.D.

Chester Swinyard, M.D.

Eugene J. Taylor, A.M.

Henry Taylor, M.D.

John Untereker, M.D.

Muriel E. Zimmerman, M.A.,
O.T.R.

With the assistance of

Associates in the Department of Rehabilitation Medicine, New York University Medical Center

PREFACE

When the first edition of this book appeared in 1958, we expressed the hope that it would serve as a basic elementary textbook. We also hoped that it would prove to be a useful reference to all physicians whether they were in general practice, specializing in rehabilitation medicine, or practicing one of the other medical specialties.

We further expressed the opinion that a majority of the medical rehabilitation procedures could and should be done by the practitioner or specialist responsible for the patient's primary medical care and that these procedures should be an integral part of such medical care. We held that this approach was essential if any substantial gains were to be made in preventing further deterioration and increased disability among the great majority of our sick and injured, for it is in the physicians' offices and general hospitals that the overwhelming percentage of our medical care is given.

We stressed, however, that over and above this group of patients are a number with severe disabilities—paraplegia, quadriplegia, severe hemiplegia with aphasia, the demyelinating diseases, and some of the other clinical problems—whose needs could not be met without a concentrated long-term program that included full utilization of the allied health disciplines and specialized rehabilitation technics.

We noted that it was obvious that no medical practitioner, general or specialist, could have at his disposal for the care of his patients all of the varied facilities, equipment, specialized skills, and allied health professional personnel needed for the management of such patients.

Now, thirteen years later, we believe these comments are as valid as when they were made in 1958. We feel this book has served its purpose as a basic elementary textbook for physicians specializing in rehabilitation medicine and as a useful reference for all physicians. We believe this has been borne out by the fact that a second edition in English was published in 1964 and that there have been two editions in Spanish, one edition in Japanese, and one edition in Croatian.

In the second edition of this book a number of important revisions were made to reflect the advances in scientific medicine, including genetics, which had been so extensive. Again, in the past seven years there have been similar scientific advances, which are reflected in this new edition. Major among these revisions are greatly expanded and rewritten sections on the rehabilitation of patients with cardio-vascular disease, cerebrovascular disease, cancer, and chronic obstructive lung disease and sections on amputee management and on the externally powered orthoses and prostheses.

As in the first and second editions, all the photographs in this book were taken by Mr. Stanley Simmons, New York University Medical Center. The artwork was done by Dr. Seymour Fradin and Mr. Robert Wilson. The editorial work was done by Mr. Eugene J. Taylor and me. All the contributors are current or former members of the faculty of the Department of Rehabilitation Medicine, New York University School of Medicine. We wish to express particular gratitude to Darrel J. Mase, Ph.D., Dean, College of Health Related Professions, J. Hillis Miller Health Center, University of Florida, for his assistance in certain revisions on the chapter on speech rehabilitation. We also wish to express our gratitude to Miss Rose Elfinbein for her editorial assistance in the preparation of the manuscript.

Just as rehabilitation services for the individual patient require the "team ap-proach," this book is a product of teamwork. The names of some members of the team appear as authors; the others are associates—medical and allied health pro-fessionals—and particularly the patients with whom we have worked during the years in developing the concepts and technics discussed herein.

Howard A. Rusk, M.D.

CONTENTS

REHABILITATION MEDICINE

CHAPTER 1

THE PHILOSOPHY AND NEED OF REHABILITATION

Two decades ago a great majority of the medical profession looked on rehabilitation as an extracurricular, adjunct activity of medicine, something dealing with social work and vocational training but something that had little concern with or held but few implications for medicine.

That trend has been reversed over the past 20 years, and today the majority of physicians, particularly those who have completed their training within recent years, now are familiar with the aims and procedures of rehabilitation and recognize that medical care cannot be considered complete until the patient with the residual physical disability has been trained "to live and to work with what he has left."

When the Rehabilitation Medicine Service (then Department of Physical Medicine and Rehabilitation) was created in New York City's Bellevue Hospital in 1946, it was the first rehabilitation medicine service, with its own personnel and beds, of any civilian hospital in the United States, and probably in the world. During the intervening years an increasing number of hospitals have recognized the importance of organized rehabilitation medicine services with an allocation of hospital beds.

Such services in hospitals, however, developed slowly in the decade following the creation of the first such program at Bellevue Hospital. There was a more rapid growth during that period of independent "rehabilitation centers." In many instances these centers were started by local affiliates of the National Easter Seal Society and United Cerebral Palsy, and were limited to the care of orthopedically handicapped children.

Most of these centers were independent, public, or nonprofit institutions designed to serve all the physicians and all the hospitals in the community, but had no direct affiliation with hospitals. Such institutions have provided valuable and needed patient care services within the community. In addition, they have contributed greatly to the recognition of the need and value of rehabilitation services by the medical profession, the public, and the hospitals.

The number of independent medical "rehabilitation centers" has not increased substantially within the past decade. The trend instead has been to focus expansion of rehabilitation services in general hospitals with the recognition that if any major attack was to be made on the problems of disability, it had to be made within the general hospitals to which practicing physicians turn for the care of their patients.

This growth also gave more recognition to the fact that within the general hospitals such services can be brought to the patient at the earliest possible time and the incidental and damaging physical, emotional, social, and vocational sequelae of the acute disease process could be alleviated or minimized. There was a growing recognition during this period that to ignore the development of rehabilitation within the general hospitals was to guarantee the continued deterioration of many of the less severely disabled persons until they, too, reached the severely disabled and totally dependent category.

It was recognized that to ignore a disability is far more costly than an early aggressive program of rehabilitation that restores the individual to the highest possible level of physical, economic, social, and emotional self-sufficiency.

Experience has shown that rehabilitation services within the hospital can best be provided by the organization of a "bed service" that has a relationship to the other hospital services similar to that of the x-ray and laboratory divisions. It is a service program of all of the other specialty services within the hospital.

STATE-FEDERAL VOCATIONAL REHABILITATION PROGRAM

Although the United States has had a state-federal program of vocational rehabilitation since 1920, this program was extremely limited in the number of persons served and the scope of services provided until 1943, when the law was amended to permit the provision of medical and medical rehabilitation services.

Prior to that time this program was limited to guidance and vocational training of the handicapped. Ironically, although a fairly substantial sum of money could be spent on vocational training for the disabled, such training had to be "training around the disability," for no funds could be spent on eliminating, alleviating, or reducing the effect of the disability on the physical capacities of the client.

The program was substantially enlarged again in 1954, with increasing funds available each year since then for patient services and for research and training of professional personnel.

The number of persons rehabilitated into employment under this program rose from 55,825 in 1954 to around 250,000 in 1970. It has been documented over and over again that for every dollar spent in vocational rehabilitation services five dollars is returned by the disabled worker in federal income taxes alone in the first 5 years of his employment after rehabilitation.

Rehabilitation during the first quarter of the century was characterized by the development of orthopedic surgery, physical therapy, occupational therapy, and the state-federal vocational rehabilitation program, all outgrowths of World War I. Under the stimulus of World War II the second quarter of the centuy has brought a recognition that the problems medicine faces in meeting the growing incidence of chronic disease and disability extend far beyond any medical specialty; they are problems that must be faced and solved by medicine as a whole. Rehabilitation is every physician's business, not merely that of the specialist in rehabilitation medicine.

There must be specialized rehabilitation centers where the "team approach," drawing on the specialized skills of the physician, physical therapist, occupational therapist, social worker, speech and hearing therapist, nurse, vocational counselor, psychologist, and prosthetic specialist, can be used to meet the needs in the more difficult cases. Our nation, however, will never have enough of these specialized centers for the rehabilitation of all the patients needing services. We will not have the funds to build and equip such facilities, the trained personnel to staff them, or the funds to pay for service. Therefore, if patients are to benefit from present-day developments in rehabilitation, the concept of rehabilitation and the basic technics must be made a part of the medical programs of all of our hospitals. The concept of rehabilitation and the basic technics must also be made a part of the armamentarium of all physicians; for, regardless of the type of disability, the responsibility of the physician to his patient cannot end when the acute injury or illness has been cared for. Medical care is not complete until the patient has been trained to live and to work with what he has left.

REFERENCES

1. Krusen, F. H., Kottke, F. J., and Elwood, P. M.: Handbook of physical medicine and rehabilitation, Philadelphia, 1965, W. B. Saunders Co.
2. Lowman, E. W.: Symposium on rehabilitation, Med. Clin. N. Amer. 53:485, 1969.
3. Licht, S. H., editor: Rehabilitation and medicine, New Haven, Conn., 1968, E. Licht, Publisher.
4. Allan, W. S.: Rehabilitation: a community challenge, New York, 1958, John Wiley & Sons, Inc.
5. Kessler, H. H.: Rehabilitation of the physically handicapped, New York, 1953, Columbia University Press.
6. Department of Health, Education, and Welfare: Areawide planning of facilities for rehabilitation services, Public Health Service pub. no. 930-B-2, Washington, D. C., 1963, U. S. Government Printing Office.
7. Rusk, H. A., and Taylor, E. J.: New hope for the handicapped; the rehabilitation of the disabled from bed to job, New York, 1949, Harper & Brothers.
8. Salmon, F. C., and Salmon, C.: Rehabilitation center planning: an architectural guide, University Park, Pa., 1959, Pennsylvania State University Press.

CHAPTER 2

THE EVALUATION PROCESS

HISTORY TAKING

In the course of medical advance a standard procedure has been evolved to facilitate the documentation of the medical history of the individual patient. Increasing importance has been properly placed on this fundamental necessity, both diagnostically and therapeutically. No format, however, can be accepted as more than a guide. Individual patients and specialties have specific problems in medical history taking that can be met only by an appropriate amendment and modification of the standard history form. Although it is always indispensable to obtain an accurate general history, the physician facing a paraplegic patient will have many questions that would be of little or no value to a surgeon taking a history of a patient with an acute abdominal condition or to a psychiatrist trying to obtain the best possible background information from a schizophrenic patient.

This discussion will be limited to those unusual elements of the medical history, which, if properly explored, may give valuable information as to diagnosis and rehabilitation procedures, as well as furnish a key to social and vocational problems.

A patient who has either suddenly or gradually lost his integrity inevitably suffers from some degree of emotional instability. He appreciates sympathetic understanding, but almost invariably he rejects an ostentatious solicitude. The physician must recognize that patience and understanding are the most important factors in his first encounter with these patients. An accurate and logical history cannot always be obtained from patients who, because of the emotional sequelae of a catastrophic illness or because of their age, are not able to think and to relate their medical histories clearly. Patience, skill, and tact are of paramount importance in these situations. The physician must guide his patient, without unduly influencing or prejudicing him, to obtain the necessary information.

Structure of a medical history

Identifying data. Identifying data include the patient's name, age, address, vocation, etc.

Chief complaints. The patient's chief complaints should always be related in his own words. It is important to resist the temptation to translate the patient's description of complaints into technical terms, for example, shortness of breath to dyspnea, pain in the chest to precordial pain, lameness into paralysis, itching or burning into paresthesia. Such an interpretation inevitably prejudices the examiner and indeed prompts him to make a tentative diagnosis prematurely.

History of present illness. This portion of the medical history must be well documented, especially in accident or compensation cases, in which the possibility of a legal procedure is always present. Chronologic sequence of significant events should be carefully noted.

History of past illnesses. Sometimes what seem like unimportant episodes to the patient may provide valuable data. For example, a history of an unidentified febrile disease many years before will give the clue to the differential diagnosis in a case of parkinsonian syndrome in a patient in late middle age. Similarly, a long-forgotten small sore on the external genitalia decades before may explain a gradually developing gait disturbance.

Family history. The family history should contain all information as to the possible hereditary or familial nature of a disabling condition. It may also give the physician information relative to the possibility of restoring the disabled patient to his community.

Habits. The effects of smoking, alcoholism, and dietary idiosyncrasies in the production of certain disabling conditions must be borne in mind and thoroughly explored.

Vocational history. The vocational history can offer important medical information, for example, industrial poisoning in a peripheral nerve paralysis. It can also offer important social information. A very detailed vocational history is needed to facilitate vocational rehabilitation and selective placement in a job that the patient can hold over a period of time.

Psychosocial history. The psychosocial history is of paramount importance, and in complicated situations the initial history should be documented by the physician and then augmented by detailed documentation by the psychiatrist, psychologist, and social worker.

Systemic review. A systemic review is a recapitulation of the history according to organ systems. It is a most valuable aid to a clear understanding of the often complicated history. Both positive and negative findings are important.

PHYSICAL EXAMINATION

The physical examination used in rehabilitation medicine contains all essential elements of the examination used in clinical practice. However, because of the nature of specific problems in physical rehabilitation, strong emphasis is laid on an extensive neurologic and orthopedic survey. The guiding principle in the physical examination of a disabled patient is that, in addition to the conventional anatomic and pathologic data, all information concerning the patient's functional capacity must be obtained. The latter distinguishes the physical examination in restorative medicine from that in other clinical disciplines. The fundamentals of the physical examination of the disabled patient to be considered are general appearance, the head and neck, the cardiorespiratory system, the gastrointestinal system, the genitourinary system, and the neuromuscular and locomotor systems.

General appearance. General appearance includes a general description of the patient's body structure, nutritional status (for example, obesity), cooperation, and apparent motivation. A simple inspection by a trained examiner can reveal a great deal of important information concerning the physically disabled patient.

Head and neck. A detailed examination of the visual, auditory, and speech apparatus is particularly important in physically disabled patients.

Cardiorespiratory system. The heart and lungs bear the brunt of the burden imposed on the patient by the increased activity outlined in a program for rehabilitation. It is important, therefore, that, in addition to the patient's anatomic and pathologic status, his functional capacity also be thoroughly evaluated.

Gastrointestinal system. A complete survey, with special attention to the function of the sphincters, is the first step to successful rehabilitation of the incontinent patient.[1] In patients with a colostomy or an ileostomy, the condition of the intestinal soma and the surrounding skin requires close attention.

Genitourinary system. The physical and functional status of the genitourinary organs is of major importance in many central nervous system diseases, especially in paraplegia. In urinary incontinence some of the patient's individual methods to initiate and regulate urination may give important clues to their ultimate management.[2]

Neuromuscular and locomotor systems. In disabling diseases of the nervous system, this is the most important part of the examination. It should always include an analysis of the patient's gait pattern.

Gait. No physical examination in disorders of the locomotor system can be considered complete without an analysis of the patient's gait pattern. To recognize pathologic gait, the examiner must be familiar with the mechanism of normal gait.[3, 4] Locomotion is a cyclic activity. The gait cycle consists of a stance phase and a swing phase, which, by rhythmic alternate repetition, produce the pattern of locomotion.

The *stance phase* starts with the foot in forward position and the heel on the ground and ends when the toe is "pushed off" and the foot leaves the ground. At this point the *swing phase* begins and lasts until the heel again touches the ground. Actually these two phases are not of the same duration. The period of support (stance phase) is longer than the swing, and there is a brief period when both legs are on the ground simultaneously. This is the *period of double support.*

The human gait is a three-dimensional activity. This means that to achieve progression in one direction several forces producing three-dimensional displacement of the body segments must come into operation. The secret of good gait diagnosis hinges on the knowledge of all the minute elements of a normal gait pattern and a recognition of any variation, omission, or addition from this normal. It must be remembered that, although the legs play the most conspicuous role in locomotion, other body elements like the pelvis and trunk and even the upper extremities have a very important role in human gait. Consequently, in gait analysis all these elements must be carefully observed.

Pathologic gait is characterized by excessive or asymmetric movement of body elements and of the center of gravity. The diagnostic value of such common abnormal gait patterns as in hemiplegia, peroneal palsy, hip fusion, muscular dystrophies, and painful weight-bearing joints is well known. With a little practice, however, the examiner can learn to recognize many more variations from normal gait that can be of great help in making a diagnosis in locomotor diseases.

In most organic diseases the conventional physical examination methods will lead to an appropriate clinical diagnosis. Restorative medicine, however, is mainly concerned with patients who suffer from a locomotor or neuromuscular disability. Therefore the commonly used tests will not give the information needed for a correct disability diagnosis. In such conditions the extent of physical limitation and the degree of resultant functional impairment must be established. To satisfy this special demand, the physician dealing with the physical disabilities must, in addition to the routine diagnostic methods, include in his armamentarium such measures as testing muscle strength and range of motion of joints and evaluating ability to perform the activities of daily living (ADL).

Testing of muscle strength. The assessment of the impaired musculature is indispensable to a good disability diagnosis. Recognizing its practical significance, both the clinician and the physiologist have been working on methods that are accurate and yet simple enough for clinical use. As a result of such widespread efforts, a great number of laboratory methods have been designed. Unfortunately, most of them are too complicated and cumbersome, and they require more instruments than are feasible in clinical practice. Today only simple dynamometers and occasionally ergometers are used. Most of the time the physician relies on subjective evaluation of muscle strength. (See Chart 1.)

Although this method possesses all the weaknesses inherent in subjective technics, its application in the course of long years has lent it such a degree of accuracy and validity that its use in clinical situations can be safely recommended. This diagnostic method was first described by Lovett in 1916 for the testing of patients with poliomyelitis.[5, 6] His method briefly consists of applying various degrees of resistance to the muscles and grading them according to their ability to overcome the given degree of resistance. This test has undergone a great deal of technical improvement as a result of its widespread use.[7, 8]

Briefly the procedure is as follows: First the patient is placed in a test position, which obviously varies with the muscle or muscle group to be tested. Then he is asked to move his muscle through its full range of motion. According to the response the following grading system is used (Lovett):

> *Zero:* No contraction felt or seen.
> *Trace:* Muscle can be felt to tighten but cannot produce movement.
> *Poor:* Produces movement with gravity eliminated but cannot function against gravity.
> *Fair:* Can raise the part against gravity.
> *Good:* Can raise the part against outside resistance, as well as against gravity.
> *Normal:* Can overcome a greater amount of resistance than a "good" muscle.

With an introduction of intermediate plus and minus grades, for example, fair plus or poor minus, further refinement can be accomplished. The testing resistance in this method is applied manually by the examiner, and the elimination of gravity can be achieved through manual assistance, powder boards, counterbalanced pulleys, or other technics.

In addition to the grading scale just described, some others were suggested to

CHART 1

MUSCLE EXAMINATION

Patient's name _____ Chart No. _____
Date of birth _____ Name of institution _____
Date of onset _____ Attending physician _____ M. D.

Diagnosis:

	LEFT				MUSCLE					RIGHT
					Examiner's initials					
					Date					
NECK					Flexors					NECK
					Extensors					
TRUNK					Flexor					TRUNK
					Extensors—Thoracic					
					Extensors—Lumbar					
					R. ext. obl. / L. int. obl. }Rotators{ L. ext. obl. / R. int. obl.					
					Elevation of pelvis					
HIP					Flexors					HIP
					Extensors					
					Abductor					
					Adductors					
					External rotators					
					Internal rotators					
					Sartorius					
					Tensor fasciae latae					
KNEE					Flexor—Outer hamstring					KNEE
					Flexors—Inner hamstrings					
					Extensors					
ANKLE					Plantar-flexors—Gastroc. & soleus					ANKLE
					Plantar-flexor—Soleus					
FOOT					Invertor—Anterior tibial					FOOT
					Invertor—Posterior tibial					
					Evertor—Peroneus brevis					
					Evertor—Peroneus longus					
TOES (4 lateral)					Flexors—Metatarsophalangeal					TOES (4 lateral)
					Extensors—Metatarsophalangeal					
					Flexor—Proximal interphalangeal					
					Flexor—Distal interphalangeal					
					Abductors					
					Adductors					
HALLUX					Flexor—Metatarsophalangeal					HALLUX
					Flexor—Interphalangeal					
					Extensor—Interphalangeal					

Additional data:
Face _____
Speech _____
Swallowing _____
Diaphragm _____
Intercostals _____

KEY

100%	5	N	Normal	Complete range of motion against gravity with full resistance
75%	4	G	Good*	Complete range of motion against gravity with some resistance
50%	3	F	Fair*	Complete range of motion against gravity
25%	2	P	Poor*	Complete range of motion with gravity eliminated
10%	1	T	Trace	Evidence of slight contractility; no joint motion
0	0	0	Zero	No evidence of contractility
S or SS			Spasm	Spasm or severe spasm
C or CC			Contracture	Contracture or severe contracture

*Muscle spasm or contracture may limit range of motion. A question mark should be placed after the grading of a movement that is incomplete from this cause.

CHART 1, cont'd

LEFT						Center		RIGHT				
						Examiner's initials						
						Date						
SCAPULA						Abductor—Serratus anterior						SCAPULA
						Adductor—Middle trapezius						
						Adductors—Rhomboids						
						Elevators						
						Depressor						
SHOUL-DER						Flexors						SHOUL-DER
						Extensors						
						Abductors						
						Horizontal abductor						
						Horizontal adductor						
						External rotators						
						Internal rotators						
ELBOW						Flexors						ELBOW
						Extensors						
FORE-ARM						Supinators						FORE-ARM
						Pronators						
WRIST						Flexor—Radial deviation						WRIST
						Flexor—Ulnar deviation						
						Extensors—Radial deviation						
						Extensor—Ulnar deviation						
FINGERS					1	Flexor profundus digitorum	1					FINGERS
					2	Flexor profundus digitorum	2					
					3	Flexor profundus digitorum	3					
					4	Flexor profundus digitorum	4					
					1	Flexor sublimis digitorum	1					
					2	Flexor sublimis digitorum	2					
					3	Flexor sublimis digitorum	3					
					4	Flexor sublimis digitorum	4					
					1	Finger extensors	1					
					2	Finger extensors	2					
					3	Finger extensors	3					
					4	Finger extensors	4					
					1	Lumbricales	1					
					2	Lumbricales	2					
					3	Lumbricales	3					
					4	Lumbricales	4					
						Dorsal interossei						
						Palmar interossei						
THUMB						Opponens pollicis						THUMB
						Adductor pollicis						
						Abductor pollicis longus						
						Abductor pollicis brevis						
						Thumb extensors						
						Flexor pollicis longus						
						Flexor pollicis brevis						
						CONTRACTIONS AND DEFORMITIES						
						SHOULDER						
						ELBOW						
						FOREARM						
						WRIST						
						FINGERS						
						THUMB						

Date_____   Therapist_____

Table 2-1. Comparison of the most frequently used grading scales*

Grade	Percent	Lovett scale	
5	100	N	Normal
4	75	G	Good
3	50	F	Fair
2	25	P	Poor
1	10	T	Trace
0	0	0	Zero

S or SS, spasm; C or CC, contracture

*From the report of the Committee on After-Effect, National Foundation for Infantile Paralysis, 1946.

express the degree of muscle strength. Since some of them are presently used in practice, a comparative tabulation of equivalent values is given in Table 2-1.

For obvious reasons spastic muscles tested by this method will not give a reliable picture of their strength. In recognition of this fact, the presence and degree of spasticity is indicated in the patient's record by S or SS and the extent of contractures is indicated by C or CC.

Another method of testing muscle strength is to apply force to "break" a test position assumed by the patient.

The common weakness of clinical muscle tests is that they test a single performance of the muscle's power and therefore do not give information about its endurance. If such information is desired, the repetitive testing technic is advised. In judging the endurance of a muscle, the effect of psychologic factors should always be taken into consideration.

Testing the range of joint movement. In many diseases involving the musculoskeletal system, the range in which the joint is able to move is impaired. Since the degree of impairment has an important bearing on the patient's functional capacities, it is essential to have reliable methods available for its measurement.

The subjective assessment of the range of motion should be used only for gross orientation; if any limitation is detected, one of the many objective methods should be used.[9, 10]

The simplest instrument for measuring range of motion is the goniometer. It consists of two shafts joined by a bolt, allowing free movement around a complete circle. The extent of movement can be read from a scale located at the jointed end of the instrument. The technic of examination is simple. The joint and shaft of the instrument are superimposed on the corresponding anatomic parts of the measured extremity, and the range of motion is read directly on the attached scale. (See Fig. 2-1.)

The main value of this instrument is its simplicity. Facility in using the goniometer, however, requires some experience; otherwise false values may be elicited. The main shortcoming of this method is that it measures the joint excursion in one plane only. For circular and other complex movements, various other devices have been suggested.

The range of the movement is conventionally expressed in degrees. The normal

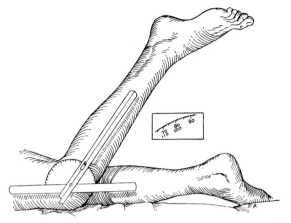

Fig. 2-1. Modified simple goniometer. Inset shows a section of scale for reading the joint excursion.

range of motion varies from joint to joint. It is determined by its anatomic structure and its relation to the surrounding soft tissues. It is essential that the examiner be familiar with the normal ranges of motion; otherwise he cannot draw any diagnostic inference from his findings. If, however, he does not have such information or charts available, a comparison with the symmetric unaffected joint may be of some help.

Recording muscle tests and range of motion findings. A reliable chart system for permanent recording of these two tests is very important. Only by having such records available will the physician have an objective picture of the patient's response to therapy. On charts recording muscle strength the previously described grading system is used. The presence of spasticity should also be indicated; otherwise a distorted picture of muscle strength might be obtained. It is customary and very useful to provide additional spaces in the muscle charts to allow for repeated follow-up testing during the course of treatment.

The recording of the joint range of motion is somewhat more complicated and often depends on the examination technic applied. Some charts indicate graphically the normal joint excursions. Other charts indicate the normal joint excursions numerically. (See Chart 2.)

Differences in nomenclature may cause misinterpretations. In trying to evaluate a joint range of motion chart from unfamiliar sources, it is important to know the method used by the examiner in making his measurements. Certain numerical values may mean entirely different ranges of joint excursion, depending on the method applied.

Meticulous, detailed history taking and physical examination with all of the additions and modifications outlined to meet the problems of rehabilitation and restorative medicine are fundamental cornerstones in diagnosis, evaluation, programming, and ultimate social and vocational placement of the disabled patient.

Text continued on p. 17.

CHART 2*

NEW YORK UNIVERSITY MEDICAL CENTER
INSTITUTE OF REHABILITATION MEDICINE
PHYSICAL THERAPY

NAME_____ DATE_____

DIAGNOSIS_____ PATIENT STATUS_____ IN____OUT_____

DISABILITY_____

RANGE OF MOTION

1. Anatomical position is starting position. Range is measured with cauda as 0°, cranium as 180°. Rotating motions are from the midsagittal plane as 0° to lateral plane as 180°.

2. All ranges are expressed as passive range of motion. Check muscle chart attached for limitations caused by tightness, weakness, spasm, or contracture.

3. The scale is divided into units of 10°. Range of motion is recorded by filling in area of range directly on attached sketch with date and examiner's initial.

4. Use of same sheet for subsequent tests is recorded in same color and dated accordingly.

5. Retrogression is marked by diagonal lines over area of previous test and dated.

6. If position is other than in sketch, indicate S for supine, P for prone.

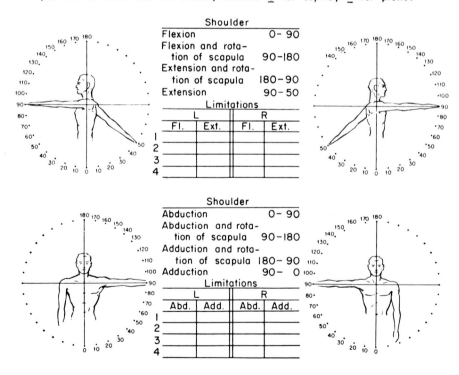

Shoulder	
Flexion	0- 90
Flexion and rotation of scapula	90-180
Extension and rotation of scapula	180-90
Extension	90- 50

Limitations			
L		R	
Fl.	Ext.	Fl.	Ext.

Shoulder	
Abduction	0- 90
Abduction and rotation of scapula	90-180
Adduction and rotation of scapula	180- 90
Adduction	90- 0

Limitations			
L		R	
Abd.	Add.	Abd.	Add.

*From White, P. D., Rusk, H. A., Lee, P. R., and Williams, B.: Rehabilitation of the cardiovascular patient, New York, 1958, Blakiston Division, McGraw-Hill Book Co., Inc.

CHART 2, cont'd

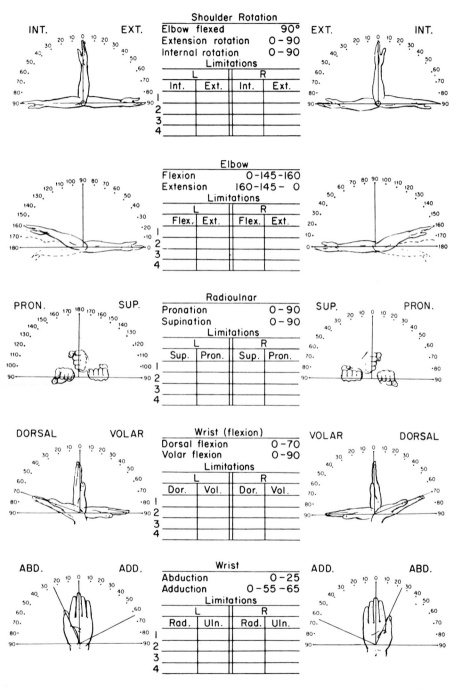

Shoulder Rotation

Elbow flexed	90°
Extension rotation	0 – 90
Internal rotation	0 – 90

Limitations

	L		R	
	Int.	Ext.	Int.	Ext.
1				
2				
3				
4				

Elbow

| Flexion | 0 – 145 – 160 |
| Extension | 160 – 145 – 0 |

Limitations

	L		R	
	Flex.	Ext.	Flex.	Ext.
1				
2				
3				
4				

Radioulnar

| Pronation | 0 – 90 |
| Supination | 0 – 90 |

Limitations

	L		R	
	Sup.	Pron.	Sup.	Pron.
1				
2				
3				
4				

Wrist (flexion)

| Dorsal flexion | 0 – 70 |
| Volar flexion | 0 – 90 |

Limitations

	L		R	
	Dor.	Vol.	Dor.	Vol.
1				
2				
3				
4				

Wrist

| Abduction | 0 – 25 |
| Adduction | 0 – 55 – 65 |

Limitations

	L		R	
	Rad.	Uln.	Rad.	Uln.
1				
2				
3				
4				

Continued.

CHART 2, cont'd

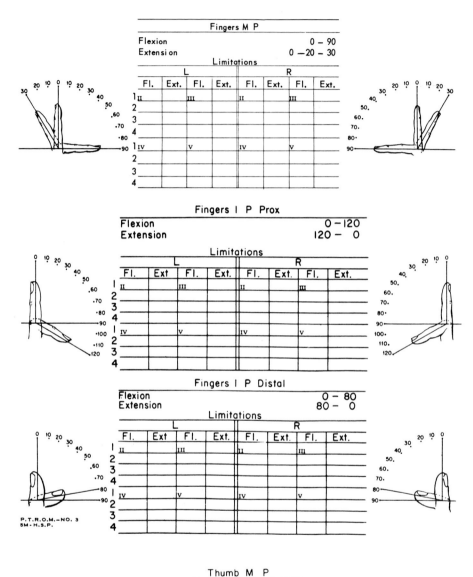

Fingers M P

		0 – 90
Flexion		0 – 90
Extension		0 –20 – 30

Limitations

	L				R			
	Fl.	Ext.	Fl.	Ext.	Fl.	Ext.	Fl.	Ext.
1 II			III		II		III	
2								
3								
4								
1 IV			V		IV		V	
2								
3								
4								

Fingers I P Prox

Flexion	0 –120
Extension	120 – 0

Limitations

	L				R			
	Fl.	Ext	Fl.	Ext.	Fl.	Ext.	Fl.	Ext.
I II			III		II		III	
2								
3								
4								
I IV			V		IV		V	
2								
3								
4								

Fingers I P Distal

Flexion	0 – 80
Extension	80 – 0

Limitations

	L				R			
	Fl.	Ext	Fl.	Ext.	Fl.	Ext.	Fl.	Ext.
I II			III		II		III	
2								
3								
4								
I IV			V		IV		V	
2								
3								
4								

P.T.R.O.M.–NO. 3
5M - H.S.P.

Thumb M P

Flexion	0 – 60 –70
Extension	70 –60 – 0

Limitations

	L		R	
	Fl.	Ext.	Fl.	Ext.
I				
2				
3				
4				

CHART 2, cont'd

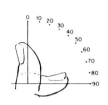

Thumb I P

Flexion	0-90
Extension	90- 0

Limitations

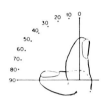

	L		R	
	Fl.	Ext.	Fl.	Ext.
1				
2				
3				
4				

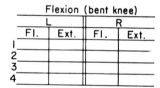

Hip

Flexion (straight knee)	0-90
Flexion (bent knee)	0-115-125
Extension	0-10-15
Extension and lumbar spine	0-15-45

Limitations

Flexion (straight knee)

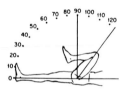

	L		R	
	Fl.	Ext.	Fl.	Ext.
1				
2				
3				
4				

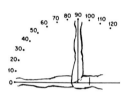

Flexion (bent knee)

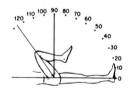

	L		R	
	Fl.	Ext.	Fl.	Ext.
1				
2				
3				
4				

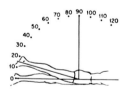

Extension

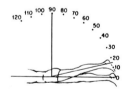

	L		R	
	Ext.	E&L	Ext.	E&L
1				
2				
3				
4				

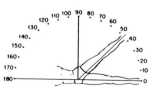

Hip

Abduction	0-45
Adduction	0-45

Limitations

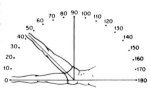

	L		R	
	Abd.	Add.	Abd.	Add.
1				
2				
3				
4				

Continued.

CHART 2, cont'd

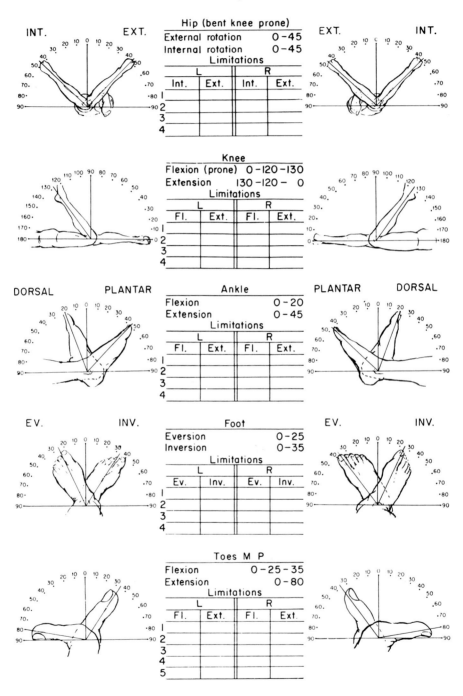

Hip (bent knee prone)

| External rotation | 0-45 |
| Internal rotation | 0-45 |

Limitations

L		R	
Int.	Ext.	Int.	Ext.
1			
2			
3			
4			

Knee

| Flexion (prone) | 0-120-130 |
| Extension | 130-120-0 |

Limitations

L		R	
Fl.	Ext.	Fl.	Ext.
1			
2			
3			
4			

Ankle

| Flexion | 0-20 |
| Extension | 0-45 |

Limitations

L		R	
Fl.	Ext.	Fl.	Ext.
1			
2			
3			
4			

Foot

| Eversion | 0-25 |
| Inversion | 0-35 |

Limitations

L		R	
Ev.	Inv.	Ev.	Inv.
1			
2			
3			
4			

Toes M P

| Flexion | 0-25-35 |
| Extension | 0-80 |

Limitations

L		R	
Fl.	Ext.	Fl.	Ext.
1			
2			
3			
4			
5			

CHART 2, cont'd

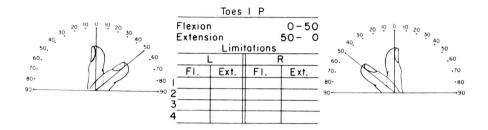

	Toes I P		
Flexion			0 – 50
Extension			50 – 0
	Limitations		

	L		R	
	Fl.	Ext.	Fl.	Ext.
1				
2				
3				
4				

TESTING IN THE ACTIVITIES OF DAILY LIVING

A test in Activities of Daily Living (ADL) serves to outline as accurately as possible how a patient functions in everyday life—in other words, how many daily activities he can perform in his own home and in connection with his work.[11, 12] (See Chart 3.)

The test forms are usually so designed that the initial test and the patient's progress can be recorded on the same sheet. The patient's deficiencies as apparent in the initial testing constitute the basis for his individual ADL program. The discharge summary on the chart should give a good picture of the patient's sufficiency in his daily activities. This will also serve as a guide for any future planning for his home and his work.

Activities tested

Most workers agree on the following major groups: bed activities, wheelchair activities, self-care activities (toilet, dressing, eating), ambulation, and elevation activities. How much each group is subdivided, how many activities are actually listed under each group, depends on the type of patient to be evaluated.

Testing technics

Just say "sit up," or "show me how you brush your teeth," or "comb your hair," or give a similar instruction to the patient. Do not merely ask the patient what he can or cannot do but have him actually perform the task, for patients may exaggerate or minimize their functional ability.

If, for example, you want to know how he can wash his hands, you must provide the sink, water, soap, and towel. The same applies to feeding and all other activities. Merely going through the motions will not substitute for a trial of actual performance. A patient may ascend and descend stairs very well in the gym, but going out dressed for the street, especially wearing heavy winter clothes, and descending the steps that lead down from the entrance door are quite a different matter.

To have a reliable and useful record of the patient's progress, there must be an objective record of performance of what he could do when he started and what he could do at discharge.

There are various ways of organizing and designing an ADL test. The test form

Text continued on p. 22.

CHART 3

NEW YORK UNIVERSITY MEDICAL CENTER
INSTITUTE OF REHABILITATION MEDICINE

TEST
ACTIVITIES OF DAILY LIVING

Mr.
Miss
Name Mrs. _____ Age ____ Room No. ____ M.D. _____
last
Date of initial test _____ Patient: In _____ Out _____

Address _____ Vocation _____

Onset date _____ Lesion _____ flaccid ___ spastic ___ Admission date _____

Disability _____

Cause _____

Decubiti _____

Surgery _____

METHOD OF RECORDING TEST AND PROGRESS

Symbols for grade

✔ Patient can perform activity independently.
S   Patient needs supervision.
A   Patient needs assistance.
L   Patient has to be lifted.
X   Activity is not indicated.

1. *At initial testing* (use *blue pencil*):
   Enter grade symbol in column G/1 and your initials in column I. The initial date appears at top of page.

2. *Progress* is recorded with *red pencil*:
   Enter grade symbol in column G/2 and your initials in column I. Date in column "Date."
   _____

If there is more than one method or item listed with an activity, circle which indicated.

BED ACTIVITIES

	G/1	G/2	DATE	I
Moving in bed: lying, sitting				
Roll to right, to left				
Turn on abdomen				
Manage: pillows, blankets				
Sit up				
Reach objects on night table				
Operate signal light				

CHART 3, cont'd

WHEELCHAIR ACTIVITIES

	G/1	G/2	DATE	I
Propel: forward, backward, turn				
Open, through, and close door				
Up, down ramp				
Bed to wheelchair				
Wheelchair to bed				
Wheelchair to straight chair				
Straight chair to wheelchair				
Wheelchair to easy chair, couch				
Easy chair, couch to wheelchair				
Wheelchair to toilet (high toilet seat, regular seat)				
Toilet to wheelchair				
Adjust clothing				
Wheelchair to tub				
Tub to wheelchair				
Wheelchair to shower (chair in stall shower, or tub)				
Shower to wheelchair				

Travel:

	G/1	G/2	DATE	I
Wheelchair to car_____ on curb				
Car to wheelchair____on curb				
Wheelchair to car____ no curb				
Car to wheelchair_____no curb				
Place wheelchair in car_____on street				

SELF-CARE ACTIVITIES

HYGIENE (TOILET ACTIVITIES)

	G/1	G/2	DATE	I
Comb, brush hair				
Brush teeth				
Shave (electric razor, safety razor), put on makeup				
Turn faucet				
Wash, dry hands and face				
Wash, dry body and extremities				
Take bath (wheelchair, walking)				
Take shower (wheelchair, walking)				
Use urinal, bedpan				

EATING ACTIVITIES

	G/1	G/2	DATE	I
Eat with spoon				
Eat with fork				
Cut meat				
Handle: straw, cup, glass				

DRESSING ACTIVITIES

	G/1	G/2	DATE	I
Undershirt; bra				
Shorts; panties				
Slipover garment				
Shirt; blouse				
Slacks; dress				
Tying necktie—bow				
Socks; stockings				
Shoes (laces, buckles, slip-on)				
Coat, jacket				
Braces, prosthesis, corset				

Continued.

CHART 3, cont'd

MISCELLANEOUS HAND ACTIVITIES

	G/1	G/2	DATE	I
Write name and address				
Manage: watch				
match or cigarette lighter				
cigarette				
book, newspaper				
handkerchief				
lights: chain, switch, knob				
telephone: receiver, dial, coins				
handle: purse, coins, paper money				

WALKING ACTIVITIES

	G/1	G/2	DATE	I
Open, go through, and close door				
Walking outside				
Walking carrying				

STANDING UP AND SITTING DOWN

	G/1	G/2	DATE	I
Up from wheelchair				
Down on wheelchair				
Up from bed				
Down on bed				
Up from straight chair				
Down on straight chair				
Up from straight chair at table				
Down on straight chair at table				
Up from easy chair				
Down on easy chair				
Up from center of couch				
Down on center of couch				
Up from toilet				
Down on toilet				
Adjust clothing				
Into car, on curb, up curb				
Out of car				
Down on floor				
Up from floor				

CLIMBING AND TRAVELING ACTIVITIES

	G/1	G/2	DATE	I
Up flight of stairs (railing, no railing)				
Down flight of stairs (railing, no railing)				
Into and out of car, taxi				
Walk one block and back				
Down curb, cross street, on curb				
Into bus				
Sit down, get up from bus seat				
Out of bus				

CHART 3, cont'd

HOME SITUATION

Note suggestions for adaptation next to each line or check when so indicated. In special instances, diagram of layout will be advisable.

Location: Urban_____ Suburban_____ Rural_____

APARTMENT: Floor _____ Rooms_____ Elevator_____ (self-service _____) None ___

Walk up_____

PRIVATE HOUSE: Floors_____ Rooms_____ Elevator_____ (self-service_____) None___

ENTRANCE: Door_____ No. steps_____ Railing: right____ left____ none____ Ramp_____

Note floor if in private home: Bedroom_____ Living room_____ Kitchen_____
Bathroom_____

BATHROOM: Door width____ Tub____ Shower over tub____ Stall shower_____

Information uncertain (explain): _____

Information unavailable (explain): _____

APPLIANCES

Column 1—Initial Summary check blue; column 2—Discharge Summary check red.

WHEELCHAIR	1	2		1	2
8″ casters			Footrests: removable		
Wheels: knobs, tape			adjustable		
Pneumatic tires			stationary		
Cushion: seat____back			Armrests: removable		
Board: seat____back			desk		
Back: snap			stationary		
reclining			Brakes____ extension		

BRACES	1	2	CRUTCHES	1	2	GAIT	1	2	DEVICES	1	2
Short leg			Axillary			Swing to					
double			Lofstrand			Swing through					
Long leg			Canes			4-point					
double			None			Other					
Pelvic band			Other			Hemi					
Knight spinal											

INITIAL SUMMARY_____

_____

_____

_____

CHART 3 (cont'd)

DISCHARGE SUMMARY

RECOMMENDATIONS (circle)

Patient is independent

Patient needs: Supervision _____ Assistance _____ Lifting _____

Patient uses wheelchair: Exclusively_____ Partly_____

Patient is ambulatory: Exclusively_____ Partly_____

ENDURANCE AND SPEED

Endurance:
    Propelling wheelchair: State number of feet _____Time_____
    Walking: State number of feet _____Time_____
    Climbing: State number of steps_____Time_____

Speed:
    Propelling wheelchair: State time for 60 feet_____
    Walking: State time for 60 feet_____
    Climbing: State time for 10 steps_____

ACTIVITIES

Grade in column G as follows: ✓, Independence; S, Supervision; A, Assistance; L, Lifting; time in column T (use red pencil).

	G	T
GETTING READY IN MORNING:_____ Total___		
Toilet needs: From wheelchair_____		
Standing_____		
Complete dressing_____		
Eating: Breakfast_____		
Lunch_____		
GETTING OUT OF HOUSE: Stairs, ramp, entrance door____Total___		
TRAVEL TO WORK: Own car (hand controls), taxi_____Total___		
Car: From wheelchair (including placing wheelchair) _____		
From standing_____		
Bus: Walk to station, cross street_____		
Date_____ Signature_____		

given in Chart 3 has been developed at the Institute of Rehabilitation Medicine, New York University Medical Center, and is currently used there. This form can be used as an example for a simple method of recording and grading.

At initial testing, a blue check (✓) is made in column G/1 to indicate that the activity is performed independently. The square is left blank if the activity cannot be performed. A cross (X), also in blue, is recorded if the activity is not indicated. Whenever the patient progresses, red pencil is used to state the date in column

"Date," and in column G/2 the following grades are used to describe the patient's performance:

1. *The patient is independent.*
   √ to indicate that he can function without help and without anybody even near him.
2. *The patient needs help.* This grading may consist of the following:
   S (supervision) to indicate that he can function without being helped, but needs somebody to stand by.
   A (assistance) to indicate that either the entire activity or part of it has to be done for him.
   L (lifting) to indicate that he has to be literally "lifted."
   X to indicate that the activity is not pertinent.
3. *Devices.* If a device is needed, it is so stated on the same line as the activity it is used for, and the activity is graded as described under 1 and 2.

This kind of grading is preferred to indicating the patient's performance in such terms as excellent or good, or in percentages, because these grades imply, without lengthy comment, what kind of help the patient may require.

At discharge the patient is tested again and the same kind of grades (as under 1 and 2) is recorded in red pencil. Since speed of performance is an important factor in the overall picture of the patient's independence, time is recorded in column T.

Wheelchair adjustments, braces, crutches, and other devices, as well as the kind of gait, must be incorporated into an ADL program, and provisions to record their use are made on the test form. A thorough investigation of the layout of the home, including furniture, entrance doors, surrounding grounds, etc., is essential in developing and adjusting an ADL program. These items are therefore listed on the test form. In regard to attendants and use of wheelchair and ambulation, recommendations are made on the basis of the initial test and the final test.

Use of the ADL test

Since Activities of Daily Living is but one aspect of the total rehabilitation program, the ADL test should be so designed that it can serve as a source of information to all members of the rehabilitation team.

The physical therapist and others working with the patient in ADL must know each detail of performance of each individual activity, for the specific difficulty must be related to specific physical therapy measures. For example, if a patient tends to fall out of the wheelchair because of inadequate sitting balance, he will need exercises for sitting balance. Similarly, any difficulty in regard to devices has to be related to occupational therapy measures. The nursing service, which must carry through whatever the patient has learned in ADL, obviously should know as many details as possible.

In addition to this detailed information it is important for all departments that the following data be stated clearly in the discharge summary of the ADL test.

1. *What degree of independence has the patient reached?*
   a. *Performance* in terms of independence or help needed (lifting, assistance, or supervision) for each major group of ADL, that is, bed, wheelchair, self-care, ambulation, elevation, traveling, and miscellaneous hand activities.

    b. Specific statements in regard to the following activities:

        (1) *Time needed* for eating (breakfast, lunch, and dinner), attending to toilet needs, and using transportation media.

        (2) *Endurance and speed* in regard to wheeling wheelchair, ambulation, and climbing steps (expressed in distance and units of time). This information is especially important when considering the possible *place of work.*

2. *Is the patient essentially a wheelchair patient or ambulatory?*

    This will depend on the foregoing, since his efficiency, which means skill, endurance, and speed, will determine whether he will use his wheelchair exclusively, only part of the time and be partly ambulatory, or be exclusively ambulatory. This again will be the basis for recommending whether any work is feasible and, if so, whether at home or outside.

    (**Special note:** Some patients may be good functional walkers, but, because of crowded space in their work situation, it may be more advisable for them to use the wheelchair while at work. For these patients it is important that they get up on crutches and braces for from 30 to 60 minutes daily for weight-bearing and exercise purposes.)

3. *Does the patient need an attendant?*

    The degree of help needed (lifting, assistance, or supervision) will depend on information under 1 and 2 and will determine whether an attendant is necessary.

4. *Overall performance*

    On the basis of the foregoing data, it is important:

    a. To consider the overall picture of the patient's performance, *especially his speed,* not in regard to groups or single activities, but in terms of big units of groups of activities connected with his possible occupation and social life.

    b. To remember that a patient has to carry out his daily activities in his *particular environment,* which must be analyzed in all its aspects in addition to the necessary motions. The following units of groups of activities and what they entail will illustrate these two points:

        (1) *Getting ready in the morning,* which includes getting out of bed, washing, grooming, dressing, and eating breakfast.

        (2) *Going out of the house,* which includes using stairs, elevator, ramp, entrance door, and outside stairs.

        (3) *Traveling to place of work,* which includes use of (1) private car (garage), placing wheelchair in and out of car (in case wheelchair is used), or (2) public transportation—taxi or bus (which also entails walking to bus station, crossing street, climbing into bus, sitting down, getting out of bus, and walking to place of work.)

        (4) *Functioning at place of work,* which includes entrance door, steps (inside and outside), ramp, elevator, distance from entrance to locker room to workroom, distance to be covered in workroom, distance from workroom to bathroom, time needed in bathroom, distance to lunchroom, time needed to eat lunch, and floor coverings (carpet, tile, wood).

    An estimate of speed of performance for units (1) through (3) will be very helpful when planning for work. Although an ADL test cannot supply the answer to the problems entailed at the *place of work,* this unit is outlined to illustrate how the discharge summary is related to work. This unit will also serve as a guide to the vocational counselor or whoever is responsible for job placement.

5. *Attitude*

    The information discussed above can be learned from the test and can be expressed in grades and units of time. In addition, however, there is the problem of the patient's attitude, which cannot be tested fully in a formal testing situation, but has to be observed carefully during his entire program.

Summary

    When summarizing a patient's proficiency in ADL, it is essential to report not only the patient's ability to be independent but also his desire to be independent.

Can he function under pressure? No matter how much tension there may be in a crowded hospital or institution, there is always more pressure when actually going to work.

In conclusion, an ADL program and the ADL test should be organized in such a fashion as to permit testing and training a patient in all essential daily activities that may have a relationship to his rehabilitation program. This means that the patient, as well as the staff, must analyze what the patient can do in terms of motion and in terms of management of furniture and equipment and what possible adaptations are necessary. Although an ADL test is not itself an end, it is an essential means toward the end goal of maximum total rehabilitation. When a complete summary is given, all observations that go beyond formal testing should be carefully reported, since no test can truly express the mysterious entity known as the "total patient."

PSYCHOLOGIC EXAMINATION
Definition

A psychologic examination is an assessment of an individual's behavior in a situation that is so contrived that his responses can be recorded, coded, compared with that of other individuals, and evaluated in a fairly unambiguous way. The situation involves tasks that may appear to be trivial to the casual observer. Yet they are capable of revealing a great deal about a person's mental life and of providing useful cues to the understanding of larger, more meaningful samples of behavior beyond the immediate test situation.

There are various types of psychologic tests. Although these types can be subdivided in many ways, we would like to consider grouping the tests under two main headings: (1) objective tests and (2) projective tests. Objective tests are designed to elicit responses that can be translated into numerical values and compared with the performances of other persons. The test stimuli and the responses have fixed meanings. There is more concern with the actual end result than with how the individual solves the problems. The most popular of all objective tests are the well-known intelligence and school achievement tests.

Projective tests are designed to assess the way in which the individual structures his environment by an analysis of how he responds to stimuli that are deliberately vague and amorphous, such as inkblots. Under such circumstances an individual's response is more a product of his own needs, ways of coping with the world, misperceptions, anxieties, and self-values than of the test materials. Although the responses can be scored, the primary aim is not to arrive at a score that will be comparable to that of other persons but to analyze the psychologic processes that are responsible for the individual response. "Projective" responses are not limited to the testing session. They occur also in daily life in situations that do not have fixed or familiar meanings or where our needs are so strong as to cause us to misinterpret objective reality. Walking down an unfamiliar, dark, deserted street and *seeing* each lamppost shadow as an ominous grotesque figure is an example of a projective response. Projective tests are powerful tools that permit a microscopic analysis of personality structure and inner processes. Popular usage equates objective tests with measures of skill and intellect and projective tests with measures of personality. However, psychologists who are flexible in their use of tests attempt to analyze the response processes in objective tests and to develop norms and scoring systems for projective tests. Nevertheless, the distinction is a useful one.

Indications

Intellectual evaluation. Since rehabilitation, fundamentally, is so related to education, an assessment of the ability to learn is important. The patient in a rehabilitation program may be required to learn verbal, motor, social, vocational, and/or personal adjustment skills. Therefore, a psychologic examination would be indicated (1) to arrive at a judgment as to how complex a task an individual can be expected to master and (2) to determine learning readiness in specific areas. Psychologists have discovered that one of the best measures of a person's ability to learn is his score on an intelligence test. This finding is consistent with common sense. But before this statement can be accepted at its face value, it should be noted that some psychologists consider intelligence as a unitary property that is manifested in all of a person's behavior. Others conceive of it as an aggregate of separate abilities, such as reasoning skill, numerical skill, and verbal skills. In rehabilitation both approaches are important. It is useful to determine patterns of specific abilities as well as general mental functioning. Our experience indicates that there are four main types of abilities in a rehabilitation program. These abilities, the tasks that are used to measure them, and their significance are presented in Table 2-2.

Table 2-2. Types of abilities, tasks, rehabilitation areas, and populations whose deficits are manifested in lower abilities

Skill	Task	Importance for rehabilitation	Populations that demonstrate deficits
Verbal	1. Vocabulary 2. Range of cultural information 3. Verbal comprehension 4. Verbal reasoning 5. School achievements	Educational, speech, and vocational goals	School failures Culturally impoverished persons Aphasic patients
Social	1. Alertness to social nuances 2. Ability to organize social experiences 3. Judgment in interpersonal situations 4. Appropriate response to demands of situation	Vocational, psychologic, and social goals	Those with psychiatric disturbances, including psychoses, neuroses, and some brain damage
Motivational	1. Ability to persevere at a task 2. Attention and concentration 3. Frustration tolerance	Vocational and psychologic goals	Those with psychiatric disturbances, particularly neuroses Those with brain damage Educational retardates
Motor	1. Ability to judge and use spatial cues 2. Speed and accuracy of movement 3. Physical grace	Physical and occupational therapy —ADL	Those with brain damage Those with poor physical coordination Aphasic patients

As Table 2-2 indicates, defects in particular abilities are selectively important for the varied goals of rehabilitation. In general a pattern of abilities takes on different meanings when considered in the light of the different goals of rehabilitation, for example, training of motor skills, vocational skills, and interpersonal skills. The profiles are most useful when they show marked deficits in one or more areas. They are used generally as supplementary evidence in helping to set goals in each of the areas under consideration. The technical considerations entering into an analysis of psychometric profiles may be found in the literature.[1, 13]

Personality evaluation. The assessment of personality is as important as the assessment of intelligence. However, the evaluation of personality structure is conceded to be more difficult and complex than is the measurement of intellect. Distortions in forming impressions of character are too well known to require elucidation. For this reason objective and accurate appraisals of personality are exceedingly helpful.

Personality structure may be described along many dimensions. Only some of these dimensions, however, are pertinent to rehabilitation. Therefore, we have found it useful to organize the data derived from projective tests according to the following outline:

1. Current problems actually facing the patient
2. Patient's perception of current problems
3. Patient's style of adaptation to the world
   a. Anxiety level
   b. Ego strength (i.e., inner resources to cope with his problems)
   c. Defenses
4. Pertinent psychologic conflicts
   a. Self-acceptance
   b. Dependency needs
   c. Body image
   d. Attitude toward disability
   e. Reaction to authority
   f. Frustration tolerance
   g. Sex role
5. Genetic sources of tensions and defenses
   a. Sources and strength of identification
   b. Level of aspiration
   c. Ability to plan
   d. Ability to relate to workers
6. Recommendations

Those parts of this scheme should be incorporated that are pertinent to the patient and to the goals that rehabilitation offers him. This framework has been particularly useful in attempts to analyze a test protocol in terms of the information helpful to the rehabilitation team.

Contraindications

Psychologic data are not always necessary in the evaluation process. This occurs when:

1. Goals are clear-cut and obvious and psychosocial problems do not appear to affect the course of rehabilitation.
2. Data are already present. This occurs when sufficient evidence can be gathered from the life history and the clinical interview to render the psychologic examination superfluous.
3. The patient is overly anxious. This occurs when a patient's emotional state is too precarious to be subjected to probing procedures. Under such circumstances, it is judicious to forego the examination or to defer it until the patient can be adequately prepared for the procedures. The psychologist must exercise his judgment in this matter.

Limitations and difficulties in administration

A physical handicap automatically reduces the range of behavior that can be sampled in many patients. For this reason some modifications of the usual instruments may be required to elicit valid test responses. The specific difficulties are listed below:

1. *Sensory limitations*—Poor hearing, e.g., may diminish responses to verbal tasks.
2. *Motor handicap*—Paralysis or weakness may inhibit or delay the patient's attempting to solve manually oriented tasks (that is, the time limits set for nonhandicapped people cannot be applied to evaluate performance).
3. *Cultural impoverishment*—Many disabled children do not have the advantage of normal environmental stimuli. For example, a study of children with muscular dystrophy who have been homebound indicates their verbal intelligence is below average.[14] Congenitally disabled cerebral palsied persons show less creative and imaginative records than do those with poliomyelitis, even when they are matched for intelligence.[15]
4. *Adequate norms*—The standards used to measure the performance of an individual are compiled generally from samplings of the performances of physically able persons who suffer from none of the preceding limitations.

Despite these difficulties, there is evidence that valid pictures of mental functioning can be obtained from as high as 85% of a population group as severely disabled as cerebral palsied persons.[16]

In addition to these factors, frequently more subtle psychologic processes are present that are related to the physical handicap or to the psychologic testing procedure. These include the length of time since onset of disability, the social and cultural background of the individual, his reactions to being away from home, and the nature of the rapport at the time of the examination. There is little systematic evidence that can help pinpoint these factors, and what exists is too scattered to be enumerated here. However, a well-trained psychologist always will take the test conditions into consideration while interpreting the results.

Special disability groups—brain-damaged persons

Psychologic testing with brain-damaged persons is carried out for three main purposes: (1) to help infer from an individual's responses to several tests whether he actually has sustained damage to the brain, (2) to evaluate the effects of brain damage on overt behavior, and (3) to assess behavioral deficits and help set up a program of reeducation.

Many patients with known brain damage do not manifest changes in external behavior. It follows, therefore, that psychologic tests, like other instruments such as the electroencephalogram, are highly reliable when they indicate positive signs of brain damage. However, the absence of indicators of behavioral disturbance in the test data does not prove that brain damage is absent. In short, testing is valid when positive signs appear, but it is not so valid when negative evidence is present.

The hemiplegic patient

Very often the physician is called on to evaluate a hemiplegic patient as a candidate for a rehabilitation program. While it is desirable to carry out thorough psychologic examinations, the physician often must make a decision about the patient's mental status from on-the-spot clinical observations. Several fairly rapid technics for assessing various areas of mental dysfunction are available for such occasions. They include tests of orientation, memory and concentration, perception, and lability. The following paragraphs describe them in more detail.

Tests of orientation. These include asking the subject his name, the name and location of the hospital, the date of examination, and the date of entrance into the hospital. The examiner should note whether the patient answers the questions correctly, as well as the manner in which he answers them. For instance, does the patient ramble? Does he pause and appear to grope for the answer? Does he burst into tears or manifest any other signs that the questions are distressing to him?

Tests of memory and concentration. These can be divided into verbal and non-verbal types of problems. If rapport is established and the patient is not upset by direct questioning, the examiner can ask the patient if he experiences any memory difficulties. Many patients will admit to difficulties. However, the patient who denies memory problems cannot always be trusted as a reliable observer. At times a patient who appears to have an adequate retention of old memories, almost to the point of dwelling on his experiences in the distant past, often may be concealing specific gaps in more recent memory and attention. Old learning frequently is retained by an individual with severe deficits in new learning. It is, therefore, not surprising to find many patients who retain memories of highly technical information but cannot carry out simple instructions in their everyday lives.

Some tests that are useful for a rapid estimate of memory and concentration are (1) repetition of digits, particularly reversed digits, and (2) serial sevens (subtracting by sevens from 100). Concentration in the motor area may be observed by watching how the patient carries out instructions to perform a sequence of one, two, or three separate acts.

Deficits in perception. These are best evaluated by intensive and careful psychologic tests. Perceptual deficits are difficult to distinguish from deficits in sensation and/or deficits in thinking. We shall use the term perception to refer to the process involved in the integration of sensations; therefore, perception deficits may transcend specific losses in given modalities. For this reason we say it is possible to perceive a musical composition as exciting or a personality as cold just as we perceive objects in the visual field. Responses to a spatial problem like the organization of an inkblot may be indicative of responses to nonvisual stimulation, for example, the stimulation provided by other persons.

A gross measure that is sensitive to very severe perceptual deficits, but not to the more subtle dysfunctions, can be elicited by having the patient copy the figures on p. 30. The mental age equivalent for each figure is also noted.[17]

The potency of perceptual measures of dysfunction is great. We have found that perceptual tests are the best indices of how well a patient can learn in a rehabilitation program. This holds true not only for the motor skills involved in activities of daily living, but also for the verbal skills in speech training of the aphasic patient. Along these same lines we may note the puzzling finding that persons with hemiplegia of the left side tend to have greater perceptual deficits than those with hemiplegia of the right side and that behavioral disturbance and aphasia may be independent of each other.[18, 19]

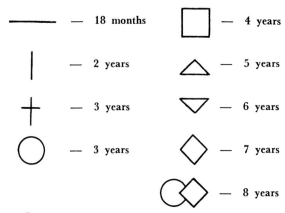

Perceptual motor development according to age

Lability. Other facets of behavior that the physician may note are emotional lability and disturbance in consciousness, for example, sleep patterns. While depression and somatic preoccupation occur with great frequency among hemiplegic patients, it is often difficult to tell whether they are due to the emotional sequelae of being faced with a loss of mastery or are due to fixations based on the rigidity of thinking brought on by the brain damage. In such cases psychologic testing may help clarify the reasons for the behavioral disturbance.

The cerebral palsied patient

Perhaps the most difficult of all groups to evaluate psychologically is that having cerebral palsy. Most of the difficulties in testing disabled persons are present and even exaggerated in those with cerebral palsy. For these reasons, special precautions should be observed. Tests should be administered by an experienced examiner. They should include a careful evaluation of the patient's intelligence, perception, and personality. Unfortunately, as we move away from the area of intelligence to other areas, the instruments are less suited to quantitative assessment and are more subject to qualitative impressions. Yet it is those areas of behavior that are most refractory to quantification that are of greatest importance, for in planning a rehabilitation program we are concerned primarily with the way the individual handles his deficits and then with the deficits per se. Therefore it is the unsophisticated equating of objectivity and quantification with practical significance against which we must be on guard.

Large numbers of cerebral palsied children have been tested and the data are available. They permit the following generalizations.

Intelligence

1. Approximately 50% to 60% of cerebral palsied patients are mentally retarded, that is, 50% do as poorly on intelligence tests as the lowest 3% of a random, normal population.[20]

2. Within the spastic group the degree of mental retardation is related to the degree of physical involvement. The intelligence test scores of paraplegic patients and those with right and left hemiplegia exceed the intelligence test scores of the triplegic and quadriplegic patients to a significant degree—18 points.[16]

Table 2-3. Proportion of diagnostic groups in mentally defective range

				Spastic group				
Spastic	*Athetoid*	*Ataxic*	*Rigidity*	*Quadri-plegic*	*Triplegic*	*Right hemi-plegic*	*Left hemi-plegic*	*Para-plegic*
0.44	0.42	0.71	0.63	0.65	0.52	0.35	0.41	0.34

3. The spastic and athetoid groups have mean IQs of approximately 72, whereas the ataxic and the rigidity groups have mean IQs between 55 and 59.[20]

4. The proportion of diagnostic groups that fall into the mentally defective range (70 or below) can be seen in Table 2-3.[20]

Consistency of the tests. Two studies indicate that test and retest results are highly consistent. In one study 153 children were tested at 6 years 3 months of age and were retested at 9 years. Only 16 (10%) of the children gained more than 10 points, while 33 (20%) decreased more than 10 points. Of those who changed more than 10 points, in 80% of the cases the examiner had suggested a retest at the time of initial examination.[16] An independent study on a totally different group revealed that 75% of cerebral palsied children maintained an intelligence quotient within 10 points of the original testing, indicating that for the group as a whole the psychometric test in the hands of an experienced psychologist is a fairly accurate predictor of later intelligence.[20]

Perception. Cerebral palsied persons tend to suffer from defects in perception[22] that can be assessed by psychologic tests. These defects include the following:

1. An inability to distinguish a figure from its masking context or background, resulting in unstable perceptions; for example, the cerebral palsied child will have greater difficulties in recognizing the bottle (Fig. 2-2) than will the noncerebral palsied. This deficit implies severe problems in reading and other learning tasks that require the individual to organize a pattern of stimuli into a meaningful whole.[20]

2. Distractibility, a corollary of figure ground defect, has been termed forced responsiveness, hypervigilance, and motoric disinhibition. The distractible child apparently is responding to irrelevant stimuli.[20, 21] Such behavior is often mistaken for negativism based on avoidance of interpersonal contacts.

3. Perseveration refers to the tendency to repeat the same response, even though the stimulus has been altered. When this tendency is present to any appreciable degree, a poor prognosis for learning is inferred, for the individual cannot shift his behavior patterns to accommodate to a changing environment.

Personality. Despite widespread opinion, there is no solid evidence that different types of cerebral palsy are associated with different personality characteristics. The incidence of emotional disturbance, however, is high. Interviews with 200 adult ambulatory cerebral palsied patients revealed emotional maladaptation in 75% of this population, 20% of which were so serious as to preclude placement in a vocational situation.[21] For the most part the motivations of cerebral palsied persons are similar to those of noncerebral palsied persons.[22]

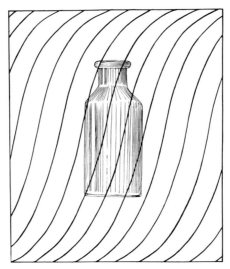

Fig. 2-2. Perceptual problem for cerebral palsied patients: extracting a figure from a masking context.

The aphasic patient

The aphasic patient poses a special problem for the psychologist. Since normal means of communication are impaired, the rehabilitation worker must modify his examination technics in an attempt to answer the following three questions:

1. Can this patient profit from a rehabilitation program, or is he too mentally disabled to be asked to learn or comprehend instructions?
2. What are the specific defects in his mental functioning?
3. What type of rehabilitation program would benefit the patient maximally?

A pretesting procedure designed to identify the specific impairment, for example, the various agnosias, apraxias, and field defects, is recommended. Sections of the Halstead-Wepman test for aphasia[23] or the Eisenson test for aphasia[24] are used frequently (Chapter 14). A qualified speech pathologist should evaluate the actual severity of the communication disorder. Indeed, the testing procedure utilized by the psychologist overlaps in part the one administered by the speech therapist. However, the psychologist is more concerned with determining the effect of the disorder on mental functioning, as well as with evaluating the effect of the mental level of the patient on his communication difficulties, than with describing the specifics of the aphasic condition.

In most instances, those in which the receptive disorder and expressive problems are mild to moderate, performances on projective tests usually coincide with those of most nonaphasic brain-damaged patients. In those cases, however, where the impairment is severe enough to interfere with an ability to respond to the task, traditional testing procedures must be suspended. Here, great flexibility and clinical skill are required. For example, patients with motor paralysis who cannot write or manipulate objects in space can be given multiple choice items that require only that the individual be able to point to the correct figure. The Full Range Vocabulary Test,[25] which instructs the subject to indicate which one of four pictorial stimuli matches a given vocabulary word, is helpful. Also, the *Handbook for Mental Examiners* provides a number of tasks that demand a minimum use of the hands, as well as a minimum use of speech.[26] Often psychometrics originally designed for preschool children are suitable for testing the aphasic patient. The Stanford-Binet and the Gesell Developmental Schedules, for instance, can elicit revealing responses, whereas the more complex tests cannot. Many items in the Gesell Schedules are similar to those included in the Activities of Daily Living Schedules.

In testing the aphasic patient, further support for the utilization of tests designed to evaluate growth patterns in children is found in the fact that many experts view aphasic speech as a regression phenomenon; that is, aphasic speech is immature, childlike speech, rather than fully developed adult speech.

Psychologic tests as tools for research in rehabilitation

As research tools, psychologic tests are useful in two ways: (1) in investigating fundamental psychologic processes of the disabled and (2) in recording behavioral phenomena pertinent for the rehabilitation process.

Historically tests have been used most often to investigate basic psychologic processes. In this tradition the implications for rehabilitation are inferred. They are not made overt, even though a disabled population may have been used. Several studies demonstrate this trend.

Paraplegic males were divided into two groups. One group included those who became sexually impotent as a result of their injuries, and the other group contained those who did not become sexually impotent. Data derived from a battery of projective tests indicated that the impotent patients felt more frustrated in their sexual drives and reacted with more depression, anger, withdrawal, and anxiety than the nonimpotent patients. In addition, the impotent patients tended to view their bodies as useless and malfunctioning organisms. The nonimpotent patients did not experience these tensions.

Another psychologic study demonstrated the importance of parental attitudes for a group of children who had had poliomyelitis. It was discovered that a significant correlation existed between a syndrome of parental attitudes and an acceptance of disability so that the children of parents who were more punitive, rigid, and authoritarian tended to be more constricted and anxious than children with less punitive parents. They failed to accept their disabilities and either denied them or used them as excuses for not adjusting to their peer groups.

Studies such as these emphasize basic psychologic processes operating among disabled persons. However, they are not concerned with the rehabilitation process per se, although their significance for rehabilitation is great. Only a handful of research studies are directly concerned with the nature of the rehabilitation process. They are designed to pinpoint specific rehabilitation practices. Some examples of such studies follow.

A comparison of patients was made between those who exhibited adequate motivation and successfully acquired skills in physical therapy and Activities of

Daily Living and those who were poorly motivated and did not acquire skills. The following differences in responses to projective tests, which were administered at the outset of the program, were revealed:

Those patients who performed well appeared to be able to direct their energies to the demands of the task (task-oriented). Those who did poorly were not able to do so (ego-oriented); their responses were marked, instead, by misdirected energy due to loss of emotional control, somatic preoccupation, and inability to organize their environment along constructive, purposeful lines. Furthermore, thought samples, elicited by having the patients make up stories to various pictorial stimuli, showed that the motivated group produced many more stories concerned with achievement and with goals and standards of excellence than did the nonmotivated group.

In attempting to investigate the various roles that psychosocial workers assumed in dealing with patients in rehabilitation, it was possible to demonstrate that those patients who were less productive and more guarded on projective tests responded differently to a psychosocial service than did the less constricted patients. The latter group, which had projected more of its own feelings into a test situation, saw counselors more than twice as often as did the more constricted group. Although both groups had vocational and personal problems, only the less constricted group was able to enter into problem-solving relationships. The counselor's role with the more constricted group, which was in many ways the more emotionally disturbed, was limited largely to furnishing vocational information and making vocational contacts.

These studies were concerned directly with the response process in rehabilitation. They are important because (1) they help provide understanding of behavior in rehabilitation, (2) they help provide practical prognostic devices, and (3) they help furnish guides to practice. For example, the last-named study, which is concerned with psychosocial processes, can guide a counselor in adopting a certain role with a patient and can provide him with a rationale for his dealings with the patient. Even modest research studies, such as these, point out the fact that psychologic tests can be used to check on the efficacy of empiric procedures and to help systematize the knowledge about these procedures.

Summary

Psychologic tests can be used to evaluate abilities (objective tests) and personality (projective tests). Since not all abilities or all aspects of personality are equally important for all rehabilitation, those features that seem particularly important for setting goals and making decisions on a rehabilitation program are stressed. Psychologic tests are not administered routinely but are prescribed for specific purposes. Some situations in which the administration of psychologic tests is not advisable have been presented. Testing severely disabled and brain-damaged patients poses special problems. Some devices for a rapid mental screening of brain-damaged patients, some important findings in regard to cerebral palsy, and some of the considerations in testing aphasic patients are discussed. Examples of the use of psychologic tests as research tools to gain meaningful insights into the behavior of disabled persons and the rehabilitation process are presented.

VOCATIONAL EVALUATION

There are many aspects of a patient's total situation that require exploration to determine whether employment or work activity should be a goal of a rehabilitation program. Medical, psychologic, social, and economic factors together with aspirations, values, and vocational potential are pertinent to this issue. Many disciplines in the comprehensive rehabilitation team may play a role in assisting the patient to define his vocational problems and to realize the goals that are decided on.

Vocational evaluation in its broadest sense is the process of gathering and understanding the implications of these vocationally significant data. The physician, psychologist, social worker, vocational rehabilitation counselor, and others may play important roles in the process of helping the patient to determine the relevance of employment as a goal and to mobilize the patient's resources and community resources to achieve his goal. Vocational evaluation, therefore, in its broadest sense is a process within the rehabilitation program that cannot be postponed until other processes are completed. It is a dynamic process wherein the patient and the counselor may periodically reevaluate vocational goals in the light of changing medical-psychologic-social factors. The early introduction of the vocational rehabilitation counselor may play a vital role in the total rehabilitation process.

There are no simple rules as to when or how the issue of vocational services should be raised with a patient. When it is the physician who decides on a patient's referral for vocational evaluation, it is important that he does not rule out referrals of severely disabled persons because he assumes they are too limited for successful vocational rehabilitation. With competent vocational services many of the most severely disabled may achieve an employment goal or a work activity. A patient's readiness, interest, and motivation for vocational services can be influenced by a variety of factors. No rules or procedures for assessing readiness and stimulating interest can be as effective as the sensitive understanding and coordinated skills of the rehabilitation team.

Vocational evaluation, in the narrow sense, refers to the evaluation of education and work history and to the use of specialized testing procedures to measure, to assess, and to discover vocational potential.

The use of specialized technics of vocational evaluation cannot be appreciated separately and apart from the entire process of vocational rehabilitation. Tests and other methods of assessing vocational potential provide data that are subject to error in predicting the vocational potential of a client. Test data when utilized unknowledgeably and arbitrarily to determine the suitability of a vocational goal may be a serious obstacle to effective vocational rehabilitation services. To emphasize the hazards inherent in the arbitrary use of vocational evaluation technics, we will deal with these technics in Chapter 17 as an integral part of the entire process of vocational rehabilitation, rather than discuss them separately in this chapter.

ELECTRODIAGNOSTIC PROCEDURES

Diagnostic evaluation in physical medicine and rehabilitation in many instances requires the investigation of the integrity of some portion of the neuromuscular system. In such a study electrodiagnostic procedures may be usefully applied to establish objectively and to evaluate (1) whether an organic neuromuscular lesion exists, (2) the nature of the defect, (3) the extent of the lesion, whether partial or complete, and (4) the prognosis for spontaneous recovery or the need for surgical intervention.

The electrical tests employed at the present time are of greatest usefulness in the diagnosis and prognosis of peripheral nerve lesions. They are, however, also of great assistance in the investigation of many other neurogenic diseases, in some myopathic disorders, and in a variety of diseases of the central nervous system with motor dysfunction. A brief introduction to electrodiagnosis follows.

Most of the procedures utilized in electrodiagnostic studies to determine various aspects of neuromuscular function may be tabulated as follows. There is no doubt that electromyogram (EMG) and conduction studies yield the most reliable and useful data.

1. Percutaneous stimulation of peripheral nerves
2. Muscle stimulation
   a. Quantitative and qualitative aspects of the response
      (1) Faradic current stimulation
      (2) Galvanic current stimulation
   b. Special tests and observations
      (1) Strength-duration curves
      (2) Galvanic tetanus ratio
      (3) Response to repetitive stimuli of varying frequency
      (4) Progressive current ratio
      (5) Neurotization time
3. Skin-resistance studies
4. Electromyography

Percutaneous stimulation of peripheral nerves. Percutaneous stimulation of peripheral nerves is a procedure that is easily performed, yet frequently neglected. Fig. 2-3 graphically depicts the subcutaneous locations of various nerves of both upper and lower extremities. At these points the nerves are easily accessible to electric stimulation. Nerves characteristically have a low threshold of excitation so that stimuli of extremely short duration may be used to produce a response. Even with relatively high voltage, this type of stimulation is not painful and is well tolerated by the patient. Percutaneous stimulation of the nerve produces visible contraction in each individual muscle innervated thereby. When the ulnar nerve, for example, is stimulated at the level of the wrist, the first dorsal interosseus, the flexor pollicis brevis, the abductor digiti quinti, and the medial lumbrical muscles may be individually observed for contraction. Partial denervation can be quickly identified in this manner. Sensory loss may be estimated by stimulation with a voltage level just below that required for motor response. In this instance the patient will or will not report a tingling sensation in the autonomous sensory zone of the nerve, depending on its status at the time.

Percutaneous stimulation also may be employed to determine whether a tendinous disruption exists. In addition, following surgical anastomosis of a severed tendon, the functional integrity of the repair may be tested by electric stimulation of the appropriate peripheral nerve from which the muscle derives its innervation.

Muscle stimulation. The methods that employ direct muscle stimulation, with the possible exception of intensity duration curves, are rapidly becoming antiquated.

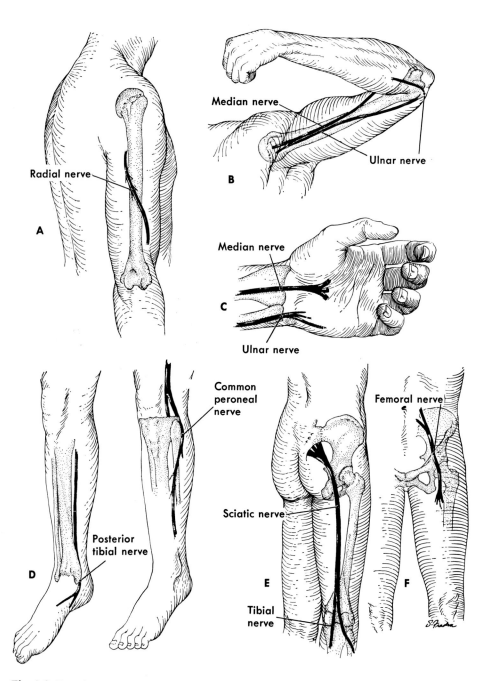

Fig. 2-3. Location of several peripheral nerves indicating sites for percutaneous electric stimulation.

Because they are still used in some laboratories, however, an abbreviated presentation follows.

Quantitative and qualitative aspects of the response. The response of skeletal muscle to electric stimulation may be studied by application of a current to the skin overlying the point of entrance of the nerve into the muscle belly, that is, the motor point.

Faradic current stimulation. The electric outflow from the secondary of an inductance coil is used. Its most important property is that it has a sharp spike with a duration of $\frac{1}{1000}$ second (1 msec. or sigma) on break of current. It is this duration of 1 sigma that lends any usefulness to faradic current in electrodiagnosis. Normal muscle has an average maximal time requirement for stimulation (chronaxie) of 1 msec.; therefore it will contract if faradic current of adequate intensity is applied. If denervation, however, is present, the chronaxie is increased and the effective duration of the faradic current will then be too short to produce a contractile response. Faradic stimulation is thus of value for the gross detection of denervation. The current, however, is painful even in moderate intensity, and patients find it unpleasant.

Galvanic current stimulation. This is a use of direct current of known polarity with which stimulation may be obtained in a most physiologic manner through the negative electrode. Most apparatus available permit accurate control of the duration of the stimulus as well as the interval between stimuli, allow for rapid change in polarity, and ·deliver a square-shaped wave with a rapidly attained maximum and minimum intensity. This current may be used to determine the muscle rheobase and chronaxie, the quality of electric response, and the effect of polar reversal.

In galvanic testing the most informative observation from direct muscle stimulation is the quality of the muscle contraction. Normal muscle contracts rapidly and sharply. Denervated muscle responds with a relatively slow contraction, a slower relaxation, and a tendency for contractions to spread contiguous fibers; the net result is a sluggish, spreading vermicular response.

In Erb's classic test the "reaction of degeneration" (RD) consists of (1) absence of response to faradic stimulation, (2) increased requirements of intensity or lack of response to faradic stimulation, (3) a sluggish-wormy contraction, and (4) reversal of the polar formula so that at least ACC = or > CCC.*

Although RD is still used in testing, it is an obsolete approach to the problem, outmoded by present knowledge of electrophysiology. The defects of this method of study will be discussed subsequently, but it may be said that the term itself is poor, since one is concerned with denervation rather than degeneration.

After nerve injury the smallest current intensity required to elicit a minimal contractile response (rheobase) varies with the time interval since injury. For the first 10 days there is no change in intensity requirement. There is then a rise of short duration that is followed by a well-marked and prolonged fall. A subsequent sharp rise in the intensity requirement frequently is a prelude to reinnervation. The observed fall in rheobase is characteristic and is in accordance with the general principle of hypersensitvity of denervated structures as postulated by Cannon.[43] Classic

---

*Intensities of the required anode closing current (ACC) and cathode closing current (CCC).

Erb testing does not take into consideration this fluctuation in rheobase, being concerned only with the increased current intensity requirement as an indication of degeneration.

The old concept of reversal of polar formula is no longer considered valid. Normally, when a motor point is stimulated with a cathode of small surface area while a large indifferent electrode is placed over an opposite part of the limb, the physiologic reaction is initiated at the motor point as the current flows across the tissues to the positive indifferent plate. If the polarity, however, is reversed, the large indifferent plate becomes the stimulating electrode; since this is far removed from the motor point, it normally elicits no response unless the current is intensified. Therefore, CCC > ACC. On the other hand, with denervation there is, in effect, no motor point, since this is the anatomic site of entrance of the now nonfunctioning nerve. Stimulation with the small cathode applicator at this point can directly affect only a small cross-section area of muscle fibers, and no response occurs. Upon reversing polarity, however, the large plate becomes the stimulating electrode and the net result is stimulation in longitudinal fashion; many more muscle fibers are affected, and a visible contraction occurs. Therefore ACC = or > CCC. It is this factor of longitudinal stimulation that also explains the often-used misconception of "distal movement of the motor point" in denervation. This is incorrect, for the motor point is a fixed anatomic landmark that cannot move.

Slowly contracting muscle fibers are a characteristic finding in denervation and are most readily observed in complete lesions. A partial lesion poses a greater diagnostic problem, and special tests may have to be done to indicate the presence of the denervation. Most applicable are stimulation away from the motor point and the use of a progressive current to demonstrate the inability of denervated fibers to accommodate.

To determine chronaxie, first obtain the rheobase—it being the smallest intensity of current of indefinite flow that will produce a minimal visible muscle contraction. This intensity of current is then doubled, and the duration of its flow required to produce the same minimal visible contraction is the chronaxie. Determination of the rheobase is subject to errors of procedure and to such variable examination conditions as temperature of the extremity, presence of adipose tissue and scars, and excessively high skin resistance. Some of these factors may be obviated by utilization of chronaxie determinations; these are accurate, reproducible studies that are highly significant in detection of and prognostication of the course of denervation states. The normal chronaxie values vary from 0.04 to 0.8 sigma, with a rough normal value of less than 1 msec. for any skeletal muscle. This variation in normal value depends on whether the muscle is proximal or distal and whether it is of an extensor or flexor group. In denervation there is characteristically an elevation of chronaxie; this may reach 50 msec. or more. Reinnervation conversely is characterized by a progressive fall toward normal time. There frequently may be a terminal lag with a somewhat elevated chronaxie persisting for a short interval after voluntary motor function has been reestablished.

Special tests and observations

Strength-duration curves. Chronaxie determinations are dependent on a formula relating muscle response, stimulus duration, and stimulus intensity as follows: $K =$

(k) (local excitatory state) (intensity × duration). The stimulus duration, of course, must fall within the utilization time limits for the inverse relationship to intensity to be valid. Strength-duration (S-D) curves are determined from this same formula; instead of a single chronaxie reading, however, multiple intensity and duration values in relationship to K are established and plotted. Such a series presents statistic advantage over unitary determinations; in addition, the shape of the plotted curve may afford prognostic information. There is a difference of opinion as to how many points should be plotted to give an accurate S-D curve, but adequate results may be obtained by utilizing between eight and sixteen determinations of K.

The normal S-D curve is essentially flat, showing a rise only at the zero end of the time abscissa. Denervation is characterized by a sharply progressing shift of the curves to the right, a fall in the rheobase, and a rise in the chronaxie. Reinnervation, on the other hand, is characterized by a shift of the curves to the left with a fall in chronaxie, a rise in rheobase, and the appearance of plateaus, the latter representing regions in which the serial determinations remain steady. S-D curves are thus a valuable means for prognosticating reinnervation. A single determination, however, indicates only whether denervation is present or absent and roughly its extent. For prognostic evaluation, serial determinations must be carried out. S-D curve determinations require care and considerably more time to carry out than does a single chronaxie; their determination, however, affords far greater accuracy as a diagnostic and prognostic tool.

In addition to those just described, the following special tests are even less commonly used.

Galvanic tetanus ratio (GTR). The GTR relates the intensity of current required to elicit a simple muscle twitch (the rheobase) to the intensity of current required to bring the same muscle into a sustained or tetanic contraction. In a normal person this ratio is approximately 3.5 to 6. The test is of value as an index of nerve growth.

Response to repetitive stimuli of varying frequency. In normal muscle the application of rapidly repeated stimuli results in a fused or tetanic contraction. The required intensity, within certain limits, is relatively stable regardless of the frequency of the applied current. Denervated muscle, on the other hand, shows an inverse relationship of intensity to frequency; early in the reinnervation phase this phenomenon reverses and returns to normal.

Progressive current ratio. This test is based on the loss of accommodation of denervated muscle. It relates the threshold of contractile response to stimulation with progressive current to the rheobase and is of considerable value in the detection of partial denervation.

Neurotization time (NT). Neurotization time is a useful index that represents a ratio of the duration of the neuropathy to the theoretic time necessary for reinnervation to take place. The formula is as follows:

$$\% \text{ neurotization time} = \frac{\text{Elapsed time since denervation} \times 100}{\text{Theoretic time for reinnervation}}$$

The theoretic time is based on the estimated rate of nerve regrowth (1 to 5 mm./day). The 100% neurotization time thus indicates that just minimal time has elapsed for reinnervation. When the neurotization time exceeds 250% with no electrodiagnostic evidence of regeneration, the prognosis is poor and surgical intervention should be contemplated.

Skin-resistance studies. The resistance of the skin to a flow of electric current is directly dependent on the integrity of submotor function. Skin devoid of sweat gland activity is electrically highly resistant, whereas skin with good sweat gland function permits an easy passage of current through the skin to the tissues below. Normal skin resistance may vary from approximately 10 thousand to 20 million ohms, depending on the local distribution of sweat glands. In a complete nerve lesion, anhydrosis results with consequent high skin resistance. Conversely, if the pathologic lesion is irritative in nature, hyperhydrosis with lowering of skin resistance may result.

The technic and instrumentation for determining skin resistance vary, but the essentials are (1) a source of voltage and (2) a variable resistance with microammeter and electrodes (Fig. 2-4). In the procedure an indifferent electrode is applied to a distant part of the body, such as the earlobe, and an exploratory electrode is used over the body areas under investigation. Normal hydrosis, anhydrosis, and hyperhydrosis may thus be mapped.

The value of this test is based on the coincidence of the area of sensory distribution of a mixed nerve with its autonomic zone of sudomotor control. The latter can be outlined by skin-resistance studies, thus permitting objective documentation of sensory disturbances in peripheral neuropathy and segmental nerve diseases. This type of investigation is known as neurodermometry.

Electromyography. Electromyography is concerned with the detection, recording, and interpretation of the electric voltages generated by skeletal muscle. Details and the technical aspects of recording electromyograms and of electromyographic equipment has been extensively described by many workers. A simple flow sheet (Fig. 2-5) may be used to illustrate the general principles of instrumentation.

Since mensuration is in milliseconds and microvolts, the apparatus must be capable of recording these minute quantities. The electric output of the muscles is detected with needle or surface electrodes and fed into the amplifier system. From here

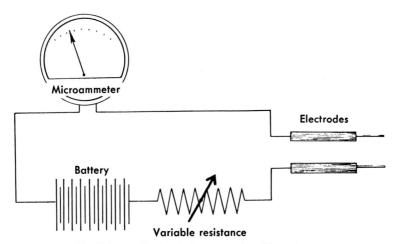

Fig. 2-4. Simple circuit for measuring skin resistance.

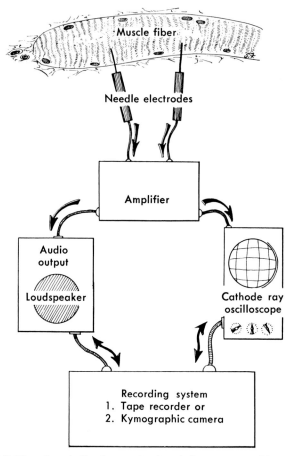

Fig. 2-5. Flow sheet indicating mechanics of electromyographic recording.

the electromotive force (EMF) is directed to two outputs: (1) the cathode ray oscilloscope and (2) the loudspeaker, in which the electric energy is converted to audible sound. The latter is feasible, since the frequency characteristics of motor unit discharge are within the confines of the audible range. One may thus simultaneously listen to and visualize the muscle potentials. The usual apparatus also includes some permanent recording device, also illustrated in Fig. 2-5. In most instances, modern day instrumentation does not require a shielded room.

Characteristic electromyographic pattern (Fig. 2-6)

Normal motor unit output. A normal voluntary muscle at complete rest is electrically silent, except for end plate activity.

With voluntary or reflex contraction, an electric potential is generated. The action of the body as a volume conductor influences the wave form, so that a diphasic or triphasic configuration may be anticipated. It has been estimated that approximately 5% of the motor units normally seen may show multiple spikes, that is, complex or polyphasic waves. The amplitude of the normal motor unit averages from

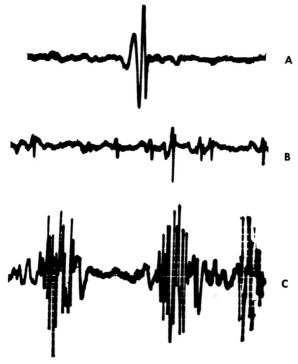

Fig. 2-6. Some typical electric potential waves. **A,** Normal motor unit potential. **B,** Fibrillation potentials of denervation. **C,** Some typical polyphasic potential waves.

100 to 2000 μv., although this is a most variable characteristic. The amplitude is not considered significantly high unless it reaches at least a level of 5.0 mv.

The duration of the motor unit action potential generally varies from 1 to 12 msec., depending on age, sex, and specific muscle. The repetitive frequency is 5 to 30/sec. or higher, depending on the intensity of the contraction. This simple wave form (Fig. 2-6, *A*), of high amplitude and long duration, is associated with a characteristic sharp popping sound emitted by the audio output.

Lower motor neuron disease

PERIPHERAL NERVE INJURIES. Electromyographic study in peripheral nerve injury presents characteristic abnormal voltages. When the nerve supply to a muscle has been disrupted, the denervated muscle, after a variable period of time depending on the animal species, will show fibrillation. Except for such an area as the tongue, these twitchings are not grossly visible. In animals there appears to be a strong correlation between the time of appearance of this denervation fibrillation and the basal metabolic rate.

In the denervated muscle, needle insertion evokes prolonged runs of electric potentials. This is in a marked contrast to the momentary "insertion potentials" observed on introduction or movement of the needle electrode in normal muscles. Furthermore, spontaneous action potentials, fibrillation, are present in the resting state. The magnitude of these waves varies up to several hundred microvolts, the duration from 1 to 2 msec., and the repetitive frequency from 5 to 30/sec. The form observed at the usual level of amplification is diphasic, with an initial positive deflection followed by a negative deflection of equal height. The sound associated with the voltage (Fig. 2-6, *B*) is similar to the clicking of radio static or the rustling of crumpled wax paper. These patterns must be differentiated from the very similar "nerve" potentials, as well as end plate activity. Spontaneous positive sharp waves are also associated with denervating lesions.

Fibrillary potentials persist until reinnervation occurs or until the muscle undergoes complete loss of contractility and fibrosis. It is of importance to note that fibrillation may be diminished by cold, ischemia, or prolonged immobilization, and that application of heat or injection of neostigmine (Prostigmin) may augment the output of fibrillation.

In a comparative study of reinnervation time, in which electromyographic data versus return of clinically observable function were used, neuronal growth was calculated at 2.6 mm./day from the former and 2 mm./day from the latter. This finding is of prognostic significance in evaluation of peripheral nerve injuries.

In the consideration of fibrillation potentials, it is an interesting hypothesis that these twitching contractions of muscle fibers represent a physiologic effort to maintain elasticity and contractility during the period of denervation. The work of Towers,[44] which showed fibrotic replacement of immobilized muscles as early as 4 months after injury, and the observation by other workers that prolonged immobilization interferes with subsequent reinnervation would favor such a viewpoint.

The motor unit voltages consistently associated with reinnervation are polyphasic in character (Fig. 2-6, *C*). In studies on the cat, polyphasic waves appeared from 9 to 16 days before perceptible movement. In human beings small polyphasic potentials appear several days to several months before functional return of movements. Similar complex waves, however, may be encountered early in acute poliomyelitis (third to fourth week of the disease) and in degenerative lesions such as amyotrophic lateral sclerosis; thus polyphasic potentials may be evidence of a denervation as well as indicative of a reinnervation process. The average physical characteristics of a polphasic motor unit potential may be summarized as follows:

Magnitude	10-1500 μv.	Number of spikes	5-20
Duration	5-15 msec.	Audio output	chug
Frequency	5-30/sec.	(chug of a Model T Ford)	

In summary, reinnervation may show only subtle differences, is electromyographically represented by a progressive decrease in fibrillation, by the appearance of polyphasic motor unit potentials, and, finally, by the appearance of normal motor unit discharges. This shift is not always so simple as indicated; a small percentage of polyphasic waves may persist for a long time after clinical recovery. Fibrillation has been observed in muscles in which no clinical weakness was evident. After reinnervation the amplitude of the action potential may be unusually high ("hypertrophied units").

PERIPHERAL NEUROPATHIES. The peripheral neuropathies exhibit changes in electromyographic records that show only subtle differences from those of the peripheral nerve injuries described. There is often complete electric silence, both at rest and on attempted voluntary contraction. Fibrillation is less frequently encountered. In the recovery phase some motor units may be seen discharging at very low frequency rates without regard to the intensity of effort, and reduplication of action potentials may occur. Large rhythmic action potentials, with voltages from six to eight times greater than normal, and fasciculations of the decrementing type similar to those described for Prostigmin-induced twitchings have also been described.

ROOT COMPRESSION SYNDROME. Irritative lesions of the roots of spinal nerves may result in a spontaneous discharge of simple or complex voltages and produce an involuntary contraction either of a single muscle fiber or of the entire motor unit. The muscles that are involved may clinically show a sustained "spasm" even at rest. Simple fasciculation voltages are of normal motor unit size and configuration and may be continuously present during rest and activity. Although the physical characteristics do not differ from the normal, their presence in the resting state and their tendency to exhibit an irregular rate and rhythm are the differential features. Complex fasciculations are similar in appearance to polyphasic potentials.

Fascicular twitches may arise from stimuli applied everywhere along the course of the motor neuron; the major number, however, arise at the anterior horn cell itself. Fasciculation potentials appear in the segmental distribution of spinal cord lesions and can be utilized in localizing the level of the lesion. It is important, however, to point out not only the significance of fasciculation waves but also the equal importance of finding fibrillation potentials. Fasciculation potentials have been reported in amyotrophic lateral sclerosis, progressive spinal atrophy, poliomyelitis, radiculitis, and spinal cord tumor.

In axonal compression, bursts of grouped motor unit potentials may serve as a differential feature in distinguishing between neuronal and axonal irritative lesions. Since they are more likely to occur in peripheral diseases, this is most common in facial nerve involvement.

POLIOMYELITIS. The electromyographic findings in poliomyelitis may be reviewed under the headings abnormal potentials, synchronization, and velocity of conduction.

Abnormal potentials: Early investigations[45] in poliomyelitis revealed a decrease in the amplitude of motor unit potentials as well as in the number of impulses fired per second. It was felt that this decrease was proportional to the decrease in muscle power and therefore represented the amount of damage sustained by the anterior horn cells. Following this line of investigation, other workers tried to correlate the incidence of fibrillation and complex motor units with the Lovett system of muscle rating. Studies were also extended in an attempt to correlate the extent and duration of denervation fibrillation with the degree of clinical recovery. In poliomyelitis it was found that none of the muscles showing 100% fibrillation between the twenty-first and sixtieth days after onset of the disease would recover motor power at the end of a year. These same investigators[46] described the presence of polyphasic potentials early in the course of acute poliomyelitis. As in the case of peripheral nerve injury, fibrillations in poliomyelitis appear between the eighteenth and twenty-first days after onset.

Synchronization: The appearance of synchronous motor unit discharges in partly functioning muscles and their antagonists as a result of poliomyelitis has been reported[47]; this was interpreted as a disturbance at the spinal cord segment of the mechanism providing reciprocal innervation and synchronization. It was also suggested that synchrony in poliomyelitis indicated an unfavorable prognosis. Later studies by another group showed that peripheral stimulation with a progressive current evoked synchronous discharge of potentials in chronic cases; this would militate against the explanation that abnormal synchronization of discharges arise at the neuron. Synchrony might also be explained as only an apparent observation, a consequence of pickup of activity by multiple electrodes of the voltages generated by one large discharging unit.

Velocity of conduction: The percutaneous application of supramaximal electric shocks to a nerve initiates an impulse that results in an action-potential output in the muscles innervated by this outflow. Accurate timing of the reaction permits estimation of the conduction velocity in the fastest nerve fibers. Normally this rate is approximately 50 to 70 M./sec. (60 M./sec. in the ulnar nerve). In patients with poliomyelitis the rate of nerve conduction is normal. In acute Guillain-Barré disease, conduction is characteristically slowed.

It is of interest to note that in acute poliomyelitis myotatic responses appear to be increased. Electromyographic studies have shown that electric activity persists in the affected muscles as long as they continue to be stretched. The etiologic basis for this finding may be the nociceptive impulses arising from meningeal irritation of posterior roots or ganglia or, as suggested by study of experimental poliomyelitis of the *Macaca rhesus,* may be due to lesions in the reticular substances of the medulla.

Upper motor neuron diseases and other central nervous system lesions. While electromyographic investigators have shown a preponderant concern with studies of the lower motor neuron system, it appears that studies of lesions at other levels of the central nervous system, whose abnormal impulse discharges affect the integrity of the motor unit, may be of equal interest and clinical significance.

In lesions, for example, associated with spasticity, electromyographic studies verify the clinical observation that the slightest stimulation triggers contractions of the stretched muscle. The electromyograph shows intense electric output in the agonist and a less regular, smaller unit output in the antagonist. It is possible, on the other hand, although at times with great difficulty, to carefully position a spastic limb so that it becomes electrically silent. Passive movement of the spastic muscle produces the same effects on the agonist and the antagonist, as well as a marked tendency to overflow into synergistic muscles. Studies with multiple-needle electrodes reveal that units of the spastic muscle seem to fire synchronously and that instead of single potentials the spastic state appears to promote group unit discharge. In addition,

Table 2-4. Comparison of neurogenic and myogenic atrophy

	Neurogenic	*Myogenic*
Size of individual unit	Normal or larger	Generally smaller
Duration of spike	Normal	Short duration
Presence of fibrillation	Present	Less frequently seen
Presence of fasciculation potentials	Present frequently	Never
Interference pattern	Single unit pattern	Total interference

tendon reflexes show an increased myotatic response. Instead of a single spike there is a run of potentials, which in some areas is followed by small recurrent discharges resembling the electric output from muscles in clonus.

Myopathies

MUSCLE DYSTROPHY AND ATROPHY. Electromyographic studies may be of value in establishing a positive diagnosis in cases of muscle disease in which clinical evaluation is inconclusive. Table 2-4 lists some differential highlights between neurogenic and myogenic atrophy.

MYOTONIA. Myotonia is a neuromuscular disease clinically characterized by a painless tonic spasm of muscles induced by voluntary or reflex movement. Electromyographic tracings reveal that the electric phenomenon associated with and following this visible contraction is output of a high-frequency potential, with a rate varying from 25 to 100 spikes/sec. The individual spikes resemble fibrillation potentials (single muscle fiber discharges) and may persist for 1 minute after voluntary effort ceases. Bizarre high-frequency discharges that are not difficult to distinguish from true myotonia may be encountered in a variety of diseases such as pseudohypertrophic muscle dsytrophy, amyotonia, and peripheral neuropathy. In myotonia the variation in the amplitude, as well as the changes in the frequency of the discharges, gives rise to a waxing and waning sound that has been compared to the sound of a dive bomber.

Myasthenia gravis. Myasthenia gravis clinically and electromyographically presents characteristics similar to those seen in a progressive partial nerve block. The EMG recorded during a weak contraction may show decrementation of the first six to ten muscle action potentials followed by fluctuation of the amplitude of a single unit potential. In a region in which unit failure has taken place, there is no recovery as long as effort is maintained, but even momentary relaxation is followed by renewal of firing and then failure again. The response of the muscle to repetitive stimulation of the nerve at specific frequencies is quite characteristic. There is an early decrementation followed by a period of facilitation, which ultimately is succeeded by a late phase of exhaustion. The defect may be restored by intravenous edrophonium chloride (Tensilon). A distinctive myasthenia-like response in the skeletal muscles may be encountered in persons with manifest or subclinical small cell carcinoma of the lung.

Metabolic disturbances. In recent years the neuromuscular manifestations of metabolic disorders have been investigated by electromyography. Although it has been known that specific ionic concentrations effectively behave like the polarizing electrodes of a galvanic current, Kugelberg[48] correlated the appearance of foci of irritation in the muscles to an index of accommodation of the nerve. This index is depressed by hypocalcemia, alkalosis, and ischemia. In hypocalcemia, for example, electromyographic findings of a spontaneous discharge of spike potentials may be demonstrated at a serum calcium level of 5.2 mg. %.

CONDUCTION VELOCITY STUDIES. The estimation of the rate of conduction of impulses in peripheral nerves may be of immeasurable aid in establishing a definite diagnosis. In considering a patient for whom differential diagnosis concerns polymyositis or polyneuropathy, for example, it is of help to know that conduction velocity in the former will be normal, whereas in the latter it is frequently profoundly slowed. If a normal nerve conducts at a velocity of 57 M./sec., the value in a severe polyneuropathy may be in the range of 20 M./sec.

Conduction studies are of value in localizing segmental neuropathy. In compression of the median nerve in the carpal tunnel, for instance, the interval between the shock artifact and the muscle response (latency) may be considerably increased, from a mean value of approximately 3.5 msec. to 7 to 10 msec.

A refinement of conduction studies is represented by the recording of nerve action potentials through the skin overlying peripheral nerves. In a peripheral neuropathy involving sensory fibers, the digital nerves may be stimulated with ringlike electrodes and recording of the nerve potential be carried out over the corresponding nerve at the wrist. The evoked potential may show a prolonged latency, may show a marked diminution of its amplitude, or may be completely absent.[49]

Electromyography is an established procedure of proved worth in the diagnosis of neuromuscular diseases. It is, however, an adjunct to such diagnosis and can in no way substitute for carefully formulated clinical judgment. Judicious application of these electrodiagnostic methods that have been only briefly reviewed as an introduction depends on awareness of such a concept.

FUNCTIONAL EVALUATION IN OCCUPATIONAL THERAPY

Two tests have been designed as guidelines to treatment planning for therapeutic measures necessary for functional performances as carried out in an occupational therapy program. These are the motor-sensory evaluation for patients with upper motor neuron disturbances and the functional motion test, which is used for patients with lower motor neuron deficiencies.

Motor-sensory test (Chart 4)

The functional motor-sensory evaluation used by the occupational therapist is primarily designed to be used with problems related to upper motor neuron lesions. Considering that these resultant disabilities primarily present problems of muscle control and sensory integration, the test is designed to evaluate gross motor, sensory, and perceptual function. It appears that in most disabilities of upper motor neuron lesion, particularly with the hemiplegic patient, a careful assessment of function in those areas mentioned reveals the presence of motor function, but that lack of sensory and perceptual integration inhibits independence. For this reason the evaluation procedure attempts to categorize the specific impairment and dysfunction to assist the therapist in treatment planning. When there are apparent problems in the area of perceptual performance, collaboration with other disciplines are essential for a comprehensive evaluation.

Functional motion test (Chart 5)

The functional motion test is used to measure the *motions* of an upper extremity in terms of active strength as available from combined muscle action. It tests movements of each arm segment separately, that is, shoulder flexion, shoulder abduction, elbow flexion, etc.

Measurement of strength of a functional motion is usually the primary need in assessment of lower motor neuron problems. However, the test also provides for

Text continued on p. 55.

CHART 4

Institute of Rehabilitation Medicine
New York University Medical Center

OCCUPATIONAL THERAPY

FUNCTIONAL MOTOR SENSORY EVALUATION

Date _____

Name _____

Address _____

Age _____ Dominance _____

Disability _____

Occupation _____

I. MOTOR PERFORMANCE

KEY

1 -- Complete impairment (0%)
2 -- Severe impairment   (1% - 25%)
3 -- Moderate impairment (26% - 75%)
4 -- Minimal impairment   (76% -99%)
5 -- No impairment   (100%)

NA -- Not applicable
NT -- Not tested

A. GROSS FUNCTIONAL PATTERNS

LEFT 5 4 3 2 1		RIGHT 1 2 3 4 5
	Hand to mouth (elbow flexion)	
	Hand to opposite wrist (elbow flexion with adduction)	
	Hand to opposite elbow (elbow flexion with adduction)	
	Hand to opposite shoulder (elbow flexion with adduction and shoulder flexion)	
	Hand to same shoulder (elbow flexion with supination)	
	Middle of waist (shoulder abduction)	
	Side reach (shoulder abduction with elbow extension)	
	Nape of neck (external rotation)	
	Behind back (internal rotation)	
	Forward reach (shoulder flexion with elbow extension)	
	Overhead (shoulder flexion with external rotation)	
	Palm up (supination)	
	Palm down (pronation)	
	Wrist extension	
	Wrist flexion	

B. HAND FUNCTION

LEFT		RIGHT
	Grasp	
	1. Type	
	2. Size	
	3. Strength	
	Release	
	Manipulation (coins in purse)	
	Consecutive finger motion	
	Prehension	

Limitations:

Flaccidity_____   Coordination _____   Pain_____ _____

Spasticity_____   Speed_____   Subluxation_____
                          Latent _____
Flexion synergy_____   Slow _____   Substitutions ___ _____

Extension synergy_____   Endurance _____   Perseveration _____

Passive ROM _____   Edema _____   Motor apraxia _____

Continued.

CHART 4, cont'd

II. SENSORY STATUS

Left		RIGHT
	Esthesiometer	
	Stereognosis	
	Gross position sense	
	Fine position sense	

III. PERCEPTUAL PERFORMANCE

Vision:
Visual hemianopsia _____   Compensates: Yes _____ No _____
Ocular pursuit _____   Other problems: _____

Visual-motor performance:
(Peg test: DI — direct placement; O — omissions; SC — self-corrects)

Type of error
Vertical   _____
Horizontal _____
Diagonal   _____
Square   _____

Visual perception:
(Frostig subtests)
Raw score
No. II Figure ground discrimination _____   (0-20)
No. V Spatial organization _____   (0-8)

Concept formation:
Ratio score
Form board: _____ (10:10)

Motor planning and self-care activities:

1. Self-care   2. General
Eating _____   Dial phone _____
Button _____   Fold paper, put in envelope ____
Grooming   Write or copy name _____
Shaving _____   Turn pages _____
Teeth ____
Cosmetics ____

Spontaneous use of extremity: Yes _____ No _____

Behavior:
Attention _____
Self-corrects _____
Comprehension _____
Confusion _____
Memory _____
Lability _____

IV. SUMMARY AND RECOMMENDATIONS

Therapist _____

CHART 5

**Institute of Rehabilitation Medicine
New York University Medical Center**
OCCUPATIONAL THERAPY

FUNCTIONAL MOTION TEST REPORT

Date _____

Name_____

Address_____

Age _____ Chart No._____

Disability_____

Occupation_____

Summary of findings

Activity limitations or problems

Goals:
        Immediate

        Long range

Recommended program

Recommended devices (if any)

Continued.

CHART 5, cont'd

LEFT (dominant)					RIGHT (dominant)		
			Examiner's initials				
			Date				
Lack of control	Loss of ROM	Active strength		Active strength	Loss of ROM	Lack of control
			SHOULDER			
			Abduction			
			Adduction			
			Flexion			
			Extension			
			Int. rotation			
			Ext. rotation			
			ELBOW			
			Flexion			
			Extension			
			FOREARM			
			Supination			
			Pronation			
			WRIST			
			Flexion			
			Extension			
			Radial deviation			
			Ulnar deviation			
			HAND			
			Flexion MCP			
			Flexion PIP			
			Flexion DIP			
			Extension MCP			
			Extension PIP			
			Extension DIP			
			Finger abduction			
			Finger adduction			
			Thumb abduction			
			Thumb adduction			
			Thumb flexion			
			Thumb extension			
			Thumb opposition			
			Grasp - palmar			
			Grasp - lateral			
			Grasp - hook			

Trunk balance _____
Sitting posture _____
Fatigue _____
Sensory losses _____
Visual problems _____
Auditory problems _____
Temperature or pain _____
Motivation _____

Coding keys:
0 = zero   1 = minimal
T = trace   2 = moderate
P = poor   3 = severe
F = fair
G = good   red = 0 to F-
N = normal   black = F to N

Activities performed   Potential   Actual
Eating
Bathing
Grooming - shaving
      - teeth
      - cosmetics
Dressing
Communications - writing
           - typing
           - telephone
           - reading
Wheelchair manipulation

Sample of
Patient's signature _____

CHART 6

Institute of Rehabilitation Medicine
New York University Medical Center
Occupational Therapy

Directions for Administering
FUNCTIONAL MOTION TEST FOR UPPER EXTREMITY

METHOD

All motions are performed by the patient independently in a sitting position in
one of the three body planes. The motion may be demonstrated by the therapist
first. It should be repeated if there is a question of fatigue. The second
positions are to be used when the first test position grade is less than fair.

GRADING

A grade of O, T, P, F, G, or N is given according to standard interpretation of
these scores as used in muscle testing.

BODY PLANES

Sagittal -- movement in the anteroposterior direction

Frontal -- movement in the lateral direction (toward and away from midline)

Transverse -- movement in the horizontal direction (dividing body in half -- top
and bottom)

SHOULDER

Abduction: Raise arm out from side to shoulder level in the frontal plane.
At end of motion apply resistance.

Adduction: Hold arm at side of body. Apply resistance at inner aspect of
upper arm.

Flexion: Raise arm forward, sagittal plane, to shoulder level. Apply resistance.

Extension: Hold arm at side of body. Apply resistance at back of upper arm.

Internal rotation:
First position -- With elbow on arm of chair and forearm in upright
position, move forearm in across front of body in frontal plane.
At end of motion, apply resistance.

Second position -- With elbow flexed, forearm resting on arm of wheelchair, move forearm in toward body in transverse plane.

External rotation:
First position -- With elbow flexed and forearm resting across front
of body (hand in lap), raise forearm out to side in frontal plane.
Hold. Apply resistance.

Second position -- With elbow flexed and forearm resting across front
of body (hand in lap), move forearm out to side in transverse plane.

Continued.

CHART 6, cont'd

ELBOW

Flexion:
  First position -- With shoulder at rest, flex elbow fully in sagittal
    position. Hold. Apply resistance.

  Second position -- With therapist holding shoulder at 90° flexion, move
    forearm in toward body in transverse plane.

Extension:
  First position -- With elbow fully flexed and shoulder flexed to 90°,
    move forearm out away from body in sagittal plane. Hold. Apply resistance.

  Second position -- With therapist holding shoulder flexed at 90° and
    with elbow flexed to 90°, move forearm out in transverse plane.

FOREARM

Supination: With forearm resting on arm of wheelchair (or table), palm
  down, rotate forearm in frontal plane. Hold. Apply resistance.

Pronation: With forearm resting on arm of wheelchair (or table), palm up,
  rotate forearm in frontal plane. Hold. Apply resistance.

Alternate position for both motions: If above grades are less than fair,
  rest elbow on arm of chair or table. Holding forearm in upright position,
  rotate in both directions.

WRIST

Extension:
  First position -- With forearm resting on arm of wheelchair, table, or
    therapist's hand, palm down, extend wrist in sagittal plane. Hold.
    Apply resistance.

  Second position -- With forearm resting in midsupinated position, extend
    wrist in frontal plane.

Flexion:
  First position -- With forearm resting on arm of wheelchair, table, or
    therapist's hand, palm up, flex wrist in sagittal plane. Hold.
    Apply resistance.

  Second position -- With forearm resting in midsupination position, flex
    wrist in frontal plane.

HANDS

If finger motions are possible, test each joint separately, same as in
  a regular muscle test. Also test following types of grasp:

Gross grasp: Open hand as wide as possible. Close hand and make a fist.

Palmar pinch: Oppose pad of thumb to pads of index and middle fingers.

Lateral pinch: Oppose pad of thumb to side of index finger and then to
  side of middle finger.

Hook: Squeeze fingers around therapist's fingers or around a 1-inch dowel.

NOTE: On gross grasp and finger pinches, test for strength by giving
  resistance with finger.

recording of gross estimates of range of motion and/or amount of control, in as much as these problems may be concurrent or superimposed on the major neurologic impairment.

This method of evaluation provides identification of each patient's specific problems, offering a rationale for physical restoration through functional activities and for selection of suitable equipment when this is necessary to start any therapeutic or functional activity. Also, through application of analysis of motions of various ADL (Chapter 9), it enables a prediction of a patient's potential performance in ADL.

Administration of test (Chart 6)

Since the purpose of the test is to check *functional motions* of the upper extremity, all testing should be done with the patient in an erect sitting position. It can be done in a semireclining or supine position but may not then reflect maximum potential, particularly in regard to shoulder motions. Grades for recording motions are the same as those used for a muscle test, as they are universally known and defined. The user should not confuse it with a muscle test, however, even though it appears similar. Note that when testing hand function, if motions are absent or minimal, gross types of grasp may be checked instead.

There are three areas for recording loss of function: one for loss of strength, a second for loss of range of motion (ROM), and a third for loss of control. When testing for strength, the ROM and loss of control problems may be detected simultaneously. If they are present, additional checking of these problems may be desirable. Note that grading of these losses is recorded only as minimal, moderate, or severe. This is felt to be sufficient for treatment purposes when they are secondary to, or superimposed on, loss of strength. When these problems become major factors, more detailed testing may be necessary. In such cases a standard ROM test or tests for upper motor neuron disabilities may be given.

Additional problems of trunk balance and sitting position may be simultaneously noted. Sensory and visual losses may be superficially noted and indications for further testing made if necessary. Fatigue may be observed or judged by the number of hours a patient is allowed to sit. Information regarding the patient's motivations, aspirations, previous hospital experiences and responses to various regimes, and types of activity interests may be obtained by informal discussion with the patient during testing of motions. The eleven activities listed on the form are graded according to (1) potential for doing them with or without equipment as based on the test, (2) the patient's verbal statement, or (3) actual demonstration.

REFERENCES

1. Brocklehurst, J. C.: Incontinence in old people, Edinburgh, 1951, E. & S. Livingstone, Ltd.
2. Buchwald, E., McCormack, M., and Raby, E.: Rehabilitation monograph II: A bladder and bowel training program for patients with spinal cord disease, New York, 1952. Institute of Physical Medicine and Rehabilitation.
3. Fundamental studies of human locomation and other information relating to designs of artificial limbs, vols. 1 and 2, Berkeley, 1947, University of California College of Engineering.
4. Saunders, J. B., Saunder, C. M., Inman, V. T., and Eberhardt, H. D.: The major determinants in normal and pathological gait, J. Bone Joint Surg. 35A:543, 1953.

5. Lovett, R. W.: The treatment of infantile paralysis, Philadelphia, 1916, P. Blakiston's Son & Co.

6. Lovett, R. W., and Martin, E. G.: Certain aspects of infantile paralysis: a description of a method of muscle testing, J.A.M.A. **66:**729, 1916.

7. Kendall, H. O., and Kendall, F. P.: Muscles, testing and function, Baltimore, 1949, The Williams & Wilkins Co.

8. Daniels, L., Williams, M., and Worthingham, C.: Muscle testing, techniques of manual examination, ed. 2, Philadelphia, 1956, W. B. Saunders Co.

9. Salter, N.: Methods of measurement of muscle and joint function, J. Bone Joint Surg. **37B:**474, 1955.

10. Moore, M. L.: Joint measurement: technique and recording methods. In Proceedings of the Second Congress of the World Confederation for Physical Therapy, New York, 1956, American Physical Therapy Association.

11. Lawton, E. B.: Rehabilitation monograph X: A.D.L., activities of daily living, testing, training and equipment, New York, 1956, Institute of Physical Medicine and Rehabilitation.

12. Lawton, E. B.: Activities of daily living for physical rehabilitation, New York, 1963, McGraw-Hill Book Co., Blakiston Division.

13. Jastak, J., and Robison, R. K.: The clinical application of factorial measures, Delaware State Med. J. **21:**169, 1949.

14. Morrow, R. S., and Cohen, J.: Psycho-social factors in muscular dystrophy, J. Child Psychiat. **3:**70, April, 1954.

15. Richards, T. W., and Lederman, R.: A study of action in the fantasy of the physically handicapped, J. Clin. Psychol. **2:** 188, 1956.

16. Cruisbank, W., and Raus, G.: Cerebral palsy, its individual and community problems, Syracuse, N. Y., 1955, Syracuse University Press.

17. Gesell, A., and Amatruda, C.: Deveopmental diagnosis, ed. 2, New York, 1947, Psychological Corp.

18. Bauer, R. W., and Becka, D. M.: Intellect after cerebral vascular accident, J. Nerv. Ment. Dis. **120:**379, 1954.

19. Reitan, R. M.: Intellectual functions in aphasia and non-aphasic brain injured subjects, Neurology **9:** 202, 1953.

20. Crowell, D., and Crowell, D.: Intelligence test reliability for cerebral palsied children, J. Consult. Psychol. **18:** 276, 1954.

21. Strauss, A. A., Kephart, N. C., Lehtinen, L. E., and Goldenberg, S.: Psychopathology and education of the brain-injured child. II. Progress in theory and clinic, New York, 1955, Grune & Stratton, Inc.

22. Glick, S. J.: Vocational, educational and recreational needs of the cerebral palsied adult, New York, 1953, United Cerebral Palsy of New York City.

23. Halstead, W., and Wepman, J.: The Halstead-Wepman screening tests of aphasia, J. Speech Hearing Dis. **14:**9, Jan., 1949.

24. Eisenson, J.: Examining for aphasia, New York, 1954, The Psychological Corp.

25. Ammons, R. B., and Ammons, H. S.: The full range vocabulary test, New Orleans, 1948, R. B. Ammons.

26. Wells, F. L., and Reusch, J.: Handbook fr mental examiners, New York, 1945, The Psychological Corp.

27. Orientation training for vocational rehabilitation counselors, Rehabilitation Service Series, no. 332, Washington, D. C., Oct., 1955, U. S. Department of Health, Education, and Welfare, Office of Vocational Rehabilitation.

28. Casework performance in vocational rehabilitation, Washington, D. C., 1955, U. S. Department of Health, Education, and Welfare.

29. Hamilton, K. W.: Counseling the handicapped in the rehabilitation process, New York, 1950, The Ronald Press Co.

30. Di Michael, Salvatore C.: Phychological services in vocational rehabilitation, Washington,

D. C., 1959, U. S. Department of Health, Education, and Welfare, Office of Vocational Rehabilitation.

31. Hull, C. L.: Aptitude testing, New York, 1928, World Book Co.
32. Crawford, A. B., and Burnham, P. S.: Forecasting college achievement, New Haven, Conn., 1946, Yale University Press.
33. Chouinard, E. L., and Garrett, J. F.: Workshops for the disabled—a vocational rehabilitation resource, Washington, D. C., 1956, U. S. Department of Health, Education, and Welfare, Office of Vocational Rehabilitation.
34. Bridges, C. D.: Job placement of the physically handicapped, New York, 1946, McGraw-Hill Book Co., Inc.
35. Hanman, B.: Physical abilities to fit the job, North Abington, Mass., 1956, Sanderson Bros.
36. Selective placement of the handicapped, Washington, D. C., 1945, U. S. Department of Labor.
37. Interviewing guides for specific disabilities (6 pamphlets: Diabetes, Heart disease, Epilepsy, Pulmonary tuberculosis, Orthopedic disabilities, Arthritis), Washington, D. C., 1954-1956, U. S. Department of Labor.
38. Rusk, H. A., and Taylor, E. J.: New hope for the handicapped, New York, 1949, Harper & Brothers.
39. Denny-Brown, D.: Interpretation of the electromyogram, Arch. Neurol. Psychiat. 61:99, 1949.
40. Kugelberg, A.: Electromyograms in muscular disorders, J. Neurol. Neurosurg. Psychiat. 12:129, 1949.
41. Licht, S.: Electrodiagnosis and electromyography, New Haven, Conn., 1956, E. Licht Publishing Co.
42. Marinacci, A. A.: Clinical electromyography, Los Angeles, 1955, San Lucas Press.
43. Best, C. H., and Taylor, N. B.: The physiological basis of medical practice, ed. 4, Baltimore, 1945, The Williams & Wilkins Co.
44. Towers, S. S.: Atrophy and degeneration in skeletal muscle, Amer. J. Anat. 56:1, 1953.
45. Hansson, K. G., Troedsson, B. S., and Schwarzkopf, E.: Electromyographic studies in poliomyelitis, Arch. Phys. Ther. 23:261, 1942.
46. Huddleston, O., and Golseth, J. G.: Electromyographic studies of paralyzed and paretic muscles in anterior poliomyelitis, Arch. Phys. Med. 29:92, 1948.
47. Buchthal, F., and Madsen, A.: Synchronism activity in normal and atrophic muscle, Electroenceph. Clin. Neurophysiol. 2:425, 1950.
48. Kugelberg, E.: Accommodation in human nerve and its significance for the symptoms of circulatory disturbances in tetany, Acta Physiol. Scand. 24 (supp. 8): 1-102, 1944.
49. Mayo Clinic: Clinical examinations in neurology, Philadelphia, 1957, W. B. Saunders Co.

CHAPTER 3

PRINCIPLES OF PHYSICAL MEDICINE

THERMAL THERAPY
Heat

Heat is one of the most commonly employed therapeutic measures. It is frequently used before massage and exercise because the effects of the heat enhance the effect of subsequent treatment measures. The physiologic effects are basically the same regardless of the method of application.

Physiologic effects. Externally applied heat brings about an increase in temperature of the tissues, vasodilatation, and in increase in circulation. The elevation of tissue temperature increases the local metabolic activity that adds to the increase in temperature and produces further vasodilatation. More capillaries become patent, and blood flow and capillary pressure increase. As a result, transudation increases. Some of the heat is conducted to underlying tissues. Excessive local heating is controlled by blood flow away from the area to other parts of the body. The skin becomes moist as the sweat glands are stimulated.

Local heating in an area of inflammation produces an increase in phagocytosis. If general body temperature is significantly elevated for a period of hours, the leukocyte count may go up to 20,000 or more.

Sedation with relief of pain and muscular tension are well-known effects of heat therapy. The exact mechanism of production is not completely understood. Wells[17] attributes the analgesic effect to the equalization of temperature from the superficial to the deep tissues.

Methods of application. Therapeutic application of heat is usually classified as radiant, conductive, or conversive heating. Radiant heat is the infrared portion of the electromagnetic spectrum. Wavelengths from 7700 to 14,000 A are called "near" or "short" infrared and those from 14,000 to 120,000 A the "far" or "long" infrared. Conductive heating is accomplished by direct application of heat in the form of hot water, heated air or heated moist air, heated paraffin, or electrically heated pads. Conversive heating is an indirect form of heating resulting from the conversion of various forms of primary energy into heat in the body tissues—medical diathermy.

The sources of radiant and conductive heating have these general characteristics in common: The heat is superficial, relatively inexpensive, and readily available. Although these heating devices are difficult to regulate accurately, there is a fairly wide margin of safety with their use.

The application of high-frequency alternating current results in deeper heating but is more expensive, is less readily available, and has a smaller margin of safety. Although the control is better with the deeper-heating devices, the safe and proper application requires a thorough understanding of the machine, its effects, limitations, and proper technics of application.

Superficial heat. The increase in temperature is maximal in the skin and progressively less in the deeper tissues. No significant rise in temperature occurs beyond 1 to 2 cm. Most sources of superficial heat require from 20 to 30 minutes to produce the desired effect. There is little danger of overdosage if the circulation and sensation are intact. Commonly used means of therapeutic application include infrared radiation and various forms of hydrotherapy.

Infrared radiation. There is no magic about infrared radiation; any object hotter than its surroundings will give off infrared rays. For therapeutic application either the nonluminous element or the luminous tungsten or carbon filament bulb is used with a proper reflector. Penetration of near infrared is about 3 mm. and that of far infrared, about 1 mm. The luminous bulbs produce a greater amount of near infrared and so are preferred to the nonluminous elements except when the glare of the visible light from the luminous bulbs would be irritating to the patient. Under identical conditions, wattage, and voltage, the tungsten filament lamps produce as much or slightly greater rise in the environmental and tissue temperature than the carbon filaments.[1]

A simple and safe device for applying near infrared to fairly large areas, such as the back of both thighs, is the so-called luminous heat "baker." The "baker" consists of several luminous bulbs mounted in the center of a semicircular metal reflector on adjustable legs. It is commercially available but can be made quite simply for home use. The "baker" is usually covered with a sheet to cut down the loss of heat due to air currents.

The intensity can be varied by changing the distance from the lamp to the skin. Smaller infrared lamps are usually placed about 14 to 18 inches from the skin; the larger ones, 24 to 30 inches. The patient should feel just comfortably warm, and at the end of the treatment period the skin should be warm, pink, and moist. If the skin is mottled and the reaction lasts several hours, the intensity should be decreased.

Open wounds and scars should be covered during treatment. Precautions should be taken where there is impairment of sensation or circulation.

Hydrotherapy. The term hydrotherapy has been used to describe the external application of water for therapeutic purposes. Water is a versatile medium for thermal application. Heat given by immersion in warmed water has some advantages over radiant heat. The buoyancy of the water provides gentle support without hindering movement, and the movement can be done with less effort than in air. The hydrotherapeutic means of applying heat that are most often used are as follows.

WHIRLPOOL BATH. The whirlpool bath is a metal container filled with water that is kept in constant agitation. It is an efficient means of conducting heat to the extremities. The usual temperature for therapeutic heating effects ranges from 110° to 115° F. except where there is an impairment of circulation. If there is any impairment of circulation, the water temperature should not exceed 105° F.

Like infrared it is easy to use and economical in time of personnel. In contrast to infrared the temperature can be controlled, and it can be used freely where there are open wounds and denervated areas. As with all forms of heating, care should be taken if there is any impairment of circulation; temperature should not exceed 105° F.

One disadvantage is that the immersed extremity is dependent, so that there is a tendency for edema to develop.

HUBBARD TANK. This full-body tank is an efficient substitute where space, money, or need is insufficient to warrant a therapeutic pool. All of the body except the head and neck is immersed in heated water at a temperature of 98° to 104° F. The water is usually agitated and aerated like the whirlpool. It provides a means of giving heat and gentle exercise and is especially useful where disease or disability affects many joints of the body.

HOT PACKS

Hot moist packs: These provide very superficial heat. Temperature control is negligible, and the weight of the pack contributes to ischemia and increases the danger of burning.

Wool packs: Such packs, if steam-heated and then spun almost dry, provide intense heat with little danger of burning. They usually have an inner hot pack, then a layer of waterproof plastic, and finally an outer layer of wool for insulation. The intense heat has a counterirritant effect. This method of applying heat appears to be the most effective in the relief of tenderness and muscle spasm such as found in acute poliomyelitis. The most effective application is a series of three or four packs left on for from 5 to 10 minutes each. Even though there are special hot-pack machines designed for the preparation of the packs, the method is costly in time of personnel.

Hydrocollator packs: These retain heat for a longer period of time and are handy for home use. Because of the weight they should not be used where extreme tenderness or possibility of ischemia is present.

CONTRAST BATH. The alternate immersion of the forearms or the lower legs in hot and in cold water produces an active contraction and relaxation of blood vessels and a significant increase in blood flow. One container is filled with water at 105° to 110° F., the other with water at 60° F. The part is immersed in the hot or cold water in the following order:

> Hot for 10 minutes, then cold for 1 minute
> Hot for 4 minutes, then cold for 1 minute
> Hot for 4 minutes, then cold for 1 minute
> Hot for 4 minutes, then cold for 1 minute
> Finish with 5 minutes in the hot water

This simple method of producing hyperemia is suitable for home use.

PARAFFIN. This simple, inexpensive, and adaptable method of applying heat produces an intense erythema, leaves the skin soft and pliable, and is well liked by most patients. It is especially useful in chronic joint diseases such as arthritis.

Melted paraffin can be used either in the home or in a thermostatically controlled container in an institution. The paraffin mixture consists of 7 parts paraffin to 1 part light mineral oil so that the melting point is about 126° F. The lower specific heat and minimal convection of melted paraffin prevent burning of the skin.

The two most common technics of applying paraffin are the bath and the pack. For a paraffin bath, the hand or foot is slowly dipped into the melted paraffin and then removed long enough for the paraffin to congeal; this is repeated until a thick coat is formed; then the part may either be left in the bath or be wrapped in towels for 30 minutes. The pack differs from the bath only in that the part, such as the knee or elbow, is painted with the paraffin mixture.

For home use the paraffin is melted in the top of a double boiler, removed from the fire when the paraffin is melted, and kept warm during treatment by the hot water in the bottom pan of the double boiler. There is little danger of burning the patient if the mixture is used when some part of the paraffin is still unmelted or when there is a film on the surface. The same mixture may be reheated and used repeatedly.

Because the effects of superficial heating are about the same regardless of source, the type selected and its application depend on (1) availability of source, (2) extent of area to be treated, (3) depth of effective penetration, (4) time necessary to produce desired effects, (5) ease of application, (6) margin of safety, and (7) individual preference.[2]

Deep heat. Direct superficial-heating devices are inadequate for the production of heat in the deeper tissues because the subcutaneous fat is a poor conductor of heat and, if the intensity of the applied heat were increased sufficiently for transfer into the deeper tissues, the skin would burn. Therefore, to obtain adequate deep heating, certain forms of physical energy must be used that can penetrate the skin and subcutaneous tissues without damage and are transformed into heat within the deep tissues. Various forms of physical energy are converted into heat in the tissues. The three types in clinical use at the present time are short wave, microwave, and ultrasound.

This use of high-frequency currents for heating is classified as medical diathermy. The Federal Communications Commission has established regulations under which diathermy machines are operated.

Short wave. Heating with short wave is by ultrahigh-frequency currents. The most acceptable method of applying short wave is by means of an induction cable that provides an electromagnetic field. The cable can be either wound around a part or coiled to make a flat applicator surface. Some machines have the cable arranged in a "drum applicator" for ease of application.

Short-wave diathermy is limited to wavelengths of 3 to 30 M. with frequencies of 10 to 100 megacycles. As the high-frequency current enters the body, it tends to spread so that a fairly large area is heated. The greatest increase in temperature occurs in the subcutaneous fatty tissue. Maximal depth of temperature rise is about 2 to 3 cm., over a period of 30 minutes. Intensity should be regulated by the comfortable tolerance of the patient.

The advantages of short wave over microwave and ultrasound are (1) a large area can be treated at one time, (2) the margin of safety is greater, and (3) skin sensations can still be used as an indication of intensity. Short wave provides more equal heating than does infrared radiation, but otherwise has little advantage over properly applied forms of superficial-heating devices used for therapy.

Microwave. The energy for microwave therapy consists of radiated electromagnetic waves. Machines used at this time radiate wavelengths of 12.2 cm. at a frequency of 2450 megacycles/sec. The various directors supplied with the machines provide focusing from several inches away from the skin surface. There should be an airspace of several inches between the director and the skin surface.

Ultrasound. The energy for ultrasound therapy consists of mechanical vibrations with a frequency ranging from 0.7 to 1 megacycle. These waves, unlike electric waves, do not travel through air, so that direct contact is required. A coupling medium such as heavy mineral oil is used between the applicator head and the skin. If used under water, a distance of from $\frac{1}{2}$ to 2 inches should separate the applicator from the skin.

Biophysical considerations of microwave and ultrasound.[3] Microwave and ultrasound are superior to short wave in production of heat in deep tissues and are more complicated in their action. Proper clinical use requires an adequate understanding of their biophysical characteristics. Instruments on the machines register the amount of energy generated by the source; how much of that energy penetrates to the deep tissues to become biologically effective depends on a number of factors.

Part of the energy of microwave is reflected back from the body surface. None of the ultrasonic energy is lost in this manner if coupling is not broken.

Depth of penetration of the electromagnetic and mechanical waves depends on the electric and acoustic properties of various body tissues. A layer of subcutaneous fat $\frac{1}{2}$ inch thick will absorb 30% of the microwave energy. The subcutaneous fat layer must be thicker than 2 inches before more than 30% of the ultrasonic energy is absorbed.

In deep tissues of high water content, such as muscle, ultrasound has greater penetration than does microwave. The reverse is true if the tissues are of low water content, such as bone. For this reason absorption and heating effect of microwave are greater in tissues of high water content than in tissues of low water content. Ultrasonic absorption and consequent heating effect are greater in the tissues of low water content.

Variation in composition of tissue affects the radiation of ultrasound to a greater extent than it does microwave. Alteration in heating effect at the interface of mediums that have different indices of refraction occurs only with ultrasound. Part of the sonic waves are transformed into transverse waves that are rapidly absorbed near the interface. When sonic or electromagnetic waves pass through a layer of material, this material may affect the wave propagation only if its thickness is comparable to or greater than the wavelength. With the wavelengths now used clinically, microwave has a wavelength in tissue of about 2 cm. and ultrasound of around 1 mm. Therefore, ultrasound produces some increased heat at the interface of different tissues in contrast to the fairly uniform heating with microwave. A portion of microwave energy is reflected from bone so that direct plus the reflected energy may heat overlying tissue to a greater extent than it does the surrounding tissue.

Clinical application of microwave and ultrasound. Adequate deep-heating effects can be obtained at a depth of 5 cm. with either ultrasound or microwave. With proper application effective rise of temperature in deep tissues occurs within 15 to 20 minutes with microwave and 3 to 10 minutes with ultrasound. Intensity of microwave varies with the distance of the director from the skin and the type of director used. Suggested dosage of ultrasound is ½ to 6 watts/sq. cm. when the applicator surface area is not more than 10 sq. cm.; intensity in any given area will vary with the speed with which the applicator is moved over the skin surface and the extent of the area treated.

Because deep temperature usually exceeds skin temperature, intensity of application cannot be based solely on subjective sensation. If unpleasant sensations of heat occur, the treatment should be discontinued or the intensity decreased. The extent of surface area that can be treated with microwave or ultrasound is small.

Present methods of producing deep heat present some specific contraindications to their use. Application of deep heat is contraindicated when there is a tendency to hemorrhage, over a pregnant uterus, when metal is present in the area treated, and over a fracture site before the callus has formed. Microwave should not be used over superficial accumulation of fluid; it should be used cautiously over growing or superficial bone, over the eyes, and over the testes. Until more information is available, ultrasound is contraindicated in or about the brain, eyes, ears, nasal sinuses, heart, reproductive organs, and epiphyses of growing bone. It should be used with care over the larger nerve structures.

General indications for heat. Heat may be used in any condition in which relief of pain, increase in circulation, and increase in local metabolic activity are desirable, if no specific contraindication is evident. It is most frequently used as an adjunct to other types of therapy. The physiologic effects of heat are most applicable in the many lesions that may affect the musculoskeletal system.

It is commonly used in the treatment of bursitis, arthritis, myositis, tenosynovitis, and tendinitis. The beneficial effects in the care of traumatic lesions such as fractures, dislocations, contusions, sprains, and strains are well established.

Precautions and contraindications. Heat is not used in acute inflammation or acute trauma until the initial reaction has subsided. Obstructed venous or arterial circulation presents definite contraindication; heat may be cautiously applied when circulation is only impaired. Known or suspected areas of malignancy should be avoided because of the increase in local metabolic rate that results from the rise in temperature.

Special precautions should be taken when sensation is absent or impaired and when an infant or psychotic patient is unable to report the onset of unpleasant sensation of heat. Careful attention should be given when applying heat to a large area of the body in the presence of impairment of the cardiovascular, respiratory, or renal systems.

Fever therapy

The production of artificial fever for therapeutic purposes has been used fairly extensively from its inception in 1919 to the recent era of the sulfonamide drugs and antibiotics. It is effective in the treatment of gonorrheal infections, ocular syphilis and neurosyphilis, rheumatoid arthritis, Sydenham's chorea, rheumatic fever, undulant fever, and nonspecific ocular infections. Its use has not been completely supplanted by more modern therapy.

The general increase in body temperature has been produced by the use of malaria, typhoid vaccine, and similar materials. Physical methods used include hot tub baths and temperature-controlled cabinets with or without short-wave diathermy devices. Hot tub baths are used when mild elevation of temperature is desired for periods of 30 to 60 minutes, the cabinets where temperatures up to 106.5° F. are to be maintained over a period of several hours.

Details of the technic, indications, and precautions can be found in the extensive literature devoted to the subject.

Cold therapy

Therapeutic use of cold is much more limited than that of heat. Local application of cold can cause vasoconstriction, decrease in blood flow, decrease in local metabolic activity, and decrease in local tissue temperature. The extent of response depends on the nature and temperature of the substance applied, the duration of application, and the area to which it is applied.

Ice bags, cold-water baths, and cold compresses have been used extensively to minimize the initial reaction of tissues to local traumatic injuries such as contusions and sprains. It has been suggested that the beneficial effects of ethyl chloride spray for painful joints and muscles are due to direct and reflex effects of the change in tissue temperature produced by the spray.[4] The destructive effects of intense cold are utilized in application of solid carbon dioxide to certain dermatologic lesions. Refrigeration anesthesia for surgical purposes is now used more frequently, but this phase of cold therapy is not within the scope of this chapter. Texts on anesthesia stress that in cases when it is indicated its value is unquestioned.

ULTRAVIOLET THERAPY

The physician should understand the physical properties of any source of radiation used for therapeutic purposes. Light is an electromagnetic wave that travels and transports energy at a speed of about 3×10^8 M./sec. or 186,264 miles/sec. Table 3-1 contains a compilation of the various ranges of the electromagnetic spectrum. The waves are rhythmic electric and magnetic oscillations that move through space

Table 3-1. Various ranges of the electromagnetic spectrum

Type of radiation	Range of wavelengths (angstroms)
Long-wave infrared	120,000-15,000
Short-wave infrared	15,000- 7,700
Visible	7,000- 3,900
Near ultraviolet	3,900- 2,900
Far ultraviolet	2,900- 1,800
Grenz x-rays	5-1
Diagnostic x-rays	0.30-0.12
Therapeutic x-rays	0.12-0.05
Gamma rays	0.1 -0.02

with identical speed, the speed of light. They differ only in the length of their waves; therefore, the most logical way to describe any of the waves comprising the electromagnetic spectrum is in terms of their wavelengths. The wavelength represents the distance from one point of the wave to the identical point on the next wave, that is, from crest to crest. Thus, as wavelength decreases, frequency increases.

For clinical purposes the range of the ultraviolet waves, from 1800 to 3900 A, produces chemical changes in tissue; the visible light range, from 3900 to 7700 A, affects the retina almost exclusively; and the range of the infrared rays, from 7700 to 120,000 A, produces superficial-heating effects.

Factors influencing quantity of absorption

Sources of ultraviolet

Natural source. The sun is the only natural source of ultraviolet. Rollier's technic for the systemic exposure of the entire body is practically the only one utilizing this source. Solar radiation, after filtration by the atmosphere, provides practically no radiation of wavelengths shorter than 2900 A. Because solar radiation varies with the season, time of day, atmospheric conditions, altitude, and latitude, it is obvious that reliance, for therapeutic purposes, must be based on artificial sources to make possible the duplication of technic.

Artificial sources. Several artificial sources of ultraviolet have been developed for therapeutic purposes. To employ a source of ultraviolet correctly, it is necessary to know which wavelengths the source produces, their quantity, and their physiologic effects.

Hot quartz mercury lamp. This lamp is the most suitable source of ultraviolet for general clinical use. Its burner is a hollow tube of fused quartz in which mercury is sealed. When voltage is impressed across the electrodes, the mercury is vaporized, forming an arc. The burner is enclosed by a metal reflector on an adjustable stand so that the ultraviolet can be applied over large areas of the body for general irradiation. Special lamps are available for local application, such as the Aero-Kromayer lamp; various types of quartz rods and disks are supplied for conduction of the rays to surfaces or cavities of the body.

Radiation from a quartz mercury vapor arc lamp is composed of 28% ultraviolet (6% of which is for ultraviolet with high germicidal action), 20% visible radiation, and 52% infrared radiation. The major radiations range in wavelengths from 2537 to 3650 A.

Carbon arc lamps. Carbon arc lamps produce radiation according to the kind of electrodes employed. These lamps are not widely used for clinical purposes. The electrodes are gradually consumed; smoke, fumes, and ash are produced; and they are more difficult to handle than the hot quartz lamps. Their chief advantage is that the carbon electrodes can be changed to vary the spectral output.

Cold quartz mercury lamps. About 95% of the radiation from these lamps has a wavelength of 1537 A. The intensity output is relatively low so that exposure of the skin is at a short distance. In lamps used for general irradiation the tube is wound into a flat, grid-type coil; for orificial irradiation the tube is straight or appropriately shaped. Cold quartz lamps are also used as germicidal lamps for the operating room and other places where sterilization of the air is desirable.

Sunlamps. A variety of lamps that are sold for home use are included in this category. They transmit radiation of 2800 A and up; the intensity is too weak for therapeutic purposes.

Distance of source from part to be treated. The *inverse square law* states that the intensity of radiation at any point from a point source is inversely proportional to the square of the distance of the point from the source. Thus, if the distance from the lamp source to the skin surface is decreased by one half, the strength of the radiation is quadrupled.

Angle of incidence of the rays. *Lambert's cosine law* states that the radiant energy per square centimeter of surface is proportional to a constant times the cosine of the angle of incidence. This means that the intensity of radiation striking the skin surface at a right angle is greatest; at an angle of 30 degrees the intensity falls to 80%; at an angle of 60 degrees, to 40%.

Dosage. The erythematous response of the patient is the best guide available for dosage at the present time.

The distance from the burner to the patient's skin and the length of exposure are the principal factors in determining dosage. In order to standardize dosage, the erythemogenic efficiency of each individual lamp must be tested periodically. This test can be accomplished by exposing, for graduated lengths of time, from a given height, six to eight small areas of skin on the flexor surface of the arm or the abdomen. From the reaction noted the erythematous dosage can be recorded for a given lamp.

In describing dosage for ultraviolet irradiation, the following terms, not to be confused with degrees of burns, have been used:

1. *Suberythematous dose (SED)*—irradiation insufficient to cause a slight reddening of the skin
2. *Minimal erythematous dose (MED)*—irradiation sufficient to cause a slight reddening of the skin, which usually disappears within 24 hours, but no desquamation
3. *First-degree erythematous dose (1D)*—irradiation sufficient to cause slightly greater reaction with slight irritation followed by fine desquamation
4. *Second-degree erythematous dose (2D)*—irradiation sufficient to cause marked erythema with itching and burning of the skin followed by free desquamation
5. *Third-degree erythematous dose (3D)*—irradiation sufficient to cause severe destructive reaction with edema and blister formation

A minimal erythematous dose is usually administered when systemic effects are desired, but dosage may be varied from an SED to a 3D, depending on the specific pathologic condition being treated.

Sensitivity of the skin. Penetration of ultraviolet varies with the thickness of the skin. The thickness of skin varies over different parts of the body, as well as among individuals. There is some correlation between skin coloring and skin thickness so that, as a rule of the thumb, the thin-skinned, red-headed individuals can be considered most sensitive and the dark-haired, swarthy individuals the least sensitive.

According to Laurens[25] increased sensitivity is found in patients with an unstable nervous system, overactive thyroid glands, elevated blood pressure, or active tuberculosis or in women during the menstrual period or in the second to seventh month of pregnancy.

Certain drugs, including sulfanilamide, quinine, and coal tar, temporarily increase sensitivity to ultraviolet rays.

Filtration factors. Dust, oil, ointment, cloth, or ordinary glass between the source of radiation and the skin serves to filter the ultraviolet.

Physiologic effects

Penetration of wavelengths in the ultraviolet range does not exceed 0.1 mm. The absorbing substance in the skin is considered to be protein or nucleic acid. The photochemical reaction produces local or distant reactions that reach their maximal effect after a latent period of several hours. The visible results are erythema and suntan.

Erythema. This effect is due to dilatation of the dermal capillaries resulting from the release of a histamine-like substance in the epidermis, which diffuses to and acts on the blood vessels just beneath the epidermis. Most effective wavelengths are around 2967 and 2537 A. Intense irradiation of wavelengths longer than 3200 A may produce rapid erythema in some individuals.

The conjunctiva is more sensitive to ultraviolet than is the skin, so that the eyes should be covered during general irradiation. The mucous membranes are less sensitive than is the skin, and the minimal erythematous dosage is about two times that of the skin. Also systemic effects from the absorption of the histamine-like substance may result in increased secretion of gastric juice, hypotension, palpation, and even prostration.

Proliferation of the epidermis with thickening of the cornified epithelium. This effect results from repeated exposure. The thickening acts as a screen to cut down the amounts of absorption of further irradiation.

Tanning. There is an increase in production of melanin and a migration of melanin granules into the prickle cell layer and the cornified epithelium. The maximal pigmenting effect occurs with wavelengths of 2537 and 2967 A. A much weaker effect occurs from 3000 to 4500 A.

Bactericidal effect. Peak bactericidal effectiveness is at 2600 A.

Antirachitic effect. This effect is produced by conversion of sterols in the skin to vitamin D. The optimum wavelength is 2800 A, with a secondary peak at 2967 A.

Carcinogenic effect. Prolonged exposure to sunrays may produce skin cancer in man and in experimental animals. Experimental evidence demonstrates the most potent carcinogenic wavelengths between 2800 and 3400 A.

Application

The eyes of both the patient and the operator should be protected. The basic dosage unit is a "minimal erythema dose." Because the maximum effect may not be anterior and the posterior surface. The dosage is increased gradually—customarily daily.

General irradiation. To expose the entire body, four exposures are usually given at each treatment. The body is divided into upper and lower portions on both the anterior and the posterior surface. The dosage is increased gradually—customarily at the rate of 1 MED each successive treatment given every other day.

Local exposure. The specially designed burner that is air-cooled (Aero-Kromayer) is placed close to the lesion; thus the intensity of irradiation is great. The time of each exposure is extremely short. If a special lamp is not available, a lamp used for general exposures can be employed if all surrounding tissue is masked to protect it from ultraviolet rays.

Special technics. Goeckerman's technic employs a coal tar ointment as a sensitizing agent to ultraviolet radiation. It is used in the control of psoriasis. The ointment is applied to lesions for 24 hours and then partially removed with lightweight oil so that a thin film remains on the lesion. A hot quartz mercury vapor lamp is then used to irradiate the area for 1 MED, increased by 1 MED daily. After the irradiation the ointment is washed off and coal tar is again applied for the treatment the following day. Marked erythema or blistering should be avoided, but tanning should be produced.

Prescription

A typical prescription for ultraviolet, in addition to the diagnosis, should include the following:

1. Source—hot quartz mercury vapor, cold quartz, etc.
2. Erythematous dose—SED, MED, 1D, 2D, 3D
3. Technic—general, local, etc.
4. Frequency of treatment—times per week
5. Dosage increment—specified by erythematous dosage
6. Total number of treatments—specified where applicable
7. Special instructions—application of coal tar, etc.

Indications

Indolent wounds. Decubitus ulcers and some indolent ulcers respond favorably to ulatraviolet. This is attributed to bacteriostasis and a prolonged increase in local circulation. Ulcers resulting from Raynaud's disease, thromboangiitis obliterans, and chronic varicosities do not respond.

Psoriasis. Goeckerman's treatment has proved valuable in cases of resistant psoriasis.

Pityriasis rosea. General irradiation at twice the MED, every other day, will usually decrease the recovery time.

Skin diseases. In combination with other therapy, ultraviolet irradiation is helpful in lupus vulgaris, acne vulgaris, and adenoma sebaceum.

Rickets. Ultraviolet irradiation is effective and was used extensively prior to the discovery of the etiology and simpler therapeutic measures.

Erysipelas. Irradiation should include at least 2 inches of the normal skin bordering the lesion. One dose sufficient to cause blistering and desquamation often cures the disease.

Tonic ultraviolet. The value of ultraviolet for its so-called "tonic" effect is unproved. However, there is no contraindication to its use if proper precautions are observed.

Contraindications

According to Krusen[20] the following constitute definite contraindications to ultraviolet: (1) progressive exudative forms of pulmonary tuberculosis involving adrenal glands; certain types of tuberculous tracheobronchial adenitis in which there may be a febrile reaction, loss of weight, and fall in blood pressure following irradiation; (2) hyperthyroidism; (3) diabetes; (4) highly nervous condition; (5) advanced cachexia or inanition; (6) advanced age in persons with acute nephritis or myocarditis; (7) acute forms of generalized dermatitis; (8) photogenic diseases of the skin, for example, pellagra, lupus erythematosus, hydroa aestivale, and xeroderma pigmentosum; (9) hyperglycemia; and (10) severe chronic nephritis.

ELECTROTHERAPY

Electric currents have a definite place among the physical agents available for therapeutic purposes. Unfortunately, the lack of accurate and uniform terminology has led to considerable confusion regarding various types of currents.

Proper and effective use of electric currents is based on physical law and physiologic response. The variation in the physiologic response of body tissue to currents of different frequency and voltage provides a basis for simple division into currents with heating effects and currents with stimulating effects.

Electric current flow may be either unidirectional or alternating. The alternating currents of very high frequency and relatively high voltage produced heating effects and are designated diathermy currents. They are discussed under methods of application (p. 63). Other types of currents, which are used mainly for stimulation of nerve or muscle, are generally considered "low-voltage currents." This group includes direct current and the low-frequency currents.

Direct current

Direct current (also called galvanic or constant) is a unidirectional current with distinct polarity, low amperage (less than 50 ma.), and low voltage (less than 100 volts). This current flows as long as the circuit is closed. When constant current passes through an electrolyte, a migration of ions occurs toward the two poles of the circuit. This polarity effect has some therapeutic application.

Therapeutic application

Ion transfer (iontophoresis). This procedure utilizes the polarity effects of direct of direct current to introduce certain ions into the body through the intact skin. The positive pole is used to introduce positive ions, the negative pole for negative ions. The ions rapidly lose their electric charge and precipitate in the superficial tissues as soluble or insoluble compounds.

Histamine and methacholine (Mecholyl) are the two drugs most frequently used. The prolonged vasodilatation produced by these drugs has been thought beneficial in arthritis, some chronic ulcers, and certain vasospastic conditions of the extremities. Other drugs that have been used include zinc, copper, procaine, penicillin, chlortetracycline hydrochloride (Aureomycin), and eserine.

A number of clinical reports have been written on the use of ion transfer for various diseases. The availability of radioisotopes has added more definitive knowledge regarding the influence of various physicochemical factors.

Electrolysis. The concentration of current when a small electrode is used may produce destructive effects. These effects are used in the removal of unwanted hair and certain skin lesions. For epilation of hair a fine platinum needle is connected to the negative pole. The needle is inserted into the base of the hair follicle and a current of 1 to 2 ma. applied. The needle is held in place until a small hydrogen bubble appears at the surface. Current flow is then stopped and the needle removed. The hair is lifted out with a small pair of tweezers.

Other forms. A number of procedures have been used in the past that are of limited value and should be considered obsolete. A partial list of these procedures includes medical galvanism, hydrogalvanic baths, the Bergonié chair, and Schnee baths.

Low-frequency currents

The variety of currents included in this classification have certain characteristics in common. Their frequency is below 10,000/sec. Their voltage is quite low and constantly changing. They can be used to stimulate motor and sensory nerves. Kovacs[26] includes the following currents in this group: (1) interrupted direct, (2) faradic, (3) surging faradic, (4) modulated alternating (rapid sinusoidal), and (5) surging uninterrupted direct with alternating polarity (slow sinusoidal).

Each impulse of interrupted direct current will produce a brief twitchlike contraction of muscle unlike the graduated tetanic contraction of normal muscle. The aim of therapeutic stimulation of muscle is to produce a contraction as nearly normal as possible. For this reason interrupted direct current has limited application. It is used when denervated muscle will not respond to other currents.

The interrupted direct and the faradic currents are used mainly for electrodiagnosis. They are discussed more fully under electrodiagnosis. The interrupted direct current may also be used for stimulation of denervated muscles.

The remaining low-frequency currents were designed to produced stimuli that evoke a more normal type of muscular contraction. The duration of impulse of faradic and modulated alternating currents is too brief to elicit a response from denervated muscle at the intensity that can be tolerated by the patient. The slow sinusoidal current has a longer duration of single impulse and is of sufficient duration to stimulate denervated muscles. These basic current forms have been modified in newer generators to provide stimuli that can be adjusted to elicit contractions closely resembling the normal contractions in both innervated and denervated muscle.

The surging faradic current is a faradic current with a smoothly increasing and decreasing rise and fall in intensity. This current will stimulate muscles having an intact nerve supply.

The modulated alternating current is simply an alternating current with a frequency of 60 cycles/sec. (time scale, 5 msec./cm.). The current may have frequency ranging up to 180 cycles/sec. This form of current, which may be either surged or modulated, is effective only in muscles with intact nerve supply.

The surging uninterrupted direct current with alternating polarity is also known as the slow (galvanic) sinusoidal current. The current form is that of a gradual rise and fall of a direct current with a reversal of flow at a rate of 2 to 90 cycles/sec. This current is the one best adapted for stimulation of denervated muscles.

Therapeutic application

Muscle stimulation. When a muscle is not able to function in a normal manner, it deteriorates. This deterioration affects the connective tissue as well as the muscle fiber. Prolonged denervation may result in a replacement of the muscle fibers by fibrous tissue. Even if this does not occur, the sclerotic changes in the connective tissue may seriously hamper the normal use of the muscle when reinnervation occurs.

The object of muscle stimulation for denervated muscles is to maintain the muscle in as normal a state as possible. Studies of the effects of electric stimulation indicate that it is helpful but that it will not completely prevent the effects of denervation. The best results have been obtained with brief periods of stimulation done frequently during the day with a slow sinusoidal type current used at a frequency of 2 to 32 cycles/sec.

Brief but frequent periods of treatment can be adequately given in the home after proper instruction by the physical therapist. There are simple, inexpensive stimulators commercially available for home use.

As reinnervation occurs, the initial voluntary contraction of the muscles will be weak. Also, if the patient has been substituting other muscles to produce the function primary to the denervated muscle, he may have some difficulty in "remembering" how to use the reinnervated muscle. Electric stimulation is beneficial during the period when anatomic reinnervation is established but maximal contraction in a coordinated manner is not yet possible.

Conduction deficit without denervation is usually a temporary condition but may last for several weeks. Stimulation during this period is of definite benefit.

Stimulation has been used in upper motor neuron lesions to reduce spasticity of muscles prior to therapeutic exercise. The mechanism of this action is obscure, but stimulation for 20 to 30 minutes at 1200 to 1500 cycles/sec. appears to be beneficial.

MASSAGE

Massage as a therapeutic measure is probably one of the oldest forms of manipulative procedures known to medicine. The Chinese described forms of massage in their records, in some cases dating back 2500 to 3000 years. Such names as Herodikos and Hippocrates have been associated with the history of the development of massage.

In the sixteenth century Ambrose Paré sought an anatomic as well as physiologic foundation for manipulative technics.

Several names are associated with the modern development of massage as a therapeutic technic useful in the armamentaria of the physical therapist and the physician.

In the last 100 years Peter Henry Ling (Sweden), Metzger (Holland), Lucas-Championnière (France), and James Mennell (England) all contributed greatly to put massage, as we know it today, on a more scientific basis. At the turn of the century Kellog, Murrell, Kleen, Mitchell, Bucholz, and others described massage with a general classification. It was used therapeutically to help mobilize a part along with other modalities of heat and exercise.

After World War II, with the impact of other forms of care, massage (tending to be a time-consuming, demanding skill) was not used as often as modalities for care. However, areas of this country and of the world exist in which it continues to be used extensively. From the books published on the subject and articles evaluating its capability during the last 10 years, massage may again be on the ascendency.

The skilled hands of trained physical therapists who are aware of the combination and capabilities of massage find it can be useful and that it has proved itself to be a valuable therapeutic tool.

Physiologic effects

Mechanical effects

Circulation. Automassage is produced by contracting muscles and moving joints. The pressure exerted on vessels tends to aid venous and lymphatic return and to prevent stasis in the capillaries. Manual massage exerts the same type of direct displacement of fluid. There is a direct as well as a reflex dilatation of the smaller vessels. Not only is there an increase in the rate of flow, but because of this increase there is an increase in the interchange of substances between the bloodstream and tissue cells.

Secondary effects of systemic massage include an increase in peripheral blood flow, an increase in the level of red blood cells, and an increment in the renal output of water.

Lymphatic flow. Lymph flow from tissue spaces back into the venous circulation is mainly dependent on external pressure. The mechanical pressure of massage helps remove the excess fluid and decreases the likelihood of fibrosis.

Edema. Excessive fluid in the tissues is decreased by massage both by direct pressure forcing fluids out of the tissues to venous and lymphatic channels and by mechanically moving the fluid along these vessels so that their pressure is lowered sufficiently to receive the excess fluid.

Connective tissue. Mechanical stretching and disruption of fibers can be effectively accomplished by friction massage when undesirable fibrosis has occurred in subcutaneous and superficial muscle layers.

Muscle. Massage improves the nutrition of muscle fibers and removes extravascular fluid. The acid-base balance is not altered by massage; this is in contrast to the alkaline tendency following active exercise. The beneficial effect of massage after strenuous exercises is probably because no additional lactic acid or other metabolites are produced so that the changes in the blood supply of the muscle permit a more rapid or more thorough removal of these byproducts already contained within the tissues.

Reflex effects

Circulation. A light stroking of the skin results in the transitory dilatation of the capillary vessels. Stronger tactile stimulation results in the triple response of Lewis— blanching, wheal formation, and a more prolonged capillary dilatation.

Muscle. The relaxation of involuntary muscle contraction as well as the transitory dilatation of capillaries has been attributed to the autonomic nervous system. These effects have been produced with stroking massage when the pressure was insufficient for mechanical effects.

Pain. Pain threshold can be increased by slowly increasing stimuli of the central nervous system. Beginning with light stroking movements and slowly increasing the intensity of the massage will result in a decrease in tenderness and increase in pain threshold.

Classification of technics of manual massage

The following classification is that of Mennell.[35] The older French terminology is included in parentheses.

Stroking (effleurage)

Superficial massage. Light stroking movements consist of slow, gentle, and rhythmic passage of the flat surface of the hand over an area of the patient's skin. Only reflex effects are expected from this type of massage, so that the pressure is just sufficient for good contact. For the same reason the direction of the stroke is unimportant.

Deep massage. The aim of this type of massage is to push along the venous blood and lymph in the direction of natural flow. Pressure in the vessels is low and the rate of flow is slow so that there is no need for rapid movement or excessive pressure. Mennell[35] has suggested a pressure of 10 mm. Hg; this is substantiated by Pollock and Wood[36] who found the venous pressure in the ankle to be 12 mm. Hg with the patient in the supine position.

The patient should be comfortably positioned to ensure maximum relaxation. Vessels compressed by contracted muscles will prevent the chief objective of the stroking movements. Proximal segments should be massaged first, then the more distal segments. The principle has been likened to "uncorking the bottle before pouring."

Compression (petrissage)

Kneading. In this type of massage a portion of a muscle or group of muscles is lifted up between the thumb and thenar eminence and the fingers, and is then compressed or kneaded from side to side alternately as the hand moves up the muscle in the direction of the venous flow. One or both hands may be used. In addition to the circulatory effects, this type of massage may be used to stretch retracted muscles and tendons and assist in stretching adhesions. For the desired effect the muscles must be relaxed and the movements done slowly and rhythmically.

Friction. Friction massage is accomplished by moving the skin and superficial tissue over the underlying tissue. The thumb, part of the hand, or one or two fingers are placed on the skin and moved over the underlying tissues in small circles. In this type of massage no lubricant is used. Pressure should be moderate and motions rhythmic. Friction is useful to loosen superficial scars or adhesions and to free adherent skin.

Vibration. Vibratory movements of the operator's shoulder and forearm are transmitted through the hand or fingers to a part of the patient's body. A larger movement of the same type is known as shaking. Such movements are rarely employed for therapeutic purposes.

Percussion (tapotement)

Percussion movements—hacking, clapping, tapping, and beating—should be used only on normal, healthy individuals. The one therapeutic use is to assist in the accommodation of tissue to pressure.

Principles of application

The therapeutic application of massage is a learned skill. As with other skilled activities, an individual variation exists in the time and practice required to attain acceptable technics.

1. The patient and the operator should be comfortable and as relaxed as possible. No clothing should constrict the body segment being treated or the area proximal to it.

2. The part being treated should be well supported.

3. All movements should be slow, rhythmic, and in the same direction. Jerky movements will result in reflex protective muscular contraction.

4. Maximum contact of the hand and skin should be maintained.

5. The hands of the operator should be warm and soft.

6. The amount of lubricant used should be just enough to allow smooth movement of the hands over the skin. The type of lubricant used depends on the condition of the part and the preference of the operator. In general, oil or lotion is better for dry or friable skin.

7. Massage should rarely be painful.

Indications

Massage is indicated when the following effects are sought: (1) relief of pain, (2) relaxation of muscle tension, (3) improvement of circulation, (4) reduction of induration or edema, and (5) stretching of adhesions.

Local application of heat is commonly used prior to massage because it enhances the desired effects. Both heat and massage are frequently used prior to active or passive exercise because the effects listed condition a patient to get maximum benefit from the exercise.

Clinical conditions in which massage is beneficial

Soft tissue trauma (sprains, contusions, strains, and lacerations). Massage is used to assist in removing the edema and hasten recovery only after the initial inflammatory reaction and tendency to hemorrhage have subsided.

Fractures. Where immobilization is added to the initial trauma, massage is one of the few methods of relieving the effects of the pain, protective muscle spasm, and impaired circulation.

Painful muscular contraction due to prolonged tension, backstrain, or strenuous exercise. Deep stroking and kneading massage are useful in relieving pain and relaxing the protective muscle spasm.

Scars and adhesions. Friction massage, often used in combination with stretching exercise, will help to free the tissue and gain mobility.

Inactivity due to disuse or peripheral nerve lesions. Massage is used to prevent or decrease fibrosis and maintain circulation.

Fibrositis. Hench and co-workers[39] defined fibrositis as "an inflammation of connective tissue." Most writers on the subject believe that these local areas of tender induration can be relieved by deep kneading massage.

Arthritis. In the subacute and chronic stages massage is usually preceded by the application of heat and followed by exercise. The goal of such treatment is the relief of symptoms and prevention of deformity.

Circulatory disorders with resulting edema. Temporary relief of the edema and prevention of the effects of stasis can be accomplished by massage.

Principles of physical medicine 75

Contraindications

Contraindications for massage include (1) acute inflammation—massage may result in systemic spread; (2) skin lesions; (3) malignant swelling; (4) acute circulatory disturbances such as phlebitis, thrombosis, or lymphangitis; and (5) the abdomen—the danger of abdominal massage far outweighs any questionable value it may have.

Connective tissue massage (Bindegewebsmassage)

Connective tissue massage is a very interesting form that yet needs further clarification. Bishoff and Elmiger[46] and also Ebner[47] have described in English the procedures and have presented the rationale for this form of massage.

REFERENCES

1. Wakim, K. G., and Krusen, F. H.: Comparison of the temperatures produced by carbon-filament and by tungsten-filament lamps, Arch. Phys. Med. **35:**508, 1954.
2. Coyne, N.: Use and abuse of heat in physical medicine and rehabilitation, Postgrad. Med. **22:**165, 1957.
3. Schwan, H. P.: The biophysical basis of physical medicine, J.A.M.A. **160:**191, 1956.
4. Borken, N., and Bierman, W.: Temperature changes produced by spraying with ethyl chloride, Arch. Phys. Med. **36:**288, 1955.
5. Bierman, W., and Friedlander, M.: Penetrative effect of cold, Arch. Phys. Ther. **21:**585, Oct., 1940.
6. Cohen, L., Martin, G. M., and Wakim, K. G.: Effects of whirlpool bath with and without agitation on circulation in normal and diseased extremities, Arch. Phys. Med. **30:**212, 1949.
7. Elter, H. S., Prudenz, R. H., and Gersh, I.: The effects of diathermy on tissues contiguous to implanted surgical metals, Arch. Phys. Med. **28:**333, 1947.
8. Engel, J. P., Wakim, K. G., Erickson, D. J., and Krusen, F. H.: The effect of contrast baths on the peripheral circulation in patients with rheumatoid arthritis, Arch. Phys. Med. **31:**135, 1950.
9. Erdman, W. J., and Stoner, E. K.: Comparative heating effects of Moistaire and Hydrocollator hot pack, Arch. Phys. Med. **37:**71, 1956.
10. Gersten, J. W., Wakim, K. G., Stow, R. W., and Krusen, F. H.: A comparative study of the heating of tissues by near and far infrared radiation, Arch. Phys. Med. **30:**691, 1949.
11. Hardy, J. D.: Physiological responses to heat and cold, Ann. Rev. Physiol. **12:**119, 1950.
12. Kendall, H. W.: Fever therapy. In Handbook of physical medicine and rehabilitation, Philadelphia, 1950, The Blakiston Co.
13. Krusen, E. M., Wakim, K. G., Leden, U. M., Martin, G. M., and Elkins, E. C.: Effect of hot packs on peripheral circulation, Arch. Phys. Med. **31:**145, 1950.
14. Martin, G. M., and Herrick, J. F.: Further evaluation of heating by microwaves and by infrared as used clinically, J.A.M.A. **159:**1286, 1956.
15. Richardson, A. W., Imig, C. J., Feucht, B. L., and Hines, H. M.: The relationship between deep-tissue temperature and blood flow during electromagnetic irradiation, Arch. Phys. Med. **31:**19, 1950.
16. Schwan, H. P., and Piersol, G. M.: The absorption of electromagnetic energy in body tissue, Amer. J. Phys. Med. **33:**371, 1954.
17. Wells, H. S.: Temperature equalization for the relief of pain, Arch. Phys. Med. **28:**135, 1947.
18. Woodmansey, A., Collins, D. H., and Ernst, M. M.: Vascular reactions to the contrast bath in health and in rheumatoid arthritis, Lancet **2:**1350, 1938.
19. Krusen, F. H., and Elkins, E. C.: Physical therapy. In Glasser, O., editor: Medical physics, Chicago, 1947, The Year Book Publishers, Inc.

20. Krusen, F. H.: Physical medicine: The employment of physical agents for diagnosis and therapy, Philadelphia, 1941, W. B. Saunders Co.
21. Bachem, A.: Ultraviolet action spectra, Amer. J. Phys. Med. **35**:177, June, 1956.
22. Blum, H. F.: Radiation; photophysiologic and photopathologic processes. In Glasser, O., editor: Medical physics, Chicago, 1947, The Year Book Publishers, Inc.
23. Forsythe, W. E., and Adams, E. Q.: Radiation: source of ultraviolet and infra-red. In Glasser, O., editor: Medical physics, Chicago, 1947, The Year Book Publishers, Inc.
24. Glasser, O.: Radiation spectrum. In Glasser, O., editor: Medical physics, Chicago, 1947, The Year Book Publishers, Inc.
25. Laurens, H.: Heliotherapy: measurement and application of solar rays. In Glasser, O., editor; Medical physics, Chicago, 1947, The Year Book Publishers, Inc.
26. Kovacs, R.: Electrotherapy and light therapy, Philadelphia, 1942, Lea & Febiger.
27. Osborne, S. L., and Holmquist, H. J.: Technic of electrotherapy, Springfield, Ill., 1944, Charles C Thomas, Publisher.
28. Kosman, A. J., Osborne, S. L., and Ivy, A. C.: The influence of duration and frequency of treatment in electrical stimulation of paralyzed muscle, Arch. Phys. Med. **28**:7, 1947.
29. Lee, W. J.: Continuous tetanizing (low voltage) currents for relief of spasm, Arch. Phys. Med. **31**:766, 1950.
30. O'Malley, E. P., Oester, Y. T., and Warnick, E. G.: Experimental iontophoresis: studies with radio-isotopes, Arch. Phys. Med. **35**:500, 1954.
31. O'Malley, E. P., and Oester, Y. T.: Influence of some physical chemical factors on iontophoresis using radio-isotopes, Arch. Phys. Med. **36**:310, 1955.
32. Shaffer, D. V., Branes, G. K., Wakim, K. G., Sayre, G. P., and Krusen, F. H.: The influence of electric stimulation on the course of denervation atrophy, Arch. Phys. Med. **35**:491, 1954.
33. Wakim, K. G., and Krusen, F. H.: The influence of electrical stimulation on the work output and endurance of denervated muscle, Arch. Phys. Med. **36**:370, 1955.
34. Wakim, K. G., and Krusen, F. H.: Comparison of effect of electric stimulation with effects of intermittent compression on the work output and endurance of denervated muscle, Arch. Phys. Med. **38**:21, 1957.
35. Mennell, J. B.: Physical treatment by movement, manipulation and massage, Philadelphia 1934, P. Blakiston's Son & Co.
36. Pollock, A. A., and Wood, E. H.: Venus pressure in the saphenous vein in the ankle of man during exercise and changes in posture, J. Appl. Physiol. **1**:649, 1949.
37. Beard, G.: A history of massage techniques, Phys. Ther. Rev. **32**:613, 1952.
38. Elkins, E. C., Herrick, J. F., Grindley, J. H., Mann, F. C., and DeForest, R. E.: Effect of various procedures on the flow of lymph, Arch. Phys. Med. **34**:31, 1953.
39. Hench, P. S., Bauer, W., Fletcher, A. A., Ghrist, D., Hall, F., and White, T. P.: The problem of rheumatism and arthritis: review of American and English literature for 1935 Ann. Intern. Med. **10**:754, 1936.
40. Kosman, A. J., Wood, E. C., and Osborne, S. L.: The effect of massage on the denervated skeletal muscle of the dog, Arch. Phys. Med. **29**:489, 1948.
41. Ladd, M., Kottke, F. J., and Blanchard, R. S.: Studies of the effect of massage on the flow of lymph from the foreleg of the dog, Arch. Phys. Med. **33**:604, 1952.
42. Mennell, J. B.: Manipuation of stiff joints, Arch. Phys. Med. **28**:685, 1947.
43. Wakim, K. G., Martin, G. M., Terrier, J. C., Elkins, E. C., and Krusen, F. H.: Effects of massage on the circulation in normal and paralyzed extremities, Arch. Phys. Med. **30**:135, 1949.
44. Beard, G., and Wood, E.: Massage principles and techniques, Philadelphia, 1964, W. B. Saunders Co.
45. Tappan, Frances: Massage techniques, a case method approach, New York, 1961, The Macmillan Co.
46. Bishoff, I., and Elmiger, G.: Connective tissue massage. In Licht, S. H., editor: Massage, manipulation, and traction, vol. 5, Physical Medicine Library, New Haven, Conn., 1960, E. Licht Publishing Co.
47. Ebner, M.: Connective tissue massage, theory and therapeutic application, Baltimore, 1962, The Williams & Wilkins Co.

CHAPTER 4

PRINCIPLES OF
PHYSICAL THERAPY

T herapeutic exercise is a science. A prerequisite to its practice is a knowledge of anatomy, physiology, physics, and kinesiology. Therapeutic exercises aim to improve the balance and stability of the body and to coordinate body movements in all aspects.

The development of therapeutic exercise as a modern science dates to the publication in 1866 in France of the now classic work of Guillaume Benjamin Duchenne, *Physiology of Motion.*[1] In describing the action of the superficial muscles of the body, Duchenne stated, "a knowledge of the muscular mechanism leads to a rational treatment of paralysis, atrophies and deformities by application of special local peripheral stimulation, physical exercise, and physiologic prosthesis."*

Emanuel B. Kaplan, in the foreword of the first translation of this French work in 1949, suggested that

> this book may be placed among the greatest books of all times—not on account of its historical significance—but, because it contains an excellent record of the kinesiology of the entire muscular system—investigated by one observer whose genius, perseverance and originality permitted him a deeper insight into the action of muscles than given to his many predecessors and, perhaps, more modern investigators.*

Each system of the body has its own particular function. The musculoskeletal system must provide motion and stability. In order that the body function with the greatest efficiency, there must be both a source of power and a control of the rate of speed and the distance through which the power is used.

Skeletal muscle makes up about 50% of the body weight. Most of these muscles bridge one or more joints of the rigid bony skeleton. Each muscle has two points of attachment—an origin and an insertion. The only physical properties of muscle are

---

*From Duchenne, G. B.: Physiology of motion. Translated and edited by E. B. Kaplan, Philadelphia, 1949, J. B. Lippincott Co.

elasticity and contractility. Muscle contraction results in (1) tension, which becomes the force used for motion and stability of the body segments, and (2) shortening, which produces the visible, rotary motion of the body segments.

TYPES OF CONTRACTION

Contractions can be divided into (1) static or isometric, (2) concentric or isotonic, and (3) eccentric types.

When the force applied is just equal to the resistance to movement, there is no alteration in distance between origin and insertion and no joint motion. This static or isometric type of contraction is the one utilized to provide stability of joint motion and prevent undesired movement.

When the tension produced during muscular contraction approximates the origin and the insertion of the muscle, the contraction is called concentric or isotonic. This is accomplished when the force applied is greater than the resistance to the movement.

The third type, the eccentric contraction, results in an increase in the distance between origin and insertion. Outside force produces the motion, and the muscular tension regulates the speed of the motion.

When a muscle is called on to overcome a resistance, the first phase of contraction is isometric until the tension in the muscle is equal to the resistance; the second phase of the contraction is either concentric or eccentric. For example, the sequence of types of contraction used as the elbow flexors assist in taking a glass of water from a table to the mouth and back again is as follows: there is isometric contraction until the tension in the elbow flexors is equal to the resistance of the weight of the forearm and glass of water and then concentric as the tension in the flexors increases, the muscles shorten, and the glass is carried to the mouth; finally eccentric contraction completes the action as the flexors regulate the rate of speed with which the glass is lowered to the table.

FACTORS INFLUENCING EFFECTIVE USE OF MUSCLE CONTRACTION

Leverage. All joint motion in the body is rotary. The tension of muscles is made useful through its action on the series of levers provided by the rigid skeletal structure. The bone acts as the lever arm and the joint as the axis or fulcrum; the force is supplied by the muscle. The leverage is useful in body motion either to favor power or to favor speed and range of motion.

The law of levers states that any lever will balance when the product of the force and force arm equals the product of the resistance and resistance arm.

As can be seen in the diagram of a lever of the first class (Fig. 4-1), the fulcrum is located between the point where the force is applied and the point where the resistance is encountered. The triceps muscle is an example of a first-class lever and also an example of the sacrifice of power in order to obtain greater speed and range of motion.

In the diagram of a lever of the second class (Fig. 4-1), the force arm is seen to be longer than the weight arm. This arrangement has an advantage of power but a disadvantage for range of motion and speed. An example of this second-class lever is the action of the gastrocnemius-soleus muscles when a person stands on tiptoe

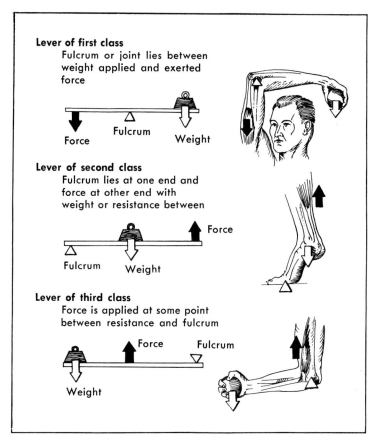

Fig. 4-1. Levers in the human body. Actions involved are (1) elbow extension, using the triceps muscle; (2) rising on toes, using the gastrocnemius and soleus muscles; and (3) elbow flexion, using the biceps muscles.

so that the body weight falls between the applied force at the attachment of the tendon of Achilles and the ball of the foot, which functions as the fulcrum.

In the diagram of a lever of the third class (Fig. 4-1), the weight arm is seen to be longer than the force arm; leverage again favors speed and range rather than power. In the body the brachialis demonstrates this class of lever.

Angle of application. Muscular force is utilized for both stabilization and rotation. The proportion used for each depends to a certain extent on the angle the mechanical axis of the muscle makes with the bone. As joint motion occurs, this angle changes. At zero degrees, with the mechanical axis of the muscle parallel to its bony lever arm, all the force would be utilized for stability. As the angle approaches 90 degrees, more of the force is utilized for rotation and less for stability. At 90 degrees the total force is rotary.

Length-strength relationship. It would appear from the preceding paragraphs that rotary strength of the biceps brachii, for example, would be greatest in full flexion of the elbow because the angle of application approaches 90 degrees. The reverse is true. The tension developed by muscular contraction decreases rapidly as the muscle shortens. The closer the muscle remains to its natural or resting length, the more tension it can generate.

Provisions, therefore, are to be found in the body for increasing the angle of application without decreasing the muscle length, thereby increasing leverage and efficiency of contraction. Examples: (1) The origin of the gastrocnemius and the insertion of the hamstring group of muscles pass over the condyle of the femur, thereby increasing the angle of insertion. (2) The patella, which is attached to the common patellar tendon, deflects the tendon away from the knee joint, thus increasing the rotary effect.

Anatomic construction of muscle. Muscles vary in length of individual fibers as well as in arrangement of fibers. Muscles with longer fibers arranged in parallel fashion have a greater range of contraction but less strength than muscles with shorter fibers arranged in pennate fashion.

Steindler[2] summarized the conditions suitable for speed and power as follows:

1. Anatomic build of muscles:
   For power: the fibers are short, more numerous, in pennated arrangement.
   For speed: the fibers are long and less numerous, in parallel arrangement.
2. Leverage of the muscle:
   For power: the muscle is attached at a greater distance from the center of motion.
   For speed: the muscle is attached at a lesser distance from the center of motion.*

An understanding of the foregoing factors is basic to the understanding of the mechanics of muscle action. A partial list of other factors that influence the resulting motion includes (1) contour of joint surfaces, (2) gravitational forces that are necessary to maintain equilibrium, (3) forces of acceleration and deceleration, (4) stability of proximal joints, (5) limitation imposed by the ligaments, and (6) action of antagonistic muscles.

All of the systems of the body are dependent on one another. The biochemical changes involved in muscle action are complex and incompletely understood. Detailed discussion of these changes can be found in standard texts.

EFFICIENCY OF MUSCLE ACTION

The body functions as a combustion engine operating at an overall efficiency of about 25%.[3] This means that only a small part of the chemical energy expended for muscular contraction is reflected in effective motion. This low efficiency is about average for a machine and is attributable in most part to the low efficiency in the muscles, which, in turn, is due to (1) the production of heat during contraction and relaxation, (2) the internal resistance of the fascia and skin, (3) the necessity of over-

*From Steindler, A.: Kinesiology, Springfield, Ill., 1955, Charles C Thomas, Publisher.

coming inertia and gravitational forces, and (4) the resistance to motion imposed by external force.

MUSCLES THAT TAKE PART IN MOVEMENT

According to Best and Taylor,[4] the following muscles take part in movement:

1. Prime movers or agonists—those muscles that are essentially responsible for the movement of the part
2. Antagonists—those muscles that oppose the prime movers (for the phenomenon of reciprocal innervation, see the following discussion on coordination of muscle function
3. Synergists—those muscles that assist the prime mover and minimize unnecessary movement
4. Fixators—those muscles that stabilize the position of neighboring segments and maintain the limb or body in a position appropriate for carrying out the particular movement

COORDINATION OF MUSCLE FUNCTION

There is no motion in the body in which only one muscle is active. Every purposeful movement of the body exemplifies the interaction of muscles as prime movers, antagonists, synergists, and fixators.

Muscular coordination might be defined as the use of the proper muscles, at the proper time, with the exact amount of force necessary. Such smooth muscular action denotes an intact nervous system controlling stimuli to an intact musculoskeletal system. The contraction of an individual muscle depends on the number of anterior horn cells stimulated and the frequency of this stimulation. If all anterior horn cells to a muscle are stimulated at once, the result is a muscle twitch. The individual muscle fibers obey the *all-or-none* law, so that a smooth contraction must arise by alternately tetanizing different parts of the muscle.

In the spinal cord occurs a mechanism whereby the effects of similar stimuli are summed whereas others annul one another. This integrative activity is present in the spinal cord, isolated from the brain. Whatever the interaction between different reflexes that compete for or cooperate in the control of a particular muscle, the final effect is exerted through the same "final common path."[5]

Voluntary or reflex contraction of a muscle is accompanied by simultaneous relaxation of its antagonist. This is Sherrington's law of *reciprocal innervation*. The inhibitory effect on the skeletal muscles is not brought about through specific inhibitory nerves but is due simply to the cessation of excitatory impulses along the motor neurons.[4]

EXERCISE AS A THERAPEUTIC MEASURE

The use of exercise as a therapeutic measure should be as carefully planned as any other type of therapy available to the physician.

The physician with his staff of physical therapists can best know what form the exercise program of a particular patient should take. The physical therapist is called on to do a muscle evaluation and a range of motion test. These tests demonstrate the areas of weakness, tightness, or whatever deviation from the "within normal limits" test. It becomes almost irresponsible and dangerous not to use the talents of all team members when an exercise program begins, the diagnosis, disability, or prognosis of the patient notwithstanding.

Essentials for planning the exercise program

The following must be taken into consideration when planning an exercise program:

1. Indication for therapeutic exercise—some alteration in body physiology that has resulted in a loss of normal movement
2. Evaluation of remaining ability of the patient—based on tests and measurements reflecting muscle strength, range-of-joint motion, functional ability, abnormal motion, etc.
3. Establishment of goal—one that will give the patient maximum independent function and efficiency within the limits imposed if damage is permanent
4. Planned, graduated exercise program—based on periodic reevaluation

Objectives of exercise

Power. Exercises for power are designed to increase strength of muscles. A common error among therapists and doctors is to attempt to use endurance-building programs to increase power. They are based on a few repetitions with maximal active effort and are especially useful where atrophy is a result of disuse. The rate and extent of hypertrophy are proportional to the amount of resistance the muscle is asked to overcome and the method of stabilization that may be used for that particular movement. It has been shown that even extremely atrophied muscles should exert their maximal effort at regular intervals.

Endurance. Exercises for endurance are designed to increase tolerance. They are based on submaximal effort with many repetitions and are especially useful following a period of convalescence.

Muscles that are yet weak and atrophied should not be subjected to endurance-building exercise until power has been restored to within normal limits.

Coordination. Exercises for coordination are designed to develop an efficient habit pattern. They are based on the principle that practice and repetition lead to precision of performance, and they are especially useful where cerebellar function is disturbed.

Range of motion. Exercises designed for maintaining or increasing range of motion are of value wherever there is limitation or potential limitations of normal range for any reason. These are particularly useful when contractures or paralysis is present. It is probably more preferable to have a limited range of motion with good power than range within normal limits with inadequate power.

Speed. Exercises for speed are designed to shorten activity time. Speed is attained by frequent repetition of functional activities until the energy expended is minimal. Practice to acquire speed is useful during the final phase of a rehabilitation program for most neuromuscular conditions.

Principles applicable to all forms of exercise

The following principles are applicable to all forms of exercise: (1) the patient should be in a position of comfort involving the least strain; (2) the proximal joints should be stabilized to eliminate undesired motion and get maximum results from segment being exercised; (3) all motions should be done smoothly throughout range and during return to starting position; (4) indications of overdosage include pain or discomfort lasting more than 3 hours and/or decrease either in range of motion or in strength; (5) short bouts of exercise repeated during the day are always preferable to prolonged period once a day; (6) effectiveness of a therapeutic exercise program cannot be evaluated unless accurate and periodic records are maintained (see discussion on muscle testing and range of motion testing); and (7) the patient should understand the object of the exercise and the manner in which it is performed.

It is of constant importance in any exercise regimen to recognize that substitution movement, or the voluntary attempt to achieve a functional goal by using movement patterns not prescribed, be recognized, minimized, or stabilized to be prevented. However, there are times when substitution patterns can be taught safely so that prescribed goals can be obtained. If used, these patterns must be continued with safety, be practical, and be esthetically acceptable.

Types of exercise

Motions can be separated into three categories: passive, active, and forced. Exercises utilizing these basic motions have been further defined as follows:

1. *Passive*—exercise accomplished by the therapist or apparatus with no active contraction of the involved part by the patient
2. *Active-assitive*—exercise accomplished by active contraction by the patient with the assistance of the therapist or mechanical device
3. *Active*—exercise accomplished by the patient without assistance or resistance
4. *Resistive*—active exercise accomplished by the patient with additional resistance, either manual or mechanical
5. *Stretching*—exercises accomplished by forced motion, either passive or active

Passive exercise. The main purpose of passive exercise is to prevent contractures by maintaining the normal range of joint motion. It is usually used when muscles are paralyzed or very weak. Passive exercise may be done by the nurse as part of bed care or may be taught to a family member, enlisting the aid of others or the patient himself in the prevention of deformity.

Principles. The therapist should (1) stabilize the proximal joint and support all distal segments; (2) keep movement within the pain-free range; and (3) maintain slow, smooth movement through complete range, avoiding pumping movement.

Effects. Passive exercise (1) prevents contracture and formation of adhesions, (2) increases proprioceptive sensation, (3) maintains resting length of muscle, (4) stimulates flexion-extension reflexes, and (5) prepares for active exercise.

Active-assistive exercise. Active-assistive exercise is the first step in a program of muscle reeducation. Strength can be attained only by active contraction by the patient with the assistance of the therapist or mechanical devices. This usually means that the therapist supports the weight of the distal segment to eliminate the resistance of its weight or the pull of gravity, so that the patient can maintain an active contraction through the greatest possible range of motion. The use of the mechanical or hydrotherapeutic modalities shown in Figs. 4-2 to 4-4 is often prescribed during this phase of treatment.

Principles. The therapist should (1) explain to the patient exactly what he wants him to do, indicating the insertion and line of pull of the muscle involved, (2) give

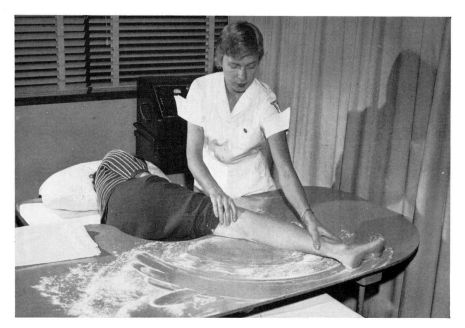

Fig. 4-2. Powdered board used for active-assistive exercise. The part to be treated is in a gravity-eliminated position for weakened musculature. The part is guided through the arc of motion by the physical therapist.

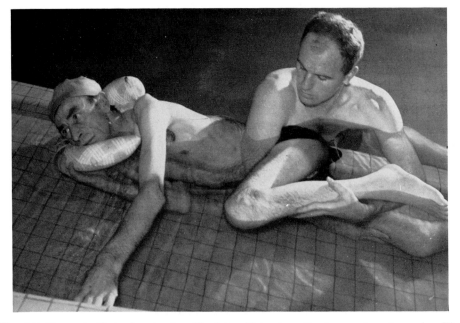

Fig. 4-3. Buoyant effect of water is utilized to aid in strengthening weakened muscles. The part is guided by the physical therapist.

only that amount of assistance required for a smooth movement, (3) avoid substitution of other muscles, (4) allow the patient a brief rest after each completed movement, and (5) gradually decrease assistance as strength increases.

Effects. Careful guidance and assistance at the beginning of a program of active exercise not only strengthens the muscles but also establishes the pattern for coordinated motion.

Active exercise. Active exercise denotes free exercise. The patient completes the total action unaided. It is a progressive step-up from active-assistive exercise that indicates assistance is no longer needed.

Principles. Principles of active exercise include the following: (1) exercise should be presented to patient, movement by movement, in proper sequence[6]—starting position, movement to the farthest point, return to the starting position, and rest; (2) exercises should be neither too easy nor too difficult; (3) patient should not be left alone during this phase of the program; the therapist checks constantly to see that the patient performs the motion smoothly and through the complete range and

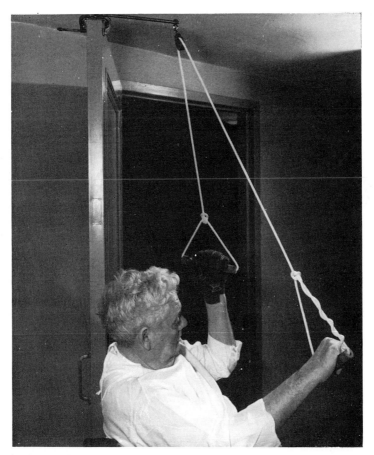

Fig. 4-4. Simple over-door pulley allows the patient to use the good right extremity to stretch the tight left shoulder.

avoids substitution movements; (4) if substitution takes place, either the exercise is too difficult or the patient is not yet ready for active exercise.

Effects. An active exercise program improves function and increases strength. It also tends to improve general body functioning by assisting in increasing cardiac reserve, respiratory reserve, and body conditioning.

Resistive exercise. In this type of exercise the patient actively carries the part through its arc of motion. External resistance, minimal as the force of gravity against the part or maximal as applied by the therapist or by mechanical devices, is added to the active exercise. The physical therapist applies manual resistance, from slight to marked, depending on the power of the muscles and the position in which the part is placed for work. The strengthening of the lower extremities must be considered in terms of function—ascending and descending stairs, getting up and down from the sitting position. "Normal" here would thus vary with total body weight. Resistance exercises are usually given when a muscle is rated "good" to "normal"; however, such exercises may be given when a muscle is rated "fair" only if the part is placed so that the forces of gravity are eliminated or assist in the movement. It may be used for muscles rated "poor" or "trace" only if gravity assists in the movement.

Principles. Principles of resistive exercise include (1) the angle of pull, the relative tension of muscle, and the amount of contraction must be considered; (2) as a general rule the least amount of resistance is exerted at the beginning and the end of the range, with most of the resistance applied in the middle one third of the range; (3) if movement is jerky, it signifies that too much resistance is being applied; and (4) resistance is applied at a point distal to the joint involved in the motion.

Effects. The major effect of any resistive exercise program is the development of strength. The significant amount of effort involved in resistive exercises differentiates them from other exercise programs; thus they must be graduated to the patient's tolerance.

Progressive resistance exercise. This method of exercise was developed during World War II by DeLorme when in the Army at Gardiner General Hospital in Chicago.[7] The aim for rapid rehabilitation of wounded servicemen was directly responsible for the development of these specific technics. Progressive resistance exercise is utilized primarily for strengthening a muscle group by means of applying progressively increased mechanical resistance. Resistance exercise with weights has the major purpose of increasing the strength and endurance of the exercised muscles.

Definitions. The following are used in progressive resistance exercises:

1. *Exercise load*—the amount of weight with which the exercise is performed, either assisting or resisting the muscles
2. *Muscle load*—the actual resistance the muscle must overcome
3. *Loading-assisting exercises*—those in which the exercise load assists the muscle (counterbalancing)
4. *Load-resisting exercises*—those in which the exercise load resists the muscle
5. *Ten-repetition maximum*—the greatest weight the muscle can correctly lift ten times
6. *Ten-repetition minimum*—the least amount of weight required by the muscle to help it to lift the extremity correctly ten times through a full arc of motion

Principles of counterbalancing. A system of pulleys is set up so that there is a

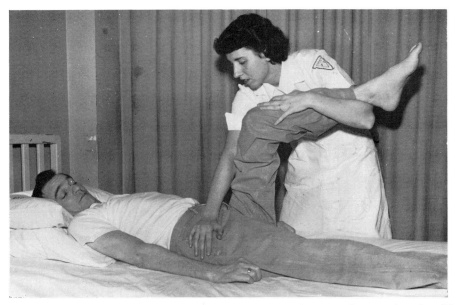

Fig. 4-5. Forced motion or stretching is aimed toward achieving normal range of motion. The physical therapist stabilizes the pelvis while stretching the hip extensors.

sufficient counterbalancing load to offset the weight of the part until such time as the weak muscle can complete the arc of motion. Weak muscles are then made to work to their maximum with no fear of overloading.

Stretching. Stretching exercises utilize forced motion. The force may be applied by the therapist or by some mechanical device (Fig. 4-5), or it may be applied by the patient, using the contraction of antagonistic muscle groups (active stretch).

Principles. These include the following: (1) motion should never be forced when acute pain is present; (2) persistent pain or decrease in range indicates overdosage, and either the force used or the duration of treatment should be decreased; (3) stretching done gently over a long period will produce the best results; (4) the distal segment should be supported so that the force is applied in the desired area; this is particularly important where a fracture has resulted in limitation of motion in the proximal joint; (5) the muscle to be stretched should be as relaxed as possible so that the stretch does not become an active-resistive exercise; (6) when stretching pluriarticular muscles, the primary joint should be stretched first and then the secondary joint, fixing the part as the emphasis changes; (7) any increase in range obtained by forced motion will be lost unless maintained either by active motion or by supportive devices if the part is weak or paralyzed.

The therapeutic use of stretching is designed to restore the normal range of motion where limitation of this range is due to loss of elasticity of soft tissue. Loss of motion due to malalignment of bone is beyond the scope of stretching procedures.

Isokinetic or accommodating resistance exercise is a new dimension in the field of resistance exercise and muscle evaluation. The underlying principle is motion at a constant speed. This prevents acceleration, which results in increased resistance.

The resistance that is developed is in proportion to the exerted muscle force and is seen on a recorder.

Isokinetic exercise is made possible by an electromechanical device that is adjustable in the anatomic axis of rotation.

It has been suggested by Hellebrandt[24] that, for increasing muscular force, work is not as important as the *rate* at which it is done. A means now exists that allows the application of resistance at a speed that produces an optional power output. The specific way in which the muscle is loaded influences the degree of effort required or the degree of neuromuscular response.

The limited amount of research so far published in this area has demonstrated several important findings with normal subjects.

1. Isokinetic exercise is an effective means with which to increase muscular tone throughout an arc of motion.

2. It increases the work a muscle can do more so than isometric or isotonic exercise.

3. The response to different systems of loading in a partial range of motion will increase significantly in this range more so than other less exercised joints.[27]

POSTURE AND GENERAL BODY MECHANICS

Posture is the position in which the various parts of the body are held while sitting, standing, walking, and lying.[8-11] Good posture is efficient posture and varies with the individual and the activity.

Good body mechanics is the correct poise and control of the body with normal functioning of every part in maintenance of proper alignment.

A recognition of what is good posture in each activity is important for all who work in physical medicine and rehabilitation. It is important in the prevention and correction of deviations.

Effects

The effects of good posture and body mechanics are (1) proper relationship between the various segments of the body, (2) minimum energy expenditure in balancing of segments, (3) ease and grace of movement, and (4) minimum energy expenditure in motion.

Special exercise programs

Commonly used special exercise programs include breathing exercises, scoliosis exercises, relaxation exercises, and some theories on the management of low back problems. The prescribed exercise program should be designed to correct deficiencies where possible or provide assistance when needed.

Breathing exercises. Breathing exercises are useful to correct or minimize a respiratory deficit or to improve posture and trunk stability. To prescribe breathing exercises correctly requires an understanding of the mechanics of respiration and etiology of the deficiency.

Mechanics of respiration. A brief summary is given in Table 4-1. A more detailed explanation can be found in standard physiology tests and special articles devoted to the subject.[4, 5, 11, 12]

Table 4-1. Mechanics of respiration

	Inspiration	*Expiration*
Type of motion	Active	Ordinarily passive rebound of elastic tissue; active if forced
Control	Reflex or voluntary	Ordinarily reflex; if forced is active
Time in cycle	Longer phase	Shorter phase
Muscles involved	Muscles that act to expand rib cage	Mainly relaxation of muscles of inspiration; forced expiration involves contraction of muscles active in retracting rib cage
	Diaphragm	Abdominals

Etiologic factors involved in deficient respiration. These include (1) habitual poor posture; (2) inadequate or asymmetric motion of the rib cage due to weakness or paralysis of the muscles active on rib cage, contracture or scar formation, pathologic changes in thoracic viscera (such as from empyema, asthma, emphysema, and tuberculosis), surgical removal of part of the soft tissue or bone, and scoliosis; (3) weakness or paralysis of the diaphragm; (4) weakness or paralysis of the abdominal muscles; (5) alterations in the normal cycle, such as *paradoxical respiration,* which is either ascent of all or part of the diaphragm during inspiration or retraction of all or part of the rib cage during inspiration rather than expiration, and *incoordinated respiration,* which implies that the normal nervous control of the sequence of events during respiratory cycle is disturbed, as often occurs in a patient with the athetoid type of cerebral palsy.

Insufficient inspiratory excursion. Whatever the etiology, there will usually be some adaptive shortening of soft tissue. The resulting tightness limits the motion of the rib cage. This should be prevented if anticipated or the tissue be stretched by active or passive means if it has already occurred.

The patient is taught to put forth maximal effort during inspiration. If the limitation of inspiratory excursion is permanent, the patient should be taught to utilize the diaphragm to its maximum extent.

Insufficient expiratory excursion. Because expiration is mainly passive, insufficient expiratory excursion is usually the result of weak abdominal muscles or a pathologic condition such as emphysema. Strengthening exercises for the abdominal muscles will help in either case. In emphysema the primary therapeutic objective is to assure, as far as is possible in any given case, adequate ventilation and thus enhance the physiologic gas exchanges in the alveoli. Since the activity of the diaphragm normally is responsible for approximately 60% of the pulmonary function, it is evident that the pathologic limitation of function of the diaphragm, which is characteristic of emphysema, must, if possible, be corrected if improvement in pulmonary ventilation is to be achieved. Planned exercise (Chapter 26) aims to restore the tonus and function of the diaphragm and its synergists.

Weakness or paralysis of the diaphragm. A weak diaphragm can be strengthened only through use. The emphasis on the diaphragm for inspiration is frequently called "abdominal breathing" because the abdominal muscles relax and the abdomen protrudes as the abdominal viscera are compressed by the descending diaphragm. If the diaphragm is permanently paralyzed, an attempt at compensation should be made by teaching the patient how to get the maximal inspiratory excursion of the rib cage by using the accessory and intercostal muscles.

Combined deficit. If the majority of the muscles of respiration are paralyzed or weakness exists to the extent that air exchange is inadequate, the patient should be taught glossopharyngeal breathing.[13-15]

Alterations in the normal cycle of respiration. Where an alteration in cycle exists, the goal of an exercise program is to attempt to establish a more nearly normal pattern.

Scoliosis exercises. There is little uniformity of opinion as to etiology or treatment of scoliosis. Before a discussion of the principles of a therapeutic exercise program for scoliosis can be given, certain basic facts must be reviewed.[2, 12, 16-18]

Motion in relation to the normal curves of the spine. Such motion includes the following.

Flexion or forward bending. This occurs in all parts of the spine but is freest in lumbar and dorsolumbar areas: (1) cervical—the anterior physiologic curve can be straightened; further flexion is due to motion between the atlas and occiput; (2) dorsal—normal curve can be increased but extent of movement not great; (3) lumbar—quite free; adults can almost obliterate physiologic curve, and children may be able to reverse the curve.

Extension or hyperextension. This occurs most in the lumbar and eleventh and twelfth dorsal vertebras: (1) cervical—limited; (2) dorsal—very slight; (3) lumbar—maximum occurs here.

Side-bending–rotation. Rotation accompanies lateral bending: (1) cervical—very free between the first and second vertebras but otherwise quite limited; (2) dorsal—lateral bending fairly evenly distributed but of slight extent; most marked in this area, especially in erect or hyperextended position; (3) lumbar—lateral bending is free movement in this area, greatest in erect position and least in extreme flexion; rotation extremely limited, especially in extreme flexion and extreme hyperextension.

Causes of scoliosis. The following factors that may cause scoliosis are adapted from Lovett[17]:

Congenital scoliosis may be caused by (1) malformation of the spine, (2) malformation of the scapula, (3) malformation of the thorax, (4) deforming intrauterine pressure, and (5) paralysis of intrauterine origin.

Acquired scoliosis may be caused by (1) anatomic, physiologic, postsurgical, or other asymmetries elsewhere in the spine—postoperative (thoracoplasty), pleural effusion (due to fibrothorax), pelvic asymmetry, pelvic obliquity (short leg), unequal vision, and unequal hearing; (2) pathologic affections of the vertebras—rickets, osteomalacia, Pott's disease, dislocation, arthritis deformans, tumors, etc.; (3) pathologic affections of the bones and joints of the extremities, causing asymmetric position—diseases of bones and joints of the leg and diseases of bones and joints of the arm; (4) distorting conditions due to disease of the soft parts—infantile paralysis, spastic paralysis, nervous diseases (hemiplegia, syringomyelia, etc.), empyema, organic heart disease, scars, throat, abdominal, or pulmonary disease, and acute or chronic inflammation of the spinal muscles (lumbago, etc.) ; and (5) habit or occupation.

Classification of scoliosis. Scoliosis is generally divided into two types—functional and structural.

Functional scoliosis. Functional scoliosis can be voluntarily corrected. There are no structural changes in the vertebras. Mild asymmetry and lateral deviation are present. There is no agreement as to the direction of the rotation of the vertebral bodies. The condition is constantly occurring and is reversible.

Structural scoliosis. Structural scoliosis cannot be voluntarily corrected. Structural changes are present in the vertebras. The degree of curvature varies with vertebral bodies rotating to the convexity of the curve. This cannot be duplicated by the normal spine and is irreversible.

Description and terminology. A description of scoliosis should include (1) the direction of deviation (whether right or left of midline), (2) involved area of spine (cervical, dorsal, or lumbar), (3) degree of deviation (mild, moderate, or severe), and (4) etiology (if causal factor is unknown, it is described as idiopathic; this applies to the majority of cases in most reported series).

Mechanics of scoliosis. The spine is a column of bones fixed to a mobile pelvis. It supports the head and trunk. Motion occurs between vertebras so that the head and trunk can move. This movement is limited by bone contact between adjacent vertebras or by their muscular or ligamentous attachments.

The spine is responsible for holding the body erect. This is accomplished by a complex mechanism that coordinates the action of the muscles in such a way that the center of gravity of the mass of the body is maintained over the base of support.

In the erect position the spinal column is vertical on posterior view but has normal curves on lateral view. Prolonged lateral deviation of the spine is abnormal. If it can be corrected voluntarily, it is considered a postural curve.

If for any reason the deviation persists, the external conformation and the internal structure of the vertebras alter in response to their new obligation (Wolff's law). Ribs, muscles, and paravertebral tissue alter their position or shape, accommodating to the changed relationship as the body continues to strive for balance and equilibrium.

It is often impossible to tell what factor caused the initial loss of symmetry. The spine, ribs, muscles, and paravertebral soft tissue are all interdependent.

Once a primary curve is established, the only way for the body to keep the head erect and shoulders level with the center of gravity over the base of support is by establishing a secondary, compensatory curve.

Principles of therapeutic exercises for scoliosis. Functional scoliosis can usually be corrected. Structural scoliosis cannot be "cured"; however, it is sometimes possible to prevent progression or to decrease that part of the curve that is due entirely to contracture of soft tissue. Treatment consists of the following: (1) cause is removed if possible; (2) symmetric exercises are given to improve general body condition; (3) mobilization is promoted where tightness of soft tissues is asymmetric or deforming; (4) asymmetric exercises are used mainly to strengthen the muscles on the side of the convexity and stretch the contracted muscles on the side of the concavity; (5) spinal elongation is promoted; traction on the spine with the Sayre head sling can be effectively combined with active exercises; (6) exercises aimed at increasing the strength of the muscles active in anteroposterior balance will prevent or decrease exaggerated curves in that plane; an example of this is the action of the abdominal muscles and hip extensors in controlling lumbar lordosis; (7) exercises aimed at derotation are performed by rotating the affected segment of the trunk toward the side of the convexity.[18] In an exercise program for patients having thoracic surgery, the best results can be obtained if the program can be initiated preoperatively.[19]

Relaxation exercises. Continuing stresses and strains, either physical or mental in origin, result in a neuromuscular hypertension that is a condition marked by reflex phenomena of excitation and hyperirritation.[20]

Principles. These include the following: (1) exercising in supine position with eyes closed; (2) developing kinesthetic sense—recognition of tenseness or muscular contraction as opposed to a release of tension or muscular relaxation; (3) learning to contract and relax muscles without moving the part; when this becomes automatic, the patient learns to relax without contracting muscles; and (4) developing habit patterns of relaxation, to be done in a quiet room at a specified time each day, since repetition is important in developing a habit pattern.

FUNCTIONAL EXERCISE

The goal of the rehabilitation process is to have the patient function to his full capacity within the limits of his disability. Many cases of failure to achieve the rehabilitation goal can be attributed directly to the fact that the patient is often taught functional activities without first laying the groundwork on which such activities must rest. The foundation for a functional training program is exercise.[21-23] These exercises may be done in the patient's bed, on an exercise mat in the gym, in parallel bars, or on crutches; they may be done individually or in groups (Chapter 6). Following is a summary of the specific aims of each exercise group.

Mat exercise (Fig. 4-6). The purposes are (1) to teach changing position, from prone to supine and supine to sitting; (2) to teach sitting balance while moving trunk and/or arms; (3) to teach movement in all directions while sitting on a level surface; (4) to teach handling of affected extremities, from higher to lower level and back; (5) to strengthen muscles—push-ups, abdominal exercise, and latissimus dorsi exercise; (6) to teach stretching—active or passive; and (7) to teach coordination and skill, for example, ball games, in preparation for standing and walking exercises in parallel bars and for ADL—bed, wheelchair, dressing, and toilet activities.

Parallel bar exercise (Fig. 4-7). These exercises are for patients with or without braces.[21, 23] The purposes are (1) to develop tolerance for standing position and weight bearing, (2) to teach standing balance while moving trunk and/or arms, (3) to strengthen upper extremities—push-ups, (4) to teach pelvic control—latissimus dorsi exercise, (5) to teach locking and unlocking braces while sitting and standing, (6) to teach coming to a standing position and sitting down while holding on to a bar, and (7) to teach basic gait pattern—even rhythm—in preparation for crutch balancing and walking.

Crutch balancing (Figs. 4-8 and 4-9). Crutch balancing is for patients with or without braces.[7, 15] The purposes are (1) to teach standing balance on crutches (lateral and anteroposterior), (2) to teach pelvic control—latissimus dorsi exercise, (3) to teach placing crutches in different directions, (4) to teach locking and un-

Fig. 4-6. Functional and strengthening activities are performed during a mat class period.

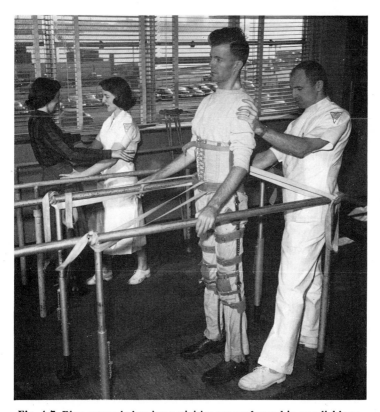

Fig. 4-7. Elementary balancing activities are performed in parallel bars.

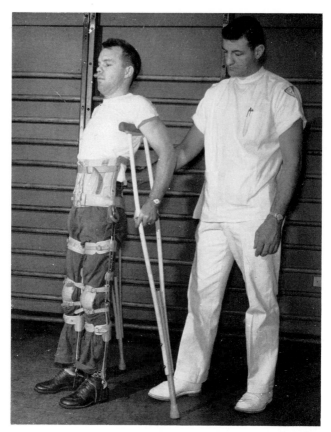

Fig. 4-8. Elementary ambulation activities progress to outside of parallel bars.

Fig. 4-9. Crutch-balancing exercises are done to improve muscle control, balance, and stability.

Fig. 4-10. Outside ambulation training is done on various types of terrain.

locking braces while in standing position, (5) to teach preclimbing exercises—push-up one crutch in parallel bars or stall bars, and (6) to teach prefalling exercises (stall bars) in preparation for crutch walking and climbing and for ADL—getting up and sitting down from wheelchair, bed, toilet, etc.

Ambulation

Basic gait technics. These technics are for patients with or without braces or crutches.[2] Their purpose is to teach good basic (safe) gait patterns—on individual prescription—in preparation for independent ambulation (Chapter 8).

Independent practice in ambulation technics (Fig. 4-10). This is for patients with or without braces, prostheses, crutches, canes, or other aids. The purposes are (1) to teach independent ambulation, one fast and one slow gait, individual selection by prescription, such as tripod, four-point, and two-point gaits, in all directions and on different floor coverings, (2) to develop endurance and speed, (3) to teach ambulation over obstacles, inside and outside, and (4) to teach ambulation outdoors on different surfaces, such as gravel, cement, and grass.

Climbing technics (Figs. 4-11 and 4-12). These are for patients with or without braces, prostheses, crutches, canes, etc. The purposes are (1) to teach independent climbing of ramps; curbs, graduated from 1 inch high to 8 inches high (standard); stairs, graduated from 1 inch high to 8 inches high (standard); and, with railings, 12 to 14 inches high (bus steps).

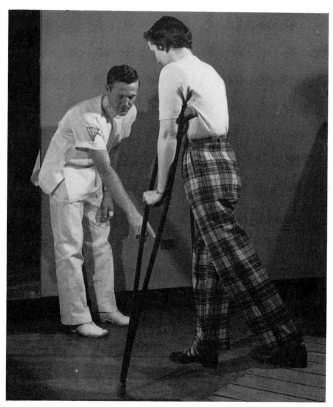

Fig. 4-11. Advanced ambulation includes instruction in all kinds of elevation activities, including going down curbs.

Falling and getting up from the floor. (Note: See Chart 7 for functional ambulation and elevation record for recording progress in ambulation.)

SUMMARY

An attempt has been made to clarify the principles underlying therapeutic exercise programs. A dynamic, successful therapeutic exercise program can be established for each patient if several factors are considered and incorporated into the program.

Patient motivation. Because a program of therapeutic exercise is a joint undertaking of the doctor, physical therapist, and patient, all the chains in this link must be sound and the forces must be pulling in the same direction. The patient who understands the goals and strives toward them is the block on which the pyramid to success is based.

Adequate prescription. Both the patient and the physical therapist depend on the physician for an understanding and interpretation of the medical problem involved. A program of exercise is based on the physician's findings. An adequate program is therefore dependent on an adequate prescription.

Performance factors. The physical therapist is responsible for supervising the exercise program and must be aware of a number of essential elements on which a

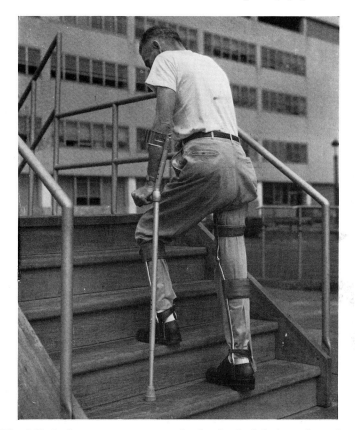

Fig. 4-12. Patient goes up steps unassisted, using both Lofstrand crutches.

successful program is based: (1) clear and concise demonstrations and explanations; (2) supervision of each exercise as done by the patient himself, with immediate correction of errors; (3) consideration of pain level and fatigue tolerance of the patient; (4) scheduling, since short, frequent periods of activity are more beneficial than long, infrequent sessions; (5) the importance of sustaining a contraction at the end of a movement and releasing that contraction slowly; (6) the policy of treating each case as a separate and distinct entity, particularly with regard to the progressions of exercise program; and (7) provision for the success of a carry-over home program, which will—to a great extent—depend on the understanding and cooperation of the patient's family regarding the total program.

Reevaluation. The goals of physical medicine, as well as the patient's pathology, must be reexamined at regular intervals. This is done under direct supervision of the physician.

• • •

All factors as presented in this summary should be constantly borne in mind when a therapeutic exercise program is to be undertaken. Each team member plays a vital role in making any program a dynamic, goal-achieving success.

CHART 7

INSTITUTION OF REHABILITATION MEDICINE

New York University Bellevue Medical Center

PHYSICAL THERAPY DEPARTMENT
FUNCTIONAL AMBULATION AND ELEVATION RECORD

Name _____ In ☐ Out ☐
Disability _____ Height of lesion _____ Spastic _____ Flaccid _____
Braces: Short—right, left; long—right, left; pelvic band; Knight spinal _____
Crutches: Axillary; Lofstrand; Canadian; Cane _____

Initial date _____ Discharge date _____

KEY

1. Each numbered block represents one week, so that progress can be recorded in the week activity has been accomplished. Fill in date above each box.
2. Fill in block 1 with blue color if activity is performed at initial date.
3. Fill in numbered block with red color when activity has been learned.
4. Fill in numbered block with X when braces have had a major adjustment (cut down). Recheck all activities and fill in with red after activity is performed again with the adjustment.

PARALLEL BAR ACTIVITIES

	1	2	3	4	5	6	7	8	9	10	11	12	13	14	15	16
Up from wheel chair holding on and with assistance																
Sit down holding on and with assistance																
Stand free in parallel bars 10 seconds 3 times																
Ambulate length of bars without stopping																

ELEMENTARY AMBULATION

Stand up from wheel chair holding on																
Sit down holding on																

Ambulate continuously with assistance
(drag-to, swing-to, swing-through,
4-point, 2-point, 3-point)*:

(a) 30 feet																
(b) 60 feet																
(c) 80 feet																

Ambulate without assistance (drag-to,
swing-to, swing-through, 4-point, 2-point,
3-point)*: (a) 30 feet

ADVANCED AMBULATION—INDEPENDENT

Stand up from wheel chair 3 times in 30-60 seconds																
Sit down 3 times in 30-60 seconds																

*Circle which.

CHART 7, cont'd

ELEVATION—INDEPENDENT

Stairs:	(a) 10 3-inch with handrail (right, left)
	(b) 10 8-inch with handrail (right, left)
	(c) 6 3-inch without handrail
	(d) 6 8-inch without handrail
Curbs:	(a) 2-inch
	(b) 4-inch
	(c) 6-inch
	(d) 8-inch
Ramp	
8-inch curb outside	
12-14 inch bus step	
Outside—down curb, cross street, up curb in:	
	(a) 2 minutes
	(b) 1 minute
	(c) 30 seconds
	(d) 12 seconds
Up from chair in gym, ambulate to elevator, out front door, down curb, cross street, up curb, and back again	

		1	2	3	4	5	6	7	8	9	10	11	12	13	14	15	16
Ambulate inside (drag-to, swing-to, swing-through, 4-point, 2-point, 3-point)*:	(a) 40 feet																
	(b) 60 feet																
	(c) 80 feet																
Ambulate inside (drag-to, swing-to, swing-through, 4-point, 2-point, 3-point)* 60 feet in:	(a) 2 minutes																
	(b) 1 minute																
	(c) 30 seconds																
	(d) 12 seconds																
Ambulate:	(a) 60 feet on gravel																
	(b) 60 feet on cement																
	(c) 60 feet on cobblestones																
	(d) 40 feet on grass																

ELEVATION WITH ASSISTANCE

6 stairs:	(a) 3-inch with handrail (right, left)
	(b) 8-inch with handrail (right, left)
	(c) 3-inch without handrail
	(d) 8-inch without handrail
Curbs:	(a) 2-inch
	(b) 4-inch
	(c) 6-inch
	(d) 8-inch
Ramp: 12-foot, 15-degree incline	
Falling and getting up from floor	

*Circle which.

REFERENCES

1. Duchenne, G. B.: Physiology of motion. Translated and edited by E. B. Kaplan, Philadelphia, 1949, J. B. Lippincott Co.
2. Steindler, A.: Kinesiology, Springfield, Ill., 1955, Charles C Thomas, Publisher.
3. Wakim, K. G.: The physiological aspects of therapeutic physical exercise. In Handbook of physical medicine and rehabilitation, Philadelphia, 1950, Blakiston Co.
4. Best, C. H., and Taylor, N. B.: The physiological basis of medical practice, ed. 4, Baltimore, 1945, The Williams & Wilkins Co.
5. Winton, F. R., and Bayliss, L. E.: Human physiology, ed. 4, Boston, 1955, Little, Brown & Co.
6. Kraus, H.: Principles and practice of therapeutic exercise, Springfield, Ill., 1949, Charles C Thomas, Publisher.
7. DeLorme, T. L., and Watkins, A.: Progressive resistance exercises: technique and medical application, ed. 2, New York, 1951, Appleton-Century-Crofts, Inc.
8. Brunnstrom, S.: The changing concept of posture, Physiother. Rev. 20:79, March-April, 1940.
9. Kendall, H. O., and Kendall, F. O.: Posture and pain, Baltimore, 1952, The Williams & Wilkins Co.
10. Lee, M., and Wanger, M.: Fundamentals of body mechanics and conditioning, Philadelphia, 1949, W. B. Saunders Co.
11. Williams, M., and Worthingham, C.: Therapeutic exercise in body alignment and function, Stanford, Calif., 1953, Stanford University Press.
12. Bennett, R. L.: Classification and treatment of early lateral deviations of the spine following acute anterior poliomyelitis, Arch. Phys. Med. 36:9, 1955.
13. Dail, C. W., Affeldt, J. E., and Collier, C. R.: Clinical aspects of glossopharyngeal breathing, J.A.M.A. 158:445, 1955.
14. Dail, C. W., Zumwalt, B. A., and Adkins, H. A.: Manual of instruction for glossopharyngeal breathing, New York (undated), National Foundation for Infantile Paralysis.
15. Dail, C. W., and Affeldt, J. E.: Effect of body position on respiratory muscle function, Arch. Phys. Med. 38:427, 1957.
16. Kleignberg, S.: Scoliosis: pathology, etiology and treatment, Baltimore, 1951, The Williams & Wilkins Co.
17. Lovett, R. W.: Lateral curvature of the spine and round shoulders, Philadelphia, 1922, P. Blakiston's Son & Co.
18. Woodcock, B.: Scoliosis, Stanford, Calif., 1946, Stanford University Press.
19. Physical therapy for thoracic surgery patients, Veterans Administration Pamphlet 10-20, Washington, D. C., 1947, U. S. Government Printing Office.
20. Jacobson, E.: Progressive relaxation: a physiological and clinical investigation of muscular states and their significance in psychology and medical practices, ed. 2, Chicago, 1938, University of Chicago Press.
21. Bennett, R. L.: Functional testing and training in physical medicine, Arch. Phys. Med. 30:263, 1949.
22. Buchwald, E.: Physical rehabilitation for daily living, New York, 1952, McGraw-Hill Book Co., Inc.
23. Hoberman, M., and Cicenia, E. F.: The use of lead-up functional exercises to supplement mat work, Phys. Ther. Rev. 31:321, 1951.
24. Hellebrandt, F. A.: Methods of muscle training, influence of pacing, Phys. Ther. Rev. 38:319-326, 1958.
25. Thistle, H. G., Hislop, H. J., Moffroid, M., Hofkosh, J. M., and Lowman, E. W.: Isokinetic contraction: a new concept of exercise, Arch. Phys. Med. 48:229, 1967.
26. Hislop, H. J., and Perrine, J. J.:The isokinetic concept of exercise, J. Amer. Phys. Ther. Ass. 47:114, 1967.
27. Moffroid, M., Whipple, R., Hofkosh, J., Lowman, E. W., and Thistle, H.: A study of isokinetic exercise, Phys. Ther. 49:735, 1969.

CHAPTER 5

PRINCIPLES OF
OCCUPATIONAL THERAPY

I n rehabilitation medicine, occupational therapy is an integral member of the total rehabilitation team. By definition "occupational therapy is the art and science of directing man's response to selected activity to promote and maintain health, to prevent disability, to evaluate behavior and to treat or train patients with physical or psychosocial dysfunction."*

The occupational therapist works with the patient, instructing him and actively involving him in a wide variety of purposeful activities common to his immediate environment and culture Thus he is stimulated to utilize his physical and emotional strengths for the purpose of achieving the maximum physical function that is needed for independence and productivity.

The therapeutic media and equipment with which the occupational therapist provides services to the patient may be simple or complex, depending on the needs of the patient and the number of other rehabilitation services available to the patient. The ultimate goal is to integrate the abilities of the patient in purposeful activities that meet everyday requirements.

The objectives of occupational therapy can best be outlined under the following principles of treatment as they apply in rehabilitation medicine:

1. Evaluation
2. Functional occupational therapy
   a. Restoration of physical function in relation to ADL and requirements of a job
   b. Training for perceptual-motor dysfunction
   c. Training in the use of prosthetic and orthotic equipment
3. Prevocational exploration
4. Supportive therapy
5. Homemaking retraining and home planning
6. Home program and follow-up care

---

*Official definition adopted by the Delegate Assembly of the American Occupational Therapy Association. In Amer. J. Occup. Ther. 24:324, July-Aug., 1970.

EVALUATION

The treatment plan to be used in occupational therapy for each individual patient is determined after a comprehensive evaluation. The evaluation procedure identifies deficits that impose limitations on functional ability, and it produces a patient profile that sets a base line for selection of treatment procedures and for subsequent charting of progress. To deal with the multiple deficits presented by patients with physical dysfunction, two different evaluation procedures are designed: one for use with patients with lower motor neuron lesions and those with joint disease such as spinal cord injury, arthritis, and multiple sclerosis; and the other for evaluation of upper motor neuron lesions such as hemiplegia and brain damage. (See Chapter 2.)

After interpretation and analysis of the deficits, the therapist can select the treatment approach and appropriate media that will realistically meet the needs of the patients and that coincide with the objectives of other rehabilitation disciplines. It is therefore important for the occupational therapist to communicate the occupational therapy treatment plans with other team members and to coordinate the program with them to achieve maximum benefit for the patient.

FUNCTIONAL OCCUPATIONAL THERAPY

Functional treatment includes three major components, and, although each requires different treatment methods, there is one goal to be achieved—to restore the patient to the maximum level of independence in relationship to his living, work, and family needs.

Restoration of physical function

Restoration of physical function in relationship to the activities of daily living and work requires the following objectives:

1. To improve muscle strength and control
2. To increase/maintain range of motion of impaired joints
3. To improve eye-hand coordination
4. To improve hand dexterity
5. To increase work tolerance and endurance

One should keep in mind that, although these objectives are physical in nature, they also may meet psychosocial needs when integrated in purposeful activity. It is not uncommon that the emotional reaction to trauma, disease or loss of limb, manifests itself in such attitudes as lack of interest, denial, and lack of motivation. Purposeful activities used in occupational therapy such as feeding, grooming, and communications skills, which emphasize independence, are stimulating and rewarding for the patient. Other activities that serve the purpose of specific exercise, and of developing work tolerance, can be presented to also provide immediate success and accomplishment for the patient. For the effective treatment of patients with physical dysfunction one should recognize that the technic or technics will differ with disabilities resulting from upper motor neuron lesions, lower motor neuron lesions, or joint disease. These differences will be apparent after the patient evaluation indicates that a combination of the aforementioned problems and secondary complications exist. Knowledge and treatment of all deficits are necessary for maximal improvement and function of the patient.

Lower motor neuron lesions. The first principle of treatment in disabilities resulting from lower motor neuron lesions is to improve muscle strength. Since active motion is imperative to achieve muscle strength, the patient is engaged in an appropriate activity. During activity, consideration is given to the frequency and duration of motion required and to the resistance to be overcome by the muscles involved in doing the work. During treatment various types of assistive equipment, such as counterbalance slings and mobile arm supports, are utilized initially to assist and to protect the weak musculature (Fig. 5-1). As muscle strength increases, the progression of activity increases from active-assistive, active, and finally to re-

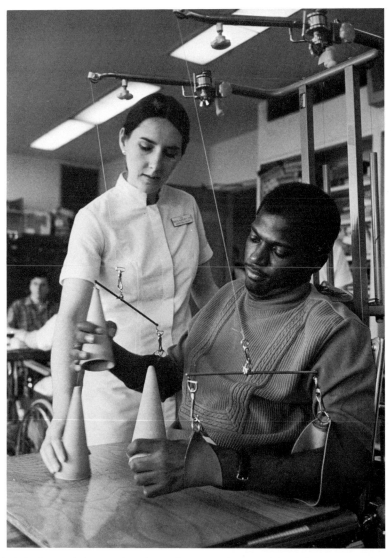

Fig. 5-1. Spinal cord injury patient in a bilateral assistive activity using counterbalance slings.

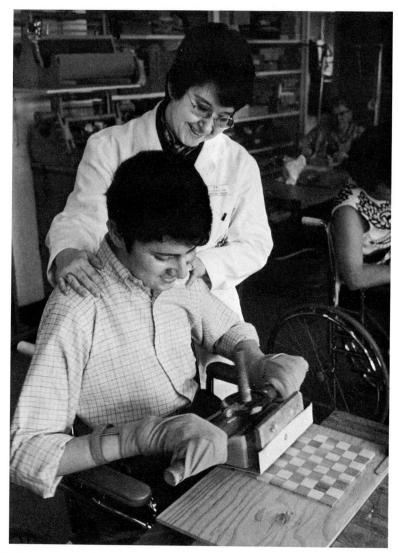

Fig. 5-2. Spinal cord injury patient strengthening upper extremities by using resistive activity.

sistive. Progression of activity is scientifically guided by the periodic and systematic increase of frequency and duration of action and by the resistance to be overcome. Grading activity for resistance can be varied through positioning of the patient in relationship to the activity, by adaptive equipment, and/or by use of weights (Fig. 5-2). For maximum improvement of muscle strength, fatigue must be carefully controlled. Intervals of rest during activity should be of an adequate length of time and frequency to accomplish maximum muscle function and to prevent overfatigue. A second principle in the treatment of lower motor neuron lesions is the prevention of deformity and contractures by maintaining freedom in joint range. This is of utmost importance for patients who will be fitted with assistive equipment and

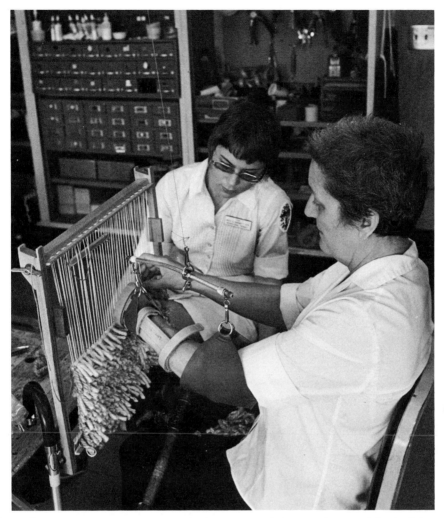

Fig. 5-3. Patient with upper motor neuron lesion retraining muscle control in the left upper extremity.

splints when irreversible loss of function requires their use to achieve maximal independence.

Upper motor neuron lesions. The major problem of restoring physical function with disabilities resulting from upper motor neuron lesions appears in the complexity of lack of muscle control compounded by neurologic deficits and in many instances deficits in perception. Muscle control is inhibited by flaccidity, spasticity, reflex patterns, synergy patterns, and sensory impairment or complete loss. Treatment goals are aimed at training for the best integrated functional use of the upper extremities (Fig. 5-3). Principles of treatment are reeducation of motor response, voluntary and involuntary; maintenance of joint mobility for prevention and/or reduction of contractures; and strengthening of flaccid musculature. Methods for motor re-education and sensory awareness include proprioceptive neuromuscular facilitation technics inherent within the activity being used. Methods for prevention of contractures include activities that require slow stretching through joint range, either static or active. When muscle control increases and voluntary isolated motion is possible, principles of muscle strengthening are employed. Since secondary complications of a visual and sensory nature can seriously impede functional use of motion, attention to these problems should be integrated in planning the treatment program.

Training for perceptual-motor dysfunction

It is not uncommon in those disabilities resulting from upper motor neuron lesions that maximal independence is seriously impeded when perceptual problems exist. These problems are not exclusive of the sensory-visual deficits that inhibit function, but often are interrelated. This is the reason why evaluation is most important as an attempt to isolate the problems interfering with function. Two of the most obvious problems are visual perceptual impairment (such as spatial disorganization, depth and vertical perception, figure-ground discrimination) and unilateral neglect. Both impose serious difficulties in training the disabled in self-care activities even when physical function is adequate. Relearning motor skills such as walking, dressing, homemaking, reading, and writing may be retarded or never achieved unless an attempt is made to treat the underlying problems. Treatment should include technics that include cross-modality stimulation, maximum cueing and diminishing cues with improvement, self-verbalization and correction, developmental sequential activities, repetition of movements, and operant conditioning technics.

Selection of activities. Selection of activities to be used for physical restoration can be considered in two general aspects: (1) physical process involved and (2) mental process involved.

1. Physical process considerations
   a. General motions—unilateral, bilateral, reciprocal—and adaption to needed motion
   b. Analysis of movement required considering position of joints and muscle action required
      (1) Primary movers of joint and type of contraction—concentric, eccentric, static, isometric, isotonic
      (2) Muscles contributing to control of action—antagonists, synergists, neutralizers—and to type of contraction
      (3) Leverage of each joint—first, second, or third class
      (4) Stabilization necessary in skeletal parts
2. Mental process considerations
   a. Intelligence and ability required—mental and chronologic age, concentration, learning
   b. Adaptability to psychologic needs; expression of attitudes such as hostility; dependency-independency; masculine-feminine identification; achievement; structure
   c. Adaptability for perceptual deficits by providing sequential development, stimulation of sensory modalities, visual-motor training, spatial organization
   d. Interpersonal relationships—contact with and/or work with other patients, with therapists, or with staff, or isolation from others

Prosthetic training

Training the amputee patient for prosthetic use begins as early as possible. Ideally, training begins immediately postoperatively for preprosthetic conditioning of the stump, muscle strengthening, and maintaining joint mobility. The program should be coordinated with physical therapy for optimum results. Since loss of limb is frequently catastrophic to the patient and may leave a psychologic impact, orientation to the purpose and usefulness of the prosthesis should begin as early as possible to prepare the patient for training and to alleviate anxiety. While the patient waits for the prosthesis, it is important to engage him in efficient use of the uninvolved extremity in one-handed activities of daily living. Successful independence at this time will lessen his feelings of despair over the loss of the limb. Depression and discouragement are common attitudes of the amputee and can inhibit motivation that is needed for participation in a use training program after receiving the prosthesis. Reaction to the loss of a limb is an individual matter and is often related to the patient's cultural and intellectual development as well as self-image. It is crucial that the therapist perceive these attitudes from the beginning and alleviate them as much as possible during the training program.

After the patient receives the prosthesis, the therapist assesses the limb for functional use, fit, and comfort (Chart 8). At the same time, the therapist teaches the patient the names and purpose of the parts of the prosthesis and discusses with him what is hoped to be achieved with the prosthesis. A cooperative relationship with the prosthetist is important and serves to solve technical deficiencies of the prosthesis if and when they arise. A comfortable, smooth-working prosthesis lends to preventing frustration and discouragement in the patient when he begins to actively use the prosthesis.

The principles of training include (1) controls training, (2) use training (ADL), and (3) vocational assessment.

Controls training includes teaching the donning and doffing of the prosthesis, necessary body movements to operate the terminal device and locking mechanisms, and the care of the prosthesis. Throughout training one should teach the patient graceful coordinated movements for the most efficient use of the prosthesis.

Use training includes the efficient use of the prosthesis in all activities of daily living (Fig. 5-4). Although most unilateral amputees will be capable of accomplishing some of these activities without the prosthesis, it is valuable to teach him to use the prosthesis in the event he may have to rely on it. Bilateral amputees must be made completely independent in ADL and for some activities, such as toileting, may require additional devices. In use training, one initially begins with teaching the concept of prepositioning, that is, anticipating the best type of approach with the terminal device and body to successfully accomplish the activity. Repetition and practice are basic for becoming proficient in prosthetic use. Avoiding body contortions and grimaces will lend to a more graceful appearance during use. It is important that the therapist provide successful experiences with every training session so as to reinforce the patient's motivation and interest.

Vocational assessment, after the patient has learned to operate the prosthesis and is confident and proficient in its use, may be included. If the patient is to return to his former job, the best approach is simulating the job situation. If the patient is not to return to his former job and new skills must be explored, then coordination

Text continued on p. 114.

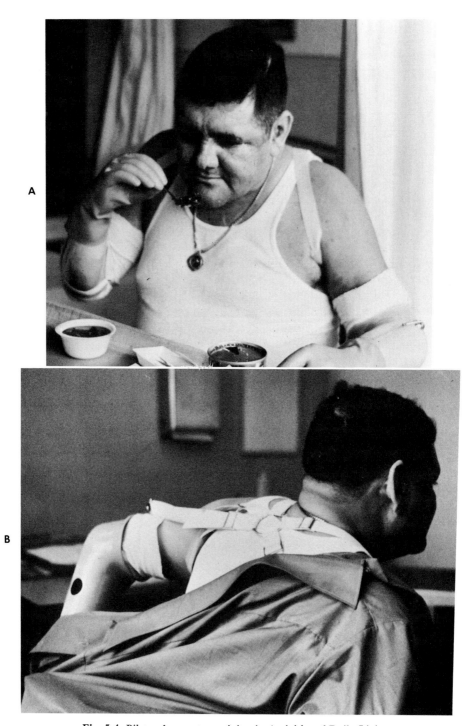

Fig. 5-4. Bilateral amputee training in Activities of Daily Living.

CHART 8

Institute of Rehabilitation Medicine New York University Medical Center OCCUPATIONAL THERAPY **PROSTHETIC CHECK-OUT** **BELOW ELBOW PROSTHESIS** Date _____	Name _____ Address_____ Age _____ Chart No. _____ Disability_____ Occupation _____

DATE OF FINAL SURGERY_____ SIDE OF AMPUTATION: Right_____ Left_____

PERCENTAGE LENGTH OF STUMP

$$\text{Unilateral B/E} = \frac{\text{Stump length x 100}}{\text{Sound forearm length}} = \underline{\hspace{1cm}}\%$$

$$\text{Bilateral B/E} = \frac{\text{Stump length x 100}}{\text{Height (inches) x 0.14}} = \underline{\hspace{1cm}}\%$$

AMPUTEE TYPE (based on percentage length of stump) _____

DESCRIPTION OF PROSTHESIS

Harness: Figure 8 or 9 _____ Shoulder saddle _____ Other_____

Harness modifications _____

Harness material: Dacron _____ Cotton _____ Other _____

Control system: Single _____ Dual _____ Biceps cineplasty _____ Other_____

Cuff: Full _____ Triceps pad _____ Other _____

Elbow hinge: Rigid: Single pivot _____ Polycentric _____ Step-up_____

             Flexible _____ Other _____

Socket: Double wall _____ Single wall _____Split _____Muenster_____

Socket material: Plastic laminate _____ Other _____

Wrist unit: Manual friction_____ Quick change_____Wrist flexion_____

             Constant friction _____ Other _____

Terminal devices: Hooks _____ Hands _____

Continued.

CHART 8, cont'd

INITIAL CHECK-OUT Date	FINAL CHECK-OUT Date
Pass ( ) Provisional pass ( ) Fail ( )	Pass ( ) Fail ( )
Patient needs further attention in the following area:	Patient needs further attention in the following area:
Medical-surgical ( ) Training ( )	Medical-surgical ( ) Training ( )
Prosthetic ( ) Other ( ) (vocational, psychologic, other)	Prosthetic ( ) Other ( ) (vocational, psychologic, other)
COMMENTS AND RECOMMENDATIONS	COMMENTS AND RECOMMENDATIONS

PATIENT'S OPINION OF PROSTHESIS

Examiner _____

CHART 8, cont'd

CHECKOUT

I. FIT

Initial CO	Final CO

HARNESS

1. Is axilla loop padded or covered and is it comfortable to amputee?

2. Is axilla loop small enough to keep cross of figure-8 harness well below seventh cervical vertebra and slightly to unamputated side?

3. Is control attachment strap below midscapular level and does it remain low enough to give adequate cable travel?

4. Can any additional harness straps be justified?

5. Does front support strap pass through deltopectoral groove?

CUFF OR TRICEPS PAD

6. Does cuff fit snugly without gapping during forearm flexion and terminal device operation?

ELBOW HINGE

7. Does elbow hinge function smoothly without pinching flesh or otherwise causing discomfort?

SOCKET

8. Is socket comfortable, especially when compression force or torque is applied?

9. Is patient's stump free from abrasions, discoloration, etc., within 10 minutes after prosthesis is taken off?

DIMENSIONS AND WEIGHT

10. Is prosthesis correct length?

11. Is weight of entire prosthesis satisfactory?

II. FUNCTION

A. RANGE OF MOTION

	Initial CO Prosthesis		Final CO Prosthesis	
	On	Off	On	Off
1. Stump rotation (total range)	o	o	o	o
2. Elbow flexion	o	o	o	o

Total rotation with prosthesis on should be half that with prosthesis off.*

Active flexion with prosthesis on should be within 10° of range with prosthesis off.

*This standard applies to medium, well-formed stumps, but is often exceeded with wrist disarticulation and long B/E stumps. Short and/or fleshy stumps may not be able to attain the 50% standard.

Continued.

CHART 8, cont'd

Initial CO	Final CO

B. CABLE SYSTEM

3. Is control cable free from sharp bends?

4. Is cable housing proper length between distal and proximal retainers, neither too loose nor too tight so that TD operates with full opening and closing?

5. Are control cables arranged so that they do not touch amputees flesh?

6. If hook is to be interchanged with hand, is hook-to-cable adapter proper length?

C. CONTROL SYSTEM EFFICIENCY

	Initial CO		Final CO		
	Hook	Hand	Hook	Hand	
7. Force applied at terminal device	___ lb.	___ lb.	___ lb.	___ lb.	
8. Force applied at hanger	___ lb.	___ lb.	___ lb.	___ lb.	
9. Efficiency = $\dfrac{\text{Force at terminal device}}{\text{Force at hanger}}$	___ %	___ %	___ %	___ %	75% or greater

D. TENSION STABILITY

10. Displacement of socket on stump with 50 lb. -- axial load

Prosthesis should not slip on stump more than 1 inch.

Initial CO	Final CO

E. WRIST UNIT AND TERMINAL DEVICE

11. If FM disconnect is used, is operating button placed in the anteromedial quadrant of the prosthetic wrist?

12. If wrist flexion unit is used, does patient operate it quickly and easily?

13. Does compression on wrist unit prevent involuntary rotation of terminal device on active opening?

14. If length adapter is used, is hand or hook fairing installed?

15. If APRL hand is used without length adapter, is back shell of hand faired to prevent sharp bend in operating cable?

16. If cosmetic glove is used, is it pulled completely onto hand and is it undamaged and properly color-matched?

17. Does TD function properly in all respects?

CHART 8, cont'd

F. TERMINAL DEVICE — OPENING AND
    CLOSING

	Initial CO		Final CO	
	Hook	Hand	Hook	Hand
18. Mechanical range				
19. Active range (forearm at 90°)				
20. Active range (fly)				
21. Active range (mouth)				

Full opening and closing
should be obtained with
forearm at 90°.

100% opening and closing
should be obtained at
mouth and fly.

Note: Children may have
less than 100%.

III. CRAFTSMANSHIP

1. Have all cut edges on cable housing been filed smooth or tipped with ferrules?

2. Is harness firmly sewn?

3. Are all strap ends sealed to prevent fraying?

4. Is neoprene lining on hook firmly bonded?

5. Are all rivets properly peened and all screws securely fastened?

6. Are all soldered joints strong, neat, and workmanlike?

7. Is cuff made of good quality leather, lined and finished in a workmanlike manner, and coated with nylon solution overall?

8. Have all trimmed edges, rough spots, and starved or uncured areas been sealed?

9. Is arm color satisfactory and consistent?

CONFORMANCE WITH PRESCRIPTION

10. Is the prosthesis as prescribed? If a recheck, have previous recommendations been accomplished?

FOLLOWING SECTION APPLIES AT FINAL CHECK-OUT ONLY:

TRAINING

11. Has the patient completed a training program in the use of his prosthesis?

12. Can the patient demonstrate effective use of terminal device and wrist unit?

(Columns at left: Initial CO | Final CO)

Fig. 5-5. Training with the use of wrist and hand orthoses.

with the counselor and a comprehensive prevocational exploration program should begin.

The length of time for prosthetic training is determined by the type of amputation, the intelligence and age of the patient, his acceptance of the disability, and his motivation. Generally the unilateral, below-elbow amputee will require from 6 to 8 hours of training, whereas the more complicated bilateral, above-elbow amputee will require from 20 to 25 hours of use training. After training terminates, a follow-up session should be arranged with the patient, if possible, to discuss problems he may have encountered during his daily routines.

Orthotic training. A too-often neglected problem is training those patients who have been fitted with orthoses, for example, tenodesis type splints and below-forearm orthosis. Since the patients who receive such equipment will rely on the use of them for independence, they should undergo training sessions to develop tolerance for

Fig. 5-6. Prevocational exploration for retraining in one-handed skill use.

wear and efficiency in use. The motivational factor, as in the amputee, depends greatly on the value of the orthosis to the individual. By this we mean where and how he is able to accomplish necessary skills with the orthosis when he would otherwise be unable to (Fig. 5-5). It is not uncommon that the most severely disabled person with irreversible damage can achieve a certain level of independence through the use of orthoses. The occupational therapist is responsible for providing these patients with activities and experiences for use of the orthoses and, at the same time, for observing the patient's comfort, his need for alteration or revision of the equipment, and his tolerance of the equipment.

PREVOCATIONAL EXPLORATION

The prevocational exploration phase of occupational therapy begins as early as possible during the rehabilitation program. As the patient approaches his maximum

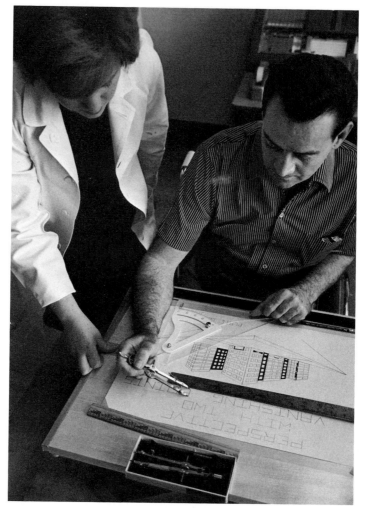

Fig. 5-7. Prevocational exploration for evaluating drafting as a vocational potential.

physical recovery, emphasis in the direction of vocational exploration takes place. Coordination with the vocational counselor is essential for maximal program planning. Vocational direction of the program is assessed by the counselor through interviews with and formal testing of the patient, whereas validation of the information gained is assessed in the prevocational program. The objectives of prevocational exploration are (1) to assess the patient's present physical and intellectual potential for returning to his former job, (2) to evaluate realistically the patient's verbalized interests and test results in terms of their compatibility with his current abilities, (3) to provide opportunity for work adjustment to regain diminished skills, work habits, and work endurance, and (4) to train and to improve communication skills in preparation for resuming academic life.

Assessment. For the patient whose disability may impose limitations on his

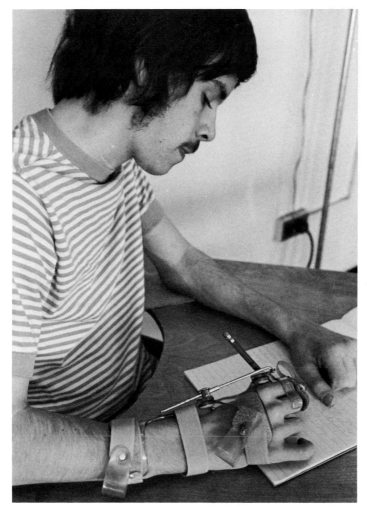

Fig. 5-8. Training patients in the use of orthosis for communication skills.

former job requirements, a simulated work environment is used for a realistic evaluation of his ability (Fig. 5-6). Cooperation with the former employer can provide necessary information and work samples to assess the patient's abilities for meeting the job requirements. When such evaluation reveals that minor changes are needed, such as adjustments of work heights and space, special equipment, and improvement in concentration and endurance, then the program is directed to solving such problems. It is ideal to gradually involve the patient to a full day's work as a true estimate of his ability.

Exploration. With the patient whose disability prevents returning to his former job, new vocational categories are explored that will be compatible with his physical and intellectual level of function (Fig. 5-7). It is not enough to rely on formal aptitude tests or verbalized interests, but rather on actual "on-the-job participation" to give a realistic picture of performance. After determining the patient's abilities and eliminating his liabilities, it is possible to predict the best vocational direction and referral for training. For disabilities resulting from brain damage, many complications present difficulties for finding suitable vocational goals. Therefore, it is critical for the therapist and counselor to keep abreast of the current trends in the occupational market that can be used in testing for those unique problems.

Work adjustment. When the disability has merely detained the patient from returning to work, then the program can provide a work atmosphere for the patient to develop work tolerance, to regain work habits, and to practice his old skills. Work adjustment is also valuable to those patients who have never had the experience of working. It offers learning and adjusting one's self to the demands and interrelationships of a working situation. This is particularly essential for the adolescent patient who is in that developmental stage of life.

Communication skills. For the patient whose academic career has been interrupted by injury, and who has potential for returning to school, the program is directed for retraining and developing skill and independence in all school related requirements from the physical and intellectual aspect. For some patients, independence in writing (Fig. 5-8) may be the only problem to be solved, but for those with other disabilities with more complex problems, including intellectual deficits, a realistic evaluation is necessary to assess the potential for returning to school at the same level or lower level. Such an assessment is organized by providing a work-study program utilizing educational material and technics.

SUPPORTIVE THERAPY

Supportive therapy is not an exclusive treatment but rather an inherent quality in all treatment of patients. The devastating effect of many disabilities, however, can cause greater distress to some patients than to others because some adjust to their illness or disability with more ease. The concept of supportive therapy provides the atmosphere in which the patient may be helped to overcome the "normal" depressed period of his adjustment and in which he may realize his abilities rather than disabilities. Through various experiences and activities the patient can begin to see himself in relationship to his environment and can begin to take a more positive approach toward his adjustment. Supportive therapy can be particularly beneficial with those patients whose physical recovery will be minimal and who must readjust themselves to an entirely new way of living. For optimum results the therapist should work in harmony with the psychosocial staff to deal appropriately with different attitudes and other complex problems.

HOMEMAKING RETRAINING AND HOME PLANNING

Principles of homemaking retraining and home planning is a comprehensive program of occupational therapy for preparing the disabled person for independent living. These principles related to the program are dealt with in Chapter 10.

HOME PROGRAMS AND FOLLOW-UP CARE

For many patients the return home after a long hospitalization period can present several problems of adjustment, particularly for those patients who are not returning to work. It is possible that in cases in which structure of the rehabilitation program kept the patient actively involved, its termination may have reverse effects on the independence that was achieved. Home programs that are planned with the patient and family prior to discharge can be directed to meet these general goals: (1) to maintain the physical status and independence gained, (2) to help reestablish previous living and work habits, (3) to provide avocational interests, (4) to assist in social participation and interests.

In planning home programs it is essential to consider the level of function that the patient has achieved and the means by which it can be maintained in a home environment. The physical requirements can most probably be incorporated in those everyday activities required for daily living. However, written instructions as a daily guide, as well as instructions to family members, will help maintain and reinforce independence at home. It is important to detail to the family members so that they can respond appropriately how much and what kind of assistance the patient will require.

In planning activities for specific exercise, a written program is essential, outlining the following: purpose of activity, how it is to be done, when it is to be done, frequency, and precautions. If materials and supplies are needed for activities, information regarding cost and where materials are to be purchased should be included. It is essential to orient the family members to the importance of maintaining the patient's interest and motivation for such programs.

A *follow-up visit* by the therapist to the patient's home, when possible, is of utmost value as a means of helping the patient overcome unanticipated problems that may interfere with establishing previous living habits. Too often these unforeseen problems arise that impede independent living and require professional consultation. Unless the patient is able to cope with them, he may withdraw to a level of dependency on the family. Home follow-up care is valuable to prevent this dependency and to deal with problems on a practical level. In addition, these visits can provide for the patient and family the reassurance and support that may make the difference between dependency and independence.

SUMMARY

As a paramedical discipline, occupational therapy provides a unique service to the patient. It provides an environment for integrating all that the patient is gaining, physically and mentally, into meaningful active experiences.

Occupational therapy is particularly concerned with man and his ability to meet the demands of his environment. The therapist administers treatment to the patient designed (1) to evaluate and increase his physical function in relation to activities of daily living, the needs of his family, and requirements of his job; and (2) to improve his self-understanding and psychosocial function as a total human being. Treatment involves the scientific use of activity procedures and/or controlled social relationships to meet the specific needs of the individual patient.*

*From American Occupational Therapy Committee: Statement of policy, Amer. J. Occup. Ther. **17:**159, July-August, 1963.

REFERENCES

1. Ayers, J. A.: Occupational therapy for motor disorders resulting from impairment of the central nervous system, Rehab. Lit. **21**:302, 1960.
2. Ayers, J. A.: Proprioceptive facilitation elicited through the upper extremities, Amer. J. Occup. Ther. **9**:121, March, 1955.
3. Dunton, W. R., Jr., and Light, S.: Occupational therapy: principles and practice, ed. 2, Springfield, Ill., 1957, Chares C Thomas, Publisher.
4. Jones, M. S.: An approach to occupational therapy, ed. 2, London, 1964, Butterworths, Publisher.
5. Willard, H., and Spackman, C.: Principles and practices of occupational therapy, ed. 3, Philadelphia, 1963, J. B. Lippincott Co.

CHAPTER 6

PRINCIPLES INVOLVED IN TEACHING THE ACTIVITIES OF DAILY LIVING

GENERAL PRINCIPLES

The purpose of the Activities of Daily Living program is to train the patient to perform, within the limits of his physical disabilities, his maximum in the daily activities inherent to his daily life in his home, at work, or at play. In contrast to vocational rehabilitation and selective placement, which is concerned primarily with the evaluation of the physical capacities of the individual in regard to his work, Activities of Daily Living (ADL) is the terminology used to denote the basic activities that are needed to carry on daily life, including getting to and from work, and how these activities are related to the particular environment of the individual patient.

As Buchwald and co-workers[1] have pointed out, the teaching of the activities of daily living is usually approached in the following manner:

1. A given ADL is broken down into its simplest motions.

2. Exercises are selected to enable the patient to perform these specific motions. The motion itself (for example, grasping a fork) may be practiced as an exercise. The exercise for this motion may be grasping a fork or similar objects. If this is not possible for some reason, such as lack of strength or lack of coordination, preparatory exercises (for example, strengthening hand and finger muscles) may have to be practiced first, or special devices may have to be made.

3. The ADL itself is practiced as a whole in a real-life situation.

The discussion of the specific activity "eating a meal while sitting in bed" will make the foregoing approach clearer, and Chart 9 may serve as an example.

The first column shows the breakdown of the activity into its simplest motions. The second column shows examples of exercises to be practiced so that the motions

121

CHART 9*

EXAMPLE OF HOW TO SELECT EXERCISES THAT ARE
BASED ON BASIC MOTIONS OF A GIVEN ADL

**BASIC MOTIONS AND EXERCISES TO BE CONSIDERED WHEN
TEACHING "EATING A MEAL WHILE SITTING IN BED"**

SIMPLEST MOTIONS OF A.D.L.	EXERCISES (TEACHING THE MOTIONS)	PREPARATORY EXERCISES (IF INDICATED)
1. Changing from supine to sitting position	Sitting up, e.g.: (a) Using hands and elbows (b) With help (c) With devices, etc.	Rolling over, strengthening of arms, etc.
2. Maintaining balance while sitting, with and without arm movements	Exercises for: Sitting-balance, e.g., with back supported, unsupported, etc.	Strengthening of shoulder girdle, back muscles, abdominal muscles, etc.
3. Grasping and holding eating utensils	Grasping and holding spoon, fork, knife (wooden spools or rubber handles may be used at first to make grasping easier)	Strengthening of finger and hand muscles Eye-hand coordination (All hand exercises in connection with occupational therapy)
4. Using eating utensils, such as getting food on utensils, cutting, etc.	Going through real motions: (a) Using regular food, or (b) Plasticine may be used for learning how to cut	Same as above
5. Putting food into mouth	Movements practiced: (a) Without food (b) With food	Same as above
6. Chewing and swallowing	Specific exercises in connection with speech therapy	

*By permission from Physical rehabilitation for daily living, by Edith Buchwald and others. Copyright, 1952. McGraw-Hill Book Co., Inc.

CHART 10*

OUTLINE OF MOTIONS AND PROBLEMS TO BE CONSIDERED WHEN
TEACHING "EATING A MEAL WHILE SITTING IN BED"

ACCORDING TO DISABILITY
(NUMBER OF MOTIONS BASED ON CHART 9)

I. Paraplegic

Involvement of both lower extremities with or without abdominal muscles (e.g., poliomyelitis, spinal cord lesions)
(a) Motions 1 and 2 have to be learned

II. Hemiplegic

Involvement of upper and lower extremity on same side, with or without aphasia (e.g., cerebral hemorrhage, cerebral palsy, etc.):
(a) Probably motions 1-6 have to be learned
(b) Performance of two-handed activities with one hand
(c) Problem of aphasia has to be considered

III. Quadriplegic

Involvement of all four extremities with or without abdominal or back muscles (poliomyelitis, spinal cord lesions, etc.):
(a) Probably motions 1-5 have to be learned
(b) Special problem: hand and finger activities, special gadgets

IV. Amputees

Loss of one or more extremities:
(a) Motions 1 and 2 for double amputation above knee; motions 3-5 for single or double amputations of hand or entire arm
(b) Special problem—management of artificial limb(s): if stump(s) too short for prostheses, special gadgets may have to be considered

V. Miscellaneous disabilities

Resulting from arthritis, poliomyelitis, cerebral palsy, multiple sclerosis, Parkinson's disease, etc., with one or more special problems of sensation, loss of strength, severe spasticity, severe tremor, and rigidity:
(a) Probably motions 1-5 will have to be learned
(b) Special problems: gadgets, hand-eye coordination, relaxation, etc.

*By permission from Physical rehabilitation for daily living, by Edith Buchwald and others. Copyright, 1952. McGraw-Hill Book Co., Inc.

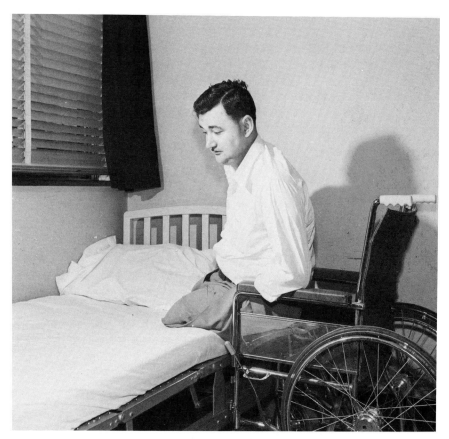

Fig. 6-1. Out of bed. Triple amputee—bilateral above-knee and left below-elbow amputations. Removal of footrests provides close approach as the patient transfers backward into the wheelchair while pushing with his right hand and his left stump on the armrests. (From Lawton, E. B.: Activities of daily living for physical rehabilitation, copyright 1963 by McGraw-Hill, Inc. Used with permission of McGraw-Hill Book Co.)

can be accomplished. The third column shows examples of preparatory exercises, if the patient for some reason or other is not ready for the exercise itself. This analysis of the motions is in the natural sequence of the ADL itself. The sequence to be practiced with the patient will depend entirely on what he is able to do. It will be best to let him start with the parts of the ADL that he can do and then gradually to teach him the parts he cannot do. For example, he may be able to put the food into his mouth only when it is put on a spoon for him, or he may be able to maintain his balance while sitting before he has learned to come to a sitting position by himself.

Since the goal is to teach him a given ADL, the starting point should be selected according to his condition and approached in such a way that he will feel that, although the task is difficult, it is not altogether impossible.

In this way a close interrelation between ADL and exercise may be seen. ADL may be regarded as exercises and the exercises as part of the ADL.

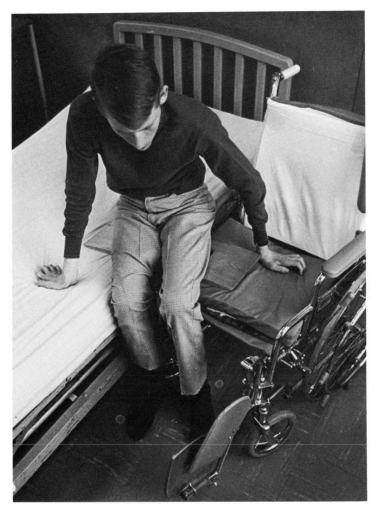

Fig. 6-2. Out of bed. Patient with involvement of all four extremities due to spinal cord injury at C-5 and C-6. The right armrest and footrest are removed for close approach as the patient transfers sideward. A sliding board is used to provide a smooth sliding surface.

In a functional training program the term exercises is used as long as motions as such are practiced, for example, sitting up or grasping eating utensils. Once the patient combines motions in a real-life situation, for example, sits up and actually eats a meal, the term ADL is used. This differentiation is arbitrary and is useful only for the purpose of simplification.

There are no rigid rules. The principle is to test the patient for what he can do, observe him carefully, and then work out with him the methods that will be the easiest or best for him; also show him what other patients in similar conditions have found helpful.

Based on Chart 9, consideration will now be given to teaching the activity, "eating a meal while sitting in bed," from the point of different disabilities. Chart 10

A 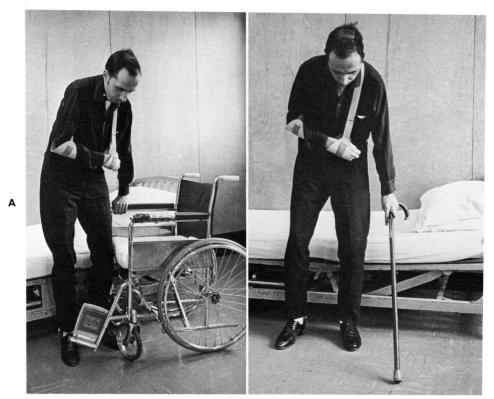 B

Fig. 6-3. Out of bed. Patient with involvement of right arm and leg due to cerebrovascular accident. **A,** Patient transfers standing up while supporting himself on his left hand. **B,** Same patient has learned to come to a standing position, as well as ambulate, using a cane.

may serve as an outline to show in general what motions may have to be learned and what problems are to be considered.

If we classify patients as (1) bed patients, (2) wheelchair patients, and (3) ambulatory patients, we find that ADL will involve different problems for the different levels. The motions necessary for "eating a meal while sitting in bed" are shown on Chart 9.

If we consider a *wheelchair patient,* we find that "eating a meal while sitting in a wheelchair at a table" will involve (1) getting dressed (which also includes getting washed, etc.), (2) getting out of bed into a wheelchair, (3) propelling the wheelchair to the table (with maintenance of balance in the wheelchair), and (4) carrying out the motions necessary for eating.

For an *ambulatory patient* the problem again is different. "Eating a meal while sitting at a table (using crutches, canes, etc.)" will involve (1) walking (which of course follows getting out of bed and getting dressed) and (2) sitting down in a chair at the table at home or in a cafeteria or restaurant (which further involves traveling to get there).

These examples illustrate that a given ADL for a given group of patients (bed,

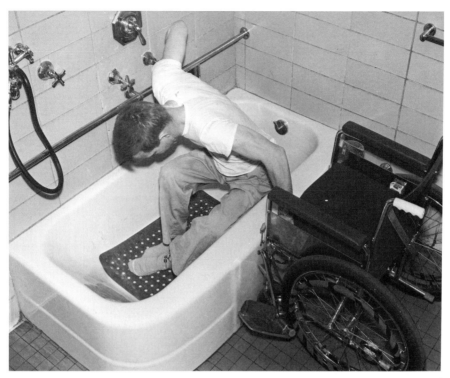

Fig. 6-4. Wheelchair to bathtub. Patient with involvement of all four extremities due to spinal cord injury at C-5 and C-6. Both footrests are removed for close approach as patient transfers, supporting himself by placing one hand on the grab bar and the other on the wheelchair seat. Transfer is very unusual for such a high lesion. (From Lawton, E. B.: Activities of daily living for physical rehabilitation, copyright 1963 by McGraw-Hill, Inc. Used with permission of McGraw-Hill Book Co.)

wheelchair, or ambulatory) is the sum total not only of all the motions of the specific activity but also of everything that was learned on the preceding level.[1]

RELATIONSHIP OF ADL TO PHYSICAL THERAPY

The different modalities and exercise programs in physical therapy are designed to maintain and/or restore range of motion and to develop strength, skill, and co-ordination to a maximum. In ADL the patient is trained to combine the different exercises learned into useful activities in real-life situations:

1. The patient has mat exercises (PT) to learn how to move about, supine and sitting. In ADL he uses these motions to learn how to get out of bed (Figs. 6-1 to 6-3), to get from a wheelchair to the bathtub and toilet (Figs. 6-4 and 6-5), and to get from a wheelchair to a car (Fig. 6-6).

2. The crutch-balancing exercises teach him how to handle crutches. In ADL he uses this skill to learn how to stand up from and sit down on a chair (Fig. 6-7).

3. The physical therapy department is responsible for the training in ambulation and climbing; in ADL the patient uses these skills to perform real activities, such

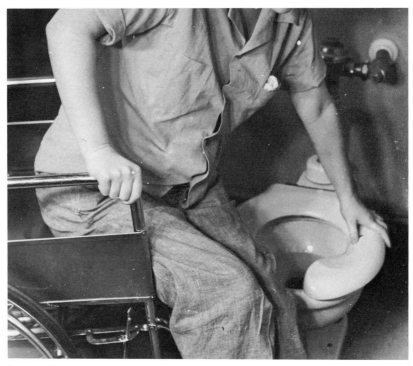

Fig. 6-5. Wheelchair to toilet. Patient with involvement of both lower extremities due to spinal cord injury at T-8 transfers by supporting herself on both hands.

as walking from room to room, standing at the sink and washing his hands, getting dressed for the street and descending a flight of stairs, or climbing a curb (Fig. 6-8).

Fortunately all ADLs have the same fundamental motions in common—those involved in changing position, sitting balance, moving in the sitting position, reaching, grasping, standing, and walking.[2]

As long as any of these motions are practiced as isolated motions, they still are only exercises and, as such, constitute most of the therapeutic exercise program in physical therapy.

Only when these motions are carried out in a real-life situation and are combined (for example, into actually "putting on shoes") do we speak of a daily activity. There is another important aspect to be considered. How does the patient get the shoes? Does somebody have to bring them to him, or can he get out of bed and take them out of the closet himself? Is the bed too high or too low for him to transfer to his wheelchair or possibly to stand up? Does the wheelchair need adjustments to facilitate the transfer? (See Figs. 6-1 to 6-3.) Can the patient wheel his wheelchair to the closet? Can he walk, although there is a carpet on the floor, or would it be safer to remove the carpet? And, finally, can he reach the shoes?

It is evident that a daily activity means more than just adding one motion to another. It also involves management of necessary furniture and equipment and, possibly, adaptations.

Fig. 6-6. From wheelchair to car. Patient with involvement of both lower extremities due to poliomyelitis. **A,** Patient transfers sideward. The left armrest has been removed for close approach. **B,** Placing wheelchair into car. The same patient can independently place the folded wheelchair into the back of the car.

ADL GROUPINGS

Activities of Daily Living are divided into the following groups:

1. *Bed activities*
   These include all the gross body motions necessary to move about in bed and to assume the desired position for a given self-care and hand activity.
   a. Changing position: rolling over, sitting up, and moving in the sitting position
   b. Maintaining sitting balance while moving trunk and/or arms in all directions
   c. Moving forward, backward, and sideward while sitting and placing legs
2. *Wheelchair activities*
   These include all the motions necessary when using the wheelchair.
   a. Wheelchair transfers—to bed, shower, tub, toilet, car, etc.
   b. Wheelchair management—handling all parts, propelling chair, etc.
3. *Self-care activities* (these are subdivided)
   a. Toilet activities
      (1) Personal hygiene (washing, bathing, and oral hygiene)
      (2) Personal appearance (care of hair and nails, use of makeup, and shaving)
      (3) Attending to toilet needs (bedpan, urinal, toilet)
   b. Dressing activities
   c. Eating activities

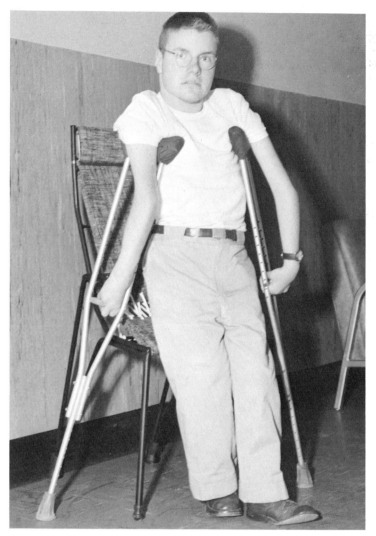

Fig. 6-7. Standing up and sitting down on chair using crutches. Patient with severe deformities of all four extremities due to arthrogryposis. (From Lawton, E. B.: Activities of daily living for physical rehabilitation, copyright 1963 by McGraw-Hill, Inc. Used with permission of McGraw-Hill Book Co.)

4. *Miscellaneous hand activities*
   The use of the hands is, of course, inherent in all ADL. The term hand activities is used here for those activities not included in bed, wheelchair, self-care, and ambulation activities.
   a. Handling the telephone, signal buttons, coins, etc.
   b. Handling eyeglasses, watch, lights, etc.
   Special considerations: A patient will carry out self-care and hand activities in bed, in the wheelchair, or in the standing position, depending on his efficiency at each of these levels and on what is most practical for him at a given time.
5. *Ambulation and elevation activities* (with or without the use of crutches, canes, braces, or prostheses)

Fig. 6-8. Climbing curb using crutches and prostheses. Triple amputee—bilateral above-knee and left below-elbow amputations. (From Lawton, E. B.: Activities of daily living for physical rehabilitation, copyright 1963 by McGraw-Hill, Inc. Used with permission of McGraw-Hill Book Co.)

    a. Ambulation activities
       (1) Inside the house on different floor coverings, such as linoleum or wood
       (2) Outside on different ground surfaces, such as cement or gravel
    b. Elevation activities
       (1) Standing up and sitting down—from wheelchair, bed, toilet, car, etc.
       (2) Climbing activities—stairs, curbs, and crossing streets
Special considerations: Ambulation activities and learning how to walk are not the same. Ambulation activities are the practical application of the basic gait training that is taught as part of physical therapy. For example, it would be rather absurd to take the patient for a walk and ask him to cross the street unless he already has learned a safe walking pattern. It should also be remembered that unless a patient can come to a standing as well as sitting position easily and without help he is not a functional walker.
6. *Traveling activities* (with or without the use of wheelchair; with or without the use of crutches, canes, braces, or prostheses)
    a. Use of private car, which includes use of garage and, in case of a wheelchair patient, placing wheelchair in and out of car (special hand controls)
    b. Use of public transportation: taxi, bus, etc.

Special note

Kitchen and household activities, although certainly part of daily living, are not included in this particular outline, or on the ADL test. At the Institute of Rehabilitation Medicine, New York University Medical Center, these activities are handled in a special department. In many other institutions, however, they are included under ADL.

THE USE OF DEVICES

Some patients need devices for certain activities. These may be permanent or temporary. They should be simple and easy to handle. The problem of independence applies in the same way as stated below in regard to any activity. It has to be stated whether the patient can handle the device independently or with help.

STANDARD OF PERFORMANCE

Once it is determined which activities compose an ADL program, the standard of performance must be defined. We say too easily of a patient "he needs a great deal of help," "he needs just a little bit of help," "he is quite independent," or "he is independent"; but such statements lack specific meaning. It is of utmost importance to define clearly what degree of independence a patient has reached. Does he or does he not need help?

1. *When does a patient need help?*
   a. If he cannot perform any part of a given activity
   b. If he can perform only part of a given activity and/or if his balance is so inadequate that he will need support to be protected from falling
2. *What kind of help is necessary?*
   a. Lifting
      If the activity *involves gross body motions* and consists in transferring the entire body weight from one place to another (for example, in transfer from a bed to a wheelchair) and the *patient cannot perform any part* of this activity, then someone must *lift* him.
   b. Assistance
      (1) If the activity consists of *small motions* (for example, eating, combing his hair, brushing his teeth) and the *patient cannot perform any part* of this activity, then someone must do the entire activity for him—feed him, comb his hair, brush his teeth, etc.
      (2) If the *patient can perform only part* of a given activity, then the help will consist of
          (a) Initiating, filling in, or completing the activity for him; for example, the patient can get from his wheelchair into a car, but someone must place the wheelchair in the car.
          (b) Support, if his balance is inadequate; for example, the patient can come to a standing position, but someone must stand by to steady him.
   The kind of help, in contrast to lifting, is designated as *assistance,* and on this basis it can be determined what kind of attendant is needed.
   c. Supervision
      If the activitity can be performed without any actual help but someone must stand by because of the patient's extreme nervousness, his inability to remember, or for other reasons, he is said to need supervision.
3. *What do we mean by independence?*
   The patient is independent when he does not need any help and can carry out a given activity entirely by himself with the necessary endurance and speed. This also means he can repeat the activity as often as may be necessary during the day without becoming exhausted.

GOALS OF ADL AND DEGREES OF INDEPENDENCE

Although ideally the goal is independence in the wheelchair, as well as in activities requiring ambulation, this is not always possible or practical. The actual goal will vary according to disability, age, occupation, and the home and work situation. The following factors must be considered:

1. There are some exceptional patients who are completely dependent on an attendant, but despite this they conduct their own businesses.

2. To function *independently* in "daily life" does not depend on the ability to walk. Many patients are "wheelchair patients" because they will never become "functional walkers." But they are certainly able to work and hold jobs if they can perform (using the wheelchair) the following activities: bed, wheelchair, and self-care activities; driving a car (with hand controls), including transfer from the wheelchair to the car and placing the wheelchair in and out of the car.

Other patients may be good functional walkers, but, because of crowded space in their work situation, it may be more advisable for them to use the wheelchair while at work.

In both instances it is important that these patients get up on crutches and braces (or in parallel bars) for at least one-half hour to one hour daily for weight-bearing and exercise purposes.

Chart 11 illustrates the different degrees of independence.

THE ADL ROOM

As long as the patient is treated on the plinth in the physical therapy department, on the bed in the ward, or in the workshop in the occupational therapy department, he is still in a somewhat artificial situation. When we want to see how he functions in a life situation, we must provide conditions that come as close as possible to real life, such as a model apartment or at least an ADL room with the basic bedroom and bathroom furniture.

Standard utensils should be placed in their logical places, such as the telephone on the desk and the electric razor in the medicine cabinet, so that activities can be tested and practiced as they would be carried out in a home.

There are many who feel that ADL can be practiced anywhere, such as on the ward or in the gym. If, however, all necessary equipment is assembled in one room, testing as well as training can be carried out more quickly and efficiently and fewer personnel are needed. The patient, as well as the staff, must think and plan in terms of a home—not of a hospital ward or a treatment area.

SUMMARY

ADL in a certain sense may be considered not as a treatment but rather as the practical application of all treatments to a life situation, and the test or proof is how the patient functions in daily life. The patient learns in physical therapy to perform "push-ups," but this becomes meaningful only if he uses this skill functionally. Gait training in the gym is of value only if the patient actually uses the gait he was taught when he walks. The best eating device is only a piece of material unless used regularly. ADL serves to train the patient to adapt to different environments,

CHART 11

CLASSIFICATION OF PATIENTS ACCORDING TO INDEPENDENCE

	FUNCTIONAL ACTIVITIES		
	PATIENT NEEDS HELP		PATIENT IS
GROUP	LIFTING	ASSISTANCE	INDEPENDENT
I	All ADL:	Patient not ready for any functional activity	
II		ADL: Bed and wheelchair activities; self-care activities	
III			ADL: Bed activities Wheelchair activities Self-care activities Eating activities Dressing activities Toilet activities Travel: Private car (from wheelchair)
IV		ADL: Travel: Placing wheelchair into car Ambulation in parallel bars: To standing position Standing and walking	ADL: Bed and wheelchair activities Self-care activities Travel: Private car (from wheelchair)
V		Ambulation: Placing wheelchair into car To standing position in 30 to 60 seconds holding on 30 feet inside Climbing: 3-inch steps with rail	ADL: Bed and wheelchair activities Self-care activities Travel: Private car (from wheelchair)
VI		Ambulation: To standing position in 30 to 60 seconds 40 feet outside Climbing: 3-inch to 8-inch steps with rail 2-inch to 6-inch curbs Up from floor in gym	ADL: Bed and wheelchair activities Self-care activities Travel: Private car (from wheelchair) Placing wheelchair into car Ambulation: 40 feet inside
VII		Ambulation (outside): 40 to 80 feet 60 feet in 1 minute Climbing: 12-to 14-inch bus step in gym 8-inch curb outside Fall and get up in gym	ADL: Bed and wheelchair activities Self-care activities Travel: Private car (from wheelchair) (standing) Elevation: Wheelchair, bed, toilet Ambulation: To standing in 30 to 60 seconds 5 times 40 to 80 feet inside 60 feet in 2 minutes, 1 minute, 30 seconds, 12 seconds Climbing: 10 to 12 8-inch steps with rail 6- to 8-inch curb in gym Get up from floor
VIII			All ADL:—Wheelchair and ambulatory Ambulation: To standing in 30 seconds to 60 seconds 5 times 80 to 120 feet continuously Cross street (60 feet in 18 seconds) while light changes, including curbs Climbing: 12 to 15 8-inch steps with rail Fall and get up from floor Travel: Private car; public transportation (bus)

but many adaptations can be made to his specific environment. Very often these adaptations will prove practical not only for a patient but for anybody, in terms of easy reach and minimum energy output. If the patient cannot be trained to function unaided, many of these adaptations will make it easier for the person who has to help him.

To stimulate and teach the patient in terms of daily necessities is not something that just happens or that is picked up "naturally" as he goes along. It must be carefully planned and practiced, and the teaching toward this end should be recognized as an important program.

REFERENCES

1. Buchwald, E., and others: Physical rehabilitation for daily living, New York, 1952, McGraw-Hill Book Co., Inc.
2. Lawton, E. Buchwald: A.D.L., Activities of daily living testing, training and equipment, Rehabilitation Monograph 10, New York, 1956, Institute of Physical Medicine and Rehabilitation, New York University–Bellevue Medical Center.
3. Deaver, G. D., and Brown, M. E.: Physical demands of daily life, New York, 1945, Institute for the Crippled and Disabled.
4. Long, C., and Lawton, E. Buchwald: Functional significance of spinal cord lesions, Arch. Phys. Med. 36:249, 1955.
5. Lawton, E. Buchwald: Activities of daily living for rehabilitation, New York, 1963, McGraw-Hill Book Co.. Blakiston Division.

CHAPTER 7

PRINCIPLES IN USE OF THE WHEELCHAIR

The wheelchair is one of the most important appliances to enable the physically disabled patient to become self-sufficient. Many patients may not be functional walkers, but they learn to carry out their daily activities independently in a wheelchair so that they can hold a full-time job, have their own business, or manage their own household. To achieve this independence a metal collapsible wheelchair with plastic seat and backrest and various adjustable and removable parts is essential (Fig. 7-1). A wooden wheelchair, which cannot be collapsed and has no removable parts, is of no help whatsoever.

The main considerations when choosing a wheel chair are:
1. Measurement of the wheel chair and its parts to fit the patient's size for best possible posture, comfort, and easy management.
2. To provide close approach for wheel chair transfer to bed, car, toilet, etc.
3. To provide close approach to tables, desks, when eating, writing, working, etc.
4. Easy storage and transportation.
5. Overall wheel chair measurements in relation to limited space, hallways, doors, and general lay-out.*

Following is a general outline of the different wheelchair parts and their adjustments. It does not present *all* solutions, but is meant as a guide when ordering a collapsible, metal wheelchair.

Sizes

The different wheelchair sizes are adult, junior, growing chair, and tiny tot (three sizes). These are all listed in detail in the various wheelchair catalogues. It

---

*From Lawton, E. B.: Activities of daily living for physical rehabilitation, copyright 1963 by McGraw-Hill, Inc. Used by permission of McGraw-Hill Book Co.

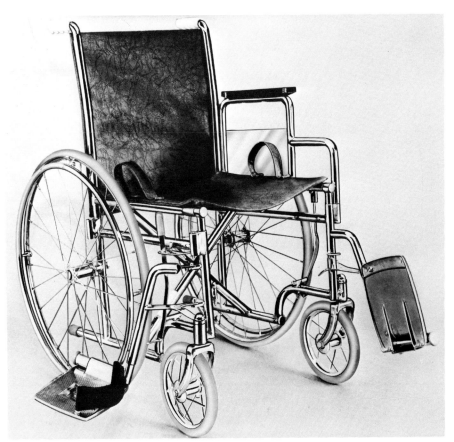

Fig. 7-1. Collapsible wheelchair with large wheels in rear, standard brakes, 8-inch casters, removable desk arms, and removable footrests with individual heel loops. (From Lawton, E. B.: Activities of daily living for physical rehabilitation, copyright 1963 by McGraw-Hill, Inc. Used with permission of McGraw-Hill Book Co.)

is of help to remember the following measurements of an adult-sized wheelchair, with large wheels in the rear and removable armrests and footrests: overall width, 25¾ inches; overall length, 41 inches (reduced 11 inches by removing the footrests); and height from floor to seat, 20 inches.

Large wheels

The wheelchair has two large wheels 24 inches in diameter and two casters 5 inches or 8 inches in diameter. The large wheels can be ordered to be placed in the front or the rear of the chair. They are preferred in the rear to facilitate close approach when transferring, etc. Another advantage of their being in the rear is that the patient has to lean backward when propelling the chair, thereby counteracting any tendency he may have to fall forward.

A *handrim* is attached to each large wheel, which is used to turn the wheel without having to place the hands on the dirty tires.

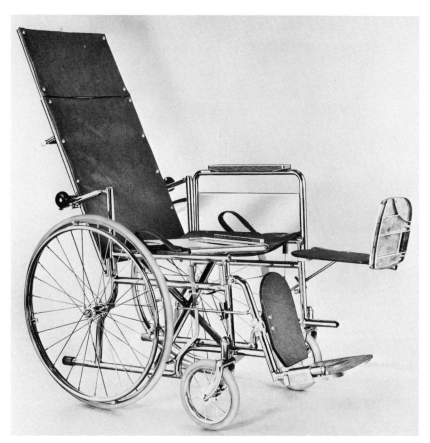

Fig. 7-2. Semireclining wheelchair with removable headrest, large wheels in rear, standard brakes, 8-inch casters, removable elevating footrests with legrests, standard removable armrests. (Desk arms are preferred; see Fig. 7-1.) Backrest adjusts to any angle from 90 to 30 degrees. (From Lawton, E. B.: Activities of daily living for physical rehabilitation, copyright 1963 by McGraw-Hill, Inc. Used with permission of McGraw-Hill Book Co.)

Standard rubber *tires* will be adequate for most patients. Pneumatic tires are helpful when using the chair outdoors. It should be remembered, however, that they make it more difficult to propel the chair indoors.

Although *brakes* should be used for the large wheels, one must keep in mind that they only prevent the large wheels from turning, but not the chair from tipping or sliding. Therefore, the patient has to be taught to prevent the tipping or sliding by not only pushing on the armrests but also always pulling the chair simultaneously toward himself while transferring.

Casters

Eight-inch casters are preferred to 5-inch casters because they give the chair a larger base and therefore more stability. They also make it easier to propel the chair over small obstacles on the ground (for example, gravel) and over door-sills.

One-arm drive

A one-arm-drive chair can be propelled with one hand only. This chair is indicated for patients who can use only *one hand* and nothing else (for example, a triple amputee or a triparetic patient). A hemiparetic patient, however, does not need a one-arm drive, since he can propel the chair with his good hand, while steering with his good foot.

Armrests

Armrests can be stationary (Fig. 7-4) or removable. Removable ones are preferred, again for close approach. *Desk arms* (Fig. 7-1) have a lower front section than the *standard* armrests (Fig. 7-2). Therefore, they make possible sitting at a table or desk without having to remove the armrests, but they need to be removable for transfers.

Armrests should be *adjustable in height* so that they can be lowered and raised according to the length of the patient's arms to provide optimum leverage for pushing on armrests when transferring or when standing up and sitting down. They should be *upholstered* for comfort as well as for facilitating grasp. *Skirtguards* on the armrests protect skirts as well as trousers from getting caught in the large wheels and also from being soiled by the tires.

Footrests

Stationary footrests (Fig. 7-4) as well as removable ones are available. *Removable* footrests (Fig. 7-1) are preferred because they allow close approach. Another essential point is that by removing the footrests the overall length and weight of the wheelchair is markedly decreased. This is important when storing the chair and especially so when placing it in the car.

The footrests should be *adjustable in length* to accommodate the distance between the knee and the foot. If the distance is too long, the patient will not be able to place his feet on the footrests. If the distance is too short, the patient's knees will be brought up too high and very bad posture will result. The footrests are ordered in different sizes according to the size of the feet.

Individual heel loops (Fig. 7-1), one on each footrest, are used to prevent the feet from slipping backward off the footrests. They are more practical than *straps* (Fig. 7-4), since they do not have to be removed when the footrests are removed.

Elevating footrests with legrests

Elevating footrests with legrests (Fig. 7-2) can be raised to any angle up to horizontal. They are indicated (1) if the knee cannot be fully flexed (for example, for an arthritic patient), (2) if the leg has to be raised because it is in a cast, or (3) if there is edema. Elevating footrests should be *adjustable in length* and also *removable* just as regular footrests and for the same reasons.

Backrest

The *standard backrest* (Fig. 7-1) reaches approximately to the patient's shoulder blades. If additional support is needed, a removable *hook-on headrest*

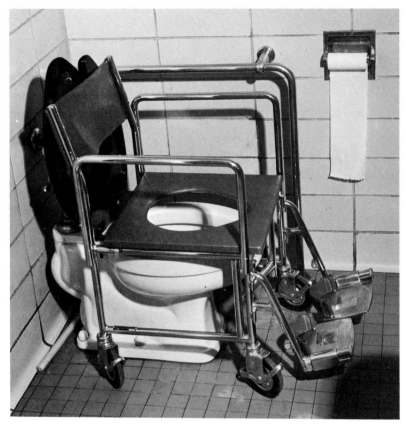

Fig. 7-3. Collapsible commode chair, 20 inches wide and folds to 5 inches, with 5-inch casters and removable armrests and footrests. The chair rolls over a standard toilet. (From Lawton, E. B.: Activities of daily living for physical rehabilitation, copyright 1963 by McGraw-Hill, Inc. Used with permission of McGraw-Hill Book Co.)

should be ordered. *Turnbuckles* on the right or left of the backrest make opening the backrest possible so that the patient can back the chair over the toilet and transfer backward onto it. This is often the only way a patient can place the wheelchair in a crowded bathroom. Turnbuckles are preferred to a zipper because they are easier to handle and also easier to attach to the backrest than is the zipper.

Reclining backrests with removable headrests support the entire back, neck, and head. They can be ordered semi- or fully reclining. The semireclining backrest (Fig. 7-2) can be lowered to any angle from 90 to 30 degrees. The fully reclining backrest can be lowered to any angle from 90 degrees to horizontal. The angle between the backrest and the wheelchair seat accommodates to the angle between the trunk and the thighs and is used (1) when there is restriction of motion at the hip joint, e.g., in arthritic patients, or (2) when there is difficulty maintaining the vertical sitting position, as in patients with a high cervical traumatic spinal cord lesion. A wheelchair with a reclining backrest always has elevating footrests with legrests (Fig. 7-2).

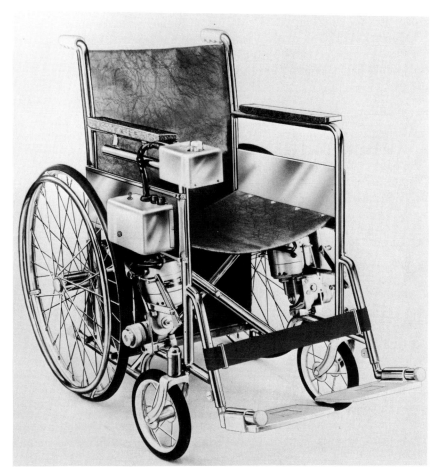

Fig. 7-4. Motorized wheelchair with large wheels in rear, standard brakes, 8-inch casters, standard stationary armrests, and stationary footrests with straps for leg support. (Removable desk arms, footrests, and individual heel loops are preferred; see Fig. 7-1.) This chair can also be collapsed if motor and battery are removed. (From Lawton, E. B.: Activities of daily living for physical rehabilitation, copyright 1963 by McGraw-Hill, Inc. Used with permission of McGraw-Hill Book Co.)

Seat

The depth of the seat should coincide with the distance from the knee to the hip. For patients with extremely short thighs and/or flexion contractures of hips and knees, a more shallow seat may be helpful.

Seat cushions and seat boards

For comfort and prevention of decubitus ulcers the patient should sit on a cushion—either foam rubber (2, 3, or 4 inches thick) or Gel pads, if special precautions have to be taken. To prevent sagging of the seat a seat board of ¼ inch plywood is placed under the cushion.

Commode chair

If space in the bathroom is too crowded, a 20-inch wide commode chair (Fig. 7-3) is suggested. This kind of chair is backed up over the toilet so that the patient does not have to transfer at all. This chair is also very helpful when traveling, since it is collapsible, has removable armrests and footrests, and can easily be put into the car.

Motorized wheelchair

A motorized wheelchair (Fig. 7-4) has a button, switch, or jog stick than can be handled with one finger. It has two batteries, which are recharged overnight (daily) at any standard 110-volt AC or DC plug.

All other parts—wheels, armrests, etc.—are selected as previously described. It must be remembered, however, that the control for the motor is attached to one armrest. And although this armrest is removable, it cannot be removed because of the wires of the control box.

A motorized chair is indicated for a patient who cannot or should not for medical reasons propel the chair by turning the big wheels. On the other hand, if propelling the wheelchair is a necessary exercise for a patient, then this chair is not indicated.

Summary

In general, a metal, collapsible wheelchair with removable armrests and footrests and the large wheels in the rear of the chair (Fig. 7-1) will facilitate the following: (1) close approach for all wheelchair transfers, (2) close approach for working at tables, desks, etc., (3) handling the wheelchair in crowded space, and (4) storage and transportation.

These removable parts are important for the completely independent patient, since he is likely to find himself in many different situations and frequently may encounter crowded space. The same removable parts are important for the partially or completely dependent patient, since they will make it easier for the person who helps him.

Specifically, each wheelchair part has to be selected carefully according to the individual patient's condition and size. (See Chapter 6 for additional illustrations of the use of the wheelchair.)

REFERENCE

1. Lawton, E. B.: Activities of daily living for physical rehabilitation, New York, 1963, McGraw-Hill Book Co.

CHAPTER 8

PRINCIPLES IN GAIT TRAINING AND THE PRESCRIPTION OF WALKING AIDS

In the prescription of suitable walking aids and the proper determination of the type of gait for patients with involvement of their lower extremities, consideration must be given to all weight-bearing joints. It is desirable that the hips, the knees, and the ankles be freed of as much stress as possible during the stance phase of ambulation. The extent to which the lower extremities are involved determines how much the upper extremities should be allowed to share in weight bearing. When the upper extremities are similarly involved (such as seen in some arthritic patients), careful attention must be paid to the weight-bearing capability of the shoulders, elbows, wrists, and fingers. The selection, measurement, and fitting of any walking aid must be considered according to the individual requirements for that patient and must be modified accordingly. Alteration in the standard gait pattern because of the addition of an assistive device such as a cane or crutch may often be necessary. In the training of any patient in ambulation, correct postural habits while sitting and standing are essentials for proper balance as ambulation progresses.

Four steps govern a gait-training program for any patient:

1. Evaluation of the patient's muscle, joint, and pain status
2. An exercise program to develop optimum range of motion, strength, and coordination to accomplish management of the walking aid as well as walking
3. Proper selection and correct fit of the particular walking aid
4. Determination of the gait pattern best suited to meet the patient's individual needs

143

EVALUATION OF THE PATIENT'S MUSCLE, JOINT, AND PAIN STATUS

Five main muscle groups share the responsibility in using walking aids for ambulation: (1) the flexors of the arm, (2) the extensors of the forearm, (3) the finger and thumb flexors, (4) the extensors (dorsiflexors) of the wrist, and (5) the shoulder girdle depressors and internal rotators. Before a program of training for walking begins, these muscles and the joints they control must be evaluated. A muscle test is done to explore the extent and locations of weakness, and the various ranges of motion are measured to determine the joint limitations. These limitations with muscle weakness could cause difficulty in the development of a comfortable, safe pattern for ambulation. In addition, the patient must be observed constantly for pain so that proper precautions may be taken to prevent undue stress on the painful area and to adjust the program accordingly.

EXERCISE PROGRAM

The exercise program for the patient being trained for the use of a walking aid is based on:
1. Limitations in joint ranges of motion
2. Extent of muscle weakness
3. Location of muscle weakness
4. Degree of pain on weight bearing

A basic prerequisite for an effective muscle-strengthening program is adequacy in ranges of motion of the joint. Muscles can be exercised through their full excursions and with optimum results only if the joints being mobilized have ranges within normal limits.

PROPER SELECTION AND CORRECT MEASUREMENT OF A WALKING DEVICE

Crutches. Wooden crutches (Fig. 8-1) with double uprights and handpiece made of lightweight wood are standard equipment at a nominal price. There are available today various types of metal crutches as well. The important consideration for selection is that the crutch be adjustable in total length and in the height of the handpiece. The weight of the total crutch assembly will be an important consideration as well.

Lofstrand crutches. These crutches are of tubular aluminum, adjustable in the total length and also in the height of the forearm piece (Fig. 8-2). Some patients prefer this crutch. Good features of the Lofstrand type include the rubber-covered armrest and the rubber-covered mold to fit the handpiece.

The tubular aluminum axillary crutch (Fig. 8-3) is adjustable in the overall length as well as in the height of the handpiece. Its special feature is the swivel capability of the handpiece, which provides adjustment in both the horizontal as well as the vertical plane. It will be found useful where supination or pronation is limited or painful.

Walkers, when selected, should be sturdy, lightweight, and measured for proper fit (Fig. 8-4).

Measurement. In determining the proper overall length of a crutch, measure the patient from the anterior fold of the axilla to a point 6 inches out from the side

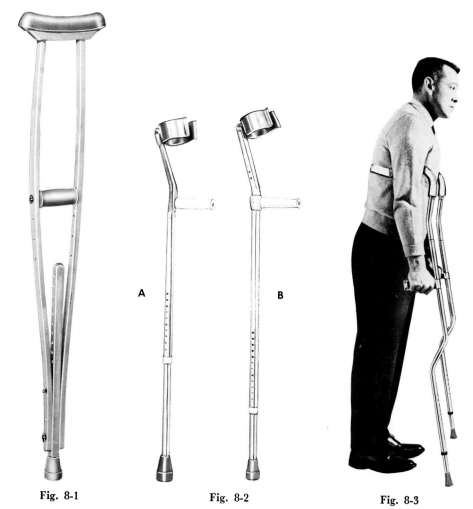

Fig. 8-1. Adjustable wooden crutch with crutch tip, crutch pad, and handgrips. (Courtesy J. A. Preston Co., New York.)

Fig. 8-2. A, Lofstrand crutch with stationary forearm. **B,** Lofstrand crutch with adjustable forearm. (Courtesy J. A. Preston Co., New York.)

Fig. 8-3. Adult ortho crutch. (Courtesy J. A. Preston Co., New York.)

of his foot, or measure the distance from the anterior fold of the axilla to the bottom of the foot and add 2 inches. Adjust the handpiece of the crutch so that the patient's elbow is approximately 30 degrees in flexion with the wrist held in extension. These last measurements of elbow and wrist can also be used for measuring of Lofstrand crutches, canes, and walkers. During stance phase, the shoulders should be in a relaxed position, neither held up nor depressed.

Canes. The conventional cane with its C-curve handle (Fig. 8-5) should be lightweight and of sturdy construction. This cane is much preferred by most patients,

Fig. 8-4. Aluminum walker. (Courtesy J. A. Preston Co., New York.)

is good-looking, comes in a variety of wood finishes, and usually costs 2 to 5 dollars. Variations of this cane are the T top and those made of metal. Recently the J-line cane (Fig. 8-6) was introduced that features a lightweight aluminum body painted black, with a biomechanical correct fit for the hand in grip as well as in ulnar deviation. Its cost is somewhat higher than the standard wooden cane, but is recommended for those patients who may require this extra support for long periods.

Walking aid adaptations. On crutches the triceps bars (Fig. 8-7), arm platforms, and the like can be adapted to accommodate the aforementioned requirements for a safe, comfortable, good-looking walking pattern. The orthotics department can advise concerning the materials to solve most problems relating to adaptations for walking aids.

Tips on the crutch or the cane (Fig. 8-8) should have a large circumference and have good suction. For crutch, tips of 2 or 3 inches in height, with a diameter of $1\frac{1}{2}$ to $1\frac{3}{4}$ inches, and of a soft white or red rubber have been found most satisfactory. Protective pads on the axillary portion of all crutches is recommended (Fig. 8-9). Handgrips are also useful in protecting the hand and are recommended (Fig. 8-10).

Fig. 8-5   Fig. 8-6

Fig. 8-5. Wooden canes. (Courtesy J. A. Preston Co., New York.)
Fig. 8-6. Functional grip cane. (Courtesy J. A. Preston Co., New York.)

DETERMINANTS OF WALKING PATTERNS

The following factors[1] must be considered when selecting a walking pattern for your patients:

1. *Step ability*—Can the patient take steps with either one or both lower extremities?
2. *Weight bearing and balance ability of the lower extremities*—Can the patient bear weight on either one or both lower extremities?
3. *Weight bearing and balance ability of the upper extremities*—Can the patient push his body from the floor by pressing down on his crutches?
4. *Direct body maintenance ability*—Can the patient maintain his body erect?

Standard crutch gaits. There are seven different crutch gaits, and the one selected depends on the patient's ability to take steps with either or both of the lower extremities.

Four-point alternate crutch gait. This gait has the following sequence: (1) right crutch, (2) left foot, (3) left crutch, (4) right foot. This is a simple gait if the patient can get one foot ahead of the other, and it is safe because there are always three points of support on the floor.

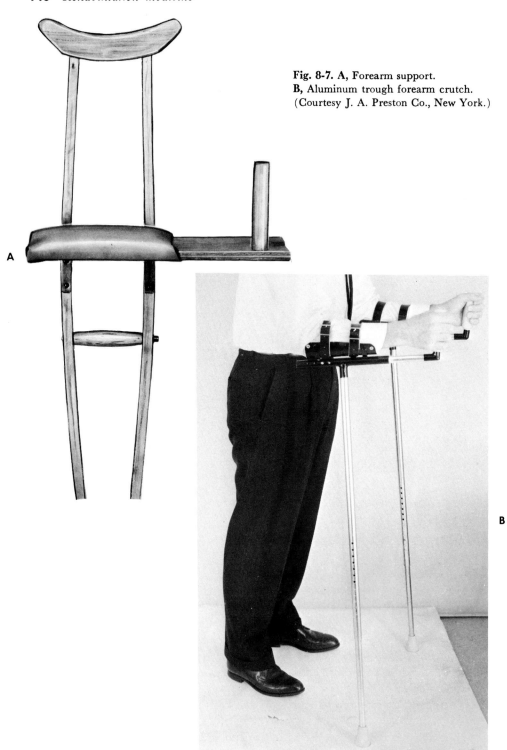

Fig. 8-7. A, Forearm support.
B, Aluminum trough forearm crutch.
(Courtesy J. A. Preston Co., New York.)

Fig. 8-8. Crutch and cane tips. (Courtesy J. A. Preston Co., New York.)

Fig. 8-9. Crutch pads. (Courtesy J. A. Preston Co., New York.)

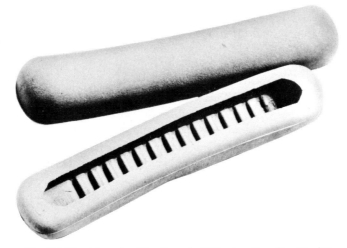

Fig. 8-10. Crutch handgrips. (Courtesy J. A. Preston Co., New York.)

Two-point alternate crutch gait. This gait has the following sequence: (1) right crutch and left foot simultaneously, (2) left crutch and right foot simultaneously. This is no more than a speeding up of the four-point alternate gait. It requires more balance control because only two points are supporting the body at one time.

Three-point crutch gait. This gait has the following sequence: (1) both crutches and the weaker lower extremity, (2) the stronger lower extremity. This gait is used when one lower extremity cannot take full weight bearing and one can support the whole body weight.

Tripod crutch gaits. When a patient has no method of placing one extremity ahead of the other, as in flaccid paralysis from poliomyelitis or spinal cord injuries, it is necessary to learn the tripod gaits for ambulation. There are two tripod gaits: the tripod alternate gait and the tripod simultaneous gait.

Tripod alternate gait. This gait has the following sequence: (1) right crutch, (2) left crutch, (3) drag the body. When the patient has learned this gait, he is taught the tripod simultaneous gait.

Tripod simultaneous gait. This gait has the following sequence: (1) both crutches, (2) drag the body. This is a sort of rocking backward and forward between crutches and feet. The one essential is that the tripod have a large base and that the body be inclined forward sufficiently to keep the center of gravity in front of the hips. If it falls behind, the patient will double up backward like a jackknife because of flexion of the hips.

Swing crutch gaits. There are two variations of the swing crutch gaits: the swing-to gait and the swing-through gait.

Swing-to crutch gait. This gait has the following sequence: (1) both crutches, (2) lift and swing body to crutches. The patient lifts both crutches and places them simultaneously in front of the body, bearing down on the crutches and lifting the body so that it moves up to the crutches, and then immediately lifts both crutches and places them on the floor ahead.

Swing-through crutch gait. This gait has the following sequence: (1) both crutches, (2) lift and swing body beyond crutches. This is difficult because it requires the patient to lift both crutches and swing the body through the crutches and then to roll the pelvis forward to get the center of gravity in front of the hips. Then he must fall forward and just at the right time raise the crutches to continue the sequence. Keeping the hips forward as far as possible with the back arched is essential for a smooth and graceful swing-through gait.

Conclusion. The crutch gait for each patient will differ according to the part or parts of the body involved. Where possible at least two gaits should be taught, one for speed and one for safety; alternating between the two gaits as the situation and the need change should be practiced.

Patients under optimum training conditions may be taught as many gaits as they can master. Each gait requires a different combination of muscles, with fatigue the earliest deterrent for the disabled. The person who becomes fatigued with one gait may change to another that allows previously used muscles to rest while others are now working maximally. The various crutch gaits can also be used as an exercise incorporating action and movement while strengthening the body musculature.

REFERENCES

1. Deaver, G. G., and Brittis, A. L.: Braces, crutches and wheel chairs: mode of management, Rehabilitation monograph 5, New York, 1953, Institute of Physical Medicine and Rehabilitation, New York University–Bellevue Medical Center.
2. Orthopedic appliance atlas: A consideration of aids employed in the practice of orthopedic surgery. I. Braces, splints, shoe alterations, Ann Arbor, 1952, J. W. Edwards.
3. Hoberman, M., Cicenia, E. F., and Offner, E.: Wheelchairs and wheelchair management, Amer. J. Phys. Med. **32**:67, 1953.

CHAPTER 9

PRINCIPLES OF SELF-HELP DEVICES

The use of self-help devices is to be considered as another adjunct of treatment available to the patient who needs rehabilitation. Devices not only can be helpful as a more dynamic approach to present treatment and resultant recovery in patients with many diseases and injuries, but also may be considered as a necessity in promoting self-sufficiency for the patient with residual permanent disability. While this is relatively old as a philosophy, a more dynamic concept has increased the potential of this field to the point that the principles and technics involved in the use of devices are worthy of special study and specific application.

Generally speaking, the main goals in use of devices are (1) to provide more dynamic treatment measures, (2) to permit early independence and the active participation of the patient in the treatment program, (3) to provide independence in daily activities despite partial or total loss of function, and (4) to help build self-esteem and achieve either partial or complete economic independence.

Self-help devices may be prescribed for the following specific purposes:

1. To provide positioning and support to help maintain body alignment against weakened segments, thus preventing either permanent loss of power in overstretched and overfatigued muscles and/or to prevent possible permanent deformity.

2. To encourage early motion and function to prevent atrophy from disease, thus maintaining muscle strength, and to prevent loss of range of motion of joints.

3. To increase function of residual muscle power and skill involved in various motions (Fig. 9-1).

4. To replace function totally and permanently lost either from muscle weakness or from loss of range of motion.

5. To assist in control of incoordination, spasm, or spasticity.

6. To replace lost body parts (as loss of hand or arm) or special senses (as loss of sensation or impairment of sight or hearing). Some of these problems are to be

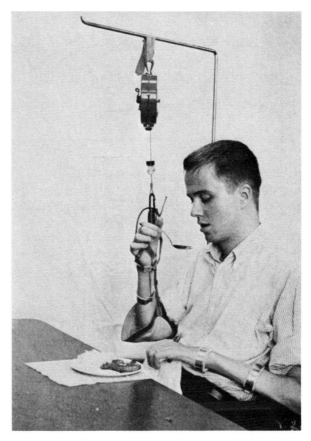

Fig. 9-1. Overhead sling and balancer used to provide early assistance to encourage use of shoulder flexion. Use of rubber band to ulnar side of wrist aids weak pronation of forearm.

recognized as belonging to other specialties—speech and hearing, training of the blind, prosthetics, etc. However, sometimes the patient refuses the assistance of a specialist, or the usual procedures are not feasible, for example, an amputee who cannot wear a prosthesis.

7. To minimize the amount of assistance needed from other persons when complete independence is not possible.

EVALUATION FOR SELECTION OF PROPER DEVICE

Although the previously mentioned reasons for devices serve to define the objectives so that suitable mechanical aids can be chosen, without further definitive approaches there can be considerable abuse resulting from improper selection. Selection of the correct self-help device depends on (1) the evaluation of the activity to be performed and the physical requirements for it, (2) the evaluation of the psychologic response to the use of equipment, (3) the evaluation of the materials and their design to be used in the construction of a mechanical appliance, and (4) the role of device in "total rehabilitation" of the *individual* patient.[1, 2]

Table 9-1. Analysis of motions used in eating (limited to use of one arm and one process only, in sitting erect position)

Body part	Motion	Purpose	Substitution	Loss	Device used to compensate
Hand	Pick-up: palmar prehension	Pick up and	1. Lacing spoon between fingers 2. Adduction of fingers	Minimal	1. Utensil interlaced in fingers (some shaping may be necessary) 2. Moleskin or tape over handle to prevent slipping
	Holding: lateral prehension to middle finger (modified lateral pinch)	Hold utensil	3. Hook grasp	Moderate	3. Built-up handle (wood, sponge, or other material) 4. Grip-shaped handles 5. Handle with horizontal and vertical dowels (pegged handle) 6. Handle with finger rings
				Severe	7. Short opponens with C-bar and utensil attachment 8. Plastic and metal holder
				Complete	9. ADL (universal) cuff 10. Prehension orthosis—manually operated or power-operated
Wrist	Stabilization (some flexion and extension or radial and ulnar deviation normally used, depending on whether grasp is hook or pinch)	Position hand for optimal function (to prevent wrist flexion)	1. Use of finger or thumb extensors	Partial stability	1. ADL wrist support, dorsal (leather with spring steel insert) 2. Flexible, adjustable nylon wrist support or Klenzac joints
				Complete	3. Tubular spring-clip (ADL) orthosis 4. Cock-up splint, rigid, palmar 5. Long opponens orthosis

Joint	Motion	Activity	Compensatory movements	Disability	Self-help devices
Forearm	Pronation	Pick up food on utensil	1. Shoulder abduction and internal rotation 2. Raise forearm to vertical position and then rotate	Partial or	1. Swivel spoon 2. Bent fork or spoon
	Supination	Keep utensil level while putting food in mouth to avoid spill	1. Shoulder adduction and external rotation	complete	1. Swivel spoon 2. Placing fork or spoon over thumb, use thumb extensors
Elbow	Flexion of forearm	Raise hand to mouth	1. Shoulder abduction 2. Trunk flexion	Partial or complete	1. Balanced forearm orthosis (ball-bearing feeder) 2. Overhead sling with "feeder" attachment 3. Overhead sling with built-up lapboard 4. Long-handled utensil 5. Functional arm orthosis
	Extension	Lower hand to plate	3. Rock forearm on edge of table		
Shoulder	Stabilization against hyperextension and internal rotation in position of slight flexion and slight abduction, plus slight active flexion and extension	Provide positioning and assist in raising hand to level of mouth	1. Trunk flexion 2. Prop elbow on table 3. Prop elbow on thigh and flex hip	Partial or complete	1. Pillow behind upper arm 2. Overhead sling 3. Balanced forearm orthosis (BFO) 4. Functional arm orthosis with hyperextension stop

Physical considerations

Assessing a patient's physical ability to use his upper extremities in the performance of various ADL is best approached by using a functional motion test[3] (Chapter 2). This is a practical application of information based on results from special analysis of motion of specified segments of various ADL as performed by the nondisabled. It is essential, therefore, first to understand or study this type of motion analysis.

There are many ways in which studies or analyses can be utilized. One approach, designed by the Gilbreths and used originally for study of the nondisabled in industry, is known as therbligs. This method consists of the use of seventeen specific divisions into which any accomplishment can be broken down for study. It includes not only the physical processes, but also the involvement of vision, mental processes, and the effects of external environment. Another, used by Lawton, emphasizes the breakdown of a given activity into a sequence of gross body motions in a particular location, with appropriate equipment, as used in each specific performance[4, 5] (Chapter 2).

Both of the above uses of motion study start with the *whole* activity and then examine or pursue those elements that indicate need for help. As the use of devices is largely confined to aiding those activities requiring the use of only the upper extremities, it is best, in this instance, to start the analysis of motion with a *limited* portion of an activity, such as the act of picking up food and putting it into the mouth or the act of writing once the pencil is held and positioned ready for use.

As the motion pattern of each portion or segment of an activity is analyzed, one finally arrives at an understanding of motions required for *total* performance. Although such an approach can be extremely time-consuming, close inspection is necessary to determine the specific purpose for which a device is required (such as holding, reaching, positioning, or stabilizing to resist forces of gravity due to body weight or weight of an object being used). This provides a rationale for need and makes it possible to set forth specific requirements for selection or construction of each piece of equipment. It also helps to detect when a device may give assistance in dual areas (not always needed) or when assistance or substitution for one motion may cause undesirable changes in the normal motion patterns.

Fortunately, it has been found unnecessary to pursue analysis of every portion of *every* activity before being able to make an initial evaluation with recommendations for specific equipment with which to start a program (either therapeutic or functional). This is discussed later in this chapter under the functional motion test. It is especially helpful with the severely disabled person who is so limited that he must of necessity start with only a small act. This will avoid frustration from poor performance and/or fatigue.

Analysis of motions as needed in ADL* performance. Tables 9-1 and 9-2 present details of the main portions of two activities—eating and writing—that can usually be managed with one hand and arm. Motion of each arm segment is ob-

---

*In this chapter the term ADL is used in its broadest sense and includes not only the self-care activities of eating, dressing, and personal hygiene and grooming, but also any daily need for living, such as various forms of communications or vocational and avocational pursuits.

Table 9-2. Analysis of motions for writing (with right hand only)

Body part	Motion	Purpose	Devices used to compensate
Hand	Opposition of thumb to index finger and lateral opposition to middle finger (modified palmar prehension)	Provides grasp	1. Built-up pencils, diameter increased 2. Spring-clip holder (used in ADL cuff) 3. Leather holding device 4. Clothespin holder with pegs 5. Holder with finger rings or finger bands 6. Adjustable holder (used with ADL cuff) 7. Fiberglas cuff
	Flexion and extension of middle phalangeal joints and/or	Moves pencil	8. Prehension orthosis—manually operated or power-operated
Wrist	Slight flexion and extension or	Moves pencil	1. ADL wrist support, dorsal (leather with spring steel insert), flexible
	Stabilization in neutral or slight cock-up position	Provides positioning	1. Anterior cock-up, rigid 2. Long opponens orthosis 3. ADL orthosis
Forearm	Stabilization in approximate midposition	Provides positioning	1. Supination assist
Elbow	Stabilization in approximate 90-degree flexion	Provides positioning	1. Overhead sling 2. Balanced forearm orthosis 3. Functional arm orthosis
	Alternating action of: Flexors	Lifts weight of arm from writing surface	1. Overhead sling 2. Foot-operated cable control to elbow of functional arm orthosis
	Extensors	Provides pressure during writing	1. Soft lead pencil (needs less pressure) 2. Certain pens, ball-point pens, felt-tip pens 3. Elastic strap from hand to forearm on anterior surface (substituting wrist flexors for elbow extensors)
	Slight flexion and extension with	Moves hand across paper	1. Ball caster support 2. Powder on board, Formica surface or Teflon-covered arm (elbow) cuffs
Shoulder	Slight internal and external rotation	Moves hand across paper	1. Overhead slings 2. Rotation mechanism on functional arm orthosis
	Stabilization in position of slight flexion and abduction	Provides positioning and	1. Overhead slings 2. Balanced forearm orthosis (BFO) 3. Functional arm orthosis
	Minimal synergistic action of shoulder girdle muscles	Minimum motion and pressure	

served and recorded separately. The same procedure can be employed similarly for the main elements of activities such as the use of telephone and typewriter; some personal hygiene measures, such as shaving, brushing teeth, and combing hair; and certain recreational or avocational pursuits, as knitting, crocheting, and piano playing. (Some of these listed activities are normally two-handed, but in each case the activity can be done with only one hand—as typing and playing the piano—or the use of a holding device can eliminate the need for two hands.)

Although the actual tasks of putting food into one's mouth and of writing (the main objectives of these two activities) are accomplished largely through the action of one hand and arm, there is, of course, need for certain other acts such as cutting of food, putting on seasonings, buttering bread, drinking from a cup or glass, picking up the pencil, and positioning and stabilizing paper. Sometimes these functions must be done by, or with the aid of, someone else. In initial attempts it is best if help is given to prevent frustration of poor or tiring performance. Later, if independence can be achieved in all areas, each of these acts can be observed and necessities of each performance carefully analyzed. Some of the factors to consider concerning the analyses of eating, writing, dressing, bathing, and vocational activities follow.

Eating. In addition to the analysis of eating given in Table 9-1, other considerations are as follows:

1. Use of other hand for cutting or stabilization of food: This may be accomplished through other technics such as use of a rocker or "one-handed" knife, which stabilizes and cuts simultaneously.
2. Handling of other equipment, such as dishes and drinking containers: Consider the use of stabilizers. For plates and cups try wet paper towels, nonskid plastic surfaces or mats, or suction bases; and for food try plate guards and scoop dishes. Handles may be used to aid lifting glasses, or lifting may be eliminated by using straws.
3. Relationship of height of chair seat to tabletop: Use cushions, adjustable tables, raised lapboards, etc.
4. Position in bed: Use pillows, backrests, Gatch beds, etc.
5. Types of food most suitable for practice: Use a spoon for puddings or mashed potatoes; a fork to cut raw apples, pears, or bananas; and a knife to cut veal cutlet or frankfurters (skinless).

Fig. 9-2 shows various devices used to aid in eating according to the losses shown in Table 9-1.

Writing. In addition to the motions of the arm (Table 9-2), analysis of writing includes the following considerations:

1. Positioning of person writing and of paper and pencil (Gardner)[6]:
   a. Body—Sitting erect, leaning slightly forward
   b. Arms—Elbows fairly close to body, slightly off front edge of desk, weight supported on belly of forearm
   c. Hands—Hand holding pencil is supported by heel of hand and tips of last two fingers or by sides of last two fingers; other hand holds paper and moves it up after each line
   d. Paper—Use one sheet at a time; place so that line drawn from lower left to upper right is perpendicular to body for right-handed person and the opposite position for left-handed person

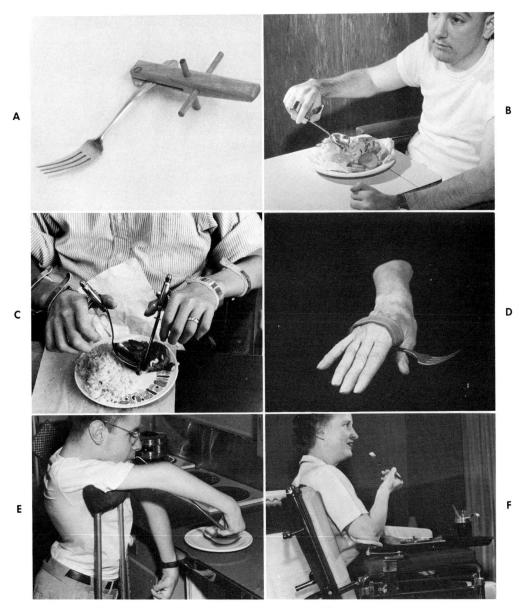

Fig. 9-2. A, Peg handle to assist weak grasp and permit holding in the usual manner. **B,** Plastic and metal holder to assist a minimal grasp yet encourage some holding. **C,** Tubular spring-clip holders with adapted eating utensils, inserted into leather pockets attached to wrist splints, provide for a "normal" positioning of eating implements. **D,** Leather utensil holder to replace complete loss of grasp. **E,** Long-handled spoon to replace loss of range of elbow motion. **F,** Ball-bearing arm support with elbow rest to assist in shoulder flexion.

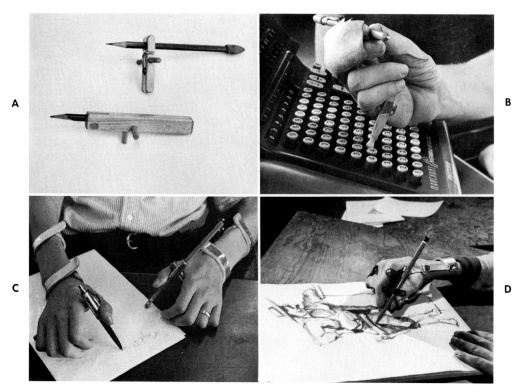

Fig. 9-3. A, Peg handles to aid weak grasp. **B,** Fiberglas cuff to replace complete loss of grasp with attachment for use of adding machine, telephone dialing, etc. **C,** Spring-clip holders for pen and pencil for writing, for stabilizing and moving paper, and for erasures (if used with a second pencil). **D,** Fiberglas cuff with wrist attachment to replace complete loss of grasp and assist in wrist flexion for greater pressure of pencil against paper.

     e. Pencil—Held by thumb and first two fingers, top part of pencil should rest next to metacarpophalangeal joint; pencil should be held loosely

  2. Other equipment used for stabilizing paper, positioning of writing implement for pick-up, positioning of work surface, etc.

Fig. 9-3 shows various devices used in writing according to the losses shown in Table 9-2.

Dressing. When motion and positioning of the whole body are involved in an activity such as dressing, and when a variety of garments adds to the complexity, start by analyzing actions such as putting on and fastening clothing, and then by observing these actions as used for each type of clothing. A sample motion analysis involving one garment only is found in Table 9-3.[7] It is possible that it may be necessary to further confine the analysis to include only one action, such as fastening; for although fastening does use mainly the hand, it is dependent on arm motions for positioning and stabilizing the hand.

Bathing. Similarly, the analysis of bathing must be broken down into its vaious actions and each component analyzed as it is used with different types of equipment.

An outline of these considerations is as follows:

1. Hand washing
   a. Use of washcloth and soap
   b. Use of hand brushes
   c. Use of washbowl (type and access to)
   d. Handling of water faucets
2. Tub bathing
   a. Use of washcloth and soap
   b. Reaching to lower extremities
   c. Drying
   d. Getting into and out of tub
   e. Handling of water controls
3. Shower bathing
   a. Use of washcloth and soap
   b. Position (sitting or standing balance)
   c. Reaching to lower extremities
   d. Drying
   e. Handling of water controls (types and placement of)

Vocational activities. Vocational objectives can also be broken down into their various elements and specific actions. However, one must remember that vocational possibilities are also dependent on the patient's potential to manage such other activities as eating in a cafeteria, using restroom facilities, and traveling to and from work.

Basic principles of motion requirements (derived from motion analysis). A compilation of the analysis of motions of various activities (eating, writing, combing hair, turning pages, typing, etc.) will show that certain motions are used more frequently than are others and thus can be designated as essential or "key" motions (Table 9-4). Until more information is available regarding participation of various muscles used to perform each motion, especially when effects of gravity and external forces are involved, it is felt best to limit observations to a given joint motion.[8] Use of motion analysis also allows for relevant application to disabilities resulting from disease or injury in which muscle weakness is not confined to specific spinal cord levels of innervation.

The three main antigravity arm motions are *shoulder flexion, elbow flexion,* and *wrist dorsiflexion* (extension). Next in importance are *external rotation* and *supination.* Some type of hand function is mandatory, as this is the grasping mechanism that is positioned by the above motions of the arm segments (or lever arms).

Further observation or study shows that elbow flexion is the most active movement; wrist motion is used mainly for stabilization, the position depending on whether the hand is used away from or toward the body and on the height of the work surface in relation to the trunk; and the shoulder acts mainly as a stabilizer to prevent hyperextension against the forces of gravity when the hand is raised from the side, and becomes active when the hand is used above the level of the chin or in extended reach. External rotation becomes important whenever both shoulder and elbow are flexed and the forearm moves out of the sagittal plane into the frontal plane across the front of the body. Supination is used mainly when the hand is engaged near or around the head. These observations apply especially to those activities used by the more severely disabled because they function generally from a wheelchair (sitting position), from which reach is limited and resistance is usually minimal. This may vary when more requirements are added.

The type of hand grasp used for different activities also provides certain basic

Table 9-3. Clothing analysis (one garment only)

Type of garment	Problem	Disability	Solution
Short-sleeved and cardigan blouses	Putting on Taking off	Loss of motion of shoulder and elbow	Loose armholes, such as raglan, kimono, or dolman Action back pleats Vertical style dress (divided into two pieces) Stick with closet hook on end (to remove garment from shoulders)
	Fastening front	Weakness of both hands	Vertical rather than horizontal button holes
		Use of only one hand	Larger buttons, size at least ⅝ inch Flat buttons Snaps instead of buttons Slipover blouse or sweater Button hook (regular or wire loop)
Long-sleeved blouse and shirts	Same as others with addition of fastening cuffs	Use of only one hand	Large enough cuff to let hand pass through Elasticized or knit cuff Elasticized thread on button Elastic loop over button
Both of above	Keeping shirt or blouse tucked in	Weakness of hands and arms	Long tails Latex gripper inside skirt or trouser band Tape from shirt tail to sock tops (for men)
	Wear of underarm from crutches		Knitted gusset piece under arms Action pleat at armhole in blouse back
		Weakness or loss of use	Material with stretch (such as jersey) Action back pleats (center back or at armhole)
	Tear or binding when wheeling chair		Three-quarter length sleeves (to prevent soiling cuffs)

principles or guidelines. It is well known that palmar and lateral pinch are the two most frequently used types of prehension for various ADL. Preference to palmar pinch was given by other investigators from findings which indicated that it could be substituted for lateral better than the reverse.[9] However, out of the total fifty-seven activities originally studied, there has been no attempt made to break them down into primary (elementary) activities, or those that are most frequently possible by the severely disabled person who functions almost entirely from a wheelchair. From this latter approach there are indications that lateral pinch is used in basic, more gross activities and palmar pinch in more skilled ones. In Table 9-4 lateral is used three times out of five (for holding motions), palmar once, and hook grasp once. Also, when hook grasp is used for eating, arm motion patterns are altered and may or may not produce desirable changes.

Functional motion test. The functional motion test was discussed in Chapter 2. Instead of resorting to a specific motion analysis of every portion of all activities with

Table 9-4. Findings and summary of analysis of motion (eating, combing hair, writing, typing, turning pages)

Body part	Motion	Active motion	Holding motion	Total
Shoulder	Abduction	0	0	0
	Adduction		5	5*
	Flexion	3	2	5†
	Extension	3		3*
	External rotation	4		4‡
	Internal rotation	4		4*
Elbow	Flexion	3	2	5†
	Extension	3		3*
Forearm	Supination	3		3‡
	Pronation	3	2	5*
Wrist	Flexion	1		1*
	Extension	1	4	5†
Hand pinch§	Palmar	1		1
	Lateral		3	3‖
	Hook grasp	1		1

*Gravity aided.
†The most frequently used arm motions not aided by gravity are (1) shoulder flexion, (2) elbow flexion, and (3) wrist extension.
‡Next in importance are (1) shoulder external rotation and (2) forearm supination.
§Some type of hand function (with or without splints) is mandatory.
‖It has been found that palmar can be substituted for lateral.

each patient, we are now in a position to test only the motions available in each patient and then to interpret the findings in terms of his potential for various activities. Resultant grading, which is used the same as in a standard muscle test, shows those deficiencies of motion that will impair function that indicate need for a device. When strength is tested, a grade of *zero* to *fair minus* in any antigravity arm or hand motion indicates need for a device, which can then be selected according to the purpose for which that motion is necessary. Moderate to severe losses in range of motion also indicate a specific need according to which motion is limited, as does loss of coordination.

Some examples of selection of equipment are as follows:

1. Loss of hand function (zero to poor-minus strength)—Use ADL cuff with various utensils, or a prehension orthosis, wrist activated, may be used instead (Chapter 12).
2. Loss of hand function (poor to fair-minus strength)—Use adapted equipment, such as universal built-up handle for spoon, fork, buttonhook, toothbrush, and adapted cup handles.
3. Loss of wrist dorsiflexion (zero to poor-minus strength)—Obtain stabilization for maintaining a neutral or slight cock-up position through use of a wrist orthosis.
4. Loss of forearm supination (moderate to severe loss of range of motion)—Use swivel fork and spoon.
5. Loss of shoulder and elbow flexion (zero to poor-minus strength)—For early stages of recovery select counterbalance overhead slings. For permanent use or regressive condition select a balanced forearm orthosis.

Psychologic considerations

Motivation is perhaps the most challenging and essential aspect in the satisfactory application of equipment. Coping with the attitudes of enthusiastic response or frank rejection requires an understanding of psychologic makeup and reasoning in regard to independence by means of this medium. Some general factors that may contribute to these attitudes follow.

Duration of disability (whether congenital or acquired, subacute or chronic). Generally speaking, the person with congenital disability or chronic disability is more realistic in recognizing the need for help. There are exceptions, however, of those who fully believe they function adequately, even though greatly impaired in speed, accuracy, and safety. The person with a subacute or acquired disability, on the other hand, is often *not* interested because of his anticipation of complete recovery.

Extent of disability (mild, moderate, or severe). Generally speaking, the patient with mild impairment is fairly well motivated because in his case devices are few and relatively simple. Patients with severe involvement are either most anxious for help or manifest total rejection as a denial of disability. Those with moderate involvement are perhaps most difficult to handle because they have "so much" and yet so little to work with.

Prognosis of disability (whether static, progressive, or improving). If the condition is static, the person is more amenable. If the condition is either progressing or improving, the person is likely to be uninterested—uninterested because he thinks he does not need such help or because his fear of increasing disability causes him to deny need for assistance.

Age as contributing factor. The young and the middle-aged patients are likely to be most accepting and the teen-aged and the elderly patients least accepting.

Type of activity to be performed. Any devices for mobility are usually desired by most patients, as are those for the self-care areas of eating and toileting. Bathing and dressing are areas in which patients seem more willing to accept help, perhaps because they need doing less frequently and because they are the more difficult to solve.

Additional physical problems and/or limitation. Other physical problems or limitations may cause pain or fear. The patient with arthritis may have increased pain on resuming a more active program. (Extreme pain and pain accompanying traumatic action are to be avoided.) Blindness may add a fear of hurting himself until he becomes secure in the use of the device.

Cultural factors. Equally important factors of influence on the disabled person are cultural and sociologic concepts of the use of special equipment by the disabled person. Cultural impacts on body image not only influence the patient's attitudes toward physical disability itself, but also carry over into attitudes toward devices that, although they minimize it, might also call attention to the disability. These values have been with man for generations and, in fact, are still inherent in much of today's thinking and doing.

Sociologic trends, with our advertising and television emphasis on bodily perfection, add to our present concept of "normal" as the anatomically perfect person. At the same time, the increasing use of mechanical equipment to extend the normal person's efficiency is minimizing the need for physical strength. These are the accepted aids to success and achievement.

The frank acceptance by the "normal" person of devices to achieve his goals and to create much of his glamor makes it imperative to supply comparable equipment adapted for the disabled person. Such specifications should result in devices that are (1) as inconspicuous as possible, (2) as near the "normal" method of doing things as possible, (3) cosmetically acceptable, (4) as little trouble as possible to handle, operate, and maintain, (5) made to allow opportunity for individual expression by permitting choice of type, design, color, and method of use, (6) flexible for easy change as indicated by physical improvement or regression, (7) adaptable to different types of disability, and (8) suitable to patients of different ages.

Mechanical considerations in fabrication of devices

In the construction of the best type of device, consideration should be given to proper fit and function, universality of use (adaptability to various disabilities and activities), adaptability to physical progress or regress, simplicity of construction (necessary personnel, tools, and time), simplicity of operation, minimum of breakdown and repair, cosmetic acceptance, reasonable cost, and place of use (indoors, outdoors, city, country, and climate). Designers must have some knowledge of tools and materials (their properties, methods of fabrication) and engineering principles (construction design, elimination of friction, levers and forces).

Materials and their properties and uses. Suitable materials available are (1) wood; (2) metals such as aluminum, Monel, and steel; (3) plastics such as Nyloplex, Royalite, Kydex, isoprene, Prenyl, Plastizote, and silicone foam; and (4) paddings such as felt, moleskin, plastisol and Reston.

Properties to be considered are those such as (1) weight, (2) strength, (3) elasticity, (4) elongation, (5) sheer strength, and (6) life expectancy.

Fabrication requirements may include (1) making cast and molds directly on body, (2) heat for bending or setting, (3) use of jigs, (4) hand tools or power tools for cutting and finishing, (5) painting or chrome plating for permanent finish, (6) method of making adjustments, and (7) different lengths of time for completion.

Each of the factors should be carefully weighed against each other in regard to the final use to which the device is put. Is strength most important? Flexibility? Weight? Availability of shop facilities and tools for construction?

Design. In construction design, consideration should be given to (1) use of the "third dimension" in stress areas to eliminate bulk and weight but provide strength, (2) use of ball bearings and certain materials to cut down on amount of friction, both for ease of operation and for reducing wear of materials, and (3) operational force requirements for mechanical levers and electronic controls.

Role of device in total rehabilitation

Although each patient may be evaluated for his physical needs and psychologic reactions and the device carefully selected and properly constructed, there still remains the necessity of fitting these requirements into the total rehabilitation needs of each patient.

The considerations are (1) medical rationale to treatment of the particular disease and/or injury and the prognosis for recovery; (2) social needs of the patient (family members and relationships, type of community in which he lives, standard of living of patient and family, economic status); (3) degree of intellectual functioning; (4) use of devices previous to present hospitalization and the results of their use; and (5) value to patient in overall goal and interests.

RELATIONSHIP WITH ORTHOSES

A self-help device is used *only* for function, but an orthosis may be used for corrective measures, stabilization, etc. Either may be used as a solution for the same problem, particularly when the main purpose of an orthosis is to enable functional activities. In any usage, however, both *should* always be cognitive of the therapeutic needs that must of necessity go hand in hand with function.

The value of an orthosis, especially the prehension orthosis, lies in the fact that it provides pinch permitting the patient to continue using standard equipment, theoretically at least. Since it is impossible for any orthosis to reproduce all types of hand prehension, some adaptations of equipment are often required in addition. Therefore, when needs are simple, a self-help device may be adequate and less cumbersome.

When the loss is in range of motion, then only a self-help device will suffice, such as a swivel spoon for loss of supination and a stocking aid for loss of hip and/or knee flexion.

When both are suitable, the decision for the use of one or the other may rest ultimately with the patient, for it is he who experiences the problem and the satisfaction or frustration of supplementary aids. When he does make the choice, he should have the benefit of exposure to each type of equipment and a full and clear understanding of the differences and rationale for use of both. Much will also depend on factors such as ease of accomplishment, ease of handling, frequency of repair, acceptance by others, and, finally, on the patient's *desires* to perform various types of activities.

TRAINING AND FOLLOW-UP

Any device, regardless of the care used in selection and/or construction, will be only as good as the amount of time and effort spent in practice with it to acquire sufficient proficiency.

Special training methods may be helpful for some devices. One may practice separate portions of a motion, in a stylized pattern, then add portions together until finally a whole or total pattern of free, smooth motion is achieved. One may start with objects such as pegs and blocks, then advance to some simple games, and finally to various ADL equipment.

Progression of skills is important. One should first start with unilateral gross activities, then progress to bilateral gross, and finally to bilateral and unilateral fine activities. Eating is a good unilateral gross activity to start with and also usually of interest to both patient and helper. Writing, a unilateral fine activity, should be attempted only after skill and control are at least fair; otherwise, results are likely to be poor and very frustrating.

Certain other factors must also be considered, such as (1) degree and type of disability, (2) complexity of motions of device and activity, (3) use of body motions for motor power, (4) age, (5) motivation, and (6) assurance of proper fit and/or function.

Proper follow-up procedures in the use of devices are necessary for consideration of (1) changes or alterations, if necessary, (2) possible discontinuance, (3) amount and frequency of use, and (4) proper care of device (cleaning and/or repair).

SUMMARY

When, through the use of self-help devices, temporary disability can be more quickly restored or independence achieved despite permanent disability, then it is the responsibility of the rehabilitation team to provide this service. There are many publications as sources of information about a host of devices.[10-15] Selection of various equipment will be improved through use of the principles discussed in this chapter. The skill of all personnel who can contribute should be brought to bear. But always it is essential to remember that the *patient* is the central figure, and maximum results are obtained with his wholehearted cooperation.

REFERENCES

1. Zimmerman, M. E.: Analysis of adapted equipment, Amer. J. Occup. Ther. **11:**229, 1957.
2. Zimmerman, M. E.: Devices: development and direction, Proceedings of the Annual American Occupational Therapy Association Conference, Los Angeles, Calif., 1960.
3. Zimmerman, M. E.: The functional motion test as an evaluation tool for patients with lower motor neuron disturbances, Amer. J. Occup. Ther. **23:**9, 1969.
4. Buchwald, E.: Physical rehabilitation for daily living, New York, 1951, McGraw-Hill Book Co., Inc.
5. Lawton, E. Buchwald: Activities of daily living for physical rehabilitation, New York, 1963, McGraw-Hill Book Co.
6. Gardner, W. H.: Left-handed writing, Danville, Ill., 1945, Interstate Co.
7. Zimmerman, M. E.: Clothing for the disabled, sect. 3, Proceedings of Workshop on Rehabilitation of the Physically Handicapped in Homemaking Activities, Department of Health, Education, and Welfare, Social Rehabilitation Administration, and American Home Economics Association, Washington, D C., 1963, U. S. Government Printing Office.
8. Basmajian, J. V.: Muscles alive, Baltimore, 1962, The Williams & Wilkins Co.
9. Keller, A. D., Taylor, C. L., and Zahm, V.: Studies to determine the functional requirements for hand and arm prosthesis (unpublished report filed in the Department of Engineering, University of California, 1947).
10. Institute of Rehabilitation Medicine, New York University Medical Center: Self-help devices for rehabilitation, Parts I and II, Dubuque, Iowa, 1958, 1965, William C. Brown Co., Publishers.
11. Bibliography on self-help devices and orthotics 1950-1967, New York, 1967, Publications Unit, Institute of Rehabilitation Medicine, New York University Medical Center.
12. Lowman, E. W., and Klinger, J. L.: Aids to independent living, New York, 1970, McGraw-Hill Book Co.
13. Klinger, J. L.. Frieden, F. H., and Sullivan, R. A.: Mealtime manual for the aged and handicapped, Essandess, N. Y., 1970, Essandess Special Editions, Division of Simon & Schuster, Inc.
14. International Centre on Technical Aids: Multilingual loose-leaves on technical aids, Fack, S-161 03 Bromma 3, Sweden, 1964 to date.
15. The National Fund for Research into Crippling Diseases: Equipment for the disabled, 1960 to date, Vincent House, Vincent Square, London, S. W. 1.

CHAPTER 10

PRINCIPLES
OF HOMEMAKING

The rehabilitation of any housewife should include the evaluation of need and/or for training in returning to her occupation as homemaker. Even if future plans should indicate some other type of vocation, either outside or inside the home, consideration should be given to the amount and kind of homemaking she will still be engaged in.

The kinds of responsibilities will depend on the homemaker's role at any given period of her life and include not only household duties but also her roles as wife, mother, mother-in-law, etc. Social and economic status may determine whether the homemaker is more of a manager or a worker. Roles may change with age, family status, or economic and social status. They may also be changed according to the type and extent of the handicap that is imposed.

It is important to remember that homemaking training is not limited to women. If a man is handicapped and cannot return to work, he may take over some household duties and so free his wife to take a job. Children also may benefit from such training, both because they are potential future homemakers and because household tasks offer excellent practice in the use of apparatus such as prostheses and wheelchairs.

The benefits or goals of homemaking training for any patient may be listed as follows:

1. Retraining or, if necessary, initial training in basic household skills
2. Reducing fatigue and amount of energy and time expended, through work simplification technics
3. Providing therapeutic exercise and/or use of prosthetic and orthotic equipment through functional tasks of homemaking
4. Providing psychologic benefits and motivation through satisfactory performance

5. Planning for adjustment of the home, if necessary, to permit disabled person to function at home as well as at the rehabilitation center

EVALUATION

The evaluation process for homemaking training is as necessary as it is for any other form of therapy. An accurate appraisal of the individual's needs will form the basis for his or her particular goals and serve as a guide for planning the treatment program. It is also a guide for other rehabilitation processes, such as vocational planning and physical retraining in physical and occupational therapy. It may be a determining factor in discharge and disposition plans.

When a patient is first seen she may be asked to discuss her home situation, providing information regarding her usual responsibilities, the type of cooking she is accustomed to doing, and a description of her home, especially the kitchen set-up. If it appears that the patient is not emotionally ready to talk about her home situation, then frequently she may just be given a simple task to perform, such as baking a prepared cake mix. This usually ensures success and helps to remove the patient's fears and uncertainty. Meanwhile, through observation, the therapist can note the patient's performance in terms of (1) physical limitations, (2) use of equipment, (3) her adeptness and ingenuity—or lack of it, (4) her concepts of work—whether haphazard and careless or careful and meticulous, (5) mental and emotional reliability (such as judgment as opposed to confusion), (6) ability to follow directions, (7) visual and perceptual deficits, and (8) safety.

When observing physical limitation, the therapist may ask herself the following questions: Is the involvement mainly in the hand or also in wrist, elbow, and shoulder? Is the patient's sitting posture good? Is manipulation of wheelchair, crutches, and prosthetic or orthotic devices a problem? Does her apparatus and/or device interfere with reaching, moving about, or manipulation of objects? Is the problem one of weakness, loss of range of motion, incoordination, spasticity, or pain? Is fatigue superimposed on these disabilities? What observations are made regarding visual, hearing, or speech deficits? Is there an intellectual impairment; does the patient understand directions and remember what she is told? Is the fear of failing in a task that was once easily managed a primary factor? What is the speed of performance? How much help does she ask for or actually need?

Some of the above evaluation factors may not be observable or forthcoming in an initial evaluation, but they are essential for consideration when planning a treatment program.

Other information basic to planning a good program are (1) type of housing—individual house or apartment, rented or owned, single story or several floors, and the number, size, and layout of rooms; (2) location—rural or urban; (3) local shopping facilities; (4) other members of family—age, sex, family role, and attitude toward disabled member; and (5) cultural, religious, and value standards.

Based on the above criteria, the evaluation may reveal need for the following: (1) practice to gain assurance of competent performance, (2) retraining in new technics, (3) development of one-handed skills, (4) selection and use of proper or suitable tools and equipment, (5) testing of work heights and arrangements, (6) testing and selection of various kinds and types of storage units, (7) use of energy-

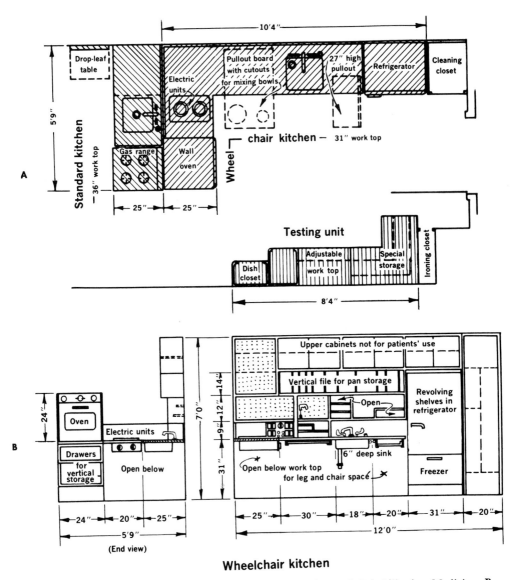

Wheelchair kitchen

Fig. 10-1. A, Floor plan of training kitchen at the Institute of Rehabilitation Medicine. **B,** Elevation of wheelchair kitchen unit.

saving methods evolved through work simplification and management-planning technics, and (8) planning for modification of the patient's living situation.

TRAINING FACILITIES OR TRAINING AREA

Although training in a hospital or rehabilitation facility does not show how the patient is going to be able to function in her own home, still it has many valuable— even ideal—advantages. For the person starting on homemaking training in the early stage of rehabilitation and for the severely disabled patient, the special so-called

Fig. 10-2. View of wheelchair section of training kitchen.

"wheelchair training unit" makes success in these activities possible. The advantage of being able to show that they *can* be done will far outweigh the objection that they are done in a special setting not at all like her own home. While in some instances major revisions in the home are necessary and in others may be desirable, if they are possible, much can be done with a few simple alterations.

In addition to a sit-down unit, it is good to include also a standard unit in the training area. This provides for a progression for some, a starting point for others, and a "normal" situation for those who cannot readily relate the principles of training in this new setting to their own kitchens at home.

If possible, a third area, for use in special problems and for additional types of storage units for demonstration and testing, makes the therapist's job easier and provides for a more complete follow-through in evaluation and training. See Figs. 10-1 and 10-2 and Zimmerman[1] for further details of a training area.

TRAINING PROGRAM

The actual steps or plan of a treatment program should never be routine but should be determined according to individual needs, as mentioned previously. Experience has shown, however, that it is usually more acceptable to the patient to start with familiar routine tasks, especially those that are of immediate concern. Several comprehensive publications on training the disabled homemaker are available for reference.[2-5]

Basic skills

Basic skills are those such as the actual operations of cooking—preparing vegetables, breaking eggs, using an eggbeater, putting together the ingredients of a cake

Fig. 10-3. Safe ways of using knives allay fears of injury, as do other safety measures.

Fig. 10-4. Breaking eggs with one hand is a new method for many women.

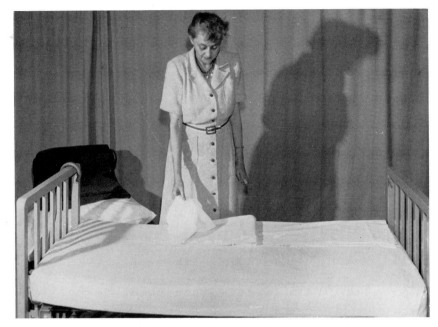

Fig. 10-5. Unfolding instead of shaking out linens is easier for persons with many types of limitations.

or casserole dish, and mixing or stirring. A few new technics may be introduced while accomplishing these tasks; for instance, using a large French chef's knife for cutting by placing the tip on the cutting surface and pressing down, instead of using the push-and-pull method (Fig. 10-3). Breaking an egg can be done with one hand after a little practice (Fig. 10-4).

Washing dishes, making beds, and ironing may be other basic skills with which to start.[6-8] Dishwashing can be managed from a wheelchair if the sink is lower and is open underneath or if a wooden rack to raise the height of a deep one is provided. Bedmaking with one hand or from a wheelchair is easier if one is taught to unfold, instead of shaking out, a sheet and if one side is made first and then the other, to save many trips or much maneuvering of the chair (Fig. 10-5).

During these operations, especially if the patient has to manage with only one hand or with two weak hands, problems of holding, stabilization, or moving of equipment may arise. One can then begin to introduce the need for special testing or training in the use of household tools.

Selection and use of equipment

The variety of household tools available usually makes it possible to manage with standard equipment. However, the ability to evaluate the different designs in terms of various problems may save a lot of frustration and wasted energy. Some of the factors to look for are as follows:

1. *Possibility of operating with only one hand.* Items such as one-handed whisks and eggbeaters, long-handled dustpans and brushes, and clip-around aprons are available.

Fig. 10-6. Casserole with two large heat-resistant handles is easier and safer to move and requires little grasp.

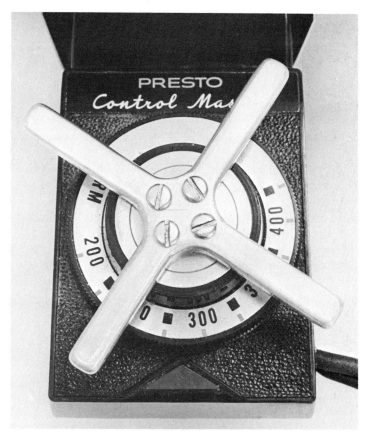

Fig. 10-7. A T-shaped handle attached to control knob enables operation with minimal grasp and also keeps hand away from hot side of frypan.

Fig. 10-8. Portable electric can opener can be used with one good hand. Note use of saucepan for stabilization of can and ease in emptying can once opened.

2. *Stabilization of tools and appliances.* There are many methods of accomplishing this. Rubber mats or wet dishcloths under bowls help prevent slipping. A new portable mixer that rests on four legs minimizes holding. Heavy iron skillets provide stability for the person with incoordination.

3. *Placement and shape of handles and knobs.* These two factors determine the (1) type of grasp necessary, (2) positioning of arms during reach and use, (3) safety, and (4) extent of energy required. A teakettle with the handle extended well down over the side requires grasp and arm positioning in a manner less tiring and also less effort to tip the kettle for pouring. A casserole with two heat-resistant handles protruding from the sides is easier to lift or slide (Fig. 10-6). An adaptation of the knob on a skillet control unit keeps the hands away from the hot metal of the pan (Fig. 10-7).

4. *Use of electric equipment to eliminate manual operation.* Such articles as electric mixers, can openers, blenders and choppers, and carving knives can facilitate work (Figs. 10-8 and 10-9).

5. *Weight, size, shape, durability, and ease of care of utensils.* Aluminum or plastic bowls may be lighter in weight than stainless steel, but it is more important that the bowl have a good, flat bottom. The new Teflon surfaces make washing

Fig. 10-9. Use and convenient storage of power equipment save energy. Pullout canisters are easy to manage and eliminate removal of lids.

Fig. 10-10. Cutting board with nails and suction cups provides stabilization of food during peeling or cutting.

Fig. 10-11. Pullout board with holes for pans or bowls provides stabilization and a lower working surface for activities that require more effort.

easier. Both the depth of bowls and pans and the type of rim and handle may be important features.

6. *Selection of containers and jars for easier opening.* Pullout or tilting canisters may be used for flour, sugar, etc.

There may be times when no standard equipment is adequate and when some adaptations may be necessary. Two of the most commonly used are the cutting board with nails to stabilize vegetables and fruits (Fig. 10-10) and the pullout board with holes for stabilizing pans and bowls (Fig. 10-11).

When considering the purchase of any equipment, either standard or adapted, in addition to the selection factors listed above, one should be guided by the following general principles: (1) the device should be really necessary, (2) it should save time, (3) it should save energy or prevent overuse of weak muscles, (4) it should provide for safety, and (5) it must meet the requirements applicable to all equipment as to durability, cleansability, ease of maintenance, and frequent enough use to justify storage and cost.[10, 11]

Work areas and storage

Almost as important as the selection of proper tools is their arrangement and storage. Basic to this is consideration of the various locations for the type of jobs to be done and the overall layout.

Every job in the home has a work area, but none is perhaps so detailed or confined within such a small space as those in the kitchen. Much has already been done toward the planning of good kitchens; these guides should be used as basic criteria,

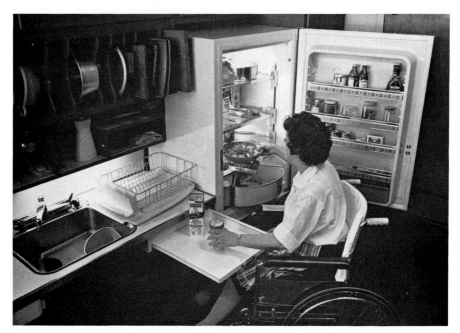

Fig. 10-12. Proper arrangement of refrigerator, work surface, and sink in area used for preparation of food.

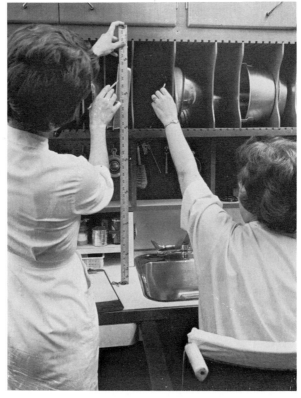

Fig. 10-13. Measuring for patient's maximum reach when planning storage arrangement.

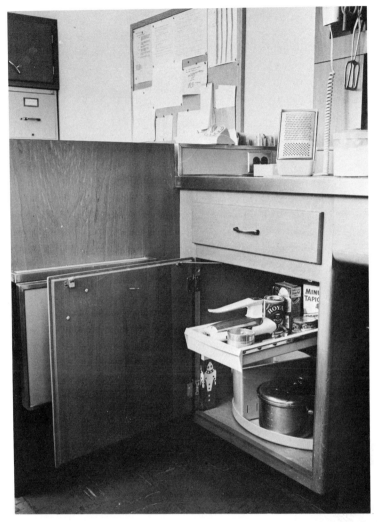

Fig. 10-14. Accessible storage devices such as pullout shelves, rotating shelves, tilt-down canisters, and magnetic racks for utensils minimize reach and effort.

with adjustments and adaptations planned as needed for the individual's physical limitations[12] (Fig. 10-12).

The two factors that most frequently need changing are the work heights and the placement and type of storage units. Proper work heights, either for standing or for sitting on a stool or in a wheelchair, will improve posture and lessen fatigue. Depth of the counter should be considered also, especially when one is confined to a wheelchair. Heights of storage cupboards and arrangement of equipment within these areas may be a major factor in achieving independence[13-15] (Fig. 10-13). Vertical filing, revolving shelves, pullout bins, and racks all bring utensils within easier reach and eliminate unnecessary lifting and bending (Fig. 10-14). The addition of extra storage arrangements by means of a midway cabinet or pegboard is an in-

Fig. 10-15. Items should be stored where they are used first, such as these requiring water.

valuable help in storing utensils at the area where they are first used (Fig. 10-15). Sometimes the use of substitute equipment, if standard items cannot be altered, may be helpful, such as the use of portable rotisseries to replace surface burners (Fig. 10-16) and ovens or drop-leaf tables to provide lower working surfaces.

A diagram such as that shown in Fig. 10-17 can be used during the recording or testing of proper heights.

Energy saving and management

Almost all patients, even those without actual orthopedic limitations (for example, those with cardiac impairment, arrested tuberculosis, or aging processes), will need a means of saving energy. This can best be provided through certain work simplification technics. These are adequately described in other literature[16] and will be mentioned only briefly here. Sitting to work; using two hands; eliminating acts of lifting, carrying, or stooping; eliminating unnecessary trips and motions; and using the assistance of gravity whenever possible are a few. One can sit to iron (Fig. 10-18) or use a wheeled cart or a wheelchair lapboard or tray to transport many objects at one time (Fig. 10-19).

Arrangement of work heights and work areas, as previously discussed, can help achieve better posture and better use of body mechanics, as well as provide shortcuts and improved methods in the performance of tasks.

Management, although a separate and unique process, is another way in which

Fig. 10-16. Portable oven on table or wheeled cart provides oven at suitable height and accessible for use.

energy is ultimately saved.[17, 18] It is economy of human resources, accomplished by the planning of work, through evaluation of the various jobs and the decisions and choices therein involved. It involves not only the homemaker but also her entire family and the use of their skills, income, and possessions. It includes the buying of furnishings, food, and equipment. It means scheduling of tasks and supervision of work done by others. It involves too the personal relationships of working with the team—the family.

If severe limitations rule out the actual accomplishment of household tasks by the homemaker herself, the role of manager may provide a place and a need for those remaining skills that involve only mental processes. Actually, management in any home, as well as in business, is a keynote to the smoothness and efficiency with which tasks involving physical effort are accomplished.

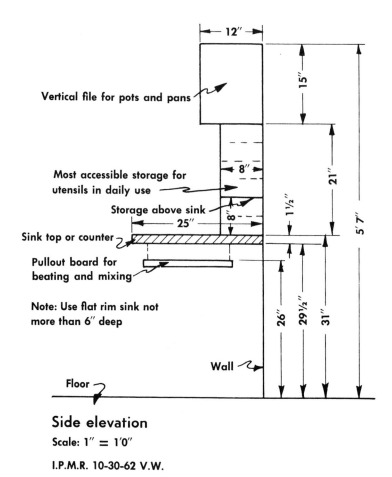

Vertical file for pots and pans

Most accessible storage for
utensils in daily use

Storage above sink

Sink top or counter

Pullout board for
beating and mixing

Note: Use flat rim sink not
more than 6″ deep

Wall

Floor

Side elevation

Scale: 1″ = 1′0″

I.P.M.R. 10-30-62 V.W.

Fig. 10-17. Suggested dimensions for wheelchair kitchen.

CHILD CARE

Although this is not an area involving every homemaker, many have infants and preschool children who demand their attention and care. Standard guidance on child care and development should be provided if it is not already available. Training in early independence is rewarding to both mother and child and is a very practical help to the disabled mother.[19]

Some of the areas in which special information and help are frequently needed are those of selection and adjustment of clothing, the bathing and lifting of infants, feeding technics, and provision for and participation in play activities[20, 21] (Fig. 10-20).

ADJUSTMENT OF THE HOME

All the evaluation and training of the disabled homemaker while in the rehabilitation center will have little meaning or value unless simultaneously a study

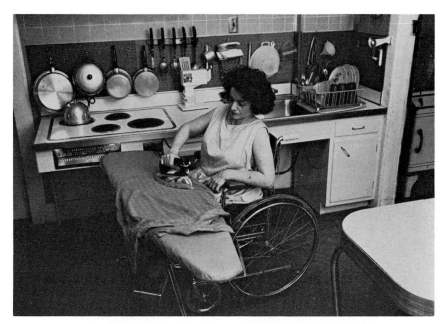

Fig. 10-18. Adjustable ironing board encourages sitting to work and makes ironing possible for the person in a wheelchair.

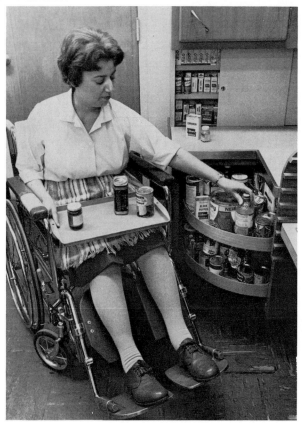

Fig. 10-19. Wheelchair tray provides a convenient and safe means of transporting items and saves many trips.

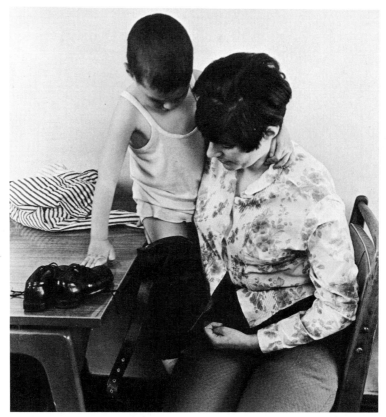

Fig. 10-20. Dressing of child is made possible by positioning child on table and/or chair so that he is within limits of reach and grasp.

of the patient's home environment is undertaken. The training program itself will undergo alteration or reemphasis according to the data regarding the existing limitations and the extent and kind of alteration possible. There may be financial problems, or the person may live in a rented house or apartment where only minor changes are allowable. Specific approaches to housing adjustments are given in Chapter 11 and in several other publications.[17-20]

Training programs in homemaking can be found in many rehabilitation centers and hospitals. In some communities, help is available in adult education programs, college or university extension services, or various health agencies. If there is no formal program, help can be enlisted from married or retired home economists, occupational therapists, nurses, or social workers.

Always, whether referred to an agency or to local help, the program and requirements for training should be under the guidance and direction of a physician, with indications as to potential, safety measures, work capacity or limitation, and possible contraindications for certain types of activity, such as standing.

SUMMARY

Homemaking training is an integral part of the rehabilitation process for any person requiring such assistance. Being able to resume former duties and to return to active participation in the home and community justifies and heightens interest in the other forms of rehabilitation measures received. To return home as a productive citizen is also of economic value.

Although training in homemaking is a specialized process and consideration should be provided for it in regard to facilities and personnel, this service should be furnished as part of the team process and be guided by the team's findings to make a contribution to the total goal.

REFERENCES

1. Zimmerman, M. E.: Homemaking training units for rehabilitation centers, Amer. J. Occup. Ther. **20:**226, 1966.
2. Rusk, H. A., and others: A manual for training the disabled homemaker, Rehabilitation Monograph 8, ed. 2, New York, 1961, Institute of Rehabilitation Medicine, New York University Medical Center.
3. Wheeler, V. H.: Planning kitchens for handicapped homemakers, Rehabilitation Monograph 27, New York, 1966, Institute of Rehabilitation Medicine, New York University Medical Center.
4. May, E., Boettke, E. M., and Waggoner, N. R.: Homemaking for the handicapped, New York, 1966, Dodd, Mead, & Co., Inc.
5. Homemaking and Homeplanning Service, Occupational Therapy Department: Training the young hemiplegic homemaker, New York, 1968, Institute of Rehabilitation Medicine, New York University Medical Center.
6. Hagman, I., and Lacy, E.: Dishwashing the easier way, Extension Division Bull. 921, Lexington, Ky., 1940, University of Kentucky.
7. Muse, M.: Seating housewives at their ironing, Agricultural Experiment Station Bull. 559, Burlington, Vt., 1951, University of Vermont and State Agricultural College.
8. Zmola, G. M.: You can do family laundry with hand limitations, Storrs, Conn., 1959, School of Home Economics, University of Connecticut.
9. Helpful homemaking hints, St. Louis, 1963, Laclede Gas Co.
10. Lowman, E. W., and Klinger, J. L.: Aids to independent living, New York, 1970, McGraw-Hill Book Co.
11. Klinger, J. L., Frieden, F. H., and Sullivan, R. A.: Mealtime manual for the aged and handicapped, Essandess, N. Y., 1970, Essandess Special Editions, Division of Simon & Schuster, Inc.
12. Heiner, M. K., and Steidl, R. E.: Let your kitchen arrangement work for you, Extension Bull. 814, Ithaca, N. Y., 1953, Cornell University.
13. McCullough, H.: Space design for household storage, Agricultural Experiment Station Bull. 557, Urbana, Ill., 1952, University of Illinois.
14. McCullough, H.: Cabinet space for the kitchen, Small Homes Council Bull. 5, Urbana, Ill., 1949, University of Illinois.
15. McCullough, H., and Farnham, M.: Space and design requirements for wheelchair kitchens, Agricultural Experiment Station Bull. 661, Urbana, Ill., 1960, University of Illinois.
16. Fitzsimmons, C., Goble, E., and Monhaut, G.: Easy ways, Extension Bull. 391, Lafayette, Ind., 1953, Purdue University.
17. Gross, I., and Grandall, E.: Management for modern families, New York, 1954, Appleton-Century-Crofts, Inc.
18. Gilbreth, L., Thomas, O. M., and Clymer, E.: Management in the home, New York, 1954, Dodd, Mead & Co., Inc.

19. May, E. E.: Child care problems of physically handicapped mothers, J. Home Economics **50:**704, Nov., 1958.
20. Boettke, E. M.: Suggestions for physically handicapped mothers on clothing for preschool children, Storrs, Conn., 1957, University of Connecticut.
21. Bare, C., Boettke, E. M., and Waggoner, N. R.: Self-help clothing for handicapped children, Chicago, 1962, National Society for Crippled Children and Adults.

CHAPTER 11

PRINCIPLES OF HOUSING

R ehabilitation falls short of its goal if it provides independence for the individual only while still a patient at a treatment center or facility and then sends him home a prisoner to home and community environmental barriers. Today most centers understand this need and include home planning in their program, although the lack of trained personnel has frequently limited its implementation.

Several facets to be considered in any such program are site and type of dwelling, ecology of housing, and architectural design features.

SITE AND TYPE OF DWELLING

Level areas are most desirable and eliminate designing to accommodate changes in ground rises. Sidewalks and curbs (removal at intersections or crosswalks) must be given attention (Fig. 11-1).

The type of dwelling depends on the needs of the occupants. Most desirable is a single dwelling or apartment located within the nondisabled community. To make this practical, access to public facilities (stores, bank, post office, library, church) must also be suitable for the disabled and aged.

Apartment complexes, partially or totally designed for the disabled, can include these facilities as well as recreation or workshop areas. Total "villages" may be necessary to provide the ultimate in independent living for the severely disabled (such as Het Dorp in Arnhem, Netherlands).

ECOLOGY OF HOUSING

Ecology of housing is a specialized derivative of the field of sociology, in which the consumer takes part in his own environmental design.[1] It expands the concept of human ecology, defined as the consideration of spatial aspects of the symbiotic relations of man and institutions. This includes not only the social, psychologic, and economic aspects of family living, but also the world of designing, construction, and industrial production. The two fields must be interrelated and their effects on each

Fig. 11-1. Ranch style house with 2-inch slope to both front entrance walk and carport.

other considered. More study of man's activities, their relationships with each other, and their needs for space and form requirements will force architects to plan with both design and function in mind. By nature they should go hand in hand.

ARCHITECTURAL DESIGN FEATURES

Architectural design features and requirements for elimination of those features creating barriers for the disabled, with consideration of ecologic and architectural demands, can be classified into three areas relating to function as follows:

1. Features providing access or egress to various areas (doors, sills, stairs, ramps, curbs, elevators)
2. Features that contribute to space enclosures (floors, walls, windows, lights, heating, ventilation)
3. Features relating to various activity work areas (kitchen, bedroom, bathroom, etc., including fixtures and furniture)

Access or egress features

Stairs (for the ambulant). Risers should not be more than 6 inches preferably. Nosings should not be abrupt or square. There should be rails on both sides, extending beyond both the top and bottom, with ends curving out of the way.

Ramps. Ramps should not exceed 1 foot rise in 12 feet of length (5 degrees) and should be at least 36 to 40 inches in width. There should also be a level entrance area extending approximately 2 feet beyond the *opening* side of the door. Rails should be available on ramps. Portable ramps may be necessary (Fig. 11-2).

Sills. All indoor sills should be eliminated and all outside ones should be confined to a minimum (Fig. 11-3).

Fig. 11-2. Portable folding ramp used for one or two steps. Made from lightweight aluminum channel and steel grid.

Fig. 11-3. Door sill from polyethylene tubing minimizes obstruction at outside doorway while offering protection from the weather.

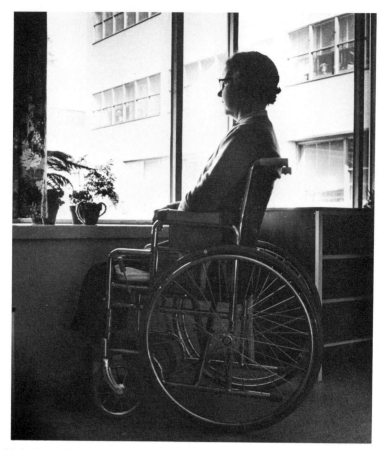

Fig. 11-4. Low window sill makes it possible to see out when sitting in a wheelchair.

Doors. A door width of 36 inches is desirable. They may be less, depending on the width of the hall or entrance area and the amount of turning space available. For small rooms, such as a bathroom, some arrangement such as a sliding door is needed to allow room for a chair and for closing the door behind the chair. Closet doors are best if they are the sliding or folding type.

Elevators. Elevators should be 3 feet in width by 4 feet in depth—the minimum interior space. The height of the operating controls should be approximately 30 to 36 inches from the floor. Doors should be self-closing, but with a time-lag.

Curbs. Curbs should be ramped. Warning features for the blind should be incorporated.

Space enclosures

Floors. Floors should be nonskid, such as terrazzo with Carborundum chips or vinyl tile. There must be *no wax*. Carpets should be eliminated, or woven mats or other coverings of firm surface should be used.

Walls. Either washable paint or vinyl coverings should be used on the walls; all surfaces should be smooth.

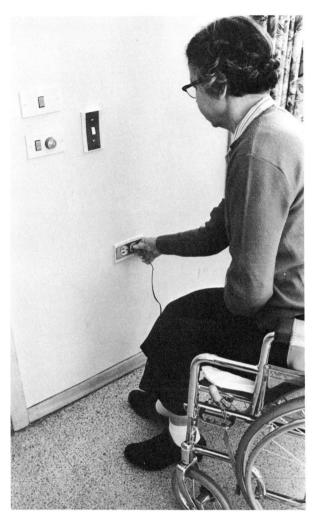

Fig. 11-5. Outlets at least 24 inches from the floor eliminate bending, require less reach, and reduce danger of falling out of chair.

Windows. Casement or awning type windows are generally easiest to open. Crank or push-bar type controls require less force and range of motion to operate. The height of the sill should not exceed 32 inches (Fig. 11-4). A bay window offers more view for those frequently confined to indoors. Pull drapes are easier to handle than venetian blinds, unless both are motorized. If venetian blinds are used, it is preferable to have the type that is enclosed between two panes of glass. Always keep a clear access to the window and controls.

Lights. Wall switches should be no higher than 36 inches and electrical outlets should be placed no less than 24 inches from the floor (Fig. 11-5). Placement should be near doors and sometimes be controlled by master switches. On portable lamps, pull chains may be easier than push or turn knobs. For persons with minimal strength some special electronic switches requiring only touch are available.

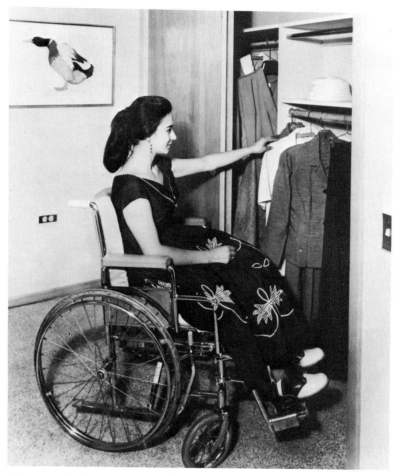

Fig. 11-6. Closet rod 48 inches from floor is suitable for many garments and reduces reach. Above storage shelf is more accessible.

Heating and ventilation. Under-floor heating requires no valuable wall space or cleaning problems, but must be supplemented to handle sudden temperature changes. Gas-fired, forced hot air provides stable, adjustable heating. It can also be combined with an air-conditioning system. Individual room thermometers may be helpful when warmer temperatures are required for only some members of a family. Good insulation also helps provide better heating.

Activity or work areas

Living, dining, and family rooms. The size of rooms depends on the number in the family and their living patterns. Arrangement should allow space for easy maneuvering of wheelchairs or use of crutches and at the same time should eliminate distances traveled to and from each area. A study of movements within these spaces (similar to a work-flow chart) should result in a truly functional space design. If well done, hallways, doorways, and separating walls will be kept to a minimum.

Fig. 11-7. Wall-mounted grab bars provide aid for transfer and also allow close access to toilet bowl by wheelchair.

Furniture should be selected according to heights of seats and of top surfaces. Chair seats should be no less than 18 inches in height and should be firm yet comfortable. Crossbars on chairs and tables are best eliminated or kept to a height of 10 inches above the floor. Sharp edges should be avoided on all pieces of furniture. Wall-hung cabinets and lights make cleaning easier and allow closer access by wheelchair.

Kitchens. For details on size, arrangement, and fixtures, see references 2 to 5 and Chapter 10. The area may include some space for eating. Otherwise, open wall areas or pass-throughs help to minimize the transport of food. When a fair amount of time is spent in the kitchen by a disabled housewife with small children, viewing access to a nearby play area saves energy and worry.

Bedrooms. Again, size and location depend on family living patterns and on the type and extent of disability. Beds, especially if double-sized, should have free access to both sides. Storage drawers can be wall hung or recessed into walls. Some under-bed storage is helpful for those who must dress in bed. Closets should be no deeper than 24 inches and are best when provided with sliding doors; rods should be no higher than 48 inches, thus eliminating unreachable, waste floor space and permitting lower shelf storage areas (Fig. 11-6).

Bathrooms. Size of bathrooms depends on the extent and type of disability and on need for assistance. The maneuvering space and the approach to fixtures deter-

Fig. 11-8. Vertical grab bar at *outside* end of tub offers support while stepping into tub and reaching for bar on adjoining wall.

mine arrangement of appliances. When individual needs are known and designed for, smaller areas may be possible and more suitable. Designing for use by persons with various disabilities may require more space to allow flexibility in approach.

Grab bars at tub and toilet are essentials (Fig. 11-7). Heights are generally suitable at 26 to 30 inches. Bars are best if horizontal and vertical, rather than diagonal. Toilet bars may need to be the fold-up or swing-away type. Tub bars should provide for standing transfer, transfer to tub seats, or transfer directly into the tub (Fig. 11-8).

The basin is best if wall mounted in a countertop not more than 30½ inches top height. The depth of the counter should allow close approach by wheelchair. It is frequently desirable to have the basin installed next to toilet so that bathing can sometimes be managed from this position. This arrangement is also helpful for persons with a colostomy.

Showers are preferred by many and are easier to manage for the more severely disabled. Sills and enclosures (except for curtains) should be avoided. A back drain and slight floor slope to the drain will prevent water from running over the bathroom floor.

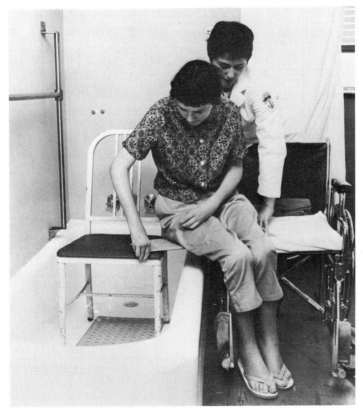

Fig. 11-9. Chair in tub at same height as wheelchair seat permits easier transfer and minimizes assistance.

Water controls should be placed on the wall adjoining the spray and within reach of the occupant. A thermostat for controlled temperature is desirable for all and is essential for some. The shower spray should be adjustable in height or should convert to a hand-held spray and is best if an on-off control is also on the hand-piece.

Tub or shower seats should be wheelchair height, if possible, to facilitate transfer (Fig. 11-9). The fold-up or easily removable types permit use of the equipment by others not desiring seats.

For further details on suitable architectural design features, see references at the end of this chapter.[6-14]

Procedures for attaining elimination of present architectural barriers in the home

Elimination of present architectural features that will create barriers for the disabled person when he returns to his home includes the following procedures:

1. *Proper timing for introduction of patient to planning.* As soon as possible,

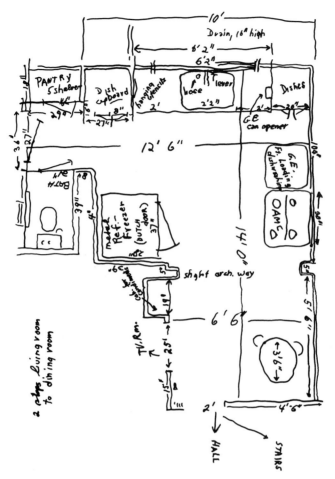

Fig. 11-10. Rough sketch of floor plan.

depending on the patient's (and family's) attitudes, the patient must be introduced to the planning.

2. *Method of acquiring information.* A rough sketch of the floor plan should be obtained from the patient or family members' description (Fig. 11-10). A home visit is preferable.

3. *Assessment of individual needs and changes.* Correlate the assessment with training programs and information from other team members. Help may be obtained from outside agencies, such as adult information programs, college or university extension services, personnel from utility companies, food and equipment businesses, editors of home magazines, and 4H Club teachers. Other helpful services are those of various professional or trade groups or individuals, such as architects, builders, carpenters, plumbers, electricians, or engineers—and neighbors and friends.

4. *Translation into scale drawings.* Scale drawings should be made suitable for

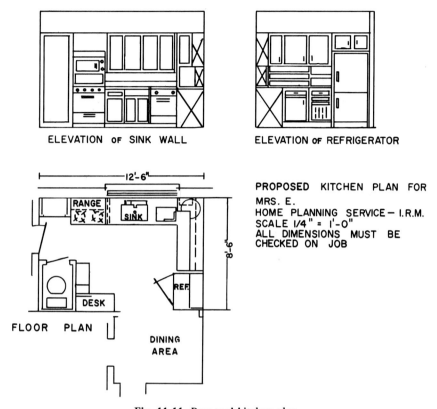

ELEVATION OF SINK WALL   ELEVATION OF REFRIGERATOR

PROPOSED KITCHEN PLAN FOR
MRS. E.
HOME PLANNING SERVICE — I.R.M.
SCALE 1/4" = 1'-0"
ALL DIMENSIONS MUST BE
CHECKED ON JOB

Fig. 11-11. Proposed kitchen plan.

conveying details necessary for the architect or builder (Fig. 11-11). (A house plan template is available from drafting or stationery stores.)

5. *Financing.* Contacts should be made with various agencies such as Division of Vocational Rehabilitation, insurance companies, and local or national health agencies or foundations. Sometimes local clubs or unions may be called on for help. (This aspect of the program may be handled by the social service department.)

• • •

Whenever possible, a follow-up visit is desirable, if not essential, to assure continuity of the program and to ascertain its results.

SUMMARY

Provision for adequate housing must be included in the concept of total rehabilitation. Otherwise, time spent on programs such as ADL and homemaking may be wasted. Vocational possibilities may be curtailed or eliminated by lack of independent access or egress to home, thus reducing economic potential.

Although the program may originate and be guided by the rehabilitation team, the value of outside assistance from other professions or individuals should be recognized and utilized.

REFERENCES

1. van Leeuwen, H.: Ecology of habitat, Wageningen, Netherlands, 1968, Department of Home Economics, Agricultural University.
2. Wheeler, V. Hart: Planning kitchens for handicapped homemakers, Rehabilitation monograph 27, New York, 1966, Institute of Rehabilitation Medicine, New York University Medical Center.
3. The therapist's guide in a homemaker's program, 1961. Available at Curative Workshop of Milwaukee, Inc., 750 N. 18th St., Milwaukee, Wis. 53200.
4. Disabled Living Activities Group of the Central Council for the Disabled: List of kitchen equipment (a guide towards choosing equipment for the physically handicapped). Available at Disabled Living Foundation, 346 Kensington High St., London, W. 14.
5. Steinke, N., and Erickson, P.: Homemaking aids for the disabled, 1963. Available at American Rehabilitation Foundation, 1800 Chicago Ave., Minneapolis, Minn. 54404.
6. American Standards Association, Inc.: Making buildings and facilities accessible to, and usable by, the physically handicapped (Specifications by American Standards), 1961. Available at Minnesota Society for Crippled Children and Adults, 2004 Lyndale Ave., S., Minneapolis, Minn. 55400.
7. Goldsmith, S.: Designing for the disabled, ed. 2, New York, 1967, McGraw-Hill Book Co.
8. International Committee on Technical Aids, Housing, and Transportation, International Society for Rehabilitation of the Disabled: The physically disabled and their environment, Svenska Central Kommittén för Rehabilitering, Fack, Bromma 3, Sweden, 1961. Available at ISRD, 129 E. 44th St., New York, N. Y. 10017.
9. Jones, S. Bramwell: The household needs of the disabled; a guide to the selection of furniture and equipment. Available at Disabled Living Foundation, 346 Kensington High St., London, W. 14.
10. Laging, B.: Furniture design for the elderly. Reprinted from Rehab. Lit. **27**:130, May, 1966. Available at National Society for Crippled Children and Adults, 2020 W. Ogden Ave., Chicago, Ill. 60612.
11. Rusk, H. A., Lawton, E., Elvin, F., Judson, J., and Zimmerman, M. E.: A functional home for easier living, Institute of Rehabilitation Medicine, New York University Medical Center.
12. Walter, F.: An introduction to domestic design for the disabled, 1969. Available at Disabled Living Foundation, 346 Kensington High St., London, W. 14.
13. Zimmerman, M. E.: A model home for the disabled, Rehab. Rec. **2**:17, Nov.-Dec., 1961.
14. Smith, C. R.: Home planning for the severely disabled, Med. Clin. N. Amer. **53**:703, 1969.

CHAPTER 12

PRINCIPLES OF ORTHOTICS AND PROSTHETICS

Man's increased longevity, his industrial environment, and the general increase in population have significantly increased the need for orthotics and prosthetics services in rehabilitation medicine. Because of this need the terms *orthotics* and *prosthetics* are widely used today, but nevertheless require definition before proceeding to the principles involved.

Ortho is a combining form derived from the Greek *orthos* denoting straight, normal, or true. The ending *tics* denotes a systematic pursuit of what the root of the word stands for. *Orthotics* then is the systematic pursuit of straightening or correcting. More specifically, orthotics deals with the application of exoskeletal devices to limit or to assist motion of any given segment of the human body. Limitation of motion may mean anything from zero degrees (immobilization) to anything less than normal range of motion, whereas assistance of motion may be throughout the normal range or through any specified range of motion. Exoskeletal devices applied for this purpose to patients suffering from neuromuscular or skeletal disorders are called *orthoses*. This is a more inclusive term than brace or splint. The *orthotist* is a professional person who designs and applies orthoses.

The term *prosthetics* is derived from the Greek prefix *pros* meaning in addition to, *tithenai* meaning to put, and the ending *tics*—the systematic pursuit of to put in addition to. Specifically, *prosthetics* deals with the addition or application of an artificial device to the body, called a *prosthesis,* to replace partially or totally missing extremities or organs. The design and fitting of prostheses is performed by the *prosthetist.*

The subjects covered in this chapter include alignment considerations of the normal lower extremity and of orthoses and prostheses; orthotic and prosthetic component indications; and fitting considerations in upper and lower extremity orthotics and prosthetics and spinal orthotics.

199

LOWER EXTREMITY ORTHOTICS

In lower extremity orthotics, the construction and alignment of a brace cannot be based solely on the condition of the disabled limb for which the brace is intended. Rather, a functionally or structurally deficient extremity must be considered as part of the body as a whole. Special attention must be given to the normal static and dynamic relationships of the hip, knee, ankle, and subtalar joints. If these normal relationships are not taken into account during fitting and alignment procedures, the brace may hinder the performance of the wearer and may tend to increase further any existing deformities.

Normally the alignment of the lower extremities is such that with a normal base of 2 to 4 inches between the centers of the heels, the ankle, knee, and hip joint axes as projected in the frontal plane are essentially horizontal (Fig. 12-1). This means that they are perpendicular to the midsagittal line. The midsagittal line is defined as a line dividing the body into two equal left and right halves. This line remains a vertical line no matter what deformity, if any, exists in the lower extremity and may be used in almost all cases to relate brace alignment to the body as a whole. This is accomplished by orienting the shoe, the brace joints, and the bands perpendicular to the midsagittal line. As a consequence, the shoe will be flat on the floor, and the joints will be horizontal and parallel to each other as viewed in the frontal plane. In the transverse plane the alignment of the segments of the leg is such that the hip and knee joint axes are perpendicular to the line of progression in normal standing with the centers of the heels 2 to 4 inches apart. The ankle axis, however, is externally rotated 20 to 30 degrees as a direct consequence of tibial torsion. The foot is toed-out by approximately 15 degrees (Fig. 12-2). This means that tibial torsion (that is, the external rotation of the ankle joint axis) is not of the same magnitude as toe-out when related to the knee axis. Therefore, the knee joint axis serves as the reference line in the transverse plane for the alignment of orthotic components. The reason for choosing the knee axis rather than the line of progression is a practical one; that is, it would be difficult to determine the line of progression and to relate all components to it, especially if there exists internal or external rotation at the hip joint. The knee axis, on the other hand, can be fairly easily determined when the knee is moved through a flexion range of about 90 degrees. The axis is perpendicular to the plane of motion of the shank and is approximately parallel to the plane of the popliteal area with the knee flexed to 90 degrees. The amount of tibial torsion (rotation of the ankle axis with regard to the knee axis in the transverse plane) and the degree of toe-out can now be related to the knee joint axis regardless of whether the patient is externally or internally rotated at the hip.

Since conventional braces do not provide motion corresponding to the subtalar joint, the correct location of the mechanical ankle axis is of great importance. To achieve this proper location, the mechanical ankle joint must be aligned in accordance with the amount of external rotation of the anatomic joint (that is, with the amount of tibial torsion). This is especially significant when free-motion ankle joints are used.

A common error is to relate ankle joint placement in the transverse plane to "toe-out." Toe-out may be defined as the relationship of the long axis of the foot

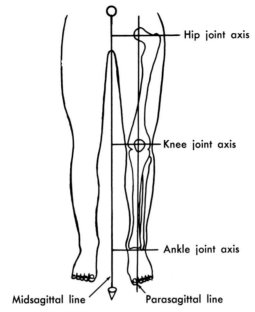

Hip joint axis

Knee joint axis

Ankle joint axis

Midsagittal line   Parasagittal line

Fig. 12-1. Frontal plane alignment.

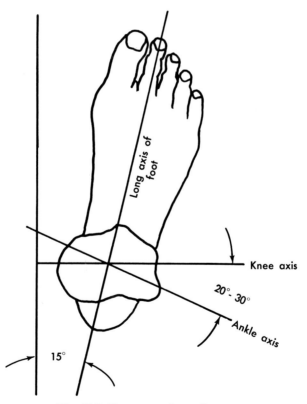

Long axis of foot

Knee axis

20°- 30°

Ankle axis

15°

Fig. 12-2. Transverse plane alignment.

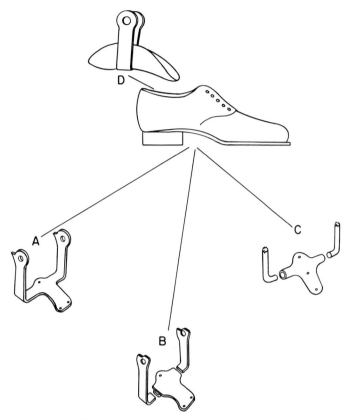

Fig. 12-3. Stirrups and foot attachment.

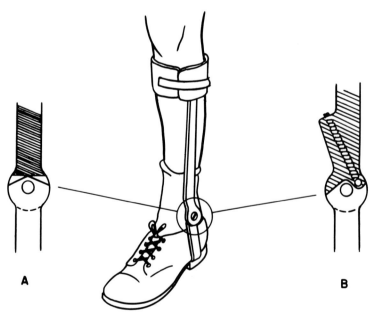

Fig. 12-4. Ankle joints.

to the line of progression. Normally, as previously stated, the foot exhibits approximately 15 degrees of toe-out. However, the amount of toe-out may be influenced by several factors other than the normal torsion of the tibia; for example, rotation in the hip or knee joints, eversion and inversion at the subtalar joint, and forefoot abduction or adduction. Furthermore, the ankle joint axis is normally rotated externally 10 to 15 degrees from a line perpendicular to the long axis of the foot. Ankle joint placement will be inaccurate, therefore, if it is solely related to the degree of toe-out. The measurement and accommodation of toe-out in orthotics, therefore, must be treated separately from that of tibial torsion. The proper relationship of the hip, knee, and ankle joints as well as toe-out in the transverse plane can best be determined by first locating the knee axis and then relating all other factors to it.

Most components used in lower extremity orthotics are prefabricated and made of either stainless steel or aluminum.

Stirrups. Stirrups may be attached either to the shoe or to a footplate. The solid stirrup (Fig. 12-3, *A*) most commonly used provides the most rigid and least bulky shoe attachment. The split stirrup (Fig. 12-3, *B*) allows the patient to transfer the brace to any shoe equipped with a flat caliper. The round caliper (Fig. 12-3, *C*) is indicated for patients, particularly cerebral palsied children, when donning the brace and shoe is a problem. With the round caliper the brace shoe and the brace itself can be donned separately and once applied can be interconnected. The solid stirrup may also be attached to a footplate (Fig. 12-3, *D*) made of metal or plastic, which also has the advantage of interchangeability of shoes.

Ankle joints. Ankle joints of most types may be used with the stirrups previously described, with the exception of the round caliper type. A limited motion ankle joint (Fig. 12-4, *A*) is indicated for positive control of the plantar or dorsiflexion angle to prevent contractures, to prevent excessive joint excursion when motor power about the ankle is inadequate, or to achieve control of a more proximal joint. For example, a plantar flexion stop can induce knee flexion, whereas a dorsiflexion stop produces a knee extension moment. A spring-loaded dorsiflexion assist (Fig. 12-4, *B*) is indicated when normal range of plantar flexion and dorsiflexion is permissible but the patient lacks adequate dorsiflexors.

Knee joints. Knee joints with a posterior offset (Fig. 12-5, *A*) are indicated when the patient has knee extensors but requires mediolateral control of the knee. The drop-ring lock (Fig. 12-5, *B*) is most commonly used when the patient lacks adequate knee extensors. A spring-loaded pull rod (Fig. 12-5, *C*) may be added to the drop-ring lock when the patient is unable to reach the knee. The spring load also provides automatic locking rather than depending on gravity. The cam lock (Fig. 12-5, *D*) provides simultaneous locking and unlocking of the double-bar uprights and provides the greatest degree of rigidity. It is primarily indicated in weight-bearing braces, or when semiautomatic unlocking is desired, since the patient can utilize the edge of a chair to trigger the bail connecting the cam locks. A plunger type lock (Fig. 12-5, *E*) provides for a more cosmetic joint, since the lock itself is concealed within the knee mechanism. It is also useful in patients who have hand weakness, since the lever attached to the plunger lock provides a mechanical advantage.

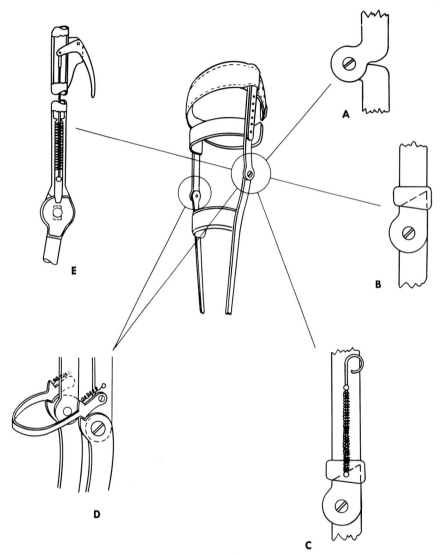

Fig. 12-5. Knee joints.

Hip joints. Hip joints providing free motion in the sagittal plane are indicated when the pelvic band is used either to control adduction or abduction or to control internal-external rotation of the extremity. A limited-motion hip joint with a posterior stop may be used to substitute for hip flexors, in addition to the indications described above for the free-motion hip joint. A drop-ring lock is indicated when the patient lacks hip extensors.

Weight-bearing devices. Weight-bearing devices provide weight-bearing relief through the skeletal system. When relief below the knee is desired, a patellar tendon–bearing brace is most commonly indicated (Fig. 12-6). The weight-bearing characteristics are similar to those of a PTB (patellar tendon–bearing) prosthesis. For

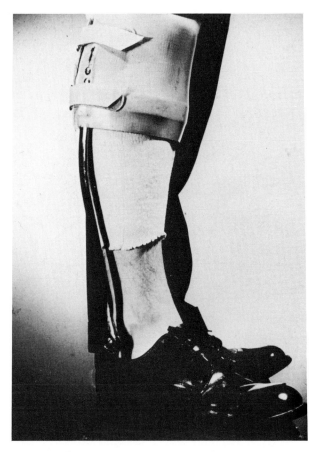

Fig. 12-6. Patellar tendon–bearing brace.

weight-bearing relief through the femur or in cases in which the PTB brace is not indicated, ischial weight-bearing devices are used. The ischial ring (Fig. 12-7, *A*) offers the least amount of comfortable weight-bearing relief, the ischial band (Fig. 12-7, *B*) is intermediate, and the quadrilateral socket (Fig. 12-7, *C*) offers the optimum and must be used when total weight-bearing relief is indicated in conjunction with a patten bottom (Fig. 12-7, *D*).

SPINAL ORTHOTICS

Although there is a great abundance of spinal brace designs, only those that are most commonly used and that present an example of motion control in various planes are described here. The two-bar lumbar (sacroiliac or chair-back) brace (Fig. 12-8) consists of a pelvic and a thoracic band and two paraspinal uprights. Its function is to control anteroposterior (A-P) motion in the lumbar area. The pelvic band, as in all braces, should be placed as low as possible without interfering with sitting at approximately the level of the coccyx. The thoracic band should extend to a point approximately ½ inch inferior to the inferior angles of the scapulas. The four-bar lumbar (Knight) brace differs from the two-bar lumbar brace in the

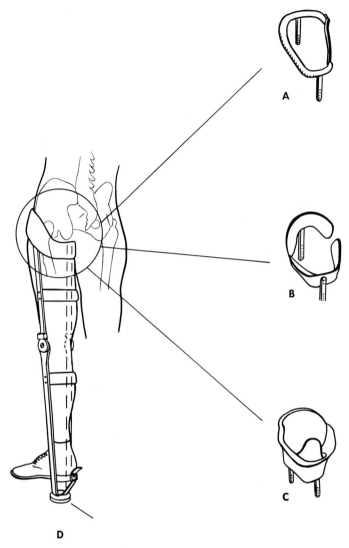

Fig. 12-7. Ischial weight-bearing devices. **A,** Ischial ring. **B,** Ischial band. **C,** Quadrilateral socket. **D,** Patten bottom.

addition of lateral uprights connecting the pelvic and thoracic bands (Fig. 12-9). In addition to controlling A-P motion, it is designed to control M-L (mediolateral) motion as well as to inhibit transverse rotation. The two-bar thoracolumbar (Taylor) brace consists of a pelvic band and two paraspinal uprights extending to just below the spines of the scapulas (Fig. 12-10). It is designed to control A-P motion in the thoracic area. A thoracic band may be added when M-L motion control is desired. The Williams flexion brace is designed to permit flexion of the lumbar spine but to resist extension (Fig. 12-11). It also controls M-L motion. In recent studies, corsets have been shown to be similar and sometimes superior in effectiveness to spinal braces. This may be explained by the fact that a well-fitted corset covers the entire

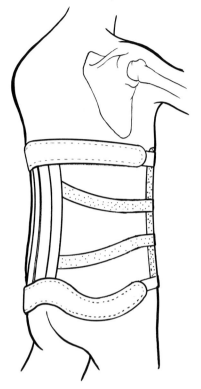

Fig. 12-8. Two-bar lumbar brace.

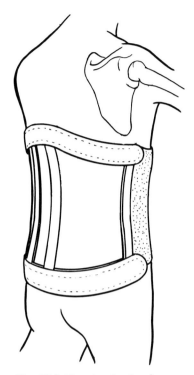

Fig. 12-9. Four-bar lumbar brace.

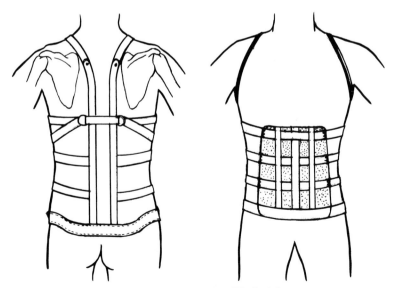

Fig. 12-10. Thoracolumbar (Taylor) brace.

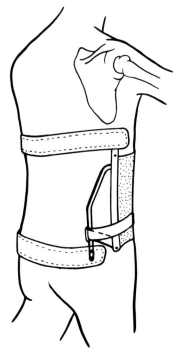

Fig. 12-11. Williams flexion brace.

abdominal area from the pubis to the xiphoid process, thereby creating a closed container. Thus an increase in the intra-abdominal pressure can be effected that, to some extent, relieves the spinal column.

UPPER EXTREMITY ORTHOTICS

Preservation of hand function, prevention of contractures, and replacement of lost hand function are one of the most important and yet difficult areas in the orthotic management of upper extremity disabilities.

Although it is impossible to include all the various designs of upper extremity orthoses, those that represent the basic principles of upper extremity orthotic applications and are most commonly used are represented in this chapter. Nearly all orthoses are fabricated from either aluminum, thermoplastic, or polyester laminates. One of the most basic functions of primates is the ability to oppose the thumb to the other fingers. Most upper extremity orthoses are based on the principle of maintaining the thumb, if not in a functional way, at least statically, in opposition to at least the index and middle fingers. The basic opponens orthosis illustrates this principle (Fig. 12-12). Although it is most often used to prevent the first metacarpal from assuming a position parallel to the other metacarpals and to prevent web space contracture (for example, in a median nerve injury), it may very well serve a functional purpose simply by placing the hand in a functionally useful position. Various components may be added to the basic opponens orthosis as shown in Fig. 12-13. Orthoses of a more permanent nature to provide for prehension

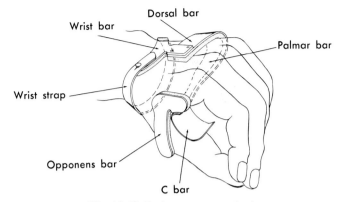

Fig. 12-12. Basic opponens orthosis.

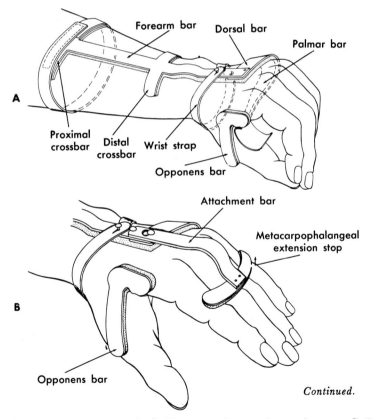

Fig. 12-13. **A**, Long opponens orthosis. **B**, Metacarpophalangeal extension stop. **C**, Finger extension assists. **D**, Activities of Daily Living—long opponens orthosis. **E**, First dorsal interosseous assist. **F**, Thumb extension-abduction assist.

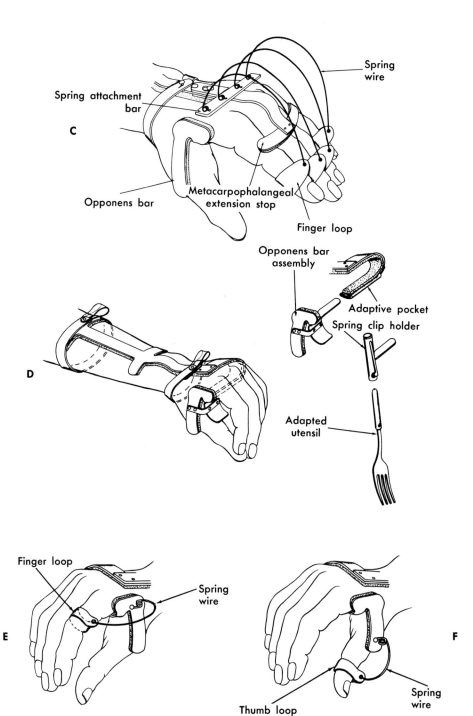

C

Spring attachment bar

Spring wire

Opponens bar

Metacarpophalangeal extension stop

Finger loop

Opponens bar assembly

Adaptive pocket

Spring clip holder

D

Adapted utensil

Finger loop

Spring wire

E

Thumb loop

Spring wire

F

Fig. 12-13, cont'd. For legend see p. 209.

utilize residual finger or wrist motion or external power (Fig. 12-14). The orthoses discussed so far are rarely used with the spastic hand. Instead, a splint that covers the greatest possible area to avoid high pressure concentration and that is designed to maintain the hand and wrist in a position of function is most commonly indicated. The volar wrist and hand splint (Fig. 12-15) is usually more comfortably tolerated in the severely spastic hand, whereas the dorsal fitting splint (Fig. 12-16) applies forces in areas where they are most effective to prevent deformity. Furthermore, exposure of the volar aspect of the forearm and the palm may have some physiologic implications. The free volar surface also provides a better friction surface that may be utilized in gross activities when holding objects against the body.

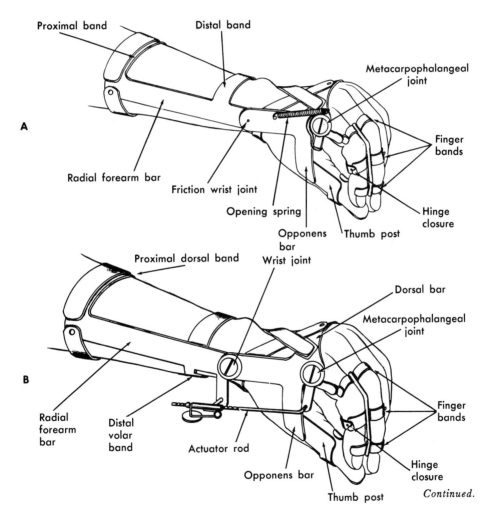

Continued.

Fig. 12-14. A, Finger-driven prehension orthosis. **B,** Wrist-driven prehension orthosis. **C,** Gasdriven (CO_2) prehension orthosis. **D,** Electrically driven prehension orthosis.

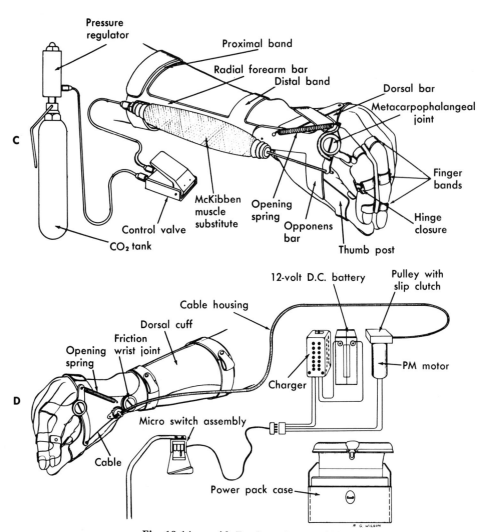

Fig. 12-14, cont'd. For legend see p. 211.

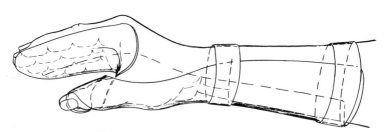

Fig. 12-15. Volar wrist and hand splint.

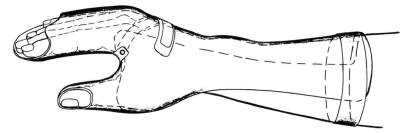

Fig. 12-16. Dorsal wrist and hand splint.

LOWER EXTREMITY PROSTHETICS

The primary objectives of lower extremity prosthetic design and components selection are to transfer body weight comfortably to the prosthetic socket, to achieve alignment stability, and to simulate normal locomotion. The degree to which these objectives can be realized depends greatly on the level of amputation. At any given level of amputation, however, there are significant variations from amputee to amputee. These relate to physical characteristics of the patient other than the amputation itself, the patient's age, and his social and psychologic adjustment to amputation. The approach used in this chapter is to relate the appropriate prosthesis and components to each amputation level.

Toe amputation. Toe amputation usually requires nothing more than a shoe filler.

Lisfranc's amputation. Lisfranc's amputation prosthesis consists of a molded arch support with metatarsal pad and a toe filler connected to it.

Chopart's amputation. Chopart's amputation and other partial foot amputations are highly undesirable from a prosthetic point of view. Although full end bearing is possible in most cases, an abnormal gait pattern results because of partial loss of the anterior lever arm during push-off. The prosthesis required must often extend to the patellar tendon and immobilize the ankle joint to provide an anterior lever arm to achieve simulated push-off. Because plantar flexion is not possible, the gait remains somewhat abnormal. Furthermore, the prosthesis tends to be quite bulky and usually requires a larger shoe size. Cosmetically it is also undesirable, especially for female amputees.

Syme's amputation. Syme's amputation has the advantage of usually permitting full weight bearing on the distal end. For female amputees, however, it is cosmetically undesirable because it produces a rather bulbous end that cannot be cosmetically accommodated in the prosthesis. Syme's prosthesis must extend to the patellar tendon for leverage at toe-off and may also be employed for partial weight bearing if full distal weight bearing cannot be tolerated (Fig. 12-17). Usually, a medial opening in the distal one third of the prosthesis or an expandable inner socket is required to accommodate the bulbous end when donning the prosthesis. The socket consists of a plastic laminate attached to a modified solid ankle–cushion heel (SACH) foot. The SACH foot does not possess a mechanical ankle joint; rather it simulates plantar flexion by the compression of a soft rubber heel wedge that can be selected in various durometers, depending on the patient's weight and gait pattern. With

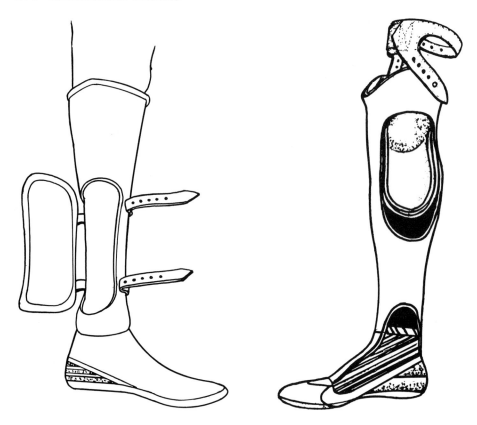

Fig. 12-17. Syme's prosthesis.   Fig. 12-18. Patellar tendon–bearing prosthesis.

the exception of the hardwood keel, the foot is made of flexible material that permits toe flexion and some adaptation to uneven terrain. Usually, there is no suspension required in Syme's prosthesis, since the bulbous end serves that purpose.

Below-knee amputation. Below-knee amputation stumps above the level of Syme's amputation cannot tolerate any significant degree of distal weight bearing. Transfer of body weight, therefore, must occur in other pressure-tolerant areas. In the patellar tendon–bearing (PTB) prosthesis (Fig. 12-18) a major portion of the body weight is distributed between the patellar tendon, counteracted by pressure in the popliteal area, and the medial tibial flare. The socket is intimately fitted to provide total contact over the entire stump area, including the distal end. A soft socket insert, although not absolutely necessary in all cases, is usually included. The socket is aligned with respect to a SACH foot in such a way as to provide mediolateral stability, and in the sagittal plane is aligned to produce knee flexion at heel strike corresponding to a normal gait pattern and to simulate push-off. This is possible by means of an alignment jig, which, during fitting, is installed between the socket and the foot. Suspension of the PTB prosthesis is provided by either a suprapatellar cuff or an extension of the anterior socket brim over the patella. Another method is to insert a supracondylar wedge between the medial portion of the socket and the medial femoral condyle. Although the PTB prosthesis provides a certain degree of M-L stability of the knee and is tolerated by most amputees, a thigh corset and side bars may be added when additional knee stability is required and/or the surface area available in the PTB socket is insufficient to allow the amputee to comfortably transfer his body weight to the prosthesis.

The conventional BK prosthesis (Fig. 12-19) with open-ended socket, thigh corset, and side joints is still indicated for a number of amputees who cannot tolerate distal end contact and who must carry a substantial amount of weight in the thigh corset. Others for whom the conventional BK prosthesis is indicated are the patient who is subject to fluctuations in stump size and new amputees who cannot readily obtain prosthetic services. Since the stump undergoes a number of changes during the first year following amputation, this latter patient can more readily maintain the fit of the prosthesis by adjustment of the thigh lacer and addition of stump socks.

In addition to the SACH foot most commonly used, there are other foot-ankle components. Of these only the single-axis foot is described here, since all others represent a very small percentage of foot-ankle components prescribed. The single-axis foot offers certain advantages over the SACH foot. The SACH foot tends to induce knee flexion, which, of course, is desirable in achieving a more nearly normal gait pattern. However, for patients who have difficulty in controlling their knees, particularly geriatric amputees, a single-axis foot reduces the tendency toward knee flexion, since plantar flexion is permitted.

Knee disarticulation. Knee disarticulation, although yielding an end-bearing stump also results in a bulbous end, and for this latter reason is not recommended for female amputees.

Gritti-Stokes amputation. Gritti-Stokes amputation and the other supracondylar amputations result in reduced bulk and length and are partial or full end bearing. Either amputee type can achieve a very good gait with the hydraulic swing phase control mechanism recently adapted to the knee-bearing prosthesis. The socket may be plastic laminated or leather molded. An anterior opening is required for donning. Often the supracondylar flares provide sufficient suspension. Otherwise, a waist belt with an elastic front suspension strap is added. A SACH or single-axis foot may be used with any of the above-knee prostheses, although the single-axis foot is more often prescribed because of its effect on knee stability.

Above-knee amputation. Above-knee amputation results in a stump that is unable to tolerate any sufficient amount of end bearing. This amputation stump is fitted with a quadrilateral socket that permits transfer of body weight through the ischial tuberosity. A total contact socket (Fig. 12-20) provides the optimum in pressure distribution but is not indicated for all amputees. Contraindications are the same as those described for the PTB prosthesis in below-knee amputations. Knee components range from polycentric units simulating the kinematics of the human knee joint and hydraulic mechanisms including those coordinating knee and ankle function for the young, active amputee, to manual knee locks for the geriatric amputee. Most commonly used, however, is the single-axis, constant-friction knee. Friction resistance in the knee can be set according to individual requirements, per-

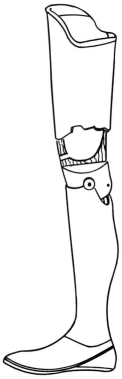

Fig. 12-19. Conventional below-knee prosthesis.

Fig. 12-20. Above-knee prosthesis with total contact quadrilateral socket.

mitting normal range of knee flexion and yet avoiding excessive heel rise. Every change in cadence, from that for which the friction is set, will result, however, in abnormal knee flexion.

The alignment of the prosthesis in the sagittal plane must be such that knee stability is achieved. The simplest method is to place the axis of rotation of the mechanical knee somewhat behind the weight line. Additional stability is achieved from midstance to toe-off by minimizing or eliminating dorsiflexion. One must realize, however, that the farther the knee axis is placed posteriorly, the more difficult it is for the amputee to initiate knee flexion. In lieu of such alignment, a friction brake in the knee mechanism that is weight-activated may be used. Alignment in the frontal plane should be such that the socket is adequately adducted to provide a supporting surface for the femur so that the gluteus medius can stabilize the pelvis when the sound leg is in the swing phase. The knee axis should be horizontal and the foot flat on the floor. A narrow base is desirable whenever possible to minimize the lateral excursion of the center of gravity.

The most sophisticated means of suspending the AK prosthesis is by negative pressure. This requires an accurate socket fit and the maintenance of a suction seal in the proximal brim area. The patient must don the prosthesis by pulling his stump

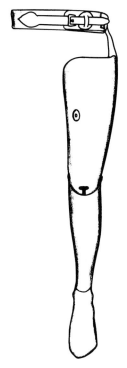

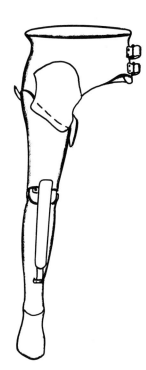

Fig. 12-21. Above-knee prosthesis with hip and pelvic belt.

Fig. 12-22. Canadian hip disarticulation prosthesis.

tissues into the socket by means of a stockinet. After the stockinet is removed, the suction valve is inserted and thus an airtight socket is created. This means of suspension is contraindicated in patients whose stumps are subject to changes in size, who suffer from excessive perspiration, or who have very short stumps or deep scars in areas where a suction fit cannot be maintained. For patients who suffer from heart disease or are otherwise physically unable to exert themselves in the proper donning of the suction socket, a hip joint and pelvic band is indicated instead (Fig. 12-21). In addition to suspending the prosthesis this system provides mediolateral hip stability. Ordinarily, stump socks are worn with this type of prosthesis. Any fluctuation in stump size can be accommodated by the addition or removal of stump socks.

Hip disarticulation. Hip disarticulation prostheses are most commonly of the Canadian type (Fig. 12-22). The unique location of the hip joint results in excellent alignment stability, yet permits the patient to walk with a free-motion hip and knee joint. This is possible because the mechanical hip joint is purposely placed anterior and distal to the acetabulum. Thus the weight line passes behind the hip joint and produces hip extension. A control strap limits the degree of hip and knee flexion. A plastic laminated socket encircles the pelvis, terminating just above the crest of the ilium for suspension.

Hemipelvectomy. A hemipelvectomy prosthesis is also of the Canadian type but

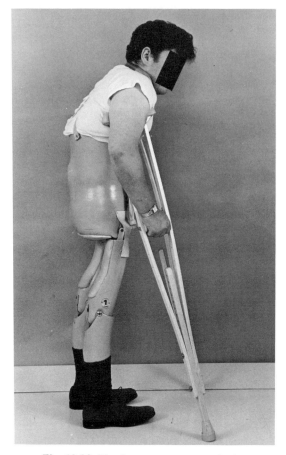

Fig. 12-23. Hemicorporectomy prosthesis.

has a socket that extends to the level of the xiphoid process to provide some weight bearing on the rib cage. Any of the knee or foot components previously described may be used with the Canadian hip disarticulation and hemipelvectomy prostheses.

Hemicorporectomy. Hemicorporectomy prostheses of the most recent design include the alignment characteristics of the Canadian type hip disarticulation prosthesis (Fig. 12-23). The hip joint, however, is modified to incorporate a stride-length control that allows the patient a reciprocal gait but requires mechanical locks at the knees. A shoulder control is utilized to unlock both hip joints to allow the patient to sit. The knee joints can be manually unlocked after the patient is seated.

UPPER EXTREMITY PROSTHETICS

Obviously, the objectives in upper extremity prosthetics are quite different from those of lower extremity prosthetics. Upper extremity function is, of course, much more important in one's pursuit of vocational and daily living activities than is

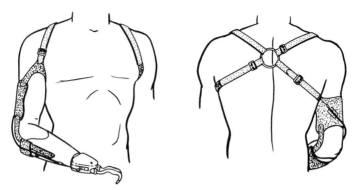

Fig. 12-24. Below-elbow prosthesis.

lower extremity function. Furthermore, an upper extremity prosthesis is much less readily concealed than a lower extremity prosthesis. The objectives, then, are the replacement of function, especially prehension, and the restoration of the body image. Unfortunately it is quite difficult to find both objectives effectively combined in the same terminal device.

Partial-hand amputation. Partial-hand amputees are rarely fitted with a functional prosthesis because of the technical difficulties involved. There is usually sufficient residual function, and most of all sensibility that makes a functional prosthesis a poor trade-off. Instead, these patients can be provided with a cosmetic glove to replace missing portions of the hand.

Wrist disarticulation. Wrist disarticulation is quite desirable from a functional point of view, since the patient retains nearly full pronation-supination in the prosthesis. The terminal devices for this and all other amputation levels are either voluntary-opening or voluntary-closing. Most common are voluntary-opening devices that are available in a functional hook or hand, as are the voluntary-closing devices. They are controlled by scapular abduction or shoulder flexion, or a combination of both. These motions are transmitted to the terminal device through a shoulder harness and Bowden cable (Fig. 12-24). Rubber bands or spring loads are used to provide terminal device closing in voluntary-opening devices. Voluntary-closing devices include an alternating locking mechanism that allows the terminal device to be locked at any angle of opening. However, a second motion is required to unlock the terminal device. Either terminal device is interchangeable in the wrist unit, which may be friction controlled or may have a quick change unit containing a rotation lock. Most amputees find a hook considerably more functional than a hand because a hook is much lighter, less bulky, and more durable. Although prosthetic hands resemble the forms and colors of human hands, they are only available in a limited number of sizes and, therefore, do not accurately match all individuals. The cosmetic glove covering the mechanical hand is susceptible to tearing and discoloration, which makes the prosthetic hand not suitable for a manual work situation. One must bear in mind that a prosthetic hook is to be looked at as a tool rather than a hand replacement, since it neither resembles the form and size of a hand

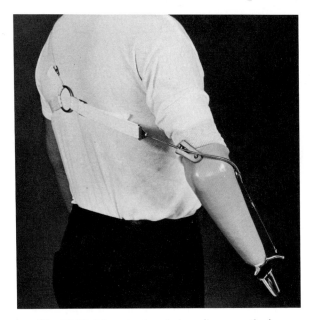

Fig. 12-25. Muenster type below-elbow prosthesis.

nor functions in the same way. Just as tools are a necessary adjunct in the performance of certain activities of nonamputees, the function of a hook is far superior to that of a prosthetic hand.

The socket, as for all the upper extremity amputations, is made of a plastic laminate and is shaped to permit maximum residual pronation and supination. A triceps cuff with flexible hinges serves as a reaction point for the control cable, as well as a link between the prosthesis and the front suspension strap of the harness.

Below-elbow amputation. Below-elbow amputees with long or medium stumps are fitted with the same prosthesis as the wrist disarticulation amputees, with the exception of socket configuration. For the short and very short BE amputee, the Muenster type prosthesis has achieved great popularity (Fig. 12-25). Here the socket extends posteriorly above the olecranon and intimately fits around the biceps tendon, thus suspending the prosthesis and eliminating the elbow hinges and triceps cuff. Although it limits elbow flexion and extension, this has been found of little consequence in the unilateral amputee. Bilateral amputees for whom maximum flexion-extension is of great necessity may be fitted with a split socket.

Elbow disarticulation. From a functional point of view, elbow disarticulation is desirable, since the patient retains nearly full range of internal and external rotation in the prosthesis. Cosmetically it is undesirable for female amputees because of the bulk produced in the elbow area by the addition of outside locking elbow hinges. The elbow unit contains an alternating locking mechanism, that is, one control motion locks the elbow and the second control motion unlocks it. Terminal device control as well as elbow flexion is produced by the same motions as discussed for the wrist disarticulation. Here, however, the housing is split in the elbow area so that when the elbow is unlocked elbow flexion occurs. When the elbow is locked, the

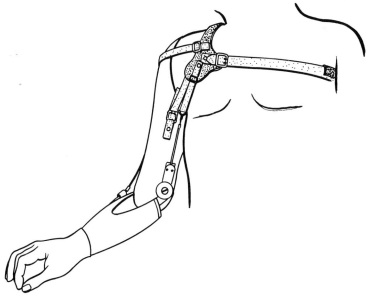

Fig. 12-26. Elbow disarticulation prosthesis.

same control motion will produce terminal device operation. The elbow lock is controlled by shoulder depression on the amputated side. The figure-of-eight harness includes a lateral and front suspension strap. The elbow lock control is attached to the latter (Fig. 12-26).

Above-elbow amputation. Prostheses for above-elbow amputees, with the exception of the elbow unit and the socket configuration, possess the same components as those described for elbow disarticulation. The elbow unit contains an internal alternating locking system as well as a turntable that permits passive control of internal and external rotation (Fig. 12-27). Depending on the stump length, the proximal socket trim lines vary. For the longest stump they are identical to the elbow disarticulation, but for a very short above-elbow stump they may extend over the acromion and contain anterior and posterior wings to stabilize the arm against externally applied rotation forces.

Shoulder disarticulation. Prostheses for these amputees require an extensive socket covering the scapula posteriorly and the pectoralis major anteriorly. Shoulder units vary from a free-motion abduction joint to friction-controlled shoulder flexion and abduction units. All other components are identical to those described for the above-elbow prosthesis. A chest strap harness with an elastic suspension strap serves to control terminal device and elbow function. Since shoulder flexion is not available, the degree of residual scapular abduction is usually insufficient to provide full elbow flexion or terminal device operation unless an excursion amplifier is incorporated into the control system. Control of elbow lock operation is obtained through shoulder elevation. The reaction point for the elbow lock cable is a waist band.

Forequarter amputation. Forequarter amputees, at the present state of the art, cannot be fitted with functional body-powered arm prostheses, although some have

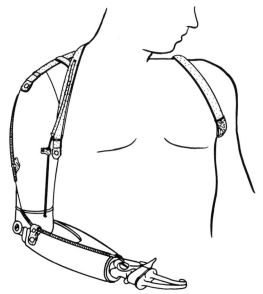

Fig. 12-27. Above-elbow prosthesis.

been experimentally fitted with externally powered devices. Instead, a lightweight shoulder cap or a passive cosmetic arm is the prosthesis of choice. More recent developments in arm prosthetics point in the direction of an increase in the application of externally powered and myoelectrically controlled devices. However, at present they are still considered experimental and have not reached the level of clinical applications.

REFERENCES

1. Lehneis, H. R.: Brace alignment considerations, Orthop. Prosth. Appliance J. **18**:110, 1964.
2. Report of the workshop panel on spinal orthotics, Committee on Prosthetics Research and Development, National Academy of Sciences–National Research Council, Washington, D. C., March, 1970.
3. Wilson, A. B.: Limb prosthetics, Artif. Limbs **2**:1, 1967.

CHAPTER 13

PRINCIPLES OF
REHABILITATION NURSING

The quality of medical care that can be provided for a patient in a designated rehabilitation service may be dependent, to a degree, on the quality of nursing care that is available constantly during the restoration process. Hence, a physician in rehabilitation and all those on the rehabilitation team will recognize the value of the numerous, intricate nursing particulars that aid in the rehabilitation of any sick or disabled person.

When nursing services are to be evaluated in the organization of a rehabilitation service, some of the factors that may deserve consideration are (1) the functions and responsibilities that are peculiar to nursing and to no other discipline, (2) the relationship of nurses to allied professional groups, and (3) the contributions of nursing toward total rehabilitation care.

In New York City, at the Institute of Rehabilitation Medicine, New York University Medical Center and Bellevue Hospital, the nursing service is recognized as a contributing force to the patient-training program; nurses participate actively in team discussions, and nursing functions and qualifications are delineated. The results may be discerned in a continuity of rehabilitation care that can come only through sustained and uninterrupted 24-hour nursing care of the patient.

It is difficult to define rehabilitation nursing with absolute exactness because the role of the nurse is a complex one. It may be said that nursing care in a specialized rehabilitation department is broadened because the nurse will find many opportunities here to render a high type of comprehensive patient care, and partly because she must work with multidisciplines within the hospital to extend the team concept of total patient care. In rehabilitation nursing the nurse is at once a nursing practitioner, a nursing educator, and, at times, a nursing coordinator. Much may depend on her relationships to other paramedical groups and her ability to see a physical disability in proper perspective.

223

ORGANIZATION AND FUNCTIONS OF NURSING PERSONNEL

Based on the belief and recognition of the worth and dignity of each person, the nursing staff endeavors to provide the highest level of nursing care to its patients. Furthermore, the nursing personnel renders an essential and unique contribution to the overall health services, both within the institution and in the community, to restore the patient to his maximal functional capacity.

All rehabilitation nursing departments must provide for administration, education, nursing care of patients, and the auxiliary personnel necessary for the care of the patient. In general, the size of the rehabilitation department may determine the size and kinds of nursing personnel. Usually one nurse will be responsible for the total program; she may have administrative and educational assistants under her direction. Nursing standards will be set, nursing care of the patient maintained, and the nursing program planned and fulfilled. In her role as administrator of the rehabilitation unit, she will analyze the nursing service required as a whole, evaluate the quality of nursing service provided by all nursing personnel, and study methods to improve patient care by encouraging the practice of rehabilitation principles in every nursing situation. The educational assistant will be responsible for the orientation of new staff, maintaining an active in-service training program, will assist in the analysis and evaluation of the educational resources within the department, will assume responsibility for continuing educational courses for nurses in rehabilitation, and will plan programs of instruction for undergraduate student nurses as part of the basic nursing curriculum. The administrative assistant assumes the responsibility to provide qualified, competent, and motivated practitioners, capable of appraising patient needs, both immediate and long term, and able to take appropriate action to meet these needs. The guidance and direction of all levels of personnel and total nursing care are the responsibility of the director of the nursing department.

The nursing director of a rehabilitation unit may be assisted in her administrative duties by clinical supervisors who supervise the nursing service and have certain administrative duties. Head nurses are directly responsible for the management of the rehabilitation unit and the overall nursing care plan for the patients. The head nurse may be assisted by team leaders whose responsibility it is to identify clearly patients' needs and to formulate a planned approach to their nursing care. Clinical instructors will work under the direction of the educational assistant, teaching nursing classes, giving nursing demonstrations, and in other ways helping to carry out the planned teaching program. Staff nurses, practical nurses, nurses' aides, messengers, unit secretaries, and volunteers may make up the remainder of the nursing service personnel. Definite functions may be assigned to all of these persons according to their degree of training skills.

NURSING ACTIVITIES ON THE REHABILITATION TEAM

In special rehabilitation centers or services where a well-integrated rehabilitation team has been assembled, rehabilitation nurses may have a dynamic part to play in their relations with the various professional disciplines on this team. One of the first responsibilities of the nurse in rehabilitation is to maintain good interpersonal relationships. This requires, on her part, the development of abilities to function smoothly with people of many and diverse professional backgrounds and training.

While this multidisciplinary approach to the rehabilitation of any patient is comparatively modern, it has its historic root in the Golden Rule, and that should help all rehabilitation workers to view each other as coprofessionals in a common cause. A deep and abiding respect for each individual's potentialities and particular skills is needed for successful rehabilitation teamwork. A statement has been made that the more seriously disabled a patient is, the more he is in need of the teamwork of many professionals and the more the individual professionals need each other. To obtain greater insight and understanding regarding patient needs, each team member must relate to all others on the team in such a way as to make possible the easy exchange of opinions and ideas, access to vital information, and discharge of mutual help or advice when these are required.

Thus, as with all other disciplines, nursing in rehabilitation must be a two-way process. There is much that the nurse can give *to* other members of the team and there is much that she can gain *from* each team member to make her nursing care more meaningful and more effective. Therefore, a desire on the parts of all team members to cooperate and to elicit cooperation is indispensable in rehabilitation.

The rehabilitation team members may discover that the nurse can render valuable service to them in her role of coordinator. The term coordinator is used in the sense that the nurse may help to unify the whole process of rehabilitation. She may collect data and information from all team members and utilize their contributions to unite every facet of rehabilitation care that can be woven into the total nursing care of the patient. While she is participating actively on the team, attending team conferences, and consulting with individual team members, she may learn much about her patients that is pertinent, advantageous, and applicable to the benefit of the patient, herself, and the team.

Following are a few examples of pertinent application:

1. An aphasic, hemiplegic patient who is undergoing speech therapy may be aided markedly by the rehabilitation nursing personnel. In the ward situation every opportunity can be utilized to promote communication for the patient and thus provide continuity of speech practice not otherwise possible.

2. Ambulation training, always a part of physical therapy and directed by physical therapists, can be carried over into the ward by a coordinating nurse. To be skilled in the technics of gait training and crutch walking is as essential for the rehabilitation nurse as for the physical therapist.

3. In the psychologic area of the patient's care the coordinating nurse or nurses, possessing insight into the aims, objectives, and case findings of the clinical psychologist, may incorporate sound principles of psychology into nursing care and thus aid the psychologist to attain some of his goals for the patient's emotional well-being.

Many more examples might be cited. Since the goal of each team member is maximum rehabilitation of the patient, the task of synthesizing activities around a 24-hour day and 7-day week must necessarily devolve on nursing personnel, who serve the patient in a rehabilitation service at all hours of the day and night.

NURSING PRACTICE IN REHABILITATION

In rehabilitation services the nurse practitioner utilizes all her previously learned skills and draws on all previously gained knowledge to develop expertise in rehabilitation. Regular hospital standards must also be met, policies enforced, and nursing service maintained.

It is true that a certain orientation is needed by a nurse who devotes her energies to rehabilitation and that a responsive acceptance of the hopeful philosophy of rehabilitation is imperative. Nevertheless, the basic principles of rehabilitation nursing in a specialized service are those of good nursing in any service of a hospital.

The fundamental idea in all rehabilitation care is that man is a total being— that he is composed of physical, mental, and spiritual entities, which taken together constitute a whole person. Therefore, a basic principle in all nursing practice is the concept that every patient is a person who must be served in many ways to aid in restoring him to dignity and usefulness. Every act or procedure of nursing must be directed toward the care of the whole patient. Nurses in special rehabilitation services and those who practice the principles of rehabilitation in all areas of basic bedside nursing care are deeply aware of the fact that skillful nursing on the physical level only is not enough—that nursing ministrations must be balanced between the technical treatments and procedures of bodily care and careful attention to the needs of the mind and spirit.

But much of the nurse's time will be given to physical care. The rehabilitation aspects of nursing care in this area are mentioned here only briefly, since they are described in greater detail in other parts of this book, as well as in nursing texts.

In general they may be said to include the following topics.

Technics related to maintenance of maximum physical and mental health. These technics are the basic procedures of all nursing care. They include measures to provide good hygiene, a sanitary environment, a wholesome and happy climate, proper rest and sleep, adequate nutrition, and diversional occupation.

Technics designed to prevent superimposed deformities. In this sphere of rehabilitation nursing care the nurse may be solely responsible for the exercise of good judgment and the application of sound rehabilitation principles. The rehabilitation of any patient most certainly should begin with methods to prevent contracture deformities, because—once developed—contractures impede the progress of a rehabilitation program. In her nursing care the nurse is therefore obliged to institute effective posture technics and to apply the principles of body alignment and correct positioning of the patient. She may use mechanical aids such as footboards, bedboards, sandbags, and pillows to position the patient properly and thus prevent foot drop, wristdrop, "claw" hand, "frozen" shoulder, and other deformities. (See Fig. 13-1.) She may initiate simple preventive exercises to maintain muscle tone and strengthen normal muscles. Often an early program of self-care activities, such as teaching the patient to wash his face, comb his hair, brush his teeth, move about in the bed, and transfer from the bed to the wheelchair, will enable the nurse to plan activities that are really exercises that help to maintain normal range of motion in the joints. Not infrequently the nurse and physical therapist may need to consult with each other and work together to achieve a degree of success in this all-important phase of rehabilitation care.

Technics that apply to management of self-care problems. These technics, classified under the general term Activities of Daily Living, have been discussed. Many of the daily living activities, of necessity, are practiced at the bedside of the

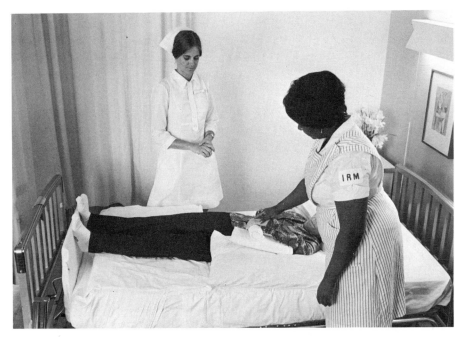

Fig. 13-1. Proper bed positioning of the patient with hemiplegia, using hand roll, pillows to elevate the affected arm, and footboard, can prevent the formation of deformities or flexion contractures.

patient or elsewhere on the ward service. Skilled nursing self-care instructors may guide a program of self-care on the rehabilitation wards. The activities that are closely related to nursing care include bed and wheelchair activities, dressing and undressing, feeding, bathing, and toileting (Fig. 13-2).

In addition to teaching skills for self-care instruction, the rehabilitation nurse must possess knowledge and adeptness in the use of mechanical devices that aid the patient to achieve self-sufficiency. Among such devices and appliances are wheelchairs prescribed for specific disabilities, braces, prostheses, crutches, canes, self-help aids and utilitarian articles of apparel.

Technics that apply to elimination of body wastes and to prevention and control of decubitus ulcers. On a rehabilitation service the urologic problems, especially those that are associated with paraplegia and quadriplegia, are managed by consulting urologists. However, certain urologic procedures in the management of the neurogenic bladder are delegated to the nurse under the direction of the urologist. Bowel management, too, may be a nursing responsibility.

Bladder management. In the management of the neurogenic bladder the following urologic examinations are done under the direction of the urologist: routine analysis, cystometry, cystoscopy, cystogram, urethrogram, and intravenous pyelogram. Attempts are made to discontinue the use of the catheter as soon as possible by trials of voiding.

Objectives. The objectives of bladder management are (1) to determine whether the patient can void after the catheter is removed, (2) to note how well the patient

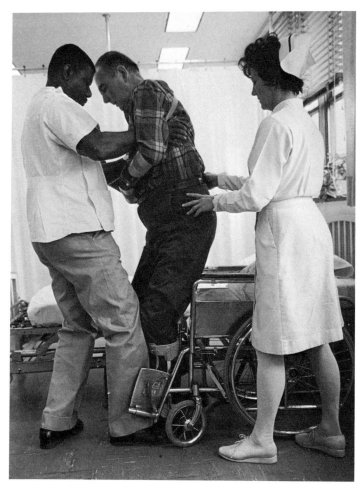

Fig. 13-2. Assisting a patient with hemiplegia to transfer from the bed to the wheelchair, the nurse is teaching the attendant how to support the affected leg, enabling the patient to do a standing transfer into his wheelchair with minimal effort.

empties his bladder, depending on the amount of residual urine left in the bladder, and (3) to obtain better urinary control after the patient has been without a catheter for a period of time, (4) to obtain maximum cleanliness for the patient, (5) to obtain maximum comfort for the patient, (6) to alleviate the emotional elements associated with incontinence, (7) to aid in the prevention of decubitus ulcers, and (8) to aid in the total rehabilitation of the patient.

Procedure for trials of voiding

1. After bladder irrigation is done, remove the catheter between 8 and 9 A.M. This is to make sure that the catheter has been working well and that the bladder is empty.

2. The patient's fluid intake should be the same as on other days (approximately 3000 ml. daily). However, some patients may drink to an excess; therefore their

fluid intake should be minimized, as an excess initiates voiding before the approximated time. The expected amount of fluid intake for 7 hours during the period of trial voiding is 800 ml.

3. Inform the patient as to signs or signals that may mean voiding. Flushing, chilling, goose pimples, and cold sweats are signs of voiding. Because each patient responds in a different manner to trial voiding, they should be observed closely and hourly while the catheter is out.

4. The use of the Credé maneuver (manual expression of the urine from the bladder with moderate external pressure, downward and backward) should also be explained and demonstrated to the patient to facilitate easy emptying of the bladder. This may also trigger reflex action in a spastic bladder to empty its urine. *Note:* The Credé maneuver is not to be used for a patient with reflux.

FOR THE MALE PATIENT. After carrying out the procedure for trials of voiding, fit the male patient with an external appliance, the condom, after removing the catheter. Explain to the patient that he should recognize signs of impending micturition so that the use of an external appliance could be eliminated. However, with a reflex type of bladder, eliminating the use of the condom is quite difficult.

FOR THE FEMALE PATIENT. Put female patients on a 2-hour voiding schedule. Place the patient on a bedpan or on the toilet every 2 hours and check to see if she can void, either by straining or doing a push-up or by the Credé maneuver. Some patients may void between the scheduled times; therefore they should be checked hourly. The patient should also be properly padded with absorbent material and should wear rubber pants to protect her from wetting clothing.

FOR BOTH MALE AND FEMALE PATIENTS. If the patient is unable to void on the first trial, carry out repeat trials on a 7-to 10-day basis or approximately the same time that the catheter is due for changing. Do this over a period of 6 months. If the patient has been unsuccessful in that period of time, the urologist may recommend surgery to help the patient to eliminate the use of a catheter. For patients with reflux, trial voiding is not carried out unless specifically ordered by the urologist.

Bowel management

Objectives. The goals of the procedures for bowel management are basically the same as for bladder management. They are (1) to establish a periodic pattern of bowel evacuation on a regular time schedule and (2) to prevent constipation and impaction of the bowel.

Procedure. The procedure to be used is as follows:

1. Start a bowel training regimen as soon as is feasible after the admission of the patient.

2. Select the most suitable time for bowel evacuation in accordance with the patient's individual needs.

3. Prepare the patient for the training and elicit his cooperation.

4. Guided by medical orders, instruct the patient concerning his diet and fluid intake.

5. Encourage mobility; all exercises aid in the process of elimination.

6. When glycerin suppositories are used, insert these regularly each day at the

prescribed hour for evacuation. Teach the patient the proper method for inserting his own suppositories.

7. It may be necessary to administer laxatives or suppositories that contain vegetable derivatives as prescribed by the physician.

8. Encourage digital evacuation on the second day if after inserting a suppository the patient fails to move his bowels.

9. After the third day of suppositories and the patient has had no bowel movement, a soapsuds enema is the last recourse, and it is advisable that the doctor be notified.

10. It should be realized that it takes a great amount of effort, patience, time, and trial and error before a person can have a stabilized bowel routine.

Decubitus ulcers. Closely related to the bladder and bowel management programs are the measures instituted by the nurse to prevent and therapeutically treat decubitus ulcers. The ulcer occurs most often as a result of prolonged pressure on the bony prominences of the body and from tissue ischemia. Early warning signs may include reddened skin areas, blister formation, and primary tissue necrosis. Infection is most common at this stage and must be guarded against.

It has been found that constant concern toward preventing the occurrence of pressure areas is the best approach to this serious problem. Of particular significance is the practice of positioning the patient from side to side and from the supine to the prone position every 2 hours or as indicated within the 24-hour period.

The purpose of the following procedure is (1) to prevent pressure sores and (2) to teach the patient self-care of the skin areas.

Procedure. The treatment of patients with decubitus ulcers is considerably more involved than is the prevention. The patient must be taught to do the following:

1. Examine the skin surfaces thoroughly every day for reddened areas.

2. Bathe daily with warm water and soap. After bathing, medicated lotions or antibacterial baby powder may be applied lightly over the body. Particular attention should be focused on the lower back, buttocks, hips, feet, and heels.

3. Use a foam rubber pillow or pad to keep pressure off any part that shows signs of redness. Report any signs of redness promptly to the doctor or nurse.

4. Change positions frequently. Do not permit pressure to remain over one part longer than 1 hour.

5. Report signs of broken skin areas or skin rash to the doctor or nurse.

6. Keep the skin dry at all times.

The *male* patient who achieves urethral voiding is usually incontinent and will require some external appliance to collect urine. A condom urinary appliance is the most practical. Change the condom daily and affix it to the phallus by skin cement. After condom application, careful inspection and examination of the genital area should be encouraged daily.

Most *female* patients have to wear specially designed pants that open in the front to enable the patient to change pads with or without braces. Examine the genital area every night for redness or ulceration.

7. Watch the braces. No part of the body should rub against the brace. Examine the body carefully for any sign of pressure from a brace that is too tight or is rubbing against the skin.

8. Do frequent push-ups in a chair. Ambulation helps to prevent pressure sores.

9. Eat plenty of protein. Protein in the diet helps to prevent pressure sores. Milk, meat and eggs, dry beans, soybeans, lentils, and cheese are protein foods.

The following points that the patient must remember must be emphasized:

1. Paralyzed parts are insensitive to pain and pressure, and serious injury can occur with no feeling to the patient. All paralyzed parts must be protected from injuries caused by bumping, sliding, falling, rubbing, or spilling hot liquids, from burns of all kinds, and from any other type of injury. The earliest sign of a pressure sore is redness on the skin, followed by a broken area. The patient must learn to recognize these signs.

2. Daily inspection of the skin areas is vital to discover signs of pressure on the parts of the body that cannot be seen easily. These parts must be observed closely.

3. Any gain in weight may cause the brace to constrict the circulation of blood by squeezing against the skin. Constriction caused by a tight brace is the same as pressure, and pressure always causes a pressure sore.

4. Pressure sores may be averted.

Some other methods for the prevention of decubitus ulcers in general use are those that relate to the proper positioning of a patient in his bed or wheelchair; the use of foam rubber mattresses; the use of Gelfoam pads, alternating pads, sheepskin, and synthetic sponge pads over pressure points; an exercise regimen; and meticulous attention to the intake of sufficient quantities of protein foods.

Use of the tilt table. Experience has shown that placing a patient in the upright position seems to diminish some of the deconditioning phenomena that may result in the occurrence of a decubitus ulcer. Accordingly, the use of mechanical or electrical tilt tables has been found helpful in treatment and in prevention.

PURPOSES. The purposes of the use of the tilt table are (1) to enable a paralyzed patient to assume the erect position, (2) to prevent contracture deformities, (3) to prevent the deconditioning effects of prolonged bed rest, such as loss of muscle tone, urinary calculi and urinary infection, interference with peripheral vascular circulation, and decubitus ulcers.

EQUIPMENT. Equipment includes (1) a tilt table, (2) three restraint straps, and (3) an overbed table.

PROCEDURE. The procedure for placing a patient on a tilt table is as follows:

1. Place the patient on the tilt table using the three-man carry (Fig. 13-3, *A*).

2. Place restraint straps at the knees, hips, and lower rib cage and fasten them securely. The angle of the tilt is easily read by the indicator (Fig. 13-3, *A*).

3. In the upright position the patient may become independent in his personal hygiene as well as engage in social and recreational activities (Fig. 13-3, *B*).

Use of the CircOlectric bed. The CircOlectric bed enables the nurse to provide maximum nursing care; the mechanical device saves time and energy.

PURPOSES. The purposes are (1) to provide immobilization when indicated and yet permit a change of position for the prevention of decubitus ulcers, flexion contractures, and other hazards of prolonged bed rest; (2) to provide vertical turning as well as tilting; (3) to make it possible for the patient to use the bedpan without being moved or lifted; (4) to enable the nurse to carry out treatments and procedures with the patient in the prone position; and (5) to enable the disabled

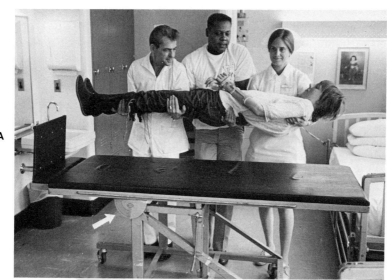

Fig. 13-3. A, Patient is placed on the tilt table using the three-man carry. **B,** In the upright position the patient can engage in social and recreational activities.

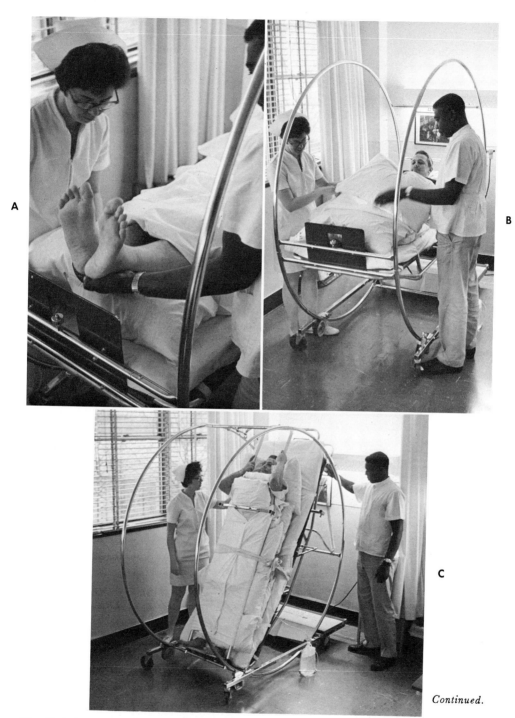

Continued.

Fig. 13-4. A, After transfer to the CircOlectric bed, special attention is focused on positioning the lower extremities. **B,** Patient is prepared to rotate to the prone position. **C,** During the turning process the nurse places her hand on the patient's shoulder for reassurance. She observes him for pallor or other signs of vertigo or nausea that can occur the first few times the total body position is changed. **D,** Patient is rotated to the prone position. Notice the urinary drainage equipment on the outside of the tubular steel frame. **E,** While in the prone position, the patient can carry out some of his Activities of Daily Living.

D
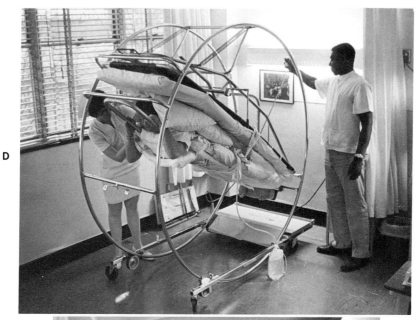

E
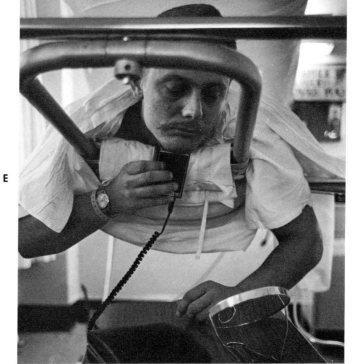

Fig. 13-4, cont'd. For legend see p. 233.

patient to become more independent in bed activities, improve his mental attitude, and facilitate his rehabilitation through the proper use of many facilities incorporated in this bed.

EQUIPMENT. Equipment includes (1) CircOlectric bed, (2) accessories, depending on illness and particular needs, and (3) an overbed table.

PROCEDURE

1. Prepare the posterior frame of the bed by placing pillows on the mattress to prevent pressure on the heels and sacral area. Then move the patient to the CircOlectric bed, using the three-man carry as used with the tilt table (Fig. 13-3, *A*).

2. Position the hips over the gatch and place the feet against the footboard. Keep the heels off the mattress by placing a pillow under the lower leg between the knee and the ankle (Fig. 13-4, *A*).

3. After 2 hours prepare the patient to rotate to the prone position. Place pillows over the anterior aspect of the lower leg, thigh, and chest area, keeping in mind the principle of protecting the bony prominences of the toes, knees, and iliac crest against pressure (Fig. 13-4, *B*).

4. Place the anterior frame into position over the patient and fasten the stud nuts at each end of the frame securely.

5. Use restraint straps for further security of the patient while turning.

6. Turn the patient gradually to prevent loss of consciousness or vertigo (Fig. 13-4, *C*).

7. Rotate the patient to the prone position (Fig. 13-4, *D*).

8. While in the prone position, the patient can carry out some personal activities such as shaving, washing, and eating (Fig. 13-4, *E*).

•   •   •

Some general and specific knowledge that is needed for the practice of nursing in a rehabilitation service or center has been described very briefly in the foregoing paragraphs. However, it is an axiom that rehabilitation for any patient should begin at the onset of his illness. Therefore, the principles of rehabilitation nursing must be made a part of total nursing care in every nursing situation. In general it may be said that these principles fall into three major categories: (1) those included in the practice of good nursing care that prepares the patient to benefit from all other rehabilitation services, (2) those that are a part of the teaching arts and that transfer knowledge of rehabilitation to all nursing personnel and to patients and their families, and (3) those principles of coordination of services that ensure continuity of rehabilitation care and aid in shortening the period of rehabilitation.

COMMUNITY IMPLICATIONS OF REHABILITATION NURSING

In her triple role of practitioner, educator, and coordinator, the rehabilitation nurse in the hospital has an excellent opportunity to lay some of the foundations of successful community living for a disabled patient. She may accomplish this in the following ways.

Educating the patient. All instruction of the patient should run concurrently with his rehabilitation training. Indeed, one might say that the rehabilitation patient is really "in school," for there are some learning situations in most of his contacts: all rehabilitation workers are teachers. The significant instruction that the nurse can give includes (1) that which is concerned with fundamental health principles, such as the necessity for cleanliness, the morale factor in wearing attractive clothing, the need for good grooming and other basic elements of mental and physical hygiene; (2) that of teaching the patient to observe the rules of good nutrition and elimination—the latter involving specific information about bowel and bladder training, skin care, and protective measures to prevent the occurrence of decubitus ulcers; (3) that which motivates the patient to persevere in performing the activities of daily living without assistance from well-meaning but overprotective families and friends; and (4) that which is necessary for the proper care of mechanical appliances, such as wheelchairs, braces, prostheses, and self-care aids.

Interpreting the patient's needs to his family. The family group must be aware of all that has gone into the education of the patient during his rehabilitation. Sometimes, unfortunately, it is found that a well-rehabilitated patient may regress when he returns to the home environment. And, often, this may occur because the members of his family do not fully understand his limitations, do not always accept his disability, and sometimes fail to comprehend the purpose of the rehabilitation program. Obviously the nurse has a responsibility to act as counselor and guide while the patient is in training in the hospital. In special, planned family-education conferences, in visiting hours, in individual interviews, etc., the nurse may impart knowledge to the family that provides some insight into the patient's social and emotional needs, his ability to perform self-care activities without assistance, the devices and appliances he may need, the cost of these articles, and how they can be obtained. Every opportunity must be utilized to build good rapport between the patient and his family in preparation for his return to the community. The education of the patient and his family in the area of nursing needs should take place in the hospital rehabilitation service under competent nursing supervision. The follow-up, then, in the community setting by visiting nurses should serve as a continuation of this instruction and not a wholly new orientation.

Transferring information to the visiting nurse in the community. It is always the hope of all rehabilitation workers that every patient will reach the ultimate goal of total independence. However, it is recognized that a certain number of disabled persons will never be capable of this achievement but that they may attain a lesser degree of rehabilitation. Consequently, on their return to the community these patients may need to avail themselves of assistance from community agencies. Since the visiting nurse service is one of the public health agencies in the community, it is clear that avenues of communication should be strengthened between the nurse in the hospital and the nurse who serves in the community. In some regions of the country, public health nurses play an active part in hospital planning for the patient's discharge. In some rehabilitation services they join the rehabilitation team periodically, and through team conferences they perceive the patient's community needs. Such a procedure places the visiting nurse at a great advantage when the patient must come under her care in his home. Her first visit to the patient's home then is not a new experience, for both the patient and the nurse recognize it as a continuation of the hospital program. But if this close relationship is lacking, the same effect may be gained by the transferral of pertinent information from hospital to community agency. It is the responsibility of the rehabilitation nurse to furnish necessary information to the visiting nurse, and the nature of the data will vary with the individual patient and his type of disability. The answers to questions such as the following are helpful to nurses who follow the disabled patient in the community: (1) How did the patient adjust to his rehabilitation program in the hospital, and what was his attitude toward his training? (2) Was he well motivated? (3) What instructions were given to the patient and his family? (4) What skills did he master? (5) How well did he assume responsibility for his own care?

SUMMARY

Rehabilitation is a medical service to the patient that is designed to prepare him for useful, productive, and happy living in his community. One of the first maxims of nursing in rehabilitation is that the patient, however seriously sick or disabled, must be regarded from the first day of his illness as one who will someday return to a normal and satisfactory mode of living; a second is that rehabilitation principles must be practiced in nursing care as early as possible, with the view of helping a patient to reach this normal state.

Perhaps the true worth of a profession can be measured in terms of its value to society. If this is true, then nursing in rehabilitation must meet its obligations. Because of the varied and consummate demands for the nurse's services in rehabilitation and because she may bring much to the success of the program, her position in a rehabilitation service and on a rehabilitation team should always be one of courteous cooperation, balanced judgment, and dynamic participation.

REFERENCES

1. Lawton, E. Buchwald, and others: Activities of daily living for physical rehabilitation, New York, 1962, McGraw-Hill Book Co., Inc.
2. Morrissey, A. B.: Rehabilitation nursing, New York, 1951, G. P. Putnam's Sons.
3. Rusk, H. A., and Taylor, E. J.: Living with a disability, New York, 1953, Blakiston Co.
4. Terry, F. J., Benz, G. S., Mereness, D., Kleffner, F. R., and Jensen, D. M.: Principles and techniques of rehabilitation nursing, St. Louis, 1961, The C. V. Mosby Co.
5. National League for Nursing: Nursing in rehabilitation, bibliography, New York, 1955.
6. Morales, P.: The neurogenic bladder in traumatic paraplegia, New York, 1969, Department of Urology and Institute of Rehabilitation Medicine, New York University Medical Center, and the New York State Rehabilitation Hospital.
7. Urological nursing procedures, Rehabilitation monograph 43, New York, 1970, Institute of Rehabilitation Medicine, New York University Medical Center.

CHAPTER 14

PRINCIPLES IN
MANAGEMENT OF
COMMUNICATION
IMPAIRMENT

Speech pathology programs in rehabilitation medicine settings generally differ from programs in university or private clinics, since the majority of patients who seek services in rehabilitation settings usually suffer from multiple disabilities. As a result, by far the greatest number of patients have verbal impairments secondary to multiple or complex conditions following cerebrovascular accidents or brain trauma, cerebral palsy, and other neurologic diseases. In many rehabilitation settings, oral deformities, stuttering, functional articulation defects, and voice problems are rarely encountered. Some programs provide complete audiologic service, whereas others only do routine audiometric screening essential to a speech evaluation, referring patients requiring more comprehensive evaluation to audiology centers. In the United States only a little better than 10% of all speech pathologists work in hospitals and rehabilitation centers. The remainder are employed by universities, clinics, and schools.

A *speech pathologist* is an individual trained to diagnose and treat speech disorders. Qualified speech pathologists have a minimum of a master's degree and hold the Certificate of Clinical Competence (CCC) from the American Speech and Hearing Association. Clinical service programs are accredited by the Professional Services Board (PSB) of the American Speech and Hearing Association. The Board has rigorously defined a set of standards for staff, physical facilities, ethics, and procedures for speech pathology and audiology programs. PSB registration should be viewed as an appropriate standard for rehabilitation medicine speech pathology services.

The term *speech therapy* is commonly used, particularly in rehabilitation medicine settings, to describe the treatment administered by a speech pathologist. When a speech pathologist is administering speech therapy, he is sometimes informally referred to as a speech therapist, although the official title is speech pathologist. Some use the term *language therapy* to refer to the type of treatment provided for patients with disorders of language.

Communication comprises all of the behaviors human beings use to transmit feelings and ideas, including gestures, pantomime, speaking, writing, and the processes of hearing, understanding, and reading that interpret visible and oral symbols. The modalities by which we express information are referred to as *expressive* or *encoding* processes and those used in the understanding and interpretation of symbols are *receptive* or *decoding* processes. These are high cognitive processes that take place in the cortex.

In order that the degree of language pathology manifested by a given patient may be identified, his performance must be measured against some standard of "normal." One may choose as the standard (1) the language common to the cultural community of unimpaired persons in which the patient lives, in which case the patient's verbal function is compared with that of others in the same community of similar age, sex, education, and achievement, or (2) the patient's own verbal behavior prior to the onset of his illness or trauma. The latter will vary from individual to individual and is based on premorbid educational achievement, specific cultural subcategory, personality, and other factors.

Accordingly, a patient is verbally impaired when he deviates in any parameter of language and/or speech processing from the "normal" communication behavior of his community or his premorbid life. If, for example, a patient was illiterate premorbidly and is unable to read and write after a cerebrovascular accident, he is not impaired, since literacy was never a part of his premorbid communication behavior.

We can think of the speech event as taking place on various levels: physiologic (neural and muscular activity), physical (generation and transmission of sound waves), and linguistic.

The descriptive natural science that deals with message transmission among speakers of any language is called linguistics. All languages differ in their linguistic characteristics, but they are composed essentially of three systems: a system of sounds (known as phonology-morphology), a lexical system (vocabulary), a system of grammar (also known as syntax), and the semantic or meaning system.

The specific processes of language organization, production, and perception are still poorly understood. It is, however, generally agreed that the elaboration of language occurs in only one hemisphere of the brain. This is designated the dominant hemisphere primarily because that is where the communication centers are located. In roughly 97% of the population the left is the dominant hemisphere, hence the close association of language pathology with right-sided motor and sensory impairments.

The smallest units of speech, called phonemes, generally divide into consonants and vowels and are the basic units from which words and sentences are constructed. Consonants combine to form syllables, which generally have a vowel as the central phoneme. The rules that determine how words are organized into sentences are

Table 14-1. Classification of English consonants by place and manner of articulation

Place of articulation	Manner of articulation				
	Plosive	Fricative	Semivowel	Liquids (including laterals)	Nasal
Labial	p b		w		m
Labiodental		f v			
Dental		θ th			
Alveolar	t d	s z	y	l r	n
Palatal		sh zh			
Velar	k g				ng
Glottal		h			

called the grammar or syntax. Word order is controlled by both grammar and the semantic value, or meaning, of words.

SPEECH PRODUCTION

Breathing is reflexive and functions differently at rest and during speech. The average resting cycle takes about 5 seconds. In speech the cycle is adjusted; inhalation is brief but adequate to accommodate the long expiratory phase while the person is talking.

During speech we change the size and shape of the vocal tract and the characteristics of the sound wave by (1) vibrating the vocal cords, thereby converting air into audible sound (phonation); (2) moving the tongue, lips, and pharyngeal wall; and (3) differentiating sounds by closing the nasopharynx with the soft palate to prevent air from escaping through the nasal cavity.

The size and shape of the vocal tract is probably most extensively changed by movements of the articulators, of which the tongue is the most flexible. Its tip, edges, and center can be moved independently by the intricate and highly specialized network of muscles that compose it. The lips are rounded and unrounded to change the shape and length of the vocal tract. These changes in the overall shape or dimensions of the oral cavity produce the various sounds known as vowels. On the other hand, consonant production is a result of interruption or modification of the airstream by the articulatory organs—the lips and tongue. (See Table 14-1.)

The process of consonant articulation cannot be ascribed solely to the action of the articulators. In addition to articulatory movements, phonemes have other characteristics that distinguish them from each other. Some sounds are voiced (combination of articulatory movement and vibration of the vocal cords); others are unvoiced, nasalized, or exploded. During speech, phonemes are rarely produced in isolation but in a continuum of infinite variation and they are affected by their "environment," that is, the sounds that precede and follow them.

How well the listener understands the end product of articulation during speech is called *intelligibility*. It depends not only on the precision of articulation but on the listener's interpretation of the loudness of the speaker's voice, his anticipation of the utterance, his familiarity with the subject matter, and even background noise. Other essential acoustic characteristics of speech include rate, phonation, resonance, pitch, rhythm, and stress patterns. All of these factors contribute to the overall quality of an individual's speech production and intelligibility.

This chapter concerns itself with those communication disorders most commonly evaluated and treated in rehabilitation centers: aphasia, dysarthria, and nonaphasic language disorders secondary to brain damage.

COMMUNICATION DISORDERS SECONDARY TO BRAIN DAMAGE

Any one or all of the linguistic systems (phonemic, lexical, syntactic, or semantic) may be impaired by damage to the cerebrum or brainstem. The most common communication disability secondary to brain damage is aphasia. By definition, aphasia is an acquired impairment of language that affects both oral and aural communication, that is, the sounds, vocabulary, or grammar, both in speaking (expression) and in understanding (reception). The term *aphasia* excludes those language disorders associated with visual or hearing deficits, generalized mental deterioration, or psychiatric aberrations. Many brain-damaged patients manifest reading and writing deficits who do not demonstrate impairment in the oral code. According to this definition, they would not be classified as aphasic.

In addition to impaired speech and comprehension, an aphasic patient may have difficulty in reading (dyslexia), writing (dysgraphia), and calculation (dyscalculia). As a matter of fact, any system of communication, whether it be shorthand, Braille, semaphoric code, or the finger spelling and sign systems of the deaf can be deficient when a person suffers damage to the dominant hemisphere. It is common for patients to have deficits of varying severity in several modalities.

Approximately one third of all stroke patients who are aphasic at the onset of the vascular accident remain so. This does not include a large number of stroke patients who are only aphasic for a few hours or days. Recent studies point up that spontaneous recovery is most dramatic during the first 4 weeks and continues for about 3 months. Many investigators have pointed out that once the condition is stabilized, few patients achieve language proficiency equivalent to the premorbid state. However, some continue to make minimal improvements with or without formal training, sometimes for years.

Types of aphasia may be distinguished from each other by their "fluency" characteristics. The *nonfluent aphasic* produces speech slowly, often with awkward articulation, overuse of high-frequency words, impaired syntax, and an absence of the small words (prepositions, articles, and conjunctions). The *fluent aphasic* produces speech effortlessly with normal intonation in well-articulated long phrases or sentences. Speech production is defective, however, since despite the ease with which words are produced, the content of speech is remarkably impoverished. The patient may substitute nonspecific words or words that are totally unrelated in meaning. This is called *verbal paraphasia*. Furthermore, patients with fluent aphasia frequently substitute one sound for another, which in some cases results in real words and at other times nonsense words. For example, a patient may say *tevelision* for *television* or *soap* for *hope*. Fluent patients often substitute one word for another. This symptom is called *literal paraphasia*.

Lesions in Broca's area produce nonfluent aphasia, whereas lesions in Wernicke's area result in a fluent aphasia. For this reason the respective aphasias are often referred to as Wernicke's and Broca's aphasia. An important characteristic of Wernick's aphasia is the disturbance in comprehending the spoken word. To a large extent this also extends to comprehension of written words. Wernicke's aphasia is only one of several types of fluent aphasia that can be distinguished from each other by the patient's ability to repeat and comprehend spoken language.

The most common type of aphasia is Broca's. Lesions in Broca's area, which is located in the posterior frontal region of the dominant hemisphere, often include the motor cortex or its radiations, hence producing an accompanying hemiplegia. In contrast, patients with Wernicke's aphasia, whose lesions are in the posterior distribution of the middle cerebral artery, are usually spared motor losses. When a patient manifests nonfluent aphasia with a severe loss in comprehension and repetition ability as well, the condition is referred to as *global aphasia.*

Most aphasic patients manifest writing deficits that mirror their speech production. For example, a patient with Broca's aphasia may have difficulty writing in complete sentences and may produce written material in a telegraphic style. A Wernicke aphasic may write with ease and use adequate syntax, but he may substitute circumlocutory phrases or words, low in informational content. Also such a patient might overuse many nonspecific or filler words.

In addition to the paraphasia, some aphasic patients sometimes produce jargon or perseverative, stereotyped phrases. Different types of jargon can sometimes be identified. Some may speak with an intonational pattern similar to normal speech with an occasional meaningful word interjected. Most jargon aphasics are unaware that they are not communicating to the listener.

In concluding this section, it should be added that some workers refer to a phenomenon called verbal apraxia in describing certain aphasic patients. These patients are usually Broca's aphasics and show great difficulty in articulation. At this stage in our knowledge of speech elaboration, it is perhaps wise to withhold judgment as to the nature of this articulatory problem and simply acknowledge that it is an occasional feature of Broca's aphasia.

Many of the words and phrases that normal speakers use are automatic (hello, goodbye). An aphasic patient may produce automatic words and phrases despite a total inability to speak voluntarily. Automatisms commonly heard are profanities, the words of a song, the recitation of serials (numbers, days of the week, months) and emotionally charged phrases, or even sentences. Characteristically these same words cannot be produced in a nonautomatic context. Verbal automatisms are analogous to the involuntary movements of dyskinesias or the reflex movements of a spastic limb.

EVALUATING LANGUAGE DISORDERS

Since language is characterized by extremely complex and highly integrative functions, the identification of symptoms, type, and severity of disorders should be carried out by persons with professional training in speech pathology. The physiatrist who functions without such a staff functions at a great disadvantage. Beyond speech pathology training, an understanding of the pathology of language and enlightened clinical judgment require intensive exposure to aphasic patients over a long period of time.

Aphasia

Although valuable clinical estimates of aphasic impairment can be made by an experienced clinician, there is no substitute for systematic and comprehensive formal testing. A typical pitfall is to attribute greater understanding to the patient than he has on the basis of cues supplied by the examiner. For example, following examination of an aphasic patient, a physician might report that he "understood everything" based on the patient's ability to follow all of the instructions given during the physical examination. If one had watched the interaction between patient and physician, it would have been clear that each of the physician's instructions was accompanied by a visual cue, usually a gesture, that helped the patient understand what was expected. When he asked the patient to walk, he motioned to him to get up and then actually got up himself.

Natural communication behavior in a verbal interaction includes a variety of visual, verbal, and contextual cues that all of us rely on for the comprehension of spoken language. Formal testing has the advantage that the stimuli are predetermined and have been tested out on patients, items are graded in difficulty, and in most tests the stimuli are designed to be as modality specific as possible. In other words, a valid test for the comprehension of spoken language is designed in such a way that speech skill and the ability to read or write will not affect the test results.

Testing the aphasic patient has many inherent problems. For one thing, since speech is an event taking place in time, which, unlike the written word, cannot be retrieved under ordinary conditions, it is thus easy to "forget" exactly how the patient expressed himself. Also, the average listener tends to "adjust" to defective speech, hence biasing his ability to evaluate critically what he hears. Intact socially appropriate affect and behavior can also prejudice one's judgment of speech behavior.

Over the years many tests have been developed and used in the measurement of aphasic deficits. Some are derived from a theoretic model of how language is organized.

In discussing the results of aphasia tests, one must be aware of a common pitfall. Many aphasia tests are tests of intelligence adapted for use with aphasic patients. Some are composed of parts of verbal intelligence tests but without the benefit of standardization on populations of aphasic patients. For this reason, among others, aphasic tests are not *diagnostic,* that is, they do not *identify* aphasic individuals. They may simply help to delineate and describe the areas and degrees of dysfunction in a given aphasic patient. A retardate, an illiterate, a schizophrenic, or even a nonnative speaker of English might fail items on an "aphasia test," but this does not identify the individual as aphasic.

The most commonly used aphasia tests are the *Minneapolis Test for Differential Diagnosis for Aphasia* (MTDDA), which was first published in 1957 by Hildred Schuell and is frequently referred to as the "Schuell Test"; the *Language Modalities Test for Aphasia* (LMTA), designed by Joseph Wepman and published in 1961; the *Token Test* (TT), a test of auditory comprehension that was first published in 1962 by DeRenzi and Vignolo; and the *Functional Communication Profile* (FCP), a quantitative rating scale designed to measure functional language use. More recently the *Neurosensory Center Comprehensive Examination for Aphasia* (NCCEA) by Benton and Spreen and the *Boston Aphasia Test* by Goodglass and Kaplan have been made available.

Dysarthria

The term *dysarthria* refers to motor speech defects that result from trauma or disease of the nuclei or fiber tracts in and adjacent to the brainstem that subserve the speech musculature. Such pathology may affect any of the so-called acoustic characteristics of speech—articulation, loudness, rate, phonation, resonance, pitch, rhythm, and stress patterns. In the majority of dysarthric speakers, articulation is the primary speech defect, usually because of weakness or incoordination of the speech musculature. However, since intelligibility is the end product of all the acoustic characteristics, any one of them may contribute to dysarthria. For example, if a patient has a paralyzed velum and air escapes through the nasal passage, those sounds that normally require velopharyngeal closure will sound abnormal: *bowl* sounds like *mowl; he's a big boy* becomes *ees a ming moy.*

Dysarthric speech may sound slurred, muddy, or halting; words may be mispronounced and indistinct. Patients usually use complete sentences or phrases except when the disorder is severe, in which case they may resort to short phrases. Vocabulary and grammar are normal. The most severe dysarthric patients may be totally speechless (anarthria) and may easily be confused with aphasic patients. One can expect some of these patients to have difficulty swallowing and sometimes chewing.

For many patients with dysarthria, articulation is optimal when speech is produced slowly. This is particularly true when certain combinations of consonants are problematic (clusters of consonant sounds like *pl* in *place, please; str* in *street, strike; br* in *bring, break*). The patient with articulatory difficulty will often substitute for, or omit, those sounds that he cannot produce easily (for example, *bing* instead of *bring*).

Disorders of rhythm and stress are peculiar to cerebellar lesions. Pitch is monotonous and each syllable receives equal stress.

Palatal myoclonus is a rare symptom of dysarthria characterized by a rhythmic variation in the sound stream produced by regular, involuntary contractions of the soft palate. It has been described as "nystagmus of the soft palate."

With the syndromes of aphasia and dysarthria described, it can be seen that aphasia may be characterized by aberrations in all four linguistic systems—the system of phonemes (sounds), the lexical system (vocabulary), the syntactic system (grammar), and the semantic system. Dysarthria, on the other hand, is a disorder of the sound system only.

Since dysarthria by definition refers to motor impairment reflected in the acoustic aspects of speech, the symptoms can be heard. Symptoms will become apparent when the examiner listens carefully to a patient speak in extended production. Since what we hear in speech is much more than the pronunciation of consonant clusters, a test such as having the patient repeat "methodist episcopal" is not adequate for an evaluation of speech production. In the course of listening, the patient's loudness, rate, pitch, rhythm, and particularly the intelligibility of his speech can be assessed. If the patient is asked to speak more rapidly, dysarthric symptoms will usually become more apparent.

Three types of dysarthria are common in the brain-damaged patient: bulbar palsy, pseudobulbar palsy, and cerebellar dysarthria. Characteristically, bulbar palsy manifests a flaccid type dysarthria, with hypernasality and misarticulation. There is also velopharyngeal incompetence with breathiness. Dysarthric patients in this category are often totally unintelligible. On the other hand, the pseudobulbar palsy group, called spastic dysarthrics, are imprecise in articulation, with monotonous pitch and monotonous loudness, and have reduced control over rhythm, but intelligibility is much less impaired. These patients remain intelligible despite having all other acoustic features impaired. The cerebellar speaker (ataxic speech) produces errors primarily related to the timing of speech. Rhythm is usually uncontrolled and the patient gives equal stress to each syllable. Intelligibility in this type of patient is rarely severely impaired.

The speech pathologist will want to assess dysarthric conditions using a variety of tests. In general, these are (1) an examination of isolated voluntary movements of the speech muscles, (2) analysis of acoustic characteristics, (3) articulation tests, and (4) ratings of intelligibility.

The performance of isolated movements (protruding the tongue, elevating the tongue tip), although useful for physical and neurologic examinations, may give very little useful information to the speech specialist, since these movements are often performed perfectly even though the movement cannot be combined into the smooth, precise, and rapid transitions required for normal speech. In severe cases one may see gross disorders of movement or incoordination.

Nonaphasic language disorders

The nonaphasic group of patients differs from dysarthric and aphasic patients in important ways: (1) they show no evidence of difficulty with sound production; (2) they have no problems in understanding speech; (3) speech is fluent and they use complete sentences; and (4) vocabulary may be impoverished, but there are no apparent phonemic or word-finding difficulties and no mixing of words or word parts. These patients are often confused and disoriented and have other symptoms of cognitive deficit. Eye contact may be poor; ability to attend is impaired.

The point of difference is that the structure and use of sounds, words, and syntax are not abnormal in these patients, although the content and manner of speaking may strike the examiner as being abnormal. The same problem often arises with psychiatric patients. The only way to be sure of the diagnosis is to rule out specific aphasic or dysarthric deficits and at the same time to identify intellectual impairment and personality as well as emotional functions by psychologic testing.

In stroke patients the nonaphasic disorders are the most common, aphasia is second, and dysarthria is a distant third. After trauma or disease, nonaphasic disorders are again most common, but dysarthria occurs more frequently than in stroke, most often as a result of brainstem damage.

TREATMENT OF THE DYSARTHRIC PATIENT

The retraining procedures appropriate for dysarthric patients will depend, naturally, on the areas of motor speech that have been impaired. In most cases, where more than one aspect of speech is affected, therapeutic efforts will focus on improving articulatory skill. This approach is based on the rationale that speech intelligibility is especially related to the accuracy of articulation. A tape recording of the patient's speech production serves as a permanent document for use in analyzing the acoustic elements of the dysarthric symptoms. The recording also serves as a baseline against which to measure progress.

Tape recorders and mirrors are used in the treatment of dysarthria so that the patient may constantly assess his own performance by listening to it on a recording or by comparing his oral movements to the therapist's in a mirror when indicated. Resistance exercises to increase muscle strength are prescribed for the patient exhibiting a weakness of a part of the speech apparatus. Exercises designed to improve rapid, alternating movements of the speech musculature are indicated for patients who show incoordination.

Many exercises can be assigned to the patient for supplementary home practice. In general, when sounds are taught, the order of presentation proceeds according to difficulty—first the bilabial and vowel sounds and then the more difficult *s, r, k,* and *g* sounds. Clusters of consonants in words and sentences are presented after the patient has mastered the individual phonemes.

TREATMENT OF THE APHASIC PATIENT

The goal of treatment for each aphasic patient will, naturally, depend on both the degree of impairment and the kind and amount of remaining language skill. Generally, it is best to set up weekly goals, with a long-term goal in mind. A long-term goal for language recovery from aphasia can never realistically be "normal" communication function. Only a small number of aphasic patients whose language residuals persist a few months beyond onset ever reach truly "normal" language proficiency, with or without retraining.

Many aphasic patients appear to improve in the course of speech therapy but do not carry over into everyday speaking situations. The ultimate goal of any language retraining program for aphasic patients should be based on a functional frame of reference; that is, the direction of treatment must be geared toward ultimately meeting the patient's daily language needs. This perspective will play a major role in determining the choice of content for language teaching. Progress in treatment must be gauged according to "functional" gains. Therefore, the *Functional Communication Profile* referred to previously can be administered periodically to measure the extent of functional improvement.

Speech therapy technics are put to a great test in the retraining of aphasic patients, since treatment of this disability—perhaps more than most others—requires a broad background of information in related fields, particularly in the areas of linguistics and behavioral psychology. Disciplined and meticulous attention to the specific modality, technic, and content of each training task in the language-training process is essential to a systematic approach.

The order for teaching the basic language elements to a severely impaired aphasic patient generally proceeds according to the following sequence of learning tasks: imitation of gross body movements, imitation of facial movements, imitation of a small repertoire of phonemes (sounds), matching identical objects, matching identical pictures, matching identical single-noun flashcards, matching a picture with its corresponding object, matching a picture with its corresponding flashcard, tracing the letters of a single-noun flashcard, copying the letters of a single-noun flashcard, listening to the oral production of a word while looking at the word in print, and attempting to imitate the oral production of the word on the flashcard after the therapist's demonstration.

After the patient has acquired a small vocabulary of nouns in this fashion, the introduction of a sentence structure in its written as well as its oral form may be made. One structure is taught at a time, through repetition, changing only one element. Many other techniques such as conversational practice and grammar exercises are used by the aphasia therapist once the patient has recovered some vocabulary. It its not within the scope of this chapter to review in depth the many methods that have been used with aphasic patients.

In any behavior as varied and complex as language there are, of course, a large number of elements that contribute toward making the task of aphasia rehabilitation one of extraordinary scope. No mention has been made here of the spectrum of psychologic and social factors in aphasia that severely affect the course of the patient's recovery and his total management. The patient's premorbid personality, education, verbal dependency, and achievement and his role in the family, motivation, and perseverance, as well as the extent of verbal impairment, naturally influence the outcome of the rehabilitation process.

It is indeed understandable that in a society in which verbal skill is a principal measure of intelligence, the general population often considers the aphasic patient to be mentally incompetent. The complex and misunderstood nature of aphasia arouses great feelings of fear on the part of the patient's family as to whether he has "lost his mind." While certain verbal skills are indeed diminished or virtually absent in many aphasic patients, it is also true that many patients, in the face of extraordinary verbal impairment, are still able to play a better than average game of bridge or chess, and remain fully aware of the many subtleties of activity taking place in their environments.

In interpreting the diverse opinions on the value of treatment that abound in the literature, the reader must bear in mind that the theories are derived from multiple disciplines and biases, not all of which are dedicated to the search for a method of treatment.

It is not surprising to anyone experienced in the study of aphasia that the outcome of language rehabilitation in patients who are neurologically stabilized is poor. Our understanding of how language is organized in the brain is primitive. We do not yet know if temporal factors play a role in language processing, to what degree inhibition and voluntary control affect aphasic behavior, and what psychologic variables contribute to the recovery process.

It is clear from the literature and from what we can observe clinically that the families of aphasic patients play a critical role in the patients' rehabilitation. This is, we believe, a result of the fact that much of what is called "successful" aphasia rehabilitation has little to do with actual improvement in communication function but is dependent on the patient's overall adjustment to life. An adequate and appropriate "speech therapy" regimen, whether based on direct or indirect methods, may indeed be a process of assisting the patient in his struggle to live with a devastating disability.

Those who have seen many aphasic patients are struck by the fact that language does not seem to be "lost" but rather that the aphasic patient is hampered in the activating and volitional processing of the elements of language and the processing which permits their use in spoken language—input and output. This fact is at once encouraging, for the patient appears to "have" language, but at the same time can be "trained" to use his stored but seemingly irretrievable vocabulary and syntax. If it were simply a matter of "other parts of the brain taking over" as some suggest, then aphasia reeducation would have had a better record of success than it has had. Similarly, it would seem that if motivation were a significant factor in recovery, we would expect a much greater number of patients to have overcome their communication deficits.

The evaluation of the result of speech therapy is fraught with difficulties that will stand in the way of our finding answers for a long time to come. For one thing, the careful selection of patients in such studies to ensure that they are indeed aphasic and that patients with other types of communication deficit secondary to brain damage are excluded is extremely difficult to establish on the basis of existing tests of aphasia. Some of this stems from lack of uniformity in test methods and disagreement over terminology. But much appears simply to be a lack of appreciation of the variety and degree of the elements that comprise the aphasia syndrome. Furthermore, comparing the effects of treatment administered to aphasic patients with untreated control patients is practically impossible except in certain unique circumstances.

The virtue of speech therapy for aphasia as a psychotherapeutic tool cannot be overemphasized. Overwhelming feelings of loss, lowered self-esteem, and fear are rampant in aphasic patients. The speech pathologist, however, must guard against the destructive effects of focusing too long on ultimate speech recovery without helping the patient to confront the fact that he may never use normal language again and that his adjustment to his residual deficits is essential to his well-being.

REFERENCES

1. Aronson, M., Shatin, L., and Cook, J. C.: Sociopsychotherapeutic approach to the treatment of aphasic patients, J. Speech Hearing Dis. 21:352, Sept., 1956.
2. Benton, A. L.: The fiction of the "Gerstmann syndrome," J. Neurol. Neurosurg. Psychiat. 24:176, May, 1961.
3. Benton, A. L.: Problems of test construction in the field of aphasia, Cortex 3:32, 1967.
4. Canter, G. J.: Neuromotor pathologies of speech, Amer. J. Phys. Med. 46:659, 1967.
5. Culton, G. L.: Spontaneous recovery from aphasia, J. Speech Hearing Res. 12:825, 1969.
6. Darley, F. L., Aronson, A. E., and Brown, J. R.: Clusters in deviant speech dimensions in the dysarthrias, J. Speech Hearing Res. 12:462, 1969.
7. Darley, F. L., Aronson, A. E., and Brown, J. R.: Differential diagnostic patterns of dysarthria, J. Speech Hearing Res. 12:246, 1969.
8. DeRenzi, E., Pieczuro, A., and Vignolo, L. A.: Oral apraxia and aphasia, Cortex 2:50, 1966.
9. DeRenzi, E., and Vignolo, L. A.: The token test: a sensitive test to detect receptive disturbances in aphasics, Brain 85:665, 1962.
10. DeReuck, A. V. S., and O'Connor, M., editors: Disorders of language, Boston, 1964, Little, Brown & Co.

11. Eisenson, J.: Aphasia in adults. In Travis, L. D., editor: Handbook of speech pathology, New York, 1957, Appleton-Century-Crofts, Inc.

12. Gazzaniga, M. S., and Sperry, R. W.: Language after section of the cerebral commissures, Brain 90:131, 1967.

12a. Geschwind, N.: The organization of language and the brain, Science 170:940, 1971.

12b. Geschwind, N.: Aphasia, New Eng. J. Med. 284:654, 1971.

13. Godfrey, C. M., and Douglass, E.: The recovery process in aphasia, Canad. Med. Ass. J. 80:618, 1959.

14. Goldstein, K.: Language and language disturbances, New York, 1948, Grune & Stratton, Inc.

15. Goodglass, H., and Berko, J.: Agrammatism and inflectional morphology in English, J. Speech Hearing Res. 3:257, 1960.

16. Goodglass, H., and Kaplan, E.: Disturbance of gesture and pantomime in aphasia, Brain 86:703, 1963.

16a. Goodglass, H., and Klein, B.: Specific semantic word categories in aphasia, Cortex 2:74, 1966.

16b. Goodglass, H., and Quadfasel, F. A.: Phrase length and the type and severity of aphasia, Cortex 1:133, 1964.

17. Grewel, F.: Acalculia, Brain 75:397, 1952.

18. Head, H.: Aphasia and kindred disorders of speech, New York, 1963, Hafner Publishing Co., Inc.

19. Lambert, W. E., and Fillenbaum, S.: A pilot study of aphasia among bilinguals, Canad. J. Psychol. 13:28, 1959.

20. Lenneberg, E.: Biological foundations of language, New York, 1967, John Wiley & Sons, Inc.

21. Marks, M. M., Taylor, M. L., and Rusk, H. A.: Rehabilitation of the aphasic patient: a survey of three years' experience in a rehabilitation setting, Neurology 7:837, 1957.

22. Millikan, C. H., and Darley, F. L., editors: Brain mechanisms underlying speech and language, New York, 1967, Grune & Stratton, Inc.

23. Osgood, C. E., and Miron, M. S.: Approaches to the study of aphasia, Urbana, Ill., 1963, University of Illinois Press.

24. St. Onge, K. R., and Calvert, J. J.: The brainstem damage syndrome: speech and psychological factors, J. Speech Hearing Dis. 24:43, Feb., 1959.

25. Sands, E. S., Sarno, M. T., and Shankweiler, D. P.: Long-term assessment of language function in aphasia due to stroke, Arch. Phys. Med. 50:202, 1969.

26. Sarno, J., and Sarno, M.: Stroke: the condition and the patient, New York, 1969, McGraw-Hill Book Co.

26a. Sarno, J. E., Sarno, M. T., and Levita, E.: Evaluating language improvement after completed stroke, Arch. Phys. Med. 52: 73, 1971.

27. Sarno, J. E., Swisher, L. P., and Sarno, M. T.: Aphasia in a congenitally deaf man, Cortex 5:398, 1969.

28. Sarno, M. T.: Aphasia: selected readings, New York, Appleton-Century-Crofts (in press).

29. Sarno, M. T.: Method for multivariant analysis of aphasia based on studies of 235 patients in a rehabilitation setting, Arch. Phys. Med. 49:210, 1968.

30. Sarno, M. T., and Sands, E.: An objective method for the evaluation of speech therapy in aphasia, Arch. Phys. Med. 51:49, 1970.

31. Sarno, M. T., Silverman, M. G., and Sands, E. S.: Speech therapy and recovery from severe aphasia: a controlled study, J. Speech Hearing Res. 13: 607, Sept., 1970.

32. Schuell, H., Carroll, V., and Street, B. S.: Clinical treatment of aphasia, J. Speech Hearing Dis. 20:43, March, 1955.

33. Schuell, H., Jenkins, J. J., and Jimenez-Pabon, E.: Aphasia in adults, New York, 1964, Harper & Row, Publishers.

34. Shankweiler, D. P., and Harris, K. S.: An experimental approach to the problem of articulation in aphasia, Cortex 2:277, 1966.

35. Shankweiler, D., Harris, K. S., and Taylor, M. L.: Electromyographic studies of articulation in aphasia, Arch. Phys. Med. 49:1, 1968.
36. Spreen, O.: Psycholinguistic aspects of aphasia, J. Speech Hearing Res. 11:467, 1968.
37. Taylor, M. L.: A measurement of functional communication in aphasia, Arch. Phys. Med. 46:101, 1965.
38. Taylor, M. L.: Understanding aphasia: a guide for families and friends, Patient Pub. no. 2, New York, 1958, Publications Unit, Institute of Rehabilitation Medicine, New York University Medical Center.
39. Taylor, M. L., and Marks, M.: Aphasia rehabilitation manual and therapy kit, New York, 1959, McGraw-Hill Book Co., Inc.
40. Vignolo, L. A.: Evolution of aphasia and language rehabilitation: a retrospective exploratory study, Cortex 1:344, 1964.
41. Weinstein, S.: Deficits concomitant with aphasia or lesions of either cerebral hemisphere, Cortex 1:154, 1964.
42. Wepman, J. M.: A conceptual model for the processes involved in recovery from aphasia, J. Speech Hearing Dis. 18:4, March, 1953.
43. Wepman, J. M.: The relationship between self-correction and recovery from aphasia, J. Speech Hearing Dis. 23:302, Aug., 1958.
44. West, R., and Stockel, S. R.: The effect of meprobamate on recovery from aphasia, J. Speech Hearing Res. 8:57, 1965.

CHAPTER 15

PRINCIPLES IN MANAGEMENT OF PSYCHIATRIC PROBLEMS

It has long been known that physical illness and disability produce emotional and mental changes. Similarly, from the very beginnings of rehabilitation as a serious endeavor in medicine, it has been recognized that emotional and mental attitudes can have a decisive effect on the outcome of the rehabilitation process. In about 50% of adults with physical disability, such factors determine the success or failure of rehabilitation. In children the figures run considerably higher, approximating 75%.

There is a body, a personality, and an external world, animate and inanimate. To fully understand a person, one must evaluate his functioning in each of these three areas as well as the relationships between them. When a person suffers a disability, whether as a result of injury or of disease, each of these areas of functioning is affected. His use of the body as a tool to experience and manipulate the environment can be impaired. His very sensation and perception of the external world may be disturbed because of blindness, deafness, sensory damage, or brain injury, so that his position in space may be disordered. His motor system may be affected so that locomotion and essential movements are disrupted. His body image may be distorted—an image that we all possess, which is ordinarily outside of awareness and which has become endowed with qualities of value or lack of value as a result of developmental and social experiences. Physical injury and deformity may threaten this image, necessitating a change or a defense against recognition of this change. Since the body image is integrated with personality organization, such a change threatens the equilibrium of the personality.

251

Personality can be defined as the sum total of a person's ideas, emotions, and behavior—rational and irrational, conscious and unconscious, defensive and learned. This personality has developed as a result of genetic and environmental factors representing life experiences. Physical disability constitutes a threat to a way of life and tends to disrupt the balance that this way of life represents. It may cause intense anxiety, depression, and/or rage. It may be interpreted as punishment for sins, real or imagined. It may represent a threat to omnipotent strivings or to normal mastery and may produce feelings of helplessness and panic. It may unloose previously controlled psychopathology such as paranoid ideas and create intolerable interpersonal relationships. On the other hand, the disability may be organized into neurotic patterns, such as dependency and fear of competition, and be unconsciously welcomed as a way out of a conflictive struggle.

Finally, the disability removes the individual from normal social experiences and from work situations—the two major sources of satisfaction and self-esteem. Disruption in family life and friendships, separation from loved ones, economic problems, shattered ambitions and dreams—all of these lead to serious threat and damage to the socially functioning human being.

The disability, then, represents a massive assault on all three of the major areas of functioning. How the individual reacts will depend on a vector resulting from the nature of the disability, the realistic problems it creates, the personality of the individual, his previous history and life experiences, the meaning of the disability, both conscious and unconscious, and the resources provided by the individual, his family, and society. It is toward an understanding of these factors and their manifestations that psychiatric evaluation in rehabilitation sets its goal.

PSYCHIATRIC EVALUATION

An adequate evaluation should provide information on the individual's intellectual status, personality, past personal history, motivation, and diagnosis.

Intellectual status. Since rehabilitation involves communication and learning, it is essential that the patient possess adequate intellectual apparatus, including learning ability, retention, memory, orientation, etc. This is of particular importance in the brain-damaged individual and influences not only the results of rehabilitation but also the organization and tempo of the procedures.

Personality. As has been indicated, physical disability may produce changes in personality, and, in turn, personality factors may influence the success or failure of rehabilitation. Those aspects of personality most pertinent to rehabilitation include the following:

1. *Ideational patterns*—disorders in perception, such as illusions and hallucinations, and disorders in thinking, such as delusions, obsessions, phobias, hypochondriasis, and thoughts of persecution, suicide, etc.
2. *Emotional patterns*—disorders in affect, such as depression, tension, anxiety, panic, mania, or inappropriate mood.
3. *Behavioral patterns*—cooperation, ingratiation, aggression, inhibition, etc.
4. *Characterologic patterns*—dependence, exaggerated independence, masochism, withdrawal, interpersonal difficulties, frustration tolerance, etc.

Past personal history. Much of the knowledge of the patient's personality is obtained through information about his infancy, childhood, and adolescence, his school, marital, and sexual adjustments, and his work history, as well as hobbies, religion, interests, etc. These factors supply the context for understanding the meaning of the disability to the patient.

Motivation. By asking a patient what his goals are, we usually can detect whether his expectations are consistent with those of the rehabilitation team. However, the private meanings behind the verbalized goals cannot be assessed so easily and require a sensitivity to nonverbalized cues. Actually, to fully understand a patient's motivation, we must know his total personality and situation, including both the conscious and unconscious meaning of the disability to him.

Diagnosis. Here we are concerned with whether the patient's symptoms are (1) reactive (and within normal limits), (2) neurotic, psychotic, or psychopathic, or (3) organic or toxic in origin. The psychiatric diagnosis in itself does not determine feasibility for rehabilitation or prognosis, although it may definitely influence the overall management.

In most cases this information can be obtained by means of a thorough medical history and examination, provided the examiner is aware of what he is looking for. However, there are cases in which the psychologic difficulties are not obvious or overt and in which the defenses blur the problems. Under these circumstances a more thorough investigation involving the use of a psychiatric history and examination, as well as the assistance of a variety of psychologic tests, is necessary.

PROBLEMS REQUIRING REFERRAL

Whether treatment occurs in a specialized setting such as a rehabilitation center, in a general hospital and outpatient department, or within the framework of individual practice, problems may arise for which psychiatric consultation is either desirable or necessary. Whether a referral is to be made, and when, depends on two factors: (1) the nature of the problem and (2) the nature of the professionally responsible individual—his knowledge, training, judgment, and sensitivity in psychiatric matters, as well as his honesty and integrity.

This is not the place to discuss the question of psychotherapy or evaluation by professionals other than psychiatrists. It is a complex problem but one to which serious consideration has been devoted by both medical and psychologic organizations and individuals. A detailed exposition of these questions can be found in Wolberg's test,[4] and the more specific role of the psychiatrist and the psychiatric team in rehabilitation can be found in the works of Grayson and co-workers[2] and Fisher.[11]

In general, patients will be considered for referral to the psychiatrist for the following reasons: (1) behavioral disturbances due to organic brain damage; (2) overt psychopathology in the form of neurotic, psychotic, or psychopathic behavior—includes addiction, alcoholism, and sexual disturbances; (3) emotional disturbances such as severe anxiety, depression, or hostility; (4) suicidal ideas or acts; (5) physical symptoms and signs that are bizarre and do not conform to a known organic pattern or are exaggerated beyond what can be explained organically; and (6) lack of improvement in rehabilitation without obvious physical or psychologic reason.

SPECIFIC PROBLEMS AND THEIR MANAGEMENT
Denial of disability

Of all the reactions to physical disability, that of denial is the most interesting and challenging. First described by von Monakow in 1885 in a hemiplegic patient, the phenomenon was observed in a blind person in 1896 by Anton and became known as Anton's syndrome. In 1914 Babinski applied the term anosognosia to those patients with left hemiplegia who completely denied or were unaware of their paralysis. For many years most observers believed that this phenomenon occurred only in brain-damaged patients and even went so far as to localize the lesion in the parietal cortex. However, recent workers have disputed this localization and have described denial in patients with all types of disordered brain function. In addition, denial has been described in disabled patients without brain damage, such as those with paraplegia and those who have had an amputation. Statistics on incidence of denial among hemiplegic patients have been presented by Nathanson and others[12] who found verbal anosognosia for the paralysis in 28 of 100 hemiplegic patients admitted consecutively to a general hospital.

Weinstein and Kahn,[1] in a study of 104 brain-damaged patients, classified denial according to explicit expression in verbal terms and implicit expression in terms of feeling or behavior. The mood of patients with explicit verbal denial is usually bland and affable during interviews, and they have been described as showing "fatuous equanimity." They may appear depressed and perplexed outside of interviews. Anxiety in the ordinary clinical sense is not present. Euphoric and hypomanic or paranoid and depressed attitudes are more prominent in the group showing implicit denial. The mood may change many times in the course of the day.

Personality studies of these patients revealed that all who presented explicit verbal denial had previously shown a marked tendency to deny the existence of illness. They appeared to have regarded ill-health as an imperfection, weakness, or disgrace. Illness seemed to have meant a loss of esteem and adequacy. The maintenance of health seemed to be a kind of moral or ethical duty, and illness represented not only unhappiness and dangers, but also a sin.

The cases described by Weinstein and Kahn represent the more seriously disturbed brain-damaged patients and are not typical of the patients seen on a rehabilitation service. They are also more characteristic of patients in the acute and early stages of illness. It is far more common to see patients with implicit denial of a mild form, in whom there may be blandness or mild euphoria, inattention, and unrealistic goals.

Denial by patients without brain injury is a nonrealistic attempt by a disabled person to maintain preinjury status. Why do some people evidence denial while others do not? Although the answer to this baffling question is at present incomplete, we can offer a number of plausible reasons.

1. Denial permits a person to carry out many of his activities without feeling that he is disabled. The disabled person can act as if he is not injured.

2. Denial can serve as an excuse for refusing help. The disabled person under such circumstances feels that disability is equated with inferiority. Because he feels inferior, he sees the donor of help as superior. He, therefore, responds with anger at the donor for making him feel inferior.

3. Denial represents an end product of an emotional way of thinking that distorts the means and ends of the situation. Such thinking is quite common in children and in adults who are under stress so that their thinking becomes primitive. The two thoughts, "I wish my injury could be forgotten" and "it can be done by actually forgetting it," become merged so that the individual mistakes the thought for the reality.

Individuals who use denial are under a constant state of tension, for they are always in danger of being contradicted by reality.

Denial, therefore, represents an important defense of the ego against overwhelming emotion and catastrophe and a means of maintaining the integrity of the personality. Since it may, at the same time, seriously interfere with rehabilitation, the management of such patients often poses contradictory and conflicting problems.

Case study. A 41-year-old white married female was admitted to the Bellevue Physical Medicine and Rehabilitation Service with a left-sided hemiplegia. She had had a routine appendectomy performed at another hospital 4 weeks previously. After an uneventful recovery in the hospital, she was discharged home. Within 24 hours she experienced a bout of dizziness and blacked out and was taken back to the hospital, where hemiplegia was diagnosed. Eventually it was proved to have been caused by carotid artery thrombosis. For 3 weeks she was observed to be emotionally labile, with wide mood swings ranging from euphoria to depression.

When she was seen in the rehabilitation evaluation conference, she seemed quite cheerful, presented minimal evidence of organic brain deficit, and appeared to be well motivated. Soon after admission to the ward she was started on rehabilitation procedures, including ambulation. However, despite her apparent cooperation, she made no progress; since it was the opinion of the staff that she possessed the physical resources to walk, the psychiatric service was asked to see her.

On examination the patient was in a severe depression. She expressed feelings of hopelessness and futility and claimed that none of the patients were being helped and that what had happened to her was too much for any human being to bear. Her history was both interesting and vital to an understanding of this patient. She had been born in Holland, where she had been a teacher of physical education. With the outbreak of World War II, the patient, being of Jewish origin, escaped the Nazis to the Dutch East Indies, where she continued her work as a teacher of physical education. When the Japanese invaded the territory, she was placed in a concentration camp, where for nearly 3 years she suffered intense privation and indignities. Following the liberation, she came to the United States and obtained a position as secretary to a physician in New York City. She was very efficient, successful, and well liked. She married one of her compatriots, who also had migrated to this country, and they had one child. The appendectomy and carotid artery thrombosis followed. At the time of the first psychiatric examination she was wheelchair bound.

The acute depression was precipitated when she first entered the ambulation room. There, confronted with a full-length mirror, she saw herself for the first time as deformed and crippled. The previous cheerfulness and euphoria were the result of denial, a defense that was unwittingly and abruptly pierced. We saw her daily and she spoke of her past and the tremendous trials she had experienced and said that this accident was the "final straw." We learned something of the meaning of her disability. There were feelings of helplessness, of having been overwhelmed by an act of fate. As she stated it, "After all I've been through, to have this happen is too much to bear." The importance of her body was evidenced by her previous interest in physical education, with its emphasis on perfection. She also expressed many doubts about ever being attractive as a woman, and she disconcerted the staff, both male and female, by talking about her sexual exploits and fantasies. The compensatory and reparative nature of these productions was clearly evident and in time disappeared.

Her treatment was one of support and ventilation. Much attention was given her by the staff. A beautician was engaged to do her hair, and she responded appropriately. Her ego resources, demonstrated adequately in her past history and confirmed by psychologic tests, came into action coincidentally with the diminution in her anxiety and depression; the need for denial was no longer pressing and she proceeded to make a rapid recovery. She learned how to walk, accepted braces and crutches, and was walking when she left the hospital.

This patient illustrates the characteristics of implicit denial, its defensive character against emotional turmoil, and an eventual rehabilitation after the emotional reaction was controlled. It emphasizes the importance of psychiatric evaluation, not only as a therapeutic aid but also as a possible preventive.

Management. In cases of complete denial, attempts at suggestion or logical argument are futile. The anosognosia itself represents a kind of psychosis, so gross is the distortion of reality. Rather than attempt to deal directly with the denial, one should treat the patient himself and his disability. Every effort should be made to engage the patient in rehabilitation procedures, which are to be introduced gently and over a long period of time. Time and milieu will be effective in most cases.

When the denial is implicit and there is considerable emotional lability, tranquilizers may be used. Whether or not tranquilizers are employed, psychotherapy is utilized. This is of a supportive nature, in which the patient can express his feelings and can experience the support of at least one interested person, plus the understanding and support of the entire team.

The brain-damaged patient

On a rehabilitation medicine service, brain-damaged patients constitute a large percentage of the total patient population. The two types most frequently seen are patients with brain damage due to cerebrovascular disease and cerebral palsy, but there are also patients in whom the damage is the result of other causes, such as head trauma, postoperative brain tumors, multiple sclerosis. These patients often present special problems, not only in presenting symptoms but also in rehabilitation management.

In a sense, the term brain-damaged patient is a misnomer, because in behavior and mental life brain-damaged persons may differ as much from each other as they do from persons without brain damage. The mere fact of brain injury is not sufficient to determine how a person thinks or how he experiences the world. Although some typical patterns of behavior have been reported for persons with brain injuries, these patterns differ widely and usually reflect group trends, which omit the many negative instances of patients who do not show the same types of reaction.

In a situation in which many different circumstances and determinants have to be taken into account to understand what is happening, it is difficult to generalize or reduce all the facts to a neat set of principles. However, from a psychiatric standpoint, the management of the brain-damaged person on a rehabilitation program requires inquiry into three overlapping areas: (1) nature of the brain injury, (2) mental makeup and behavior of the person, and (3) services that a rehabilitation medicine program can render and the social environment that the individual must face.

Concerning the brain injury itself, the following factors have been noted:

1. Age at which the injury occurred. If the injury occurs early in life, as in cerebral palsy, habit patterns to cope with the defects of the disease are developed that are difficult to break.

2. Site of the lesion. For example, temporal lobe lesions tend to lead to more disruption of function than do lesions elsewhere in the brain.

3. Size of the lesion. Often this is misleading; we have seen patients with hemispherectomy who showed no striking impairment of rehabilitation ability.

4. Nature of damage. A clean gunshot wound may leave virtually no sequelae, whereas a cerebrovascular accident may produce gross changes in behavior.

5. Length of time since damage occurred. Many World War II brain-injured patients were able to function at their preinjury level when reexamined 10 years later, despite extensive brain wounds.

The knowledge provided by these facts alone may be insufficient to plan a treatment program. For this reason information about the patient's premorbid life history, particularly with regard to his psychologic status, is of crucial importance. For example, the successful hard-driving businessman who took great pride in his independence may be depressed, not as a direct consequence of brain damage incurred in a cerebrovascular accident but by his loss of independence.

Intimate knowledge of the rehabilitation program, including the skills that can be taught, the personnel involved in the teaching, and the other patients on the program, is necessary. In addition, knowledge of the meaningful environmental alternatives that confront the patient on his discharge is important.

As indicated previously, the behavioral reactions to brain injury are varied. From the point of view of the rehabilitation program, the patients may be divided into two groups: (1) those who show limited, specific, or no deficits following brain damage ("nonorganic" mental syndrome) and (2) those who demonstrate profound global impairment following brain damage ("organic" mental syndrome).

"Nonorganic" mental syndrome. This group contains those with limited impairment of sensory or motor modalities and may include those with aphasia but with the structure of their mental apparatus remaining essentially intact. Patients in this group may be depressed, may show exaggerated dependency reactions, or may show any of the gamut of behavioral disturbances. However, when these disturbances occur, they generally are not the result of the brain damage per se but are reactions to internal and external stresses.

"Organic" mental syndrome. This group of brain-damaged patients generally demonstrates profound behavioral disturbance, anxiety, and disorders in mentation. These patients show evidence of an organic mental syndrome characterized by (1) perceptual insecurity and distortions, (2) thinking disturbances, (3) regression, and (4) disorders of a body image.

Perceptual insecurity. This refers to the profound difficulties that some brain-damaged patients have in apprehending and integrating their experience. Although perceptual insecurity might be associated with or due to sensory deficits, the two are not identical. At times the perceptual difficulty may be manifested in such diverse phenomena as visual inattention or fluctuations in memory, orientation, and concentration. Perceptual anomalies are reported, for example, displacement and projections of sensations, reversals of commonly perceived shapes, and denial of experiences.

Thinking disturbances. These refer to the lost powers that some brain-damaged persons have in categorizing their experiences properly. They no longer can grasp essential similarities in their surroundings. When this occurs, we say the person cannot think abstractly but reasons in a concrete way. Logical relationships are either difficult to establish or are based on irrelevant aspects of stimuli. For example, an object may be identified by its use and not by its more general connotation that transcends its specific uses, so that an apple may be referred to as something you eat and not as fruit or a chair as something you sit in but is not known as a piece of furniture.

Regression. This is characterized by the fact that the individual shows childlike responses. It is manifested by a profound dependency as well as tendency to see the world not in an objective way but according to the person's needs, wishes, or fears. Such a person has an exceeding low frustration tolerance. When the psychologic test performances of "organic" patients are compared with those of different populations, the responses in some ways resemble those of children and in other ways those of normal adults. While the behavior may be childlike in some respects, evidences of mature functioning may also appear. This is typical of regression.

Brain-damaged children who sustained their injuries at birth or in early life do not show phenomena of regression as often as they show uneven development of skills and aptitudes. They tend to cover up defects in adaptation through ordering their world in special ways. Such children may build up deviant habit systems and show impoverished fantasies even when their intelligence is not affected.

Body image disturbance. This disturbance is one of the signal characteristics of brain damage. The individual's body schema shows profound disturbance and distortions. These disturbances may be elicited by a variety of neurologic and psychologic tests.

• • •

These four characteristics generally overlap each other and are seldom found to exist independently. They are associated with states of acute and profound anxiety, low frustration tolerance, stereotyped thinking, impulse disturbances, fluctuating attention span, or extreme debilitating loss of energy and lack of motivation.

The management of the brain-damaged patient must take into account a careful evaluation of behavior in terms of whether it can be classified as nonorganic or organic. If it belongs in the nonorganic category, then the patient can be treated like any other patient on the rehabilitation service. The program should be geared to help the patient work with the deficits that he has. Untoward emotional reactions are probably due to reactions to the disability or other factors rather than to the brain damage itself. Such patients can be helped by insight as well as supportive psychotherapy.

Patients with organic brain damage require special handling. The rehabilitation program and goals are generally more limited. Fewer activities are prescribed. Special attempt is made to keep the patient's environment constant so that therapists and classes remain stable. A great effort is made at repetition and drill in teaching. Many trials are necessary to attain mastery in a skill, and these trials are spaced rather than massed together. The patient is informed of his progress and is always made aware of his immediate goals. Staff members are encouraged to build warm, personal relationships with the patient. They are told about the patient's psychologic problems to prevent their interpreting as personal any aggressive or abusive behavior that may be directed at them. Despite the gentleness that is necessary in handling the patient with organic brain damage, firmness is important. A firm therapist presents a fixed, predictable stimulus figure in a world that has become otherwise disordered.

Psychotherapy for patients with organic brain damage must generally be limited to the supportive variety, for these patients have difficulty in absorbing insights. Our experience with drug therapy, especially the use of tranquilizers, is limited.

Case study. A 56-year-old male suffered a cerebrovascular accident with left-sided hemiplegia 6 months prior to admission. He had worked for a utility company and in his spare time functioned as a local political party boss, exerting considerable power locally. After the stroke, he insisted on remaining at home and having his wife tend to his every need. Because of his inordinate demands and because of a developing wrist contracture, he was referred to the rehabilitation service.

During the evaluation conference his dependency and controlling behavior were noted, and we anticipated difficulty in management. However, because of his excellent work history and obvious capabilities, he was accepted on service. The anticipated difficulty began immediately. He was demanding of attention, cried like a child, and soon had the patients and staff "down on him." Staff conferences were concerned as much with staff hostility as with the problems of the patient. A firm course was decided on. The psychologist saw him frequently, utilizing specific episodes on the ward and interpersonal reactions to make him aware of what he was doing. The ward patients, without psychiatric briefing, also managed to exert pressure, and attitudes were noted. He made considerable progress.

The social worker consulted with the family, and as a result the patient's wife went to work and continued to work after he returned home. He returned to his previous employment, as a result of consultations between the vocational counselor and his employer.

Anxiety

It is probably safe to say that every patient who suffers physical disability experiences anxiety. By anxiety is meant a condition of heightened and often disruptive tension accompanied by a vague and disquieting feeling of prospective harm or distress. It is really a signal of anticipated danger. Psychiatrists have distinguished between fear and anxiety, calling the former a reaction to actual, present, external danger, and the latter a reaction to unknown threats, repressed impulses deep within the personality, or repressed feelings striving toward consciousness. Physiologically and emotionally there is little difference, if any, between the two, and for the purposes of this chapter no distinction need be made. It is important to recognize the existence of anxiety, to seek its sources, to determine the ways in which each individual copes with it, and to elaborate the effects of both the anxiety and the nature of the defenses used against it. It is apparent that if anxiety is a signal of danger, it must be an important biologic and psychologic factor in the preservation of the individual and cannot be lightly dismissed. As a matter of fact, much of modern psychiatry and psychiatric theory is based on a recognition of the central role that anxiety plays in the biologic and psychologic equilibrium of the individual.

Through its effect on the autonomic nervous system, anxiety is particularly likely to disrupt physiologic functions. In acute forms it may produce generalized visceral tension, with spasms of the cardiac and pyloric ends of the stomach, intestinal symptoms (such as diarrhea or constipation), palpitation, tachycardia, extrasystoles, vasomotor flushing, and respiratory distress. The hands and face may perspire, the patient may assume tense posture, show excessive vigilance, and make fidgety movements of the hands or feet, the voice may be uneven or strained, and the pupils may be widely dilated.

When the anxiety is intolerable or unacceptable, a variety of defenses are brought into play to minimize the painful feelings, and very often it is these defenses that present themselves as psychiatric symptoms and that often determine the diagnosis attached to the anxiety. Thus anxiety may be repressed from conscious awareness. It may be displaced onto an object or a situation to produce a phobia. It may be converted into a hysterical symptom. It may be controlled and walled off by means of obsessions or compulsions. There may be utilization of behavioral mechanisms, such as apathy, boredom, hyperactivity, or compulsive flow of speech, as means of expressing or containing the anxiety. It may lead to psychosomatic disorders or to hypochondriasis.

The diagnostic label attached to a person very often is based on the sources of anxiety and the ways in which he copes with it. Thus, depending on the qualitative and quantitative aspects of anxiety and on its defenses and modes of expression, the person's behavior is called reactive (realistic), neurotic, or psychotic. In the reactive types the anxiety is related to reality factors and may be estimated by the observer to be appropriate to the situation. In the neurotic patient, even in the face of a real threat such as a physical disability, the anxiety is considered to be quantitatively greater than the situation calls for, or it is expressed in neurotic mechanisms such as phobias and hysteria. The root of neurotic anxiety will be found in earlier unresolved life situations and adaptations, so that the person organizes the present real situation into a previous one and reacts to it on the basis of these earlier perceptions. In the psychotic patient the anxiety is so overwhelming that it disorganizes mental functioning, leading to a gross break with reality.

Fortunately, most patients with physical disability manifest realistic or mildly neurotic anxiety that does not interfere significantly with rehabilitation. Such patients respond well to support, to a sympathetic milieu, and to rehabilitation procedures. The program and the physical procedures provide an outlet for the release of tensions. Rehabilitation in a group setting also has the advantage of lending the support of other patients and helping to remove the feeling of estrangement and uniqueness so often associated with the disability.

Anxiety is not always a detriment to rehabilitation. In some patients it may lead to compensatory activity that facilitates rehabilitation, although one must be careful lest the patient engage in activity too strenuous and too rapid for his own welfare.

In many patients one can observe that anxiety interferes with rehabilitation because motor, perceptual, or ideational performance becomes either disorganized and fragmented or rigid and inhibited. So much concentration and energy are expended as a defense against the anxiety that very little remains for dealing with realistic problems. Where the anxiety interferes with rehabilitation, special attention must be given to the patient.

Supportive measures in a warm, empathic type of relationship are often very helpful. The patient should be allowed to ventilate his fears, should never be made fun of for exhibiting "foolish" fears, and should not be argued with concerning them. Where protective support is not effective, exploration of the roots of the anxiety—whether realistic or in unconscious conflicts—should be attempted by a trained professional following adequate psychiatric evaluation.

In addition to psychotherapy, use of sedatives or tranquilizers may be indicated, but only as part of the total management plan rather than as an isolated therapeutic measure.

Case study. A 65-year-old paraplegic patient had a long-standing history of marital discord. Six years before he sustained his disability, he went to a doctor with a complaint of dizziness and crying spells. The doctor told him he had a cardiac condition, and the patient interpreted that as a sign that he was going to die. He stopped working and confined himself to his house because he was sure that if he went past the street corner of the block he lived on he would never return home alive. His tension became so great that one day he decided he could not stand the situation any longer. He got drunk and left the house, walking beyond the corner, all the way to the other end of town. He felt suddenly free of the anxiety and went back to work.

The day before he sustained his disability he had observed an automobile accident and the same feeling of anxiety and impending death had come over him. The next day he returned to his job as a glazier and did not take adequate precautions in fastening his safety belt. He fell and sustained his paraplegia. With the paraplegia came the same feeling of relief that he had attained years earlier when he was afraid of dying. "I made up my mind I was over the worst."

On admission to the rehabilitation center, several months after his injury, his fear of death had disappeared. However, he cried easily and complained of headaches. He was very irritable, and when he became upset, he lost his power of speech for brief periods. While on the rehabilitation program he worked very hard, but, because of his poor physical coordination, his progress was very limited. He became increasingly agitated and depressed. The depression mounted when he felt that he would not be able to ambulate. For the patient this meant that when he returned home he would be forced to be with his wife all of the time. Should he become angry with her, he no longer would be capable of fleeing from the situation. He was trapped.

Psychotherapy, largely of a supportive nature, was carried out with this patient. The patient found considerable relief in talking about his feelings. As the therapy progressed, it became clear that he was filled with a tremendous amount of anger. He was basically a passive, dependent man who had secretly been hoping that the rehabilitation center would solve all of his problems, including his marital discord. When he realized that his excessive demands could not be fulfilled, he became very angry at all the rehabilitation personnel. He failed to attend classes, shouted at therapists, and became a management problem. In psychotherapy it was possible to show him that because he was angry at himself he was trying to provoke anger in the staff so that he would be sent home. While the patient's anxiety level was diminished sufficiently so that he could work on the program, his understanding of the effects of his behavior on other people increased to the point where his uncontrollable anger was sufficiently tempered to enable him to participate and gain at least partial self-sufficiency.

The therapeutic strategy was aimed at reducing the anxiety by permitting the patient to label the source (himself) and the effect of his anger (displacement) and to distinguish between the two. Although no attempt was made to have him understand his basic need to return to the woman he disliked, the patient completed rehabilitation and chose to return to live with his wife. His stay at home has been fairly successful in view of the difficulties in the past marital history.

Depression

Undoubtedly the same must be said of depression as of anxiety—that all patients with physical disability experience it. It may vary from a state of mild sadness to the extremes of psychotic delusions and suicide. The degree of depression is determined not only by the nature of the disability but also by the particular individual who suffers it and the special meaning the disability has for him.

Depression is both a symptom and an adaptation to life experience and may present itself in a variety of ways. In the milder forms the patient is quiet, inhibited, unhappy, self-depreciative, discouraged, and uninterested in his surroundings. In more severe depression there is a constant unpleasant tension. Every experience is mentally painful. The patient's interests are very limited and always melancholy. Conversation is difficult, and he is dejected and hopeless in attitude and manner. He may be so preoccupied with depressive rumination that attention, concentration, and memory are impaired. Insomnia, especially that due to early awakening, is common. (Anxious patients usually have difficulty in falling asleep.) Bodily symptoms such as constipation, loss of appetite, fatigue, tightness of the head, and headache are frequent complaints. Delusions may appear, and patients usually express ideas of guilt, unworthiness, and self-accusation. Suicidal thoughts are frequently entertained. When elements of anger and anxiety are expressed, the patient may then become agitated.

Patients have a need to defend themselves against the depression, and one of the common defenses is acting cheerful, elated, and even euphoric. In extreme instances this progresses to states of hypomania and mania. The cheerful patient is often deceptive because the observer usually reacts favorably to such a trait, sees it as courage (which it is, of course), and thus comes to admire it. However, there is always at least some element of depression behind such an outward appearance.

Depression is always associated with loss and is allied with feelings of grief. It is not uncommon for the physically disabled patient to personalize his body, or parts of his body, and feel toward it as he does toward a cherished possession. Loss of function of a limb or loss of the limb is not infrequently seen as loss of a loved one. Thus a paraplegic patient, in response to a sentence completion test question, "To be without . . .," responded, "friends is like having no limbs."

Guilt feelings usually lead to depression. The guilt may be for real or imagined sins, and all kinds of experiences are dredged up from the past to explain the accident and to heap blame on oneself. One patient, a young girl, went sleighing against the wishes of her father, crashed into a tree, and suffered a cervical injury with a resulting quadriplegia. Her depression was profound, and she connected the accident with her disobedience. When she was ready to go home, her greatest source of anxiety was her father. She feared his disapproval and his restriction on her activities. This depression became more apparent just before discharge. Another patient, a male who had suffered a cerebral thrombosis with resulting hemiplegia, was in such a state of depression as to inhibit his ability to work. He blamed his stroke on earlier extramarital relations and saw it as a punishment for his sins, even though the stroke had occurred 8 years before his admission to the rehabilitation center.

Depression is also due to loss of mastery resulting from the disability. Much of our self-esteem is derived from accomplishments that have social value. In the maturing process we learned to walk, to talk, to control our bladders and bowels, to dress ourselves, to wash ourselves, and to perform a host of other activities. These are skills that are acquired in a social matrix involving parental authority. Sometimes they are acquired in a punitive context, sometimes in a permissive context.

Parents may be patient or impatient, start independence training too soon or at the right time. In any case, out of these skills comes a feeling of mastery that leads to confidence and self-esteem. When a physical disability strikes, many of these skills and controls are lost, some permanently. The patient is projected back into a child-like, helpless state, and this regression leads to depression, particularly if the acquisition of these skills was associated with traumatic earlier experiences.

Fortunately, the patient with a mild depression does not often present a management problem. Usually the milieu, the personnel, and the activity will provide sufficient support and outlet so that the rehabilitation process will not be disturbed. Indiscriminate psychiatric intervention should be avoided in those patients who are progressing and are gaining support from the program.

Patients with more severe depression usually reveal an inhibition in their activity, and because the rehabilitation process is impaired the depression is more likely to be recognized. Supportive therapy is indicated, and insight therapy is to be avoided unless the patient has been in a continuous psychotherapeutic relationship.

When the depression is severe, when it assumes bizarre forms, and particularly when suicidal thoughts are being expressed, psychiatric consultation should be sought. Further therapy includes use of drugs such as dextroamphetamine (Dexedrine) or amobarbital (Amytal). The decision to employ shock therapy must be made by a qualified psychiatrist, and such treatment must be administered by him.

Case study. A 40-year-old coal miner was admitted to the Institute of Physical Medicine and Rehabilitation following paraplegia (at the level of the twelfth thoracic vertebra) that resulted from an accident in which a slab of slate fell and crushed his back. The patient came from a small mining town in Pennsylvania. He was married and had two children. He had worked in the mines for 22 years, although when he was a child he had always wanted to do office work. On admission he was acutely depressed. He did not accept his disability and believed that he would be working again. He told the psychologist that he felt sometimes that he was trapped in his bed and that rocks were coming down to crush him. He spoke about the uselessness of life and stated that he might be better off by going home and being taken care of. Clinical examination gave the impression of a man who felt helpless in the face of a situation that he could not master. The history indicated he had been a hard-working, conscientious, and sincere person who had suffered intermittent feelings of depression and had been quite dependent on his wife.

The therapeutic strategy from a psychiatric standpoint was focused about several points: (1) the staff was alerted to this man's continual need for support and reassurance; (2) he was placed on a program that permitted him to advance at his own pace; (3) he was encouraged to develop his arithmetic and spelling skills, and a program of training in book-keeping was worked out in order that he might pursue goals that were meaningful to him and that would permit him to achieve a sense of mastery; (4) he was permitted to return home on medical furlough for 2 to 3 weeks at a time several times during the year when his homesickness grew too strong; (5) he was seen frequently for intensive psychotherapy by the psychologist; the interviews were practically on a daily basis throughout most of his stay at the center, and they were often conducted on the ward, particularly when the patient was grounded with bouts of fever.

In the sessions he repeatedly recounted his worries about his family. He blamed himself for not taking care of them and spoke about how unworthy he felt, particularly when people offered to help him. An attempt was made to get him to see that he was expecting too much of himself and that if he did not achieve a magic cure it was not because he was not trying. During this period, which lasted about 6 months, the patient made slow but steady progress in mastering motor skills. Arrangements were made for him to complete a correspondence course and qualify for a high school diploma. Organizations in his home community were notified of his vocational aims so that a job would be waiting for him on his return.

Toward the end of his rehabilitation the patient's outlook showed a decided change. When he returned 6 months after discharge for a medical follow-up, he was doing very well. He had continued to make gains in motor skills through diligent practice at home. He was hoping to be placed by the local office of the Division of Vocational Rehabilitation.

The psychiatric management was based on the premise that this patient's extreme depression could be treated by helping him attain a mastery over his environment, which his disability had taken away from him. The therapist in the role of a reassuring, sympathetic, interested observer played the part of an authority figure on whom the patient could rely for support and who helped him plan the various steps in his future.

Suicide

Psychiatric problems pose relatively few threats to patients' lives. Suicide is one of the most prominent, and the proper evaluation of a suicidal threat always demands considerable experience and judgment.

Patients with physical disability, especially in the early stages, will often make remarks such as "I'd be better off dead" or "I wish I were dead." Where such statements lack a tone of conviction and where there are no other evidences of psychiatric disturbance, it is best not to make too much of these remarks. At most one might listen to them and then move on to the business at hand; otherwise, the physician reinforces the patient's lack of self-confidence.

There are, however, situations that must be taken as serious or potential suicidal risks: (1) depressions of a psychoneurotic or psychotic nature; (2) regardless of diagnosis, a past suicidal attempt or a history of a severe depression; (3) insistent threats of suicide; (4) anorexia, severe weight loss, insomnia, listlessness, apathy, persistent expressions of discouragement and hopelessness, continuous weeping, and general motor retardation that cannot be explained on the basis of the patient's disease and disability; and (5) dreams of death, mutilation, and funerals.

It is always a source of surprise that the problem of suicide appears so seldom on the rehabilitation service. In our experience this is true for both the Bellevue Rehabilitation Medicine Service and the Institute of Rehabilitation Medicine Service, where cultural, social, and economic differences are evident. Considering the enormous stress imposed on patients by their disabilities, one might expect suicide to be a more serious and distressing problem. Yet the fact remains that, in the many years of our experience, there has never been a case of suicide, although depressions, as well as threats of suicide, have been common.

While no controlled study of this problem in physically disabled persons is available, several comments can be made on this subject.

1. While grossly disturbed psychotic patients are not usually accepted for rehabilitation, some psychotic patients have been. A fair number who have developed their physical disabilities as a result of suicidal attempt do well on a rehabilitation program. This group may have satisfied masochistic and self-destructive impulses and, having done so, find their physical disability then subserves their emotional needs and does not cause the kind of conflict one would see in a relatively emotionally healthy person.

2. We do see a large number of elderly patients with arteriosclerotic and senile changes. This group has been found to be particularly dangerous as suicidal risks

in the general population, yet we have not seen any attempts by them on our services.

3. Sociologic studies[7] have indicated a correlation between incidence of suicide and social disintegration and isolation. It may be that the rehabilitation setting, representing a well-organized and emphatic social structure, provides even the most deprived patient with a feeling that someone really cares. This is a potent antidote to the isolation that apparently is a predisposing factor in suicide. In addition, the presence of a psychiatric team on the rehabilitation service may serve a preventive function in forestalling the acting out of suicidal impulses. Whatever the reason, the fact that so few suicides occur attests to man's basic tenacity for life.

Aggression and hostility

The third of the triad of basic emotions that are so often a cause for concern (anxiety and depression being the others) is that of hostility or anger. Of all the emotions, anger is most difficult to manage because it is socially the least acceptable. Anger also produces angry reactions in others and, therefore, may be very threatening to a person who is dependent on them.

The hostile, aggressive patient may be exhibiting a reaction to his disability, seeing it as an unfair act of fate, and his anger may simply reflect his need to strike out at this unknown force that has so changed his life. Because it is more difficult to deal with the unknown than the known, anger may be displaced and directed at doctors, nurses, or other staff members.

However, the tendency to display hostility and aggression may have developed as a result of earlier deprivations and injuries in childhood, emotional or physical. A disability may reactivate these earlier feelings and intensify their expression. This is what happens typically in a neurotic type of reaction.

Because of dependency needs or earlier conditioning, a patient may be unable to express hostility overtly. In such a case he will appear ingratiating or overdependent. He may express his anger by using the dependency to get everyone to pay attention to him. Or if he is suffering from guilt feelings or fear of expressing an unacceptable impulse such as anger, the anger will be retroflexed and turned inward, leading to a depression.

The patient who has carried angry feelings within himself throughout his life may have been able to crystallize them in socially acceptable and constructive channels, such as work, athletics, or organizational activity. However, the physical disability may temporarily or permanently prevent such activity, and the anger and tension may then mount to overwhelming proportions and may burst out into antisocial and sometimes destructive acts. A patient may throw food or a bedpan, shout in a wild fashion, or race down a ward in his wheelchair, crashing into others. The behavior may assume psychopathic or psychotic degrees.

The management of the hostile and aggressive patient can be a sorely trying experience. The first impulse on the part of the staff is to react in kind, and this serves to aggravate the situation. The best way to deal with the problem is to be tolerant and attempt to understand what forces are driving the patient. This is not always easy, especially when the welfare of other patients is concerned.

The team approach proves useful in the management of this problem for the following reasons: (1) it provides a complete evaluation, which permits the staff to see not just the hostility but also the forces behind it; (2) it provides an opportunity for each member of the staff to check his feelings against those of others; (3) it permits the shifting of responsibilities among the staff, thus allowing for personality differences; and (4) it provides opportunity for a unified, consistent, and rational approach to the patient.

When the anger is a reaction to the disability in an individual with a reasonably mature and integrated past performance, support and understanding will help the patient overcome these feelings. Participation in a rehabilitation program, with its orderliness, structure, and outlet for tension, will often reduce the hostility.

When this is not effective, psychotherapy is often helpful, even if all it provides is an outlet for angry impulses. It may serve to siphon off these feelings and relieve the tension in other areas of rehabilitation. With insight the patient can become more realistic and cooperative and then proceed with his program.

In patients who have had an angry type of personality as part of their character structure, the problem is a more difficult one. Psychotherapy will have to be more intensive and extensive, and all that can be accomplished may be to permit the physical rehabilitation to proceed. Use of sedatives or tranquilizers is at times effective in cutting down the acting out, thus facilitating the rehabilitation program.

Case study. A 17-year-old youth, with a past history of delinquent behavior, was admitted with traumatic paraplegia. He and several other boys had stolen a car. While being chased by the police, the car crashed and this boy received a concussion, fractured ribs, punctured lung, and severed spinal cord. After 5 months of critical illness he was transferred to rehabilitation for paraplegic training. His behavior on the ward followed a pattern of previous behavior in schools. He was aggressive, destructive, and defiant, terrorized the other patients, and generated such hostility on the part of the staff and patients that they insisted he be discharged. He was most provocative, lying in bed until 1 P.M., and he was careless about routine and self-care.

After psychiatric evaluation and psychologic testing, a conference was held, and it was most dramatic to observe the hostility of the staff change to understanding and sympathy when they saw how much self-destruction and self-hatred was involved in his aggressive behavior. As they learned of his life history, the disturbed aspects of his development, and his responses on the Rorschach and the Thematic Apperception Test, a feeling of poignancy took hold. Even though he continued his behavior on the ward, the staff was most patient. However, he resisted any psychotherapy. He complained, "I'm not crazy, I'm crippled." We worked indirectly through the staff. One day he asked to see the psychiatrist. He had been reading the autobiography of the bank robber Willy Sutton and in it had read some sympathetic lines about the prison psychiatrist. "If it's good enough for Willy Sutton, it's good enough for me," he decided. During the ensuing sessions he obtained some insight into his role in provoking the dislike of others. He talked about courage and wondered why he thought of robbers and gangsters as courageous, but not men like Einstein or Roosevelt. The patient then rejected further discussion and was taken home by his mother. However, he had reached the point of wearing braces and had begun vocational training in watch repairing and silver work. Our efforts were effective in creating an environmental situation that permitted some degree of rehabilitation, although we did not consider that the patient's personality had changed.

Dependency and exaggerated independency

Among the characteristics that distinguish human organisms from those lower in the evolutionary scale is the long period of dependency necessary for maturation from infancy to the adult state. This dependency is physical, emotional, and social. The infant is absolutely dependent on the mother for the satisfaction of his physical needs, and without this the infant cannot survive. The infant is also dependent on the mother or a substitute for the satisfaction of his emotional needs, and without it the infant either dies or suffers impairment of emotional development. As the child grows, he becomes dependent on the family, the school, and society in general for the acquisition of intellectual, technical, and social skills. It is not surprising then that dependency plays such an important role in human personality and functioning.

Under normal and healthy circumstances, barring psychologic or physical obstacles to growth and maturation, the child developing into adulthood gives up many of his dependencies and becomes an independent, mature person. This does not mean that the mature person functions as an isolate. On the contrary, a normal human being always has need for other human beings, is dependent on them for many satisfactions, and can depend on them when in need. Under normal circumstances, independence does not mean isolation, nor does dependency mean helplessness. Maturity, then, implies interdependence.

Unfortunately many persons experience emotional injuries during the dependency period due to deprivation or overindulgence, excessive authority, or overpermissiveness. These may be associated with feeding, with bowel and bladder training, with discipline and the learning of controls, or with an overall attitude of rejection or ambivalence toward the child. As a result of his helplessness and "copelessness," the child may grow up with a variety of adaptations and defenses against such experiences. He may "fixate" at the dependency level and fail to mature adequately. He may repress these experiences or feelings associated with them and proceed to mature while unconsciously carrying on a struggle to repair the injuries suffered in childhood.

Physical disability deprives the individual of his mature skills and projects him into a state of actual dependency. How he reacts will depend on what the dependency means to him, as determined by his earlier experiences. And since the goals of rehabilitation often are in direct contradiction to those of dependent reactions, the very outcome of rehabilitation may rest on the management of the dependency needs.

Dependency may be manifested in a variety of ways:

1. *Overt dependency or behavior that is reminiscent of a helpless child.* The patient is demanding and constantly calls the staff to do something for him, to pay attention to him, and to recognize him no matter what the circumstances. He constantly complains of symptoms, the behavior of other patients, and the personnel. He resents any attention being given to other patients and usually maintains a mental bookkeeping system to record the time spent on him and the time spent on others. He will often interrupt activities of others to complain of himself.

2. *Pseudocooperation.* The patient will appear like the ideal patient, do everything asked of him, follow the program rigidly, and seem pleasant and friendly to everyone. The clue to his difficulty lies in the fact that he is *too* cooperative and *too* uncomplaining, as well as in the fact that he makes no appreciable progress even though he seems to participate in the program.

3. *Reaction-formation.* The patient may react to dependency by means of

"reaction-formation" and behave in precisely the opposite fashion—by exaggerated independent action and by perfectionistic goals. Such a patient seems to have considerable drive and will participate in the program with an enthusiasm that engenders great pleasure in the staff. However, he refuses help when he needs it, insisting on "doing everything himself," or he may move along at such a rapid pace as to be harmful. The perfectionism may keep him at one task so long that it blocks any progress in rehabilitation.

These three types of reaction do not exhaust the means whereby a patient can manifest dependency. There are many others too numerous to mention.

The overtly dependent patient often creates a problem with the staff, engendering a great hostility and impatience because of his inordinate demands. He also generates hostility in his fellow patients because of jealousy, as well as outright social disapproval by them of such behavior. The staff reacts like rejecting parents and the patients like jealous siblings.

The pseudocooperative patient is most deceptive, and a long time can pass before anyone suspects anything is wrong. Like the compliant "good" child, he is taken for granted and often neglected.

The patient who displays exaggerated independence often generates admiration and, because he seems to be well motivated, is erroneously encouraged in his actions. Because of his overt behavior, his dependency needs are often missed.

The management of dependency reactions is based on an understanding of the dynamics behind them. The patient whose past history reveals a reasonably mature and well-integrated pattern may bend in the face of the acute problems presented by the disability and may fall back on dependency reactions for purposes of security. Dependency in such an individual is a reaction to anxiety and a feeling of helplessness and, as such, serves as a psychologic crutch. Just as we would not take away a crutch from a patient with a fractured leg until the leg heals, so we should not attempt to remove the psychologic crutch from the acutely disabled individual. The dependency in such a patient should be satisfied by means of support, sympathy, and warm acceptance, and generally such a patient will give up the dependency and proceed on the rehabilitation program. Dependency is not a major portion of his personality makeup. Rather, it is a temporary adaptation, and under favorable circumstances his basic health and maturity will emerge. (This assumes that there is no extensive brain damage or overwhelming physical disability. Even in the latter, such as the poliomyelitic patient with quadriplegia and respiratory difficulty, it is amazing what healthy personality resources and sound rehabilitation milieu can accomplish.)

Where dependency behavior has played an important role in the life history of the patient and has become a basic part of his character structure, the management is very much more difficult. The disability simply fits into his infantile and immature outlook on life and is used not only as a crutch for anxiety-relieving purposes but also as a crutch for an entire way of life. To attempt to satisfy the demands of such patients is futile because they are insatiable; it leads not to relief but to more demands. To attempt to deny the demands leads to bitter complaints and to overt hostility. It is of the utmost importance to evaluate such patients early and to plan a mode of approach early. In general the attitude of the staff should not be an overtly hostile one, but it must be firm. Such a patient must not be given an opportunity to manipulate the staff. At the same time, it may be of value to engage such a patient in psychotherapy to provide an outlet for feelings, to permit direction, and eventually, to encourage self-direction.

Under the best circumstances, even without disability, the infantile personality is very difficult to treat. Partial and limited goals are usually necessary, and, in the presence of a disability, these should be oriented to the rehabilitation process.

In a sense the perfectionistic patient who shows exaggerated independence poses a different problem and usually is easier to manage. The fact that he does not tend to generate hostility and rejection on the part of the staff in itself makes the task easier. The consistent cooperation of the staff is essential to help these patients tone down their activities and to accept certain compromises in their goals. Pointing out the dangers of "too much, too fast, too soon" can be helpful, although this must be patiently repeated day in and day out. Psychotherapy is recommended in such patients, for it helps them to gain insight into the reasons for their behavior.

Case study. A 30-year-old paraplegic patient sustained his injury in an automobile accident. During the six years following the injury the patient had been home and had received treatment at a local hospital for 18 months so that he became partially self-sufficient. About 3 years ago the patient began to develop decubiti. He was very careless and did not pay much attention to keeping off pressure areas, thus causing numerous sores to erupt. He was forced to return to the hospital for ten skin grafts and remained there for some 16 months. At the end of this period he was referred for rehabilitation. He repeatedly suggested to the doctors that they amputate his legs because of the difficulty in treating them. In addition, he stated that "they would be less weight to carry around."

When he was first seen by the social worker, he described himself as "lazy." He painted a self-depreciating picture, saying that he gambled and was interested in "bad things." He was lukewarm about the whole idea of rehabilitation; if he was admitted, it was all right with him, and if he was not admitted, that was all right too. At home his mother did many things for him and he never resisted her attempts. He told the social worker that nothing ever bothered him and told the psychologist that he wanted to finish rehabilitation as soon as possible because "I'm taking up room that someone else is entitled to."

Because of his decubiti he was not able to participate in a full rehabilitation program. It was noted soon after his admission that he began to request drugs, to drink, and to precipitate problems on the ward. Because of the difficulties and the reports of the psychologist and the social worker, he was referred to the psychiatrist and a program of psychotherapy was established.

In the course of therapy the patient told the psychiatrist that he (the psychiatrist) was out to trick him. When he was asked why he felt this way, he said it was because people are like that, and then went on to wonder what makes people with deformed bodies look "ugly" to other people. On further exploration, it turned out that he felt that many people, particularly doctors, because they represented authority figures, were trying to run his life for him and tell him what to do. When a doctor, for example, urged him to stop drinking or work harder on the program, he would become furious and react for the sheer sake of being defiant rather than to further his own progress. These aggressive attitudes were puzzling because they contradicted another series of attitudes that appeared. When the vocational counselor began to help the patient formulate future plans, he noted a reluctance on the part of the patient to become deeply involved in planning. When the psychiatrist explored the possible reasons for this reluctance, he found that the patient was very anxious about employment because he was afraid to compete with others on an equal basis.

The contradiction between the patient's wish to control other people and his fear of his powers was strikingly exemplified by an incident that he reported to the psychiatrist. During his stay at the rehabilitation center he became involved in a love affair with a nurse. Following his pattern of continuously testing other people, he suggested to her that she prove her love to him by supplying him with drugs. When she complied with this request, he became terrified at his own strength in being able to exercise such strong authority over another human being. This incident and several others were used to provide the patient with some insight into the fact that his baiting of authority figures was a form of testing his relationships with people. He anticipated and provoked rejection to demonstrate to himself that people wanted to reject him. Beneath his aggressive, antisocial attitudes existed profound feelings of low self-esteem that triggered off self-punitive behavior (provoking others to be angry, wanting his legs to be cut off, "doing bad things," "I used to pass a crap game—I felt I was being dragged in by both hands and feet and I'd play until I lost every dime I had," "maybe God did this [paraplegia] to me to slow me down"). A corollary of these self-punitive needs were the strong dependency needs that proved to be the potent underlying theme of his "acting out" behavior. (When he was home, he let his mother take care of all of his needs. He described himself as lazy—"would like to sit out in the Florida sun all day." He verbalized at the beginning of rehabilitation that he had not been too interested in walking because of fear of competing for a job.)

This case illustrates how the combination of self-punitiveness and dependency underlies a variety of behavioral disturbances and emotional upsets that could easily destroy the efforts of a rehabilitation team. The dependency in this case was manifested in some instances by overt behavior; in others, by the defenses erected against it (defiance of authority, pseudoindependence). The therapeutic strategy was aimed at permitting the patient to ventilate his feelings to the psychiatrist rather than have him direct aggression at the nearest authority figure, as well as providing him with some insight into the needs behind his provoking, antisocial behavior. The patient proved to be a rather formidable psychiatric challenge because the psychotherapist was forced to deal with many instances of behavior that were designed to test the therapeutic relationship. It was believed that the limited goals of helping the patient maintain sufficient self-control and develop a desire to work on the program were accomplished. It should be pointed out that basic characterologic changes, for example, the strong underlying dependency needs, could not be completely resolved in the 6-month rehabilitation period. In this case it was recommended that psychotherapy be continued when the patient returned to his own community. The psychotherapy was not continued and the patient lost all the ground he had gained on the rehabilitation service. He requested that his legs be amputated, and the surgeons in his home town complied with his request.

The psychotic patient

There are two types of psychotic patients who are encountered on a rehabilitation service. The first is the patient who becomes physically disabled as a consequence of his psychosis, such as the patient who attempts suicide and suffers his disability as a direct result of the attempt. The second is the patient who becomes psychotic *after* his disability, in which case the disability is not a consequence of the psychosis.

Disability due to psychosis

The first category is not uncommon. Severely depressed patients attempt suicide by violence, such as throwing themselves out of windows, slashing their wrists and severing nerves and tendons, or by carbon monoxide poisoning, which produces cerebral anoxia with brain damage and neurologic sequelae. Where the psychiatric picture is dominant, as in psychotic depressions or schizophrenic reactions, management on a general or rehabilitation service usually is not feasible. Such patients require hospitalization in a psychiatric setting, where treatment for the primary disorder can be given. The physical disability must, however, not be neglected, and rehabilitation consultation may be necessary to prevent more serious impairment, such as contractures, urologic complications, and decubiti. When the acute psychiatric problem is under control, transfer to the rehabilitation service can then occur. The following cases are two examples of such problems.

Case study 1. A 17-year-old white female, a brilliant college student, entered into a suicide pact with a friend. Both girls turned on the gas and went to bed. The friend died, but the patient was discovered in coma. She was rushed to a general hospital, where she remained in coma or semicoma for nearly 6 months. As she came out of her comatose state, she became very disturbed, screaming and shouting uncontrollably, to such a degree that she could not be managed and was, therefore, transferred to Bellevue Psychiatric Hospital, where she was cared for on the medical service. In addition to her behavioral problem, it was noted that she was unable to see, was paralyzed in both lower extremities and in one arm, and was paretic in the other arm. It was decided that if the behavioral problem could be controlled, she could be transferred to the Physical Medicine and Rehabilitation Service. This was set as the psychiatric goal. Communication was established, and by means of suggestion and training she managed to learn to control her outbursts. (This was before the availability of tranquilizers.) She was then transferred to the Bellevue Physical Medicine and Rehabilitation Service, where an energetic program was instituted, involving quadriparetic training and a reading program for her visual agnosia. Because of the severity of her brain damage, a plateau was reached, but she did make moderate advances in a physical sense. The original psychiatric diagnosis was that of schizophrenia.

Case study 2. An 18-year-old college student threw himself from a third-story window in response to hallucinatory commands and suffered fracture of his thoracic vertebras with severance of the spinal cord at the tenth thoracic vertebra. After a laminectomy in a general hospital, he was transferred to a state mental hospital for treatment of paranoid schizophrenia. He responded to insulin therapy and made a fair recovery, after which he was accepted by the Bellevue Physical Medicine and Rehabilitation Service. On admission he still presented symptoms of a serious nature. His contact with reality was transitory and tenuous. He was obsessed with the idea that he could not eat because the food was poisoned. However, because there were no overt behavioral difficulties, he was kept on the service and seen frequently by the psychiatric team; he continued to make a good recovery psychiatrically. The paraplegia was managed in the usual way, although there were complications of decubiti and renal calculi. After about 18 months he was discharged home on full self-care and with his future educational plans worked out.

Psychosis following disability

The second category, which includes the patient who develops signs of psychosis after disability, is not so common as might be supposed. While this may be because of a selection of patients, who are screened on evaluation, it may also be because of the tremendous support provided patients in a rehabilitation milieu. It may be due also to the fact that the disability serves a valuable function in helping maintain integrity of his personality and prevents an acute psychotic break with reality. However, when such episodes occur, they require psychiatric consultation and decision regarding management. In general the same principles hold as for the first category; that is, it must be decided whether it is feasible and safe to manage the patient on the rehabilitation service or necessary to transfer him to a psychiatric hospital. Following are examples of this problem.

Case study 1. Seven years previous to our contact, a 48-year-old Negro woman had attempted suicide by jumping from a window while under the influence of alcohol. As a result she suffered a spinal cord injury resulting in paraplegia. After many years in several hospitals she was transferred to our service for paraplegic training. Because of her past history, she was thoroughly examined psychiatrically, and, although there were evidences of schizophrenia, her contact with reality was good and it was decided to keep her for training. She did very well until arrangements had to be made for her discharge. There were several major obstacles to overcome that delayed the process. The patient became increasingly apprehensive and complained that patients were talking about her and that the doctors had spread a rumor that she was to be transferred to a mental hospital. These were delusions, and despite attempts at reassurance the symptoms increased in intensity and she was transferred to the psychiatric division. There she quieted down immediately, and after only 3 days without specific therapy she was returned to the rehabilitation service. The delusion and the tension had disappeared, and she continued on her program until discharge. (While on service, it should be stated that she had enormous support from the staff, who took a very special interest in her. Among pertinent reasons for this, the patient happened to be an excellent baker, and she kept the staff well supplied with delicious pies.)

Case study 2. A 28-year-old Puerto Rican male was referred for evaluation by a manufacturing company. While he was at work, a small object weighing about 6 ounces fell on his thigh. He developed pain and difficulty in walking and was transferred to another department, where he developed back pain. Complete examination by a variety of specialists and hospitals failed to reveal any organic basis for his difficulty. On admission the patient walked with a bizarre limp, dragging one foot behind him. Reexamination revealed no evidence of any organic basis for his symptoms. However, on clinical psychiatric examination and on psychologic testing, evidence of a paranoid schizophrenic process was discovered. He had marked bodily preoccupation with somatic delusions and paranoid trends based on homosexual fears, as well as other signs of severe personality disorder. It was felt that the physical disability was on a conversion basis and that it served to prevent an acute psychotic break. Because of language difficulty, low intelligence, and legal implications, no treatment was attempted, and the patient was referred back to the original source. It was our opinion that to take away the symptom at this time, whether by suggestion, narcosis, or hypnosis, would probably precipitate a psychotic episode that we were not prepared to deal with because of the reasons mentioned.

It must be emphasized that in speaking of psychosis we distinguish between overt psychosis and latent psychosis. In a sense it is like the difference between tuberculous infection and tuberculous disease. There are many individuals functioning in society who have psychotic personality structures. Under sufficient stress the psychosis may manifest itself, but not always. Thus there are patients with physical disability who may be schizophrenic but who do not manifest it overtly. The label of psychosis does not automatically militate against rehabilitation, although experience indicates that such patients often experience difficulties in the rehabilitation process.

The unmotivated patient

The rehabilitation program provides a setting in which a disabled person can help himself. Patients who refuse to help themselves generally are referred to as "unmotivated."

The patient who is not motivated for the rehabilitation program is one whose goals are not congruent with those of the rehabilitation effort. His energies are not directed to the tasks at hand but are taken up with other, more private goals. Usually such an individual does not share in the goals of the larger group of which he is a member; for example, a patient who is unattractive physically will not attempt to improve her appearance because her ugliness gives her an excuse not to mingle with other people. She may be considered "unmotivated" to change her looks. However, it would be fallacious to note this without adding that in a sense she is motivated, but her motivation does not coincide with that of most people. The "unmotivated" patient can be understood only if we keep this supplementary observation in mind.

Why does a patient cling to goals that sabotage his chances for success in rehabilitation? This is not easy to answer. A painstaking analysis of each patient's behavior and circumstances is required. Before any attempt can be made to understand some of the conditions that result in poor motivation, we should note that the goals set up in a rehabilitation program are not absolute or given. They are determined by the team. It is possible, therefore, that on occasion the team, because of faulty judgment or misinformation, decides on inappropriate goals for a given patient. When this occurs, we can say that the patient is not motivated because the team has set up the wrong goals. In some cases this occurs because the team members assume that the patient sees the world as they do and that he has the same desires and wishes as they have. The fallacy behind this assumption can be illustrated by the following example. A group of severely disabled cerebral palsied and poliomyelitic adolescents were asked which of these two conditions was more unpleasant—cerebral palsy or poliomyelitis. Each group answered that they would rather have their own condition than the other one. The patients with cerebral palsy explained that they were more fortunate than those with poliomyelitis because they were born with their handicaps and did not become sick later in life; the patients with poliomyelitis, on the other hand, thought that they were more fortunate because at least they were born "normal" and were, therefore, normal people who had become sick, whereas the same could not be said for the cerebral palsied patients. In short, each group interpreted the same facts differently and perceived the world within the framework of its own needs. Such differing patterns of needs in the various patients can be overlooked all too easily.

In most instances, however, the very existence of a team and the fact that the goals of rehabilitation are set by a team rather than by one individual provides a system of checks and balances. The goals decided on, therefore, are usually realistic.

Which forces within the patient contribute to his lack of motivation to fulfill the goals set by the team? We can answer the question by citing some of the reasons for an absence of motivation.

1. The patient fears that he will not achieve complete recovery and refuses to accept partial fulfillment of this goal.

2. The patient wishes to punish himself by not improving.

3. The patient wishes to punish or control others by making himself dependent on them.

4. The patient's goals are irrelevant to rehabilitation; for example, the patient wants rehabilitation to cure shyness and anxiety by some magic means.

5. The patient is still in a state of psychologic mourning or shock so that he can-

not set up reality-oriented goals; for example, such a case often is seen among patients who exert great effort on the physical medicine part of the program but refuse to participate in the psychosocial aspects of it.

6. The patient has been misinformed about his physical condition; for example, the patient may have been told to allow 2 years to pass before he begins to think of walking.

7. The patient has an irrational investment in his body or his bodily skills so that a disability means not only a physical handicap but also a total destruction of powers of self-defense. A coal miner sustained a paraplegia and was depressed, not because he was not learning to walk but because he could no longer defend himself physically. He told our psychosocial service staff that his only hope lay in the possibility of carrying a gun with him at all times.

8. The patient built up habit patterns resulting in strong secondary gains from being disabled. This is seen most easily in patients who become reliant on financial subsidies during a long course of illness. Gradually the subsidies become incorporated into the personality structures as major defenses. Any attempt at vocational rehabilitation threatens financial security and may therefore be strongly resisted.

9. The patient's need for achievement appears to be exaggerated. Such a pattern is found among people who have built up successful careers at the expense of other aspects of life. Very often a way of life oriented only about hard work makes a physical disability unbearable. The way of life has been shattered. Originally, this way of life may have represented a defense against a strong underlying wish to be taken care of. With a physical disability, this wish, which had previously been unacceptable, now becomes socially permissible. An example of this may be seen in the case of a railroad employee who had worked 364 days in each of the last 3 years. (He had worked for the same company for 17 years.) When he became paraplegic, he refused to consider employment, on the grounds that his financial subsidy would be jeopardized. His great devotion to his job probably was rooted in a fear of his wish to be taken care of so that he had to prove how hard he could work and how self-sufficient he was.

10. The environmental situation may seem hopeless to the patient, and he therefore has no incentive to rehabilitate himself. He may feel that the process is futile, because there is nothing of worth in his future. This may involve an unhappy home situation or an environment that offers no vocational opportunities.

11. Where the disability has occurred in adolescence, before the acquisition of either intellectual, vocational, or emotional skills, the problems of rehabilitation often seem overwhelming and may appear as poor motivation.

Although we have been speaking of motivation as if it were a unitary whole, we must point out that patients may be motivated for some parts of the program but not for others. In our experience, there are several vital areas to explore in discussing the goals of rehabilitation with a patient who has just entered on a course of treatment. These include discovering what the patient anticipates and how realistic he is with regard to (1) complete physical restoration to a predisability status, (2) motor skills for self-care, (3) vocational and social planning, and (4) psychologic and adjustment problems.

What can be done about the unmotivated patient? Two basic steps are involved here. One is preventive, and the other is therapeutic. To prevent a decrease of motivation from occurring, it is suggested that (1) all medical information that is given to the patient be weighed carefully in advance; (2) the patient be helped in setting up unambiguous and realistic goals; (3) bad habits and secondary gains be prevented from forming by asking for psychiatric evaluation as early as possible in those cases that augur trouble; (4) the team be cautious about the timing of the program; for example, different patients require different lengths of time to adjust to the fact that they are disabled; some patients are not ready for future planning until several months have elapsed; and (5) the staff try to recognize the patient's specific needs and utilize the personnel of the team accordingly.

Therapeutic recommendations for the unmotivated patient are, of course, dependent on the reasons for the lack of motivation. At times, challenging the patient directly about his lack of motivation is helpful. Such a challenge is best carried out by confronting the patient with the reality of his situation rather than by punitive measures or coercion. Challenging a patient serves to mobilize the patient's anxiety. In those instances in which a patient is anxious already or in which his ego resources are weak, such a challenge must be handled very carefully. An example of appropriate challenging is that of a 21-year-old paraplegic patient with a high school education and a background in farming who persistently avoided vocational planning by telling the counselor that he wanted to go to college to become a high school teacher. This goal seemed unreasonable in the light of information derived from psychologic tests. When the patient was challenged with these data and told that he seemed to be "pulling the counselor's leg," he readily admitted that the goals verbalized to the counselor were designed to ward off anxiety associated with thinking about the future realistically. Constructive planning then became possible.

In many instances the nonmotivated patient does not respond to challenge or to confrontations of reality. Under such circumstances, several approaches are used: (1) those areas in which the patient does show positive motivation are emphasized and the other phases of the program are discontinued temporarily; (2) identification with positive forces and with other patients in rehabilitation is built up by encouraging participation in group and recreational activities; (3) psychotherapy with a special emphasis on building up a relationship with the therapist is undertaken; (4) authority figures are used to lend help to the patient who has a magic sense of self; authority figures, such as the medical director of the hospital, are called on to help the patient accept his dilemma and channel his energies appropriately.

Pain and addiction

A frequent accompaniment of physical disability is the symptom of pain. While the source of the pain is often tangible, determined by the anatomy and pathology of the disability, on occasion it is not. Similarly, even though the source is tangible, the nature and degree may not seem to be consistent with the disease. This has led to a separation of pain into organic and psychogenic, which may be a convenient clinical classification but often leads to errors in management.

In the final analysis all pain is a psychic phenomenon, in that its presence is dependent on the functioning of the higher nervous centers, and is subjective in nature. Unless this is emphasized, a dichotomy in thinking and practice may result.

There may be a tendency to treat the pain rather than the patient. Since pain can inhibit rehabilitation activities and lead to interpersonal difficulties between patient and staff, its proper management may spell the success or failure of the entire process.

Pain produces emotional changes even in well-adjusted individuals. It increases dependency needs, self-centeredness, and secondary gain. The personality of the patient and his reaction to the disability and to pain will influence his subjective sensation and behavior. Environmental factors also play a role. Pain tends to appear worse at night because of anxiety, aloneness, and absence of external interests. The patient is more aware of himself at such times. Similarly, pain may be forgotten during recreational activities, yet complained of while the patient is in the presence of his physicians.

Patients vary in their threshold to painful stimuli. Some tolerate a great deal of pain, whereas others carry on hysterically in the presence of objectively mild stimuli. Tolerance of painful stimuli may be constitutionally determined, but there are certain personality differences that can be observed. The dependent, immature person and the anxiety-ridden patient are more reactive to pain. The detached, emotionally controlled patient exhibits a higher tolerance, as does the depressed, inhibited patient. Hysterical patients may show conversion phenomena that may be revealed as pain, paresthesia, or anesthesia—with or without organic pathology.

Certain disabilities present special problems. In paraplegic patients, pain *below* the level of the cord lesion is common and can be quite severe. It is often difficult to explain such pain, especially in patients in whom laminectomy has revealed complete transsection of the cord. Psychiatric studies have failed to explain this pain on a purely psychogenic basis.

Pain may affect rehabilitation in a variety of ways: it (1) may prevent necessary physical activities, such as transfer from bed to wheelchair or ambulation; (2) may lead to sleeplessness and fatigue, which can hinder rehabilitation; (3) may necessitate surgical or pharmacologic intervention, which may delay or prevent rehabilitation; (4) may lead to interpersonal difficulties, arising out of demands on the staff and resentment of complaints by other patients; (5) may lead to bodily preoccupation on the part of the patient, with subsequent withdrawal from activities; and (6) may be used for secondary gain to maintain dependency and to manipulate the staff.

Successful management of pain stems from a thorough medical and psychosocial examination. Accurate medical diagnosis, as well as a careful evaluation of the character and duration of the pain, is essential. A detailed analysis of the environmental, personality, and social factors should be carried out. The role of insurance, compensation, and welfare must be considered.

When the evaluation is complete, means of relieving the pain should be exhausted, one by one. Such measures include the use of nonnarcotic drugs, physical therapy, radiation therapy, anesthetic block, and surgical procedures. At the same time, adjustments in the patient's life setting should be an intensive part of the therapeutic plan. Decrease of physical activity, use of rest periods, simplification or decrease of responsibilities, and exhaustive inquiry as to the presence of possible sources of painful feelings in the environment—past, present, and future—should be considered.

A major factor in this type of approach is the attitude and performance of the physician. The physician who views pain traditionally as merely a symptom of physical disease, and fails to see a personality experiencing the pain, will find it necessary to resort to drugs, frequently narcotic drugs, to control the pain. On the other hand, the physician who is able to offer his patient sympathetic understanding, steady support, and realistic and frank perspectives can obtain similar results while resorting only infrequently to habit-forming drugs. He prescribes drugs as a specific part of the relationship between physician and patient and not as a substitute for that relationship.

The physician has three types of responsibilities in the management of chronic pain: (1) he must treat incapacitating pain, (2) he must provide management for long-term rehabilitation or elimination of the condition causing the painful syndrome, and (3) he must do no harm as he attempts to rehabilitate.

One of the dangers in the management of chronic pain is that of addiction to narcotic drugs. Approximately 20% of individuals admitted to the United States Public Health Hospital at Lexington, Kentucky, can be termed "medical" addicts, in whom addiction has stemmed from administration of drugs by physicians for therapeutic purposes. Three phenomena characterize the addiction syndrome.

1. *Physical dependence.* Long-term administration of narcotics in large enough doses inevitably results in a syndrome that requires repeated administration of these drugs to prevent the appearance of objective physiologic reactions. Morphine, in doses of ¼ grain 6 times a day for 3 weeks in nonaddicts, produces physical dependence in most persons.

2. *Psychologic dependence.* This is concomitant with and often precedes the development of physical dependence. Several predisposing hazards should be looked for. These are preexisting alcoholism in the patient, evidences of neurotic behavior, extreme degrees of anxiety, overdependent attitudes toward the physician, and previous psychotic episodes.

3. *Tolerance.* This means that, as time passes, an increasing need for the drug is manifested to achieve the same effect. When this is noted, it should be a clue that the management plan is not effective, and a review should be instituted; for, if addiction develops, it means that the therapeutic plan has failed.

Since tolerance usually precedes either physical or psychologic dependence on drugs; special attention should be given to it. Tolerance is probably more dependent on frequency of dose than on size, so that careful choice of dose to effectively manage the particular pain is necessary. Choice of drug should be guided by the nature of the pain: an intermittent pain requires a short-acting drug like meperidine (Demerol), whereas continuous pain requires a long-acting drug like levorphanol tartrate (Levo-Dromoran) or methadone hydrochloride. Because drugs of the morphine group, including dihydromorphinone hydrochloride (Dilaudid), heroin, and pantopium (Pantopon), produce a relatively high euphoric state, they must be used with care in patients suffering from psychologic tension as well as pain. Codeine can be used freely if it is used with discretion because it is rare for one to become addicted to it. Finally, all narcotic drugs produce cross-addiction, and the development of tolerance to one results in increased tolerance for all, which precludes switching from one to another to avoid addiction.

Adjunctive drugs in the management of pain should include chlorpromazine (Thorazine) and barbiturates to control the anxiety occurring at night that may intensify the pain. A mixtures such as secobarbital sodium and amobarbital sodium (Tuinal) or secobarbital sodium, butabarbital sodium, and phenobarbital (Ethobral) to carry the patient through the night is usually more effective than the short-acting drugs.

Conversion reactions

The neurotic individual, suffering from anxiety arising out of earlier unresolved conflicts, copes with the anxiety and its causes in a variety of ways that serve to minimize or deny the existence of the painful emotion. Of the many mechanisms at his disposal, that of *conversion* is most pertinent in physical disability and rehabilitation. Conversion produces an apparent physical impairment that prevents the unacceptable impulse from being acted out. Thus it paralyzes the hand that might strike, blinds the eyes that might peep, silences the voice that might talk out of turn. In conversion the physical symptom counteracts the anxiety, permitting the patient to be consciously comfortable. He thus relieves his anxiety at the cost of incapacity.

Although many symptoms may result from conversion, disturbances of sensation and motion are the most common. The disturbances of sensation include anesthesia (absence of sensation), paresthesia (distorted sensation), and pain. Superficial or deep sensation or the special senses may be affected. The disturbances of motion include paralyses, usually of limbs, digits, or the speech mechanism, and uncontrolled movements, as in specific types of tics or convulsions.

Since many of the phenomena of conversion simulate the syndromes seen in rehabilitation medicine, differential diagnosis is of the utmost importance, although at times very difficult. This is particularly true in those cases in which combinations of organic disease and conversion reactions are present. The patient with a conversion reaction, however, usually demonstrates the following characteristics:

1. The neurologic signs follow the patient's idea of anatomic distribution. Hysterical anesthesia, for example, does not follow segmental or peripheral nerve distributions but involves areas such as the hand and foot—so-called glove and stocking anesthesia. Inconsistent and physiologically impossible combinations are seen, such as hysterical paraplegia without loss of bowel or bladder control or without expected atrophy of muscles.

2. The conversion symptoms have a symbolic meaning and are related to the unconscious conflict or the precipitating situation. Thus conflict over the impulse to look at sexual activity may lead to hysterical blindness. Conversely, the symbolically significant area may be spared, as in hysterical hemianesthesia (loss of sensation of one half of the body), which customarily in males spares the genital area.

3. The patient shows far less anxiety than organic illness of the same degree would be expected to produce. Often, this absence of anxiety leads to an expression called *la belle indifférence,* a term introduced by Charcot, who first described this condition.

4. Patients who present this reaction display features of a neurotic personality. Their history indicates previous tendencies to evade or deny unpleasantness. They are very suggestible and usually are prone to therapeutic suggestions such as faith cures, spin adjustments, and placebos.

Treatment may include authoritative suggestion, narcoanalysis, and hypnosis—all of which may succeed in removing the symptom. There is danger, however, in thereby attacking the symptom because of the anxiety that lies behind it. Relief of the symptom may release overwhelming anxiety and even precipitate a psychotic reaction. Therefore, only a trained person should undertake such treatment. Often, removal of the symptom is not sufficient—treatment of the total personality is required. This may involve intensive psychotherapy of a psychoanalytic nature.

Case study. A 21-year-old nurse was admitted with paraparesis following an infection diagnosed as poliomyelitis. After careful neurologic examination and electromyographic and neurophysiologic studies, it was decided that she had neither anterior horn cell nor muscle damage.

Before her illness the patient had been working very hard, laboring under the responsibilities of a supervisor's job in which she felt inadequate. She went to a resort for a weekend, where she met a young man who showed more than casual interest in her. While dancing one evening, she became dizzy and went to her room. Upon returning home, she was hospitalized, and a diagnosis of poliomyelitis was made. She was discharged home, treated by a visiting physiotherapist, and cared for by a resentful mother for almost a year with no progress. At the urgent suggestion of the family doctor she was admitted to the Rehabilitation Service.

The absence of organic damage plus the presence of positive psychologic factors led to a diagnosis of hysterical paraparesis. She was seen three times a week by our psychiatric resident. During this time she continued in a wheelchair and participated in ambulation exercises and physiotherapy. No progress was made in either sphere. Because of the limitation of time and the pressing necessities of the patient's future, a two-therapist type of treatment was decided on. The attending psychiatrist acted as the authoritarian figure, firmly confronting the patient with the facts, removed the privileges of a wheelchair, and granted her use of one cane. The role of the resident was to deal with her feelings, of which for the first time there were many, as well as significant productions relating to her mother. The latter had been utterly dominated by the patient's helplessness.

This course was effective. Four months after admission the patient left the hospital walking, and arrangements were made for continued intensive psychotherapy at an outside clinic. The patient's motivation had been changed from emphasis on her physical symptoms to emphasis on her psychologic problems.

It must be stated that conversion reactions occur outside the awareness of the patient. They are not conscious or deliberate and in this way differ fundamentally from malingering.

Sexual problems

Of all the problems confronting the disabled patient, probably least has been written about those concerned with sexual functioning. There are good reasons for this. The disabled patient seldom ventures to discuss them, and the examiner frequently fails to question the patient on this emotionally charged subject. Both are anxious; both are protective. The disability is often so overwhelming and the reality problems so difficult that discussion of sexual problems seems superfluous, if not carrying with it the feeling that one is adding insult to injury.

Yet the problem does exist, and it often is the source of considerable shame, anxiety, and depression. The problem exists not only for the disabled patient but for spouse or lover as well, and the latter's attitudes and behavior can markedly affect the patient's motivation.

Sexual behavior is dependent on both biologic and psychologic factors. Biologic, neurologic, and endocrinologic factors are necessary for adequate physiologic sexual functioning. From a behavioral standpoint early psychosexual development, along with numerous social and cultural experiences, leads to attitudes, both conscious and unconscious, that affect sexual fantasies, impulses, and behavior. Sex then, in human beings, represents more than a biologic phenomenon. It is a profoundly emotional experience, carrying with it a host of emotional attitudes, both positive and negative, and often reflecting the very deepest sources of self-esteem and satisfaction.

In physically disabled persons sexual function may be interfered with in two ways: (1) anatomic injury to the genitalia or the spinal cord, which may make normal sexual functioning impossible, and (2) psychologic changes resulting from the disability, such as doubts about adequacy, attractiveness, and potency.

The outstanding example of physical disability in which anatomic injury plays a determining role interfering with sexual function is spinal cord injury, with resulting paraplegia or quadriplegia. From a physical standpoint it is important to distinguish spinal cord damage that results in diminution of motor power without impairment of sensation from spinal cord damage that impairs motor powers as well as sensation. The sexual experience in each type is different.

The problems of the male are somewhat different from those of the female. In the male, cord injury leads to partial or complete impotence. Studies have shown that those paraplegic males who are impotent are considerably more depressed, withdrawn, and emotionally overwhelmed than those who are potent. They regard their own bodies as useless, and they present more difficulties in constructively dealing with problems of readjustment and rehabilitation. Many experience dire apprehension at the thought of returning home to wives and family. Some of these patients establish a much better relationship with paraplegic females than with nondisabled women. The demand on their adequacy as a masculine sex partner is greatly reduced in such an attachment.

In the female, not only are attractiveness and acceptability problems, but also the ability to bear children is a source of concern. Since anatomically and physiologically the role of the female sexually is more "passive" than that of the male, physical disability does not impair her receptivity. Because of sensory difficulties, vaginal and clitoral orgasm may be impossible, but this does not preclude sexual satisfaction. Nor is fertility always impaired. We have seen paraplegic females conceive and go on to term with delivery of normal children.

A serious problem for both male and female paraplegic patients is an esthetic one. Since control of bladder and bowel function is lost and since assistance is often needed from the husband or wife, the sexual experience becomes entwined with excretory functions, which can produce a markedly inhibiting effect on the spouse or lover. It requires a most mature and understanding partner to cooperatively participate in these experiences. It is no surprise, therefore, that some marriages in which one partner becomes paraplegic end in divorce. Many of these problems might be prevented if adequate attention were given to sexual problems in rehabilitation, not only to the patient but to the family as well.

Case study. A 37-year-old attractive female, married and mother of three children, developed poliomyelitis resulting in quadriplegia. Her physical rehabilitation proceeded satisfactorily, and with the use of gadgets she was able to maneuver in a wheelchair and engage in some hand activities. However, serious family problems developed. The husband, an intelligent man, had taken to alcohol and the family was on the verge of collapse. The patient became increasingly anxious and afraid to go home, since she was so utterly dependent on her husband.

Interviews with the husband elicited the fact that he could not tolerate having sexual intercourse with his wife, even though he found her attractive. When she was home weekends, the housekeeper was usually away. He had to assist his wife in bowel and bladder functions, as well as in changing her menstrual pads. As he put it, "The whole business has gotten to be so clinical that any idea of sexual relations is revolting."

He voluntarily admitted to some kind of immaturity or weakness in this attitude but could do nothing about it. He was reluctant to tell his wife about it for fear of hurting her. Instead he blotted out the whole business by drinking himself into insensibility when his wife was home.

The problem was tactfully discussed with his wife. This spurred her to greater efforts in self-care. The husband agreed to make an attempt to work out his own feelings and problems, and arrangements were made for him to receive psychologic counseling. He stopped his drinking and the possibility of renewed family integrity was thus established.

The majority of diabled patients do not suffer from anatomic or physiologic interference with sexual function, yet many do have sexual problems. In some this may be of a mechanical nature, involving difficulty of positioning due to muscular weakness. Usually whatever difficulties exist arise out of psychologic reasons, many of which overlap those discussed elsewhere in this chapter. Problems of cosmetic appearance and vanity, self-esteem, inferiority, depression, and anxiety all may be reflected in sexual attitudes and functioning. Depression usually inhibits sexual desire. Anxiety may inhibit but may also increase sexual fantasy and desire. Aggressive impulses often are channelized sexually, and there may be frightening drives. As with attitudes toward excretory functions, sexual attitudes are often associated with experiences in childhood, and these may involve unpleasant attitudes on the part of parents and of society generally.

It has been our impression that sexual perversions are uncommon, although this statement is not based on any special study. However, clinical experience with thousands of disabled patients in closed settings, based on observations by physicians, nurses, and attendants, as well as on psychiatric interviews and psychologic testing, would seem to support this view.

It should be emphasized that some patients have had sexual problems and anxieties before the disability. In such patients sometimes one sees a relief from these anxieties as a result of the disability. However, in our experience, this represents the exception rather than the rule.

The management of sexual problems is based on their recognition. As pointed out, they are often overlooked because of anxiety in both the patient and the examiner. If a patient is to be questioned about sexual matters, it must be done tactfully and appropriately. As a general rule, sexual questions should not be asked unless and until other areas are explored. Since the disability is what most patients are concerned with, this must be the starting point. As time passes and a trusting relationship is established, sexual matters may then be discussed in a matter-of-fact manner that will be reassuring and supportive to the patient. If resistance appears, the issue should not be pressed. It will be found that if a favorable environment and relationship exist, in time the patient himself—if it is a problem—will bring up the subject.

It is often helpful to a patient to be able to air his problems with a sympathetic person. Once they are aired, reassurance, clarification, explanation, and suggestion may be employed. Psychoanalytic interpretations are appropriate only in psycho-therapeutic settings and must be used judiciously and by a knowledgeable person. Otherwise they will either pass in one ear and out the other or lead to a justifiable disruption in the relationship.

In our experience, patients who are often reluctant to discuss their feelings about sexual problems may discuss their attitudes more openly in group therapy. We have observed that patients on the ward, who have lived together and come to know each other, may discuss their feelings about sex and even the technics they use to overcome mechanical difficulties. Sometimes these are discussed in a flippant, wisecracking, joking way to cover up the great anxiety that the group feels. Group therapy, under the leadership of a person who is skilled in interpersonal relations and psychotherapeutic technics and who has some knowledge of the problems of paraplegic patients, is a potent vehicle for getting patients to communicate in this area. Group discussions also seem to bring out some difference in attitude between married and unmarried patients and between older and younger patients. It is a recommended form of management but requires great skill. In our setting, successful group therapy was achieved only after several failures.

CONCLUSION

While we have unduly, if understandably, emphasized the "problems" in physical disability and rehabilitation, this chapter should not be concluded without comments on the positive and healthy aspects of both patients and the rehabilitation milieu. The fact is that, despite what appear to be shattering problems, many patients do well. Physical disability imposes a profound challenge to the individual because of the disease itself, as well as the necessities demanded by the rehabilitation process. The patient is sharply confronted with himself, and he is also confronted with inter-personal relationships and the social structure of the rehabilitation milieu. He is en-gaged in a struggle, out of which can come change—real change—in his personality, attitudes, feelings, ways of thinking, awareness, and perceptions and in the direction and goals of his life. It represents a true crisis, and the outcome may lead to more meaningful and mature, more constructive and creative functioning as a human be-ing than was possible before. We have seen this happen, and we should know more about how it happens and why. We have been deeply absorbed in the effect of the personality on the rehabilitation process. What we need to know more fully is the effect of rehabilitation experience on personality.

REFERENCES

1. Weinstein, E. A., and Kahn, R.: Denial of illness: symbolic and psychological aspects, Springfield, Ill., 1955, Charles C Thomas, Publisher.
2. Grayson, M., Levy, J., and Powers, A.: Psychiatric aspects of rehabilitation, Rehabilitation monograph 2, New York, 1952, Institute of Physical Medicine and Rehabilitation, New York University-Bellevue Medical Center.
3. Aldrich, C. K.: Psychiatry for the family physician, New York, 1955, McGraw-Hill Book Co., Inc.
4. Wolberg, L.: The technique of psychotherapy, New York, 1954, Grune & Stratton, Inc.
5. Hoch, P., and Zubin, J.: Anxiety, New York, 1955, Grune & Stratton, Inc.
6. Hoch, P., and Zubin, J.: Depression, New York, 1955, Grune & Stratton, Inc.

7. Sainsbury, P.: Suicide in London, Maudsley monograph 1, New York, 1956, Basic Books, Inc.
8. Barker, R., Wright, B., Myerson, L., and Gonick, M.: Adjustment to physical handicap and illness: a survey of the social psychology of physique and disability, New York, 1953, Social Science Research Council.
9. Garrett, J.: Psychological aspects of physical disability, Office of Vocational Rehabilitation Service Series 2100, Washington, D. C., 1953, U. S. Government Printing Office.
10. Noyes, A.: Modern clinical psychiatry, ed. 4, Philadelphia, 1953, W. B. Saunders Co.
11. Fisher, S. H.: The psychiatric team in a physical rehabilitation service, Amer. J. Orthopsychiat. **27:**19, 1957.
12. Nathanson, M., Bergman, P. S., and Gordon, G. G.: Denial of illness: its occurrence in one hundred consecutive cases of hemiplegia, Arch. Neurol. Psychiat. **68:**380, 1952.

CHAPTER 16

PRINCIPLES IN MANAGEMENT OF SOCIAL PROBLEMS

To help the disabled person receive the careful, skilled treatment he needs in adjusting to group living and social contacts while undergoing rehabilitation procedures and in later life, social work services given by a professionally trained social caseworker are essential. In the following discussion the term management refers to a democratic way of approaching critical human problems.

All social problems arise from the way human beings share with each other, the distribution of material goods, and the availability of medical, social, and educational resources. Social problems such as those that result from illness can happen to anyone and, for this reason, blame and accusations are useless in finding the cure and prevention.

SOCIAL NEEDS OF THE PATIENT

Wherever the physically disabled person lives, he needs an opportunity to work at the prevailing wage scale so that he may meet his basic needs for food and shelter for himself and his family. He needs suitable transportation to and from work. If he is confined to a wheelchair or has difficulty climbing stars, his housing has to be such that he can get in and out and move through the house easily and must be at a rent level that he can afford. The nondisabled person can manage in a walk-up apartment, but a physically disabled person confined to a wheelchair cannot. Religious services and church activities are essential to many patients and certainly need to be available to them. Recreational facilities are as necessary for them as for anyone else. Last, but of equal importance, is the need for satisfactory relationships with family and friends.

In too many instances the family, the individual, or his community has not made adequate provisions to meet these needs. Not all of these needs can be met by the individual or his family without help from the community. If adequate housing is not available at a rent level the patient can afford and if the opportunity to work is denied him, he is forced to depend on his family and the state to support him. A rehabilitation center's contribution to the physically disabled person can be completely destroyed when the community to which he returns assumes no responsibility.

The rehabilitation center, in its broadest concept, is a community agency whose main purpose is to enable the physically disabled person to resume as productive a role as possible in his family group and community. Although the center is located in one community, it can and will serve patients from many communities. In addition to serving the patient's needs, it also trains greatly needed personnel, in all disciplines and from all over the world, who, in turn, serve the needs of physically disabled people in their home communities. The patient's eligibility requirements, unlike those of many state- and city-sponsored facilities, can be flexible. If his physical and emotional status make rehabilitation treatment feasible and if funds can be secured for his treatment, he can be admitted. There are no racial, religious, or geographic boundaries that prohibit his acceptance. Patients are accepted on the inpatient or outpatient services, depending on the degree of their disabilities and on their social situations. In either service all the facilities of the program are available to them and their families.

Direct service to patients is only one aspect of the social worker's function. Of equal importance is her work with the families and the sharing of her findings with the other team members so that the best possible service can be given the individual patient.

REHABILITATON CENTER ENVIRONMENT

The environment of the rehabilitation center has manifold ramifications for both patients and staff. The physical plant, which is designed to meet the needs of the physically disabled person, is new and strange to the average patient and his family. It is physically designed to encourage the patient in self-care, which is of vital importance, but it also has a negative component that must be recognized by all members of the staff. Where else can the physically disabled person find such a physical environment, a unique community designed to meet his needs? Thus many patients and their families will be reluctant to have the patient return to a house that is physically inadequate and to a community in which the patient's physical, vocational, and social needs are negated by the attitudes of many of his fellowmen.

In no other setting are there so many different disciplines housed together to meet the needs of one patient as in a rehabilitation center, and it is in this interaction that the emotional environment of the setting arises. Although the center cannot be considered a traditional hospital, it is nonetheless a medical setting, and the welfare of the patients within this small community is intimately related to the quality of relationships existing between various team members. If positive relationships exist between staff members, the emotional climate provided by the setting is of vast therapeutic value.

It is important to understand this specific community historically. Although the center is basically a medical facility, it recognizes the need for contributions from other professions carrying out its broad services to physically disabled persons. The very presence of social workers and members of other professions is sufficient evidence of this fact. Since human beings, members of the staff as well as patients, live through their feelings, not intellect, this can be an area of conflict between team members. In any hospital setting, life-and-death factors control the environment, and the medical profession must, of necessity, assume the role of authority. "This authority, coupled with the prestige of the medical profession, represents formidable hurdles for other professional groups in the hospital."*

Although such authority may not be as necessary in a rehabilitation center, since the services offered are not feasible for critically ill people, the tradition of medicine can prevail. "It is recognized that the profession of medicine is extremely structure- and status-oriented. This is brought about in part by the standardized training in medical schools and the rigidity of the American Medical Association professional code."* Hubert Bonner,[3] a social psychologist, has pointed out that the physician's domination by strict codes tends to make him conservative in his professional attitudes and behavior. There are, of course, exceptions to this observation ". . . but the cultural fact is significant since it presents a contrast to the profession of social work in which there is a much higher degree of professional equalitarianism and a strong liberal philosophy based on identification with basic human needs in our society."* The historic difference between the training of the physician and the social worker, if resolved in the interest of the patient, can produce a healthy balance. In all fairness it must be stated that the personality makeup of the physician and the social worker will, in reality, determine how each will utilize his training and philosophy.

The hospital and its environment represents a subculture for patients and staff alike, and it is necessary to understand and deal with this subculture in building a professional service as a part of the hospital setting.[1] The medical function of a rehabilitation center, which is its primary purpose, has to be the central focus of all the activity. Thus the social worker, to be useful and realistic, has to work with patients and their families within this framework. This situation can cause difficulty and conflict to the social worker because she has to find the best means possible for carrying out her work. It can make an insecure worker feel defensive and robbed of status if she cannot accept and understand the primary purpose and historic roots of a medical setting. The social worker is not, of course, the only nonmedical worker who has to understand this situation. A "status race" can become contagious and infectious, since all human beings—professional staff, patient, and the cleaning personnel alike—have varying degrees of status needs. It is any disproportionate need that each team member needs to be aware of in himself and, by his awareness, avoid. Only in this manner can a truly democratic team approach be realized and a basically friendly environment be provided for the individual patient.

---

*From Blackey, E.: Social work in the hospital: a sociological approach, Social Work **1:**43, April, 1956.

THE SOCIAL WORKER AS A MEMBER OF THE REHABILITATION TEAM

The specific functions of the social worker in a rehabilitation center are not so easily defined as they may be in a traditional hospital setting. There are reasons for this. The traditional hospital team consists of physicians, nurses, and social workers, whereas in rehabilitation the psychosocial team includes psychiatrists, psychologists, and vocational counselors. Since members of all three of these disciplines use the patient-worker relationship to help patients achieve changes, the social worker cannot claim that this method of helping is her exclusive contribution. But she has an important service to offer, a service that depends greatly on her skills and on other agencies in the community. She is an "enabling person." Her contribution is primarily to enable the patient to use other services within the hospital or to use the appropriate resources outside. The social worker represents the community within the hospital and the larger community outside. Her job, from the intake interview on, is to help the patient relate to the immediate society of the center and to his own community to which he will return. If the patient can adequately relate himself to these two societies, he may have no need or desire for social service.

The social worker's unique service lies in her ability to do environmental manipulation. Her areas of special emphasis are concerned with adequate income, housing, recreational outlets, availability of transportation, and the understanding and acceptance of the patient by his family.

Since the social worker represents the hospital community and is responsible for "social management" therein, it is necessary for her to establish good working relations with other team members. Thus understanding the roles of the other team members is essential. When necessary there will be an interchange of roles frequently among professional team members just as there is in the roles of parents in child care. Rigid adherence to roles serves no constructive purpose.

The rehabilitation physician is in charge of his patient just as he is in a standard patient-doctor relationship. He decides on admission and discharge with help from other team members.

The physical therapist has to work closely and intimately with patients; the nature of his work causes the patient pain and anxiety and emphasizes to the patient his physical inadequacy. This situation calls for skill and understanding on the part of the physical therapist. It becomes clear at times why the patient misses his program—to avoid pain and anxiety, not just treatment. The same situation can exist in occupational therapy when a patient is learning to feed himself or to write again.

The nurse helps the patient to increase his physical adequacy. This is a new concept to most patients and families, who have seen nurses in the role of doing. For example, during acute illness a nurse feeds a patient. In rehabilitation the nurse may help and teach the patient to feed himself. The ward becomes the patient's "field placement" or "internship," where he is expected to practice what he has learned in class. In such a situation—when the nurse is helping the patient to do for himself rather than doing for him—the patient may feel that there is a lack of understanding for his pain and anxiety. For this reason there may be excessive complaints about nurses.

The psychiatrist is responsible for the diagnosis of mental and emotional illness. He is a consultant for all team members and agency representatives, and he engages in psychotherapy. Also, he is available to the social worker for supervision in the management and understanding of patients she may be seeing for intensive casework help.

The vocational and psychologic services offer counseling to patients to help them adjust to the services, as does social service, until they are able to plan vocationally. Since patients come primarily for physical reasons, timing is essential in prevocational planning. When a patient is lying on a stretcher and is asked to think about his vocational future, he may become antagonized and feel very much as a child who is asked by a grownup, "What do you want to be when you grow up?"

The psychologist has specific tools to use in and contributions to make from his psychologic testing. He can ascertain the present intellectual functioning and evaluate the intellectual endowment of the patient. His projective technics are diagnostic aids aimed at uncovering unconscious conflicts. Understanding gained from an evaluation of these areas is of vast importance in determining the personality limitations and potentialities of the patient.

The speech and hearing services offer a unique contribution. If the patient cannot communicate through speech and has severe hearing losses, he is greatly limited in his ability to adjust. The anxieties that arise from speech and hearing losses present serious problems of frustration and misunderstanding between the patient and his family.

The recreational service gives the patient an opportunity to participate with others in various social activities. If the patient does not use this service, it is important to understand why he does not. Very likely he will avoid social contacts at home if he does so at the center.

The prosthetic service also plays an important role in rehabilitation, since the patient's use of a prosthesis indicates his ability to come to terms with his loss. At times the service is called on to meet a social problem of magnitude. If a patient cannot learn to walk with artificial limbs, should they be procured for him? From a social worker's point of view, the dignity and independence of the human being are violated, perhaps for no other reason than cosmetic reasons, if the legs cannot be procured. Some governmental and city agencies that are responsible for the financial costs will not purchase the prostheses unless they can be utilized in walking.

The self-help devices and activities of daily living service offer the only ways in which a patient can learn to feed himself, write, put on clothing, apply lipstick, etc., and to take care of his other daily needs. In this area families need to accept and understand this service, so that they will not destroy the patient's hard-earned gains and initiative when he returns home.

The multiple professions and varied services of the professional team, as outlined, make the rehabilitation center a highly specialized community. The new patient rarely has advance information concerning the services that are available to him in this community. He usually has been referred by private physicians, union health and welfare funds, local chapters of national voluntary health agencies, state divisions of vocational rehabilitation, insurance companies, or social agencies.

In all instances he has been referred because it is thought that a rehabilitation service is the appropriate resource for him. In the majority of instances the patient is willing, if not eager, to come. The referral source may have certain goals or expectations, but the patient and his family are usually seeking a medical cure for the disability.

THE CONTRIBUTION OF SOCIAL WORK

The social worker meets with the patient and his family as soon as feasible after his admission. She is the patient's first contact with the psychosocial team. The social service intake interview has many purposes: (1) to give the patient and his family information and understanding regarding the services that are available to him, (2) to evaluate from the patient's social history his environmental resources and his abilities to utilize the services, (3) to refer the patient and his family to the appropriate resource, outside of the center, for alleviating the environmental problems, (4) to refer the patient to the appropriate discipline within the psychosocial service, and (5) to share these findings with members of the rehabilitation team informally and at scheduled meetings.

The casework focus, from the intake interview to discharge plans, will be on the specific reality problems of the individual and his family and on finding ways to reduce the pressure or relieve undue hardship.[4]

Many patients may have problems in social functioning that are caused by a personality disorder. In other patients the disability that has disrupted the social situation drastically may have reactivated a personality disturbance in both the patient and his family. It is in determining these complex factors that the psychiatric evaluation, the social history, the psychologic evaluation, and the medical diagnosis are essential in deciding whether casework help can be beneficial or is a suitable choice.[1] The choice of treatment that will be used and the discipline that will carry it out are decided at the psychosocial team conferences. The psychiatrist, psychologist, vocational counselor, recreational worker, and social worker represent the various disciplines at the psychosocial team conferences.

Frequently it is necessary to refer a patient's family to various agencies in his home community while he is still at the center. These agency referrals will be varied, such as public assistance, housing authorities, visiting nurse services, family agencies, medical clinics, recreational centers, and mental hygiene clinics. People usually have some resistance to seeking any kind of help, and such resistance must be worked through before the family will act on the referral.

The focus used in casework help in a rehabilitation center has been mentioned. The goals of the casework treatment will vary with the individual.

The selection of patients for casework help is decided on at the psychosocial team conferences in conjunction with the psychiatrist, psychologist, and vocational counselor. The pathology of the patient is not necessarily the decisive factor in considering the advisability of casework help. The reality problems of the patient and the estimated length of his hospitalization are essential factors in deciding what ways help can be given.

The multiplicity of the patient's psychologic and social problems, the degree of his physical helplessness, the positive and negative aspects of his environment, his intelligence, and his readiness and capacity to examine his problems must all be defined before it is decided what casework treatment will be most beneficial to the patient.

The following histories were selected to exemplify the social casework approach in understanding and assisting the disabled individual in rehabilitation.

Case study 1 *("Nobody wants you when you are old and gray")*. Mrs. J., a 65-year-old widow with compound fractures of both hips, came to the Institute of Rehabilitation Medicine to learn how to walk. That she was inadequate in all activities of daily living seemed unimportant to her on admission. She needed desperately to be able to walk so that she could be admitted to a church home for the aged. This had been her lifelong plan; she felt that if she were not accepted at the church home there would be no place for her to go. She had no relatives. She worked very hard on her program and became adequate in wheelchair transfers and in dressing and undressing. But she remained adamant in her wish to walk, and she did not want to discuss her disposition problems. With her consent, her plan for admission to the church home was discussed with the administrator of the home and it was discovered that it was not feasible for her to be admitted in a wheelchair. This was discussed with her and she was assured that a home could be found for her. It was then that she poured out her fears of what would become of her if her plan failed. From then on she was able to plan realistically and constructively.

Case study 2 *("Please take care of Mom")*. A teen-aged boy who sustained a disability of paraplegia as a result of an automobile accident just a few weeks before his admission to college came to the Institute of Rehabilitation Medicine in a state of deep depression. He did not care to discuss his social situation or vocational plans with anyone. He was polite but insisted that he came for physical treatment and that he would make his own plans. He had no financial or housing problems. In view of his opposition it seemed advisable not to antagonize him by insisting that he have a psychologic evaluation or be seen regularly by the social worker. He therefore was seen very casually in the halls and wards by the social worker for many months. He eventually consented to come to the worker's office. There he spent his time complaining about the physician's neglect and lack of interest in him. These complaints, along with the story of his life, continued for many weeks. He was the oldest of three siblings; he had a brother 2 years younger and a sister 4 years younger. His father had died when he was 12 years old. They had had a close relationship, and the father once remarked, "If anything happens to me, you must help your mother." The patient had not wanted such a burden, yet he felt a kind of pride that his father made such a request. To face the impact of his permanent physical disability of paraplegia meant to him consciously that he had failed his father, his mother, and himself. How could he take care of himself, much less "Mom"?

It became understandable why this young adolescent could not talk with the social worker or psychologist and why he complained about his medical care. He longed for understanding and felt unworthy. A bladder infection made it necessary for his rehabilitation to be temporarily interrupted and he was transferred to a local hospital, where the social worker visited him one evening. The visit was brief but friendly. When the patient was readmitted to the rehabilitation service, he asked for an appointment with the social worker. He said that her visit made a profound impression on him. It meant someone knew how he felt, did not judge him, and wanted to help him plan. He now wanted this help. He asked the social worker to talk with his mother. The social service contact with the patient became less frequent as he was now able to use the vocational and psychologic services and work out his admission to college.

In these two case illustrations, we were dealing with fairly healthy individuals who had environmental resources that could be mobilized, and, from a physical point of view, both of them were completely adequate in taking care of their needs from a wheelchair at the time of their discharges.

The next group of patients present more complicated problems from many points of view.

Case study 3 *("He will not attend his program")*. An 18-year-old intelligent boy presented a management and disposition problem. A year prior to his admission he had sustained an injury in a boat accident that resulted in quadriplegia. He complained that pain in all four extremities was so intense that he was unable to work on his program and that the usual medication did not relieve the pain. On admission, he had no awareness of the degree of his disability.

Prior to his disability the patient's sole interests were concerned with physical mobility and manual skills. His vocational goal was to become a jet pilot. He had frequent mood swings, vacillating between cheerfulness and deep despair, verbalizing suicidal thoughts during the latter. At all times he was anxious. His frustration tolerance was low. He had extreme difficulty in expressing anger directly for fear of retaliation. When he was angry, he had difficulty in talking and attributed this to his problem in breathing. He referred to himself as "lazy" and used this as an excuse not to attend classes. He became angered when he felt the staff did not treat him as an adult. He disliked being told what to do instead of being asked. He had many bodily fears and anxieties about his wasted body.

Both of his parents were tense, frightened people whose emotional problems had interfered in their marriage and the rearing of their three children: the patient, a sister 2 years younger, and a brother 4 years younger. The mother felt responsible for the patient's accident, believing that if she had given him sufficient security, he would have been more easily satisfied and less interested in dangerous pursuits such as flying. She also felt that if she had been a better wife her husband would be happier. The physical environment of the home was unsuitable for a person confined to a wheelchair. This presented another problem, since the financial status of the family was modest.

The anxiety of the patient's father expressed itself in his impulsive need to do something: move to a new city or buy a new house, irrespective of how suitable this might be for the entire family. This particular form of activity on the part of the father frightened the mother, since she could make no moves that were not carefully planned ahead.

The patient's mother felt the patient was closer to his father than to her. The father was described as having been inconsistent in his treatment of the patient. He would be "busy" and "strict," then relax all rules. This placed the mother in a difficult situation. If she complained to the father, he would accuse her of being "bossy" and of trying to dominate him. She had more or less withdrawn into the background, not knowing what to do.

The father, quite an immature person, but intelligent, went through a phase of demanding an answer to why the accident happened to his son. He could see no way of helping the patient except through buying him whatever he mentioned. He was asked to think of things he and the patient could do together in the future. This helped to channel his need for activity into exploring possibilities instead of acting impulsively.

The mother was given as much reassurance as possible that she was a good mother and could still help her son. Both parents were urged to consider changes in the home, on the basis of actual money they had.

The patient was seen daily by the social worker for many weeks and twice by the psychiatrist. Dreams that were related by the patient to the social worker were helpful in understanding underlying conflicts. The significance of the dreams was elaborated on by the psychiatrist, but their interpretation was not used by the social worker in her work with the patient.

Very gradually the patient was able to express some of his pent-up anger to the social worker. Some he expressed indirectly by failing to keep appointments. He at last said that he could not go to his rehabilitation program and work because it meant accepting that he was a "quad," and that he would prefer to die. Since his relationship seemed fairly stable with the social worker at this point, he was confronted with the consequences. He finally said he did not want to learn to use his hands. The social worker said that she did not believe this, rather that he feared he would fail. He was silent. He was asked if he would be interested if he could be given a "guarantee" that he could learn to write and feed himself. He thought he might. He was told that his occupational therapy worker felt certain he could succeed if he wanted to do so. From then on the patient began to work on his program. Daily contacts with the social worker were no longer essential. Planning with the patient's family in the area of changes they can make in the home and the amount of help they can give the patient in terms of vocational goals is being discussed. The patient will complete his high school education while at the rehabilitation center, and before he leaves the center prevocational exploration can begin.

   Case study 4 (*"All I ever wanted was a family"*). The last case concerns a 40-year-old man who was separated from his wife. He presented a management and severe disposition problem. He had a physical disability of triplegia and organic impairment, including visual problems, and he suffered from severe depression. Several years prior to his admission he had undergone surgery for a brain tumor. He considered his admission to the rehabilitation service as his last chance and hope.

   The patient was well educated and excessively polite. However, he constantly made complaints against attendants and was suspicious of everyone's intentions toward him, but he was unable to recognize his anger and suspiciousness. He was supersensitive and very easily offended. He had deep fears about rejection and abandonment; however, he did not verbalize these for some time; rather, he acted them out. When he finally expressed his fear that his rehabilitation was failing, his thoughts about this were explored. He then said for the first time that he wished that his mother would keep him.

   On admission, he had been very bitter toward his mother. He talked about his parents' failure in their marriage, his lonely life as a little boy in boarding school, and the terrible blow the failure of his own marriage had been—a deep unresolved hurt. It was to have been his family, the one he felt he never had.

   The casework relationship given to the patient was a supportive one to help him be as comfortable as possible so that he could attend his classes. He developed acute pneumonia and had to be transferred to another hospital. At that time his mother and his wife were available to him.

   These few case examples by no means represent the vast number of "social problems" arising daily in a rehabilitation center. They are used to illustrate the caseworker's approach to the "management" of the patient's "social problems."

COMMUNITY RESOURCES

   Community resources needed by the physically disabled person are not always available to him after he leaves a rehabilitation center. Even basic needs such as adequate housing, work, transportation, and continual nursing care are frequently very difficult to find in too many communities. "All social institutions develop in relation to areas of human need. They represent the organized response of the community to the needs of its members."*

   While the foregoing statement is no doubt true, social awareness and knowledge of the facts known only to organized social workers and not made known to the citizens in a given community will all too slowly develop social institutions in relationship to human needs. Community organizers who work with lay groups, not organized institutions, are needed. It is the "unorganized community" that needs to become aware of the basic unmet needs of the physically disabled person. It, in turn, can change the eligibility laws and can insist on a broader interpretation of existing laws that have a bearing on the dignity and welfare of physically disabled persons and other citizens with unmet needs. It can even develop new resources not presently in existence but sorely needed. The local community cannot know of the isolated needs of its citizens without education and communications from the rehabilitation workers.

---

*From Woodward, L. E.: Findings of the Commission in Social Work, Ann. N. Y. Acad. Sci. 63:365, 1955.

In most communities local housing projects paid for by the taxpayers operate on rigid economic eligibility requirements. These eligibility requirements need to be broadened for physically disabled persons on the basis of need and the economic eligibility worker out on an individual basis.

Many physically disabled persons who can work, want to work, and have skills with which to work cannot always find work. Too large a number of them have no means of solving their transportation problems. Liberal insurance companies have taken risks and have advanced money for cars, and it has "paid off." Why can't some way be found for organized institutions to take risks on their citizens?

It is generally conceded that institutional living of the severely disabled person seems to keep the "body alive" but adds "little to the soul." Perhaps families could assume more responsibility if nursing relief were considered as essential as food and shelter by "organized society."

In the final analysis, all rehabilitation workers have this responsibility to the community in which they work—to make known to the community the unmet needs of the people they serve. There is no other way to solve the problem.

REFERENCES

1. Austin, L. N.: The place of social casework in psychotherapeutic treatment methods, Ann. N. Y. Acad. Sci. 63:383, 1955.
2. Blackey, E.: Social work in the hospital: a sociological approach, Social Work 1:2, April, 1956.
3. Bonner, H.: Social psychology: an intradisciplinary approach, New York, 1953, American Book Co.
4. Woodward, L. E.: Findings of the Commission in Social Work, Ann. N. Y. Acad. Sci. **63:**365, 1955.

CHAPTER 17

PRINCIPLES IN MANAGEMENT OF VOCATIONAL PROBLEMS

Rehabilitation medicine is concerned with the individual—with his total needs. In its efforts to assist each person to achieve his fullest potential, rehabilitation medicine is keenly aware of the importance of developing the sensitivities, skills, and services that recognize the uniqueness and diversity of each person's needs. The comprehensive rehabilitation medicine program brings together a group of disciplines that provide services in a milieu that mobilizes the patient's strengths and skills toward overcoming his limitations and realizing his fullest potential. For many patients, realizing their fullest potential is possible only if they attain suitable employment. To provide effective vocational services for this group, the comprehensive rehabilitation team must first appreciate the significant and complex role work may play in the person's total adjustment.

WORK AS A FACTOR IN TOTAL ADJUSTMENT

The goal of rehabilitation may be seen as the promoting of ego integrity and feelings of self-worth. For some people—perhaps most, but not all—work and employment are crucial in achieving this goal. In our society, employment may determine both economic and social status. Income, economic independence, and occupational level can be potent factors affecting self-concept and determining feelings of personal adequacy, dignity, and worth.[1-4]

Disabling conditions may induce intense feelings of loss of identity, worthlessness, and dependence. The total rehabilitation effort seeks to encourage the patient to develop fundamental life values that challenge the basis for this destructive self-image.[5] The opportunity to consider and to engage in work may, when properly and sensitively timed, play a vital role in facilitating the total rehabilitation process.

Successful adjustment to disability, to family, and to society may be jeopardized when vocational and educational problems are dealt with inappropriately and ineffectively.

The inability to achieve employment may be a threat to a person's mode of life and that of his family. This problem can be compounded when the lack of employment is due to physical disability because the person may regard himself and be regarded by others as inferior, not only with respect to his specific limitation but as a total person.[2] In the person for whom severity of disability necessitates considerable physical dependence, the achievement of vocational and economic independence may be a goal of tremendous importance.

Work is generally more than a means of earning a livelihood. Neff,[3] discussing the wide range of needs that are fulfilled or frustrated by different kinds of work, points out a number of the more important motivations for work. Several of these have already been noted: material needs, self-esteem, and respect from others. Activity and the need for creativity are also significant. Activity refers to the need "to have things to do." Work activity appears to play a role in combating boredom, but what is boring for one individual may be stimulating for another. Thus both the nature of a work activity and the characteristics of the individual are factors in evaluating the relationship between work and boredom.[6] Some psychoanalysts feel that work serves as a means for sublimating aggressive feelings and that lack of work results in anxiety because of the pressure of inacceptable impulses.[7] Perhaps activity should not be considered separately from the other aspects of work, since whatever the activity, its impact is conditioned by its significance to the individual in terms of income, self-esteem, creativity, etc.

Work may be a means of satisfying the need for creativity in that work activity may be part of a process of self-realization and fulfillment. The meaning of "creativity" in this context need not be limited to those with exceptional or outstanding talents. Work may provide the opportunity of being creative in this broad sense if it allows the expression and satisfaction of the person's unique qualities. The work of the artist, poet, novelist, researcher, business innovator may be obviously creative. But the work of the carpenter, industrial worker, and clerical worker may also be creative or creatively satisfying.

The problem of creative job satisfaction is related to the question of work alienation—the meaningfulness or meaninglessness of the job to the worker. According to Wilensky,[8] work alienation appears to be anchored in the social and technical organization of work and its threat to personal identity. Positive attachment to work is fostered by opportunity for social contacts on the job, control over work pace, and an orderly career. Work alienation is fostered by limitations in freedom (for example, close supervision and lack of opportunity to exercise judgment and intelligence), a blocked career, lack of opportunity to use one's skills, and consumer pressures outrunning income.

Technologic and industrial development appears to be increasing the number of workers who are alienated from their jobs. This phenomenon seems related to mechanization and automation leading to "subdivided tasks, bereft of initiative and responsibility, to despiritualization of work."[9]

We have discussed some of the ways in which employment may be of crucial significance as a determinant of an individual's total adjustment. Employment, however, is not always a possible or a preferred goal. There are situations in which the quality of an individual's total adjustment may be enhanced by his ability to develop the awareness that employment is not a necessary or sufficient requirement for a meaningful and fulfilling life. With the development of an appropriate value system, we may envisage a variety of activities—other than employment—that offer the opportunity for fulfillment. Social and recreational activities; reading, education, and conversation; avocational interests; and family activities—these and other pursuits can provide satisfactions that for some are preferable to employment.

In recent years technologic and economic developments have forced a reconsideration of the role of work for the individual. Predictions have been made that because of the population increase and the advances in industrial technology, a decreasing percentage of the available work force will be able to meet the need for goods and services.[10-13] This would lead to a large reservoir of technologically unemployed. To meet this problem, economists, welfare specialists, and presidential commissions have raised the issue of guaranteed annual income and the negative income tax. The National Commission on Technology, Automation, and Economic Progress submitted a report to the President in February, 1967, recommending that for those unable to work there be a guaranteed annual income.[14] In an economy of abundance, and with increasing unemployment, the role and ethics of work may undergo a radical transformation, especially if income and standard of living become less dependent on employment.

If our interest in a disabled person's total adjustment is to include concern for improving the quality of the satisfactions and enjoyment derived from his life activities and relationships, our understanding of the role of work and employment must be strengthened and enriched. Rehabilitation medicine, especially vocational rehabilitation, has tended to equate "successful rehabilitation" with the achievement of employment. This equation is frequently arbitrary and invalid—the job unsuitable or meaningless; and increasingly so, perhaps we can no longer assume that a job should be the measure of a person's total adjustment.

THE PROFESSION OF REHABILITATION COUNSELING

Rehabilitation counseling, a profession with a history of less than 30 years, is still in the process of defining its professional boundaries. The physician, and other members of the rehabilitation team, responding to the profession's youth and tenuous image, may expect too little or too much from the rehabilitation counselor. Inaccurate perception of a team member's professional role and competency can be highly disruptive to the team's effectiveness. It is our purpose, therefore, in this section to give the other members of the rehabilitation team an understanding of the rehabilitation counselor's background, his training and philosophy, and what can be expected from him optimally.

Vocational guidance or vocational counseling is a profession familiar to all of us. Schools and agencies provide these services to assist students and others to select an occupational goal congruent with their interests and abilities. Since World War II there has been increased concern for the welfare of the sick, the disabled, and the aging, and more recently for the poor and the socially and educationally disadvantaged. To meet the vocational needs of these groups, it was felt necessary to establish a new profession, generally known as rehabilitation counseling. Over sixty universities in the United States have graduate programs in rehabilitation counseling, a 2-year program at the master's level, with some programs providing further education on the doctoral level. The counselor's academic training includes: principles of counseling, dynamics of human behavior, vocational development, medical and psychologic aspects of disability, psychologic and vocational assessment, occupational information, and placement technics.

The profession of rehabilitation counseling focuses on the occupational adjustment of the person with a handicapping condition. The handicap may be the result of one or more of the following types of disabling conditions: medical-physical, mental, emotional, aging, poverty, and social-educational deprivation.

To appreciate the role of the rehabilitation counselor, we must be aware that he sees occupational adjustment—the ability to work—as a psychologic-behavioral phenomenon. Not only is this phenomenon intrinsically complex, but also it may be influenced by interdependent physical, social, family, and economic factors. The readiness to become involved with vocational services, the selection of a vocational goal, the capacity to pursue a program of education or training, the ability to get and hold a job—all of these can be influenced by the individual's feelings and values, by family relationships, by social experiences, and by the attitudes of employers and fellow employees.

When vocational services are based on these concepts, it becomes evident that the rehabilitation counselor can seldom be of optimal help to a patient if his role is limited to routine vocational testing and to providing occupational information. Vocational rehabilitation counseling is part of a process wherein the patient and the counselor enter into a partnership that seeks to provide the patient with experiences and services that can help him develop the attitudes, values, strengths, skills, and opportunities required to make optimal use of his potential for work.

THE PROCESS OF VOCATIONAL REHABILITATION

The process of vocational rehabilitation in its broadest sense refers to the total spectrum of services that may be required to assist a handicapped individual to achieve an optimal vocational adjustment. This process may include medical, psychologic, social, and vocational services, that is, the total rehabilitation process. The vocational rehabilitation process may be characterized as "action coupled with interaction."[16] Interaction is the counseling relationship, the face-to-face interaction between counselor and patient. Action refers to a variety of experiences and services utilized by the counselor to assist the patient to achieve a suitable vocational adjustment. Through counseling interaction the counselor seeks to develop a relationship-partnership with the patient wherein there is a joint endeavor to improve the patient's abilities to understand and to deal constructively with his vocational problems. When this joint endeavor includes action, the counseling relationship is the vehicle through which the action is determined and through which consequences of the actions are clarified and assimilated.[17-21]

The wide variety of "actions" that the counselor can introduce into the process of vocational rehabilitation can in terms of their purposes be classified into the following categories: evaluation, treatment, training, placement, and follow-up. These services, however, should not be viewed as discrete and consecutive. Rather they are overlapping, intertwining activities.

Evaluation may include consideration of the vocational significance of medical, social, psychologic, educational, and leisure-time activities. In addition, specific vocational data may be obtained through a variety of technics to be discussed in more detail later. These vocational evaluation technics may include activities that assess not only work skills but also attitudes, feelings, and behavior that may affect success in employment. Evaluation and treatment may be part of the same process, that is, the same activity may serve both to evaluate and to overcome vocational problems.

Training may include a variety of programs: education in regular schools—academic, technical, vocational, commercial—and colleges; training in vocational rehabilitation centers and workshops; individual tutoring; and on-the-job training. A training program may provide opportunities for reevaluation and for treatment.

Job placement may be the final stage of the rehabilitation process when, after evaluation, treatment, and training, the patient is ready for employment. Job placement may, however, also be utilized in the evaluation, treatment, and training processes themselves. A job can provide an opportunity to evaluate interests, skills, behavior, and personality factors related to employment. It can also under certain circumstances, with cooperation between employer and counselor, be a treatment and/or training experience.

Follow-up, our final classification of the services in the vocational rehabilitation process, refers to the counselor's activity in assisting the client to deal successfully with job problems. This may require counseling sessions with the client concerning relationships with supervisors and fellow employees, job performance and job satisfaction, need for further training, social and recreational needs, housing, and transportation. Follow-up may also include contacts with employers, a service that can be effective in clarifying and resolving job problems.

The focus of the process of vocational rehabilitation should be on treating each client *individually,* and yet it is this sensitive and skilled individualization of services that is generally the most difficult goal to achieve. Failure to attain this goal may be due to a variety of factors. The rehabilitation counselor may be required to make many decisions as to the types and the timing of services, a task that is difficult because decisions must frequently be based on subjective judgments concerning complex data and complicated problems. To add to the counselor's difficulties, he may have the authority to deprive a client of funds and services, unlike the situation when a person seeks help and counseling from physicians, psychotherapists, social workers, guidance counselors, and others and is generally free to determine for himself whether to follow the advice given.

The problem of individualizing the process of vocational rehabilitation and of client freedom of choice is aggravated when agencies, through their counselors, make decisions that reflect bureaucratic agency needs. Statistical concerns, financial factors, and large case loads may limit flexibility, innovation, and creativity—leading to inadequate services and to curtailing the clients' right to participate in decision making. Both client and counselor become prisoners of a system demanding conformity and caution rather than encouraging flexibility and experimentation.

VOCATIONAL EVALUATION

Vocational evaluation is the process of collecting vocationally significant data and of analyzing and understanding the implications of these data for vocational decision making and planning. The technics utilized in the process of vocational evaluation include: counseling, psychologic-vocational testing, and work evaluation.

Counseling. The counseling relationship permeates the entire process of vocational rehabilitation.[17, 19, 20] It is the vehicle through which the patient and counselor deal with occupationally significant information. The vocational implications of medical, psychologic, economic, social, educational, and occupational information are discussed. The patient's feelings toward himself and his disability, toward dependence and independence, his involvement with insurance and litigation, his values, and his fears and anxieties are all factors that may have important vocational consequences.

Educational history is explored and evaluated. Relevant data may include type and level of academic achievement, interests and aptitudes, adjustment to school, and patterns of behavior in dealing with school and with social demands. A review of work history provides data on vocational skills and aptitudes, work habits, job adjustment (attendance, relationships to co-workers and supervisors, quality of work, productivity, etc.), ambitions, job satisfactions, and values.

Psychologic-vocational testing. The need for testing and the nature of the testing program should generally be worked out with the patient through the counseling relationship. The predictive value of tests is subject to error, and consequently the evaluation of the significance of test results is a challenging task.[22] Too often there is a tendency to ascribe an excessive degree of importance to test scores. Test scores must be used cautiously, bearing in mind that vocational potential is an amalgam of aptitudes, abilities, interests, needs, and self-concepts whose development may be influenced positively or negatively by a variety of environmental factors. Test data viewed statistically without the recognition of the complex and dynamic nature of vocational potential can be a dangerous obstacle to facilitating an optimal vocational adjustment.

Standardized tests of intelligence, personality, aptitude, interest, and achievement provide scores that usually indicate how the individual compares with "norm" groups. Interpreted sensitively and cautiously, these data may sometimes be useful in pointing out areas of strength and weakness. Cues and suggestions for further vocational exploration and evaluation may be gleaned from the test results. Important also are the observations made during testing as to the patient's fatigability, distractibility, anxiety, ways of handling a variety of problems, perseverance, frustration tolerance, flexibility, etc.

Work evaluation technic. Because of the inadequacies of the standardized testing technics described, a great deal of effort has gone into development of new methods for evaluating and maximizing the vocational potential of individuals with medical-physical, emotional, social, educational, or economic disabilities or disadvantages. There is considerable variability in the terminology used to describe these technics. For our present purpose the term *work evaluation* is used to denote evaluation of vocational potential through the utilization of real or simulated work. Work evaluation technics include prevocational evaluation, sheltered workshops, and job tryouts.

Prevocational evaluation. The term *prevocational evaluation* has been used for a variety of work evaluation procedures.[23] Here we use this term to refer to a program usually in a rehabilitation center or in a hospital, staffed by occupational therapists or vocational evaluators. Job tasks and work samples, simulated and real, are utilized to gain a better understanding of vocational potential. Some prevocational or work sample programs favor a standardized structure in which everyone is processed through identical evaluations.[24] The emphasis is on evaluation as a statistically predictive instrument. This approach has been criticized for failing to appreciate the complexities of work behavior and giving undue importance to data that have serious predictive limitations.[25]

Individualization of vocational evaluation technics may be superior to uniformity. The prevocational unit, if utilized innovatively, may contribute significantly to the rehabilitation process when emphasis is on individualized services and on providing tasks and experiences tailored to the individual's needs.[26, 27] Work samples and work projects can be developed for specific purposes. Study programs and training programs can be introduced. Employers can be involved to develop projects for evaluating a patient's ability to return to his previous employment. In this approach the patient participates actively in developing his own prevocational program. The program focuses on assisting the patient to understand and to develop his vocational abilities and potential. This type of prevocational program is not grounded in statistically validated scores. Its usefulness depends on the staff's sensitivity, ingenuity, flexibility, knowledge, and skill. Utilized with job tryouts, training, and educational opportunities, and as part of a sophisticated, flexible, and comprehensive program of vocational rehabilitation, it may be an extremely valuable tool.

Sheltered workshops. Sheltered workshops generally provide unskilled and semi-skilled job tasks with remuneration in a setting conducted not for profit but for the purpose of carrying out a program of rehabilitation.[28] The sheltered workshop can be utilized for a number of purposes. For some, who cannot meet the demands of competitive employment, it provides the opportunity for permanent, paid employment in a protected work environment in which the demands of the job can be modified or reduced to meet the needs of the handicapped person. Workshops are also utilized for work evaluation and for diagnosis, modification and treatment of work problems such as interpersonal relationships, underutilization of abilities, adjustment to work pressures, work satisfaction, and the concept of self as a worker.[29]

Job trials. Job trials or job tryouts are the utilization of actual jobs in a real employment situation for the purpose of work evaluation. This technic, which has received little attention, may possess important advantages over the traditional methods of work evaluation used in vocational rehabilitation. If vocational rehabilitation agencies and specialists join forces with business and industry, it may be possible to develop on-the-job work trials that are much more realistic and effective than evaluation technics now used. A greater variety of job opportunities could be made available to the handicapped, and employers through their involvement in the evaluation program would have an opportunity to develop more enlightened attitudes toward hiring the handicapped. An additional and very important advantage of a job trial program might be the opportunity for the handicapped to solve their vocational problems without referrals to agencies and programs serving only the handicapped, an experience that for many is negative and unnecessary. In the past few years, programs for meeting the educational and employment needs of the socially and economically disadvantaged have experimented with new methods of delivering services and have involved employers in new programs for recruiting, evaluating, and training the disadvantaged. This type of innovative, employer-involved program may offer new and richer perspectives for the physically handicapped. Neff,[25] in discussing the problems of vocational assessment, describes the issue well: "Although it might need something of a social revolution to bring about, the site of the vocational evaluation ought to be in the work-place itself, so that the potential employee can receive the kind of intensive and personal assistance and attention he receives in a well-designed rehabilitation facility. . . . If this sounds like something of a millennium, I believe we can make an intelligent start."* A start has been made in a number of hospitals and medical centers[26, 27] where the large range of jobs found in these settings have been utilized for purposes of work evaluation and on-the-job training.

Final word on vocational evaluation. Vocational evaluation—the process of collecting, analyzing, and understanding the implications of vocationally significant data—should be perceived as more than the technologies of counseling, testing, and work evaluation. Many activities or experiences may contribute to vocational evaluation. This perception of vocational evaluation is consistent with the concept that medical-physical and psychosocial-vocational factors are determinants of vocational potential, of vocational goals, and of vocational adjustment. From this point of view, medical evaluation and treatment, psychosocial evaluation and treatment, training, education, and job placement are all part of or may contribute to the vocational evaluation process.

VOCATIONAL GOALS FOR THE SEVERELY DISABLED—WHAT IS UNREALISTIC?

Untold numbers of persons with severe disabilities have had their occupational goals and aspirations shattered by rehabilitation counselors and other professionals who respond with "That's unrealistic." Refusing to accept this assessment, some persons have through drive, ingenuity, and perseverance realized their ambitions and proved the professionals to be wrong.

Vocational rehabilitation agencies have traditionally evidenced reluctance to provide funds and services to persons considered to have poor prospects for successful employment. Despite the advocacy of ability—rather than disability—as the measure of vocational potential, vocational rehabilitation personnel all too frequently perceive severity of physical limitations as an index of residual vocational potential. Severely disabled persons have been deprived of education and training and of other services on the basis of arbitrary policies and invalid indices of vocational potential. Several recent studies[20, 26, 31, 32] of the vocational achievement of individuals with quadriplegia report that with effective rehabilitation services substantial numbers of quadriplegics can enter into competitive employment. These studies demonstrate that

---

*From Neff, W. S.: Vocational assessment—theory and models, J. Rehab. **36:**27, Jan.-Feb., 1970.

failure to attain employment is often attributed to severity of disability, when in reality the failure is due to inadequate services.

Concern for "being realistic" in selecting a vocational goal permeates the field of vocational rehabilitation. The rejection by counselors and agencies of a vocational goal as being unrealistic is frequently an arbitrary decision reflecting bureaucratic policy that stresses "sure bets" and smothers initiative and innovation. Wright,[33] in an eloquent article on being realistic, raises a number of issues that deserve serious consideration by those concerned with broadening the vocational perspectives of the handicapped. She reports studies showing that even the best judgment by experienced professionals as to what reality is may not be a good reflection of that reality. Prejudice, myths, hopes, and fears fashion a reality that may only exist in the minds of men. Reality as predicted may become reality as actualized, whereas being unrealistic can be a source of hope, achievement, and redefinition of the boundaries of new realities. Not all mistakes that can be avoided ought to be avoided, for in avoiding them we also avoid learning from them. If we are to enlarge and enrich the vocational perspectives of the handicapped, we must exert our energies to changing reality rather than to accepting reality.

THE STATE VOCATIONAL REHABILITATION AGENCY

Each state has a vocational rehabilitation agency that is responsible for providing a variety of services to persons with a vocational handicap to develop, preserve, or restore their ability to perform remunerative work. The federal government, through the Rehabilitation Services Administration, Social and Rehabilitation Services, Department of Health, Education, and Welfare, participates with the states in financing the program. The services available from the state agencies include diagnostic services; medical, surgical, psychologic, and physical restoration services; vocational counseling and evaluation; education and training; job placement; and appliances, tools and equipment, supplies, and stock for self-employment.

Since the state rehabilitation agency is the primary source of funds for vocational rehabilitation services, the private rehabilitation centers must seek to develop an effective liaison with the state agency. To accomplish this it is generally essential for the private center to have its own staff of rehabilitation counselors, who have the responsibility of establishing a working relationship with the state agency that is successful in obtaining the needed services. The practice of utilizing the services of state agency rehabilitation counselors in lieu of hiring its own staff rarely proves successful for a private center. To develop an optimal program of vocational services the rehabilitation counselor must be an integral member of the center's team. Responsibility to another agency is an obstacle to achieving this goal.

JOB PLACEMENT

Some patients may be able and may prefer to find a job through their own efforts; others may require a variety of special services. Dealing with job interviews, for example, can be a challenging and anxiety-provoking experience. The counselor, through discussion and role-playing technics, can assist the patient to deal more effectively with the job interview. The counselor can also play an important role in preparing the employer to be more receptive. The placement process can be viewed therefore as "readying" the job applicant for job seeking and "readying" the employer for hiring.

Obtaining suitable employment for the disabled, especially the severely disabled, often requires skills, know-how, ingenuity, and perseverance that may challenge the most experienced counselor. However, it has been increasingly demonstrated that success in employment can be achieved even for the most severely disabled when the resources of the professional, the business community, and the disabled person are utilized effectively and innovatively.[26] As experience and knowledge accumulate, disabilities once considered insurmountable obstacles to employment become less formidable.

Because of the nature or severity of their disability, some patients may require an intensive, flexible, and time-consuming placement service. A sizable number of placement failures result from the unavailability of this type of placement service. Although the vocational problems of the severely disabled are receiving increasing attention, there is still a great need for expanded and improved services.

EMPLOYER ACCEPTANCE

The job applicant who is handicapped by chronic illness or disability may be confronted by obstacles to employment despite the fact that he possesses satisfactory job skills. Employer attitudes and personnel policies may deprive him of the opportunity of demonstrating his ability to perform successfully on a job. However, great progress has been made in this area in the past 25 years. Through extensive programs of education and through improved services to the handicapped, increasing thousands of handicapped have been rehabilitated into employment.

Both public and private endeavors have contributed to the expansion of job opportunities for the handicapped. The President's Committee on Employment of the Handicapped and similar programs on the state and local levels have, with business organizations, labor unions, and community organizations, overcome employer resistance to hiring the handicapped. The U. S. Civil Service Commission, in cooperation with federal agencies, has pioneered in demonstrating that the severely disabled and the mentally handicapped, when properly placed, make successful employees.

Studies have established that disabled persons, when properly placed, are competitive with unimpaired workers in productivity, attendance, job stability, safety, and cost of employment. There is no flawless method for selecting, training, and placing employees, either for those with disabilities or for those without. Improved employment opportunities for the persons with handicaps will follow after there is improved understanding and cooperation between business, industry, the agencies providing services, and the handicapped themselves.

THE CHALLENGE OF THE SEVERELY DISABLED

Successful rehabilitation requires the coordinated services of many disciplines. Each participating profession brings to the rehabilitation team its own body of professional skills. To achieve optimal goals with the severely disabled, each profession must acquire the specialized knowledge and skills required for the effective resolution of the problems of the severely disabled. As Walker[34] points out: "In reality, the inabilities of patients to achieve appropriate goals do not represent failures by the patient, but by rehabilitation personnel who have not developed technics and programs which tap patient resources. Patient failures cannot be justified on the basis of rationalization that such severely disabled patients cannot work."*

Severely disabled persons are demonstrating in increasing numbers that when effective rehabilitation services are available, they can successfully enter into employment. With our current knowledge and technics many more could achieve success in employment if they were given the opportunity to make use of an effective rehabilitation program. Rehabilitation, health, and educational personnel in every profession providing services to the disabled must in increasing measure become aware of the vocational potential of the severely disabled and of the specialized services required to ensure that the severely disabled are given the opportunity to realize their full potential for work and for employment.

---

*From Walker, R. A.: Vocational rehabilitation of the quadriplegic, Arch. Phys. Med. **42:**716, 1961.

REFERENCES

1. Allan, W. S.: Rehabilitation—a community challenge, New York, 1958, John Wiley & Sons, Inc.
2. Dembo, T., Leviton, G. L., and Wright, B. A.: Adjustment to misfortune—a problem of social psychological rehabilitation, Artif. Limbs **3:**4, March, 1956.
3. Neff, W. S.: Work and human behavior, New York, 1968, Atherton Press.
4. Roe, A.: The psychology of occupations, New York, 1956, John Wiley & Sons, Inc.
5. Wright, B. A.: Physical disability—a psychological approach, New York, 1960, Harper & Row, Publishers.
6. Vroom, V. H.: Work and motivation, New York, 1964, John Wiley & Sons, Inc.
7. Neff, W. S.: Psychoanalytic conceptions of the meaning of work, Psychiatry **28:**323, 1965.
8. Wilensky, H. L.: Varieties of work experience. In Borow, H., editor: Man in the world of work, Boston, 1964, Houghton Mifflin Co.
9. Friedmann, G.: Industrial society, New York, 1955, The Free Press.
10. Carroll, T. E.: The ideology of work, J. Rehab. **31:**26, July-Aug., 1965.
11. Hansen, C. E.: Rehabilitation and the guaranteed annual income, J. Rehab. **34:**13, Nov.-Dec., 1968.
12. Hayes, H. E.: Work without wage—a need for the future, J. Rehab. **34:**20, Sept.-Oct., 1968.
13. Theobald, R.: Guaranteed income: next step in economic revolution? New York, 1966, Doubleday & Co., Inc.
14. U. S. News and World Report: Job or subsidy for everybody? Feb. 14, 1967.
15. Muthard, J. E., editor: The profession, function and practices of the rehabilitation counselor, Gainesville, Fla., 1969, Regional Rehabilitation Research Institute, University of Florida.
16. Malikin, D., and Rusalem, H., editors: Vocational rehabilitation of the disabled—an overview. New York, 1969, New York University Press.
17. Hamilton, K. W.: Counseling the handicapped in the rehabilitation process, New York, 1950, Ronald Press.
18. Lofquist, L. H.: Vocational counseling with the physically handicapped, New York, 1957, Appleton-Century-Crofts, Inc.
19. McGowan, J. F., and Schmidt, L. D.: Counseling: readings in theory and practice, New York, 1962, Holt, Rinehart & Winston, Inc.
20. Patterson, C. H.: Counseling and psychotherapy, New York, 1959, Harper & Row, Publishers.

21. Tyler, L. E.: The work of the counselor, ed. 2, New York, 1961, Appleton-Century-Crofts, Inc.
22. Super, D. E., and Crites, S. O.: Apprasing vocational fitness by means of psychological tests, New York, 1962, Harper & Row, Publishers.
23. Muthard, J. E., editor: Proceedings of the Iowa Conference on Prevocational Activities, Washington, D. C., 1960, U. S. Department of Health, Education, and Welfare, Vocational Rehabilitation Administration.
24. The job sample in vocational evaluation, New York, 1967, Institute for the Crippled and Disabled.
25. Neff, W. S.: Vocational assessment—theory and models, J. Rehab. **36:**27, Jan.-Feb., 1970.
26. Specialized placement of the quadriplegic and other severely disabled, U. S. Department of Health, Education, and Welfare, Vocational Rehabilitation Administration Project RD 509 (mimeographed report), April, 1963, Institute of Rehabilitation Medicine, New York University Medical Center.
27. Vocational rehabilitation in a suburban community hospital, Vocational Rehabilitation Administration Project 359, Sept., 1963, Long Island Jewish Hospital.
28. Chouinard, E. L., and Garrett, J. F.: Workshops for the disabled—a vocational rehabilitation resource, Washington, D. C., 1956, U. S. Department of Health, Education, and Welfare, Office of Vocational Rehabilitation.
29. Gellman, W. G., Gendel, H., Glaser, N. M., Friedman, S. B., and Neff, W. S.: Adjusting people to work, Chicago, 1957, Jewish Vocational Service.
30. Siegel, M. S.: The vocational potential of the quadriplegic, Med. Clin. N. Amer. **53:**713, 1969.
31. Siegel, M. S.: Planning for employment for the quadriplegic, Proceedings of the Seventeenth Spinal Cord Injury Conference, New York, Sept. 1969, U. S. Veterans Administration.
32. Geisler, W. O., Sousse, A. T., and Meagan, W.: Vocational reestablishment of patients with spinal cord injury, Prosth. Int. **2:**29, 1966.
33. Wright, B.: The question stands: should a program be realistic? Rehab. Counseling Bull. **11:**291, 1968.
34. Walker, R. A.: Vocational rehabilitation of the quadriplegic, Arch. Phys. Med. **42:**716, 1961.

CHAPTER 18

PRINCIPLES OF PRESCRIPTION WRITING

The importance of proper prescription writing in physical medicine and rehabilitation is not so well recognized as it is in the other clinical specialties. This limitation in communication between the physician and the therapist can be eliminated only by using the same principles in prescription writing in restorative medicine as are practiced between the physician and the pharmacist. One of the major obstacles is that dosage of many essential therapeutic activities in restorative medicine are not so well defined quantitatively as are the elements in pharmacotherapy. A good prescription must be concise and at the same time specific so that any therapist with sufficient technical knowledge should be able to carry out the orders precisely the way the physician prescribed them. It also has to be unequivocal to prevent misinterpretation of the instructions.

Generally, it is considered poor practice to have prescription forms printed on which the physician indicates his orders by the simple use of checkmarks. This leads to schematization of medical function, with the inherent danger that essential instructions that are not included in the printed form could easily be omitted. The prescription blank, although containing all essentials of specific information, must allow for a great deal of flexibility in communication between the physician and the therapist.[1]

The form shown on p. 307 should suffice for the prescription of most physical treatments and, with minor modifications, for most therapeutic exercises also. For more specialized exercises, special forms may be needed; such forms can be found in books and articles describing the exercises. In the following discussion some special considerations in prescribing individual therapeutic activities will be given.

THERMOTHERAPY

In addition to specifying the heat source in thermotherapy, it is important to give the exact temperature desired for the treatment. Whenever this cannot be done,

Patient's name_____ Age _____ Date _____

Clinical diagnosis_____

Disability diagnosis_____

Form of treatment_____

Frequency of   Duration of
    treatment_____   treatment_____

Electrode (type, size, etc.)_____

    Active_____

    Indifferent_____

Site to be applied_____

Special instructions_____

Precautions_____

               _____M.D.

utmost care is indicated. It is suggested that not too much reliance be placed on the measuring gauges on the short-wave machine. They do not indicate the degree of heat produced in the body. The size and type of the electrode should be specified. When noncontact electrodes are used, their distance from the body must be indicated.

If a radiant heat source is prescribed, its type and distance from the body should be indicated, in addition to the other data contained on the prescription form. Care should be taken to protect those parts of the body that are not to be treated with radiant heat.

Microwave equipment, though commercially available, is still in the stage of development. It is a potent source of heat and can be localized easier than some of the other modalities. The type of the electrode and its distance from the body, together with the treatment period, must be specified. Care must be taken to avoid deep and superficial burns that might occur as a result of improper application. The general area of the eyes should be avoided.

The principles of ultrasound therapy are different from those of all other thermotherapeutic agents. Its clinical effect, however, according to the prevailing concept, is still primarily thermal. It can be applied through water or by direct contact. Since there are no practical methods to measure the intensity of heat produced in the tissues, one has to rely on empiric specification of the mode and duration of application. When the contact technic is used, some suitable material (oil, water) to be placed between the treatment head and the skin should also be prescribed.

ULTRAVIOLET

The main representatives of this group of therapeutic tools are the conventional ultraviolet and cold quartz light lamps.

In prescribing ultraviolet therapy, the minimal erythematous dose (MED) is a convenient method of determining dosage. The minimal erythematous dose is not a constant value but rather one that depends on the patient's skin pigmentation, the thickness of the superficial skin layers, and vascularization. It depends also on the type and age of the lamp used and, most importantly, on the time of application and distance from the body. The minimal erythematous dose has to be determined for every patient individually. The technic of this procedure is as follows[2]:

A grid of about 10 by 2½ inches is made out of cardboard, heavy canvas, or other suitable material in which six circular openings ½ inch in diameter are cut. Provision is made so that these apertures can be covered by individually hinged flaps. This grid is placed on the site to be treated, and by a system of covering and exposing the openings the skin can be exposed to ultraviolet light for increasing lengths of time. Within a few hours one or more of the exposed areas will become erythematous. The minimum erythema dose is determined by the blanching of the skin after 24 hours.

HYDROTHERAPY

In writing a prescription for hydrotherapy, the first thing to decide is whether the physical, chemical, hydrostatic, or simple cleansing effect of the water is desired.

When the temperature of the water is intended for treatment, such vague terms as cold, tepid, hot, and warm should be avoided. Instead the exact temperature or temperature range should be prescribed. Whirlpool baths entail a combination of mechanical and thermal forces. Treatment should be prescribed according to the specific need of the patient. For example, in Buerger's disease the temperature of the whirlpool bath should not exceed 100° F. In peripheral nerve injuries, in which causalgia is present, an excessive amount of pressure in the stream of water may produce unwarranted pain.

The use of carbon dioxide baths is not so popular in this country as in some parts of the world. In prescribing such therapy it is important that in the precautionary remarks instructions be given to protect the patient from massive inhalation of carbon dioxide.

When water is combined with galvanic current or when some active drug is added to it, the exact strength of the current or the exact concentration of the added drug should be clearly indicated.

Specific instructions to watch the patient constantly (color, respiration, pulse rate, of the required procedure.

Hot packs and similar methods of treatment require a step-by-step prescription blood pressure) for signs of circulatory complications should always be included when extensive or prolonged hydrotherapeutic measures are prescribed.

ELECTROTHERAPY

Electrotherapy is perhaps the easiest to control quantitatively, since the current is delivered by electronic equipment that is usually provided with sensitive measuring devices. When constant current is used to introduce drugs into the body (ion transfer), it is important that, in addition to the intensity and duration of the current, the concentration of the drug be specified. It is also important to remember that the drug thereby introduced produces, in addition to a local response, generalized systemic reactions. Therefore, it is desirable that the first few treatments be given by the physician himself, and only after the final dosage and the patient's reaction have been established should the patient be turned over to the therapist.

In muscle and nerve stimulation the exact location of the active electrode not only should be specified in writing but also should be indicated on a schematic anatomic picture.

THERAPEUTIC EXERCISES

Therapeutic exercises can be defined as purposeful and graded movements of the body or its single elements, with the goal of improving its (1) strength, (2) endurance, or (3) skill. To achieve the therapeutic goal, the exercise must be selected carefully. An exercise regimen designed to improve skill will, as a rule, do very little to increase the strength of the muscles.

In the prescription the selected exercise, together with the duration and frequency of the treatment, must be described. It is important to specify the rest periods between phases of activity.

Since all therapeutic exercises by their very nature impose an increased demand on the patient, special instructions should be given to watch for fatigue or adverse reaction on the part of the cardiorespiratory system.

The prescription for an exercise regimen is often so lengthy that it is impossible for the physician to write detailed orders for each and every patient. However, if prepared instructions are used, they should be designed to allow the widest latitude for individual modifications to suit the particular patient's need. In prescribing specialized exercises it is advisable that the instructions of the original authors be carefully adhered to.[3, 4]

MASSAGE

Massage is the most popular but at the same time the most abused form of physical therapy in general clinical practice. If its indications and execution are carefully observed, it can be a very efficient therapeutic tool. By the same token, however, if no such care is exerted, it can serve only as a psychotherapeutic placebo or, even worse than that, it may do definite harm to the patient. Massage should be prescribed only by a physician who has sufficient knowledge of muscle physiology and pathology to anticipate its therapeutic effect. The various forms of massage, which are discussed in Chapter 3, must be selected carefully and indicated in the prescription. If massage is given in combination with some other therapy, such as heat or ion transfer, the sequence of the measures should be indicated. Mechanical devices for massage are unphysiologic and should not be prescribed. Massage for local or general weight reduction is based on erroneous physiologic considerations, and prescription for such a purpose is to be discouraged.

OCCUPATIONAL THERAPY

Occupational therapy can be a most valuable tool in the treatment of physical or mental disabilities.[5] However, it can be entirely useless unless the prescription is given after careful analysis of the patient's needs. This is the only therapeutic activity that results in the production of an object by the patient. In prescribing and administering occupational therapy, it is important to remember that the object produced is essentially a by-product and that it is more important for the patient to go through the mechanics of production than to produce a perfect object. The prescription should be as specific as possible. In addition to the information given in the standard prescription, it is desirable to indicate the reason for which the activity is prescribed and the anticipated result. If the physician feels that to achieve his therapeutic goal some technical modification of the conventional tools or equipment is desirable, he should specifically describe this in the prescription. In prescribing occupational therapy he must consider the patient's general health and work tolerance in addition to the disabling condition. It is essential also that the duration and frequency of treatments be indicated. If a gradual increase in the work load is desired, the physician must decide and specify the tempo of increasing the work load. Only with such precautions can occupational therapy be an efficient therapeutic tool in the hands of the physician.

PROSTHETICS

As a result of the constant development in the field of prosthetics, the prescription of artificial limbs has become increasingly complicated. However, the physician should never refer his patient to a prosthesis maker without a prescription. The best prosthetist will be unable to supply a suitable prosthesis without a great deal of information that only the physician can supply. The prescription should specify every element of the prosthesis—the joint mechanism, suspension apparatus, material to be used in manufacturing, and many other minute details.[6] With such data the prosthetist can proceed to take measurements and produce precisely the type of prosthesis the physician prescribes for the patient. If instead of a conventional leg some of the newer modifications are used (for example, suction socket or soft socket) more technical knowledge will be required before the prescription can be written.[7] Since an artificial arm is intended to perform more intricate functions than an artificial lower extremity, its prescription is more complicated. In addition to the material used in manufacturing, all other elements (for example, suspenders, cables, joints, utility hooks, and dress hands) must be prescribed in minute detail. Special attachments for holding tools or utensils can also be prescribed in connection with utility hooks.[8]

The successful use of an artificial leg depends to a large extent on the postoperative regimen. Shrinking and toughening of the stump is one of the most important parts of the postoperative care. It is not enough simply to prescribe a shrinker or elastic shrinker bandage. It is essential that the physician or an experienced therapist demonstrate and personally supervise the first few applications. The number and type of stump socks must also be prescribed.

BRACES

Nothing should be taken for granted or left to the bracemaker's decision. The bracemaker is a technically trained person, who, without a medical prescription, has no knowledge of the patient's clinical needs.

In prescribing a brace, not only the affected limb but also the patient's general condition should be considered. No braces should be ordered for patients who cannot use them. The prescription must be as precise as possible and must include every elements of the brace. Uprights, joints, locks, and straps must be clearly described.[9-11]

Wheelchair prescription

Date_____

Name_____ Age_____ Address_____

Clinical diagnosis_____ Weight_____

Disability diagnosis_____

Size { Adult / Junior / "Tiny Tot" }   Type { Universal / Traveler / Amputee }

Driving mechanism: Manual < One-wheel < Right / Left ; Two-wheel

Electric

Other, specify_____

Wheels: Large, 20-inch, 24-inch, 26-inch   Casters: 5-inch, 8-inch

Tires: Regular   Pneumatic < Semi / Full

Brakes: Regular   Extension < Bilateral / Unilateral

Handrims: Regular   Rubber-covered   Other_____

Seat: Hard   Soft   Cushion   inches_____

Back: Standard   Reclining < Semi / Full   Raised   Detachable

Headrest

Arms: Regular   Detachable   Desk   Wooden / Upholstered

Legrest: Standard   Elevating / Swinging   Fabric legrest panel

Footrest: Fixed   Removable   Parallel   Toe loops   Heel loops
          Heel straps

Special accessories: Arm supports and slings
                        Utility tray
                        Commode

Remarks_____

_____M.D.

Encircle the desired specifications.

An observance of the following criteria laid down by Bennett can be of considerable assistance:

(1) The device must serve a real need. Applying unnecessary apparatus can be just as dangerous as not applying necessary apparatus. (2) The device prescribed must be of a design that can be constructed and, as necessary, be repaired by any good orthotist. (3) The device should be as light weight as possible but must also be capable of standing up under expected wear. (4) The device must be reasonable in cost. (5) The device must be sufficiently simple so that it can be properly

applied by the patient or his family. (6) The device must be acceptable in appearance, and (7) must in no way endanger the structural security of bodily segments through its use.*

WHEELCHAIRS

A wheelchair should never be thought of simply as a chair with four wheels under it but rather as a piece of complicated prosthetic equipment. It is also important to remember that it is often not the most expensive wheelchair that will answer all requirements. All major elements of a wheelchair (for example, wheels, casters, seats, back, arms, and footrests) can be changed or modified. Such changes and modifications, together with the exact measurements of the essential parts, and sometimes the material they are made of, must be specified in the prescription. Brakes are essential to the wheelchair's efficiency and to the patient's security and should never be omitted from the prescription.

Special types of wheelchairs (for example, one-wheel drive or electric) can be useful in selected cases. They should be prescribed only if there is a real need for them; otherwise they can become expensive and cumbersome gadgets that the patients often do not use. There are some books and catalogues that offer useful specimens of wheelchair prescriptions similar to the form shown on p. 311. Such printed forms will prevent the physician from overlooking important elements in this prescription.

SELF-HELP DEVICES

This is the area in which the physician's ingenuity and creative imagination will be most helpful. In contrast to prostheses and braces, these devices are often not well defined.[12-14] They may range from complicated mechanical or electric devices to enable the quadriplegic person to turn pages in his book to a plain long nail in a board to help the hemiplegic housewife to peel potatoes. Some of them will have to be made to the physician's exact specifications by skilled engineers; others can be homemade or be bought in any variety store. However, simple as they are, they make the handicapped patient's life easier, and that is why it is the physician's responsibility to look around, often outside his conventional orbit, to discover such devices for his patients. This activity should not be dismissed as "gadgeteering," since it can sometimes offer more practical help than all conventional medical knowledge. However, it should not be overdone by making the patient an operator of a battery of mechanical gadgets.

---

*From Bennett, R. L.: Orthotics for function. I. Prescriptions, Proceedings of the Second Congress of the World Confederation for Physical Therapy, New York, 1956, American Physical Therapy Association, pp. 147-156.

REFERENCES

1. Gordon, M. M.: Prescription writing. In Krusen, F. H.: Physical medicine and rehabilitation for the clinician, Philadelphia, 1951, W. B. Saunders Co.
2. Kovacs, R.: Manual of physical therapy, Philadelphia, 1949, Lea & Febiger.
3. DeLorme, T. L., and Watkins, A. L.: Progressive resistance exercise, New York, 1951, Appleton-Century-Crofts, Inc.

4. Smith, O. F. G.: Rehabilitation, re-education and remedial exercises, Baltimore, 1949, The Williams & Wilkins Co.

5. Dunton, W. R., Jr., and Licht, S.: Occupational therapy principles and practice, Springfield, Ill., 1957, Charles C Thomas, Publisher.

6. Daniel, E. H.: Amputation prosthetic service, Baltimore, 1950, The Williams & Wilkins Co.

7. Klopsteg, P. E., and Wilson, P. D.: Human limbs and their substitutes, New York, 1954, McGraw-Hill Book Co., Inc.

8. Deaver, G. G., and Daniel, E. H.: The rehabilitation of the amputee, Arch. Phys. Med. **30:**638, 1949.

9. Deaver, G. G., and Brittis, A. L.: Braces, crutches, wheelchairs—mode of management, Rehabilitation monograph 5, New York, 1953, Institute of Physical Medicine and Rehabilitation, New York University-Bellevue Medical Center.

10. Orthopaedic appliances atlas: A consideration of aids employed in the practice of orthopaedic surgery. I. Braces, splints, shoe alterations, Ann Arbor, 1952, J. W. Edwards.

11. Bennett, R. L.: Orthotics for function. I. Prescription, Proceedings of the Second Congress of the World Confederation for Physical Therapy, New York, 1956, American Physical Therapy Association.

12. Institute of Physical Medicine and Rehabilitation, New York University-Bellevue Medical Center: Self-help devices for rehabilitation, Dubuque, Iowa, 1958, Wm. C. Brown Co.

13. Manual of cerebral palsy equipment, Chicago, 1950, National Society for Crippled Children and Adults.

14. Lowman, E. W.: Self-help device for the arthritic, Rehabilitation monograph 6, New York, 1954, Institute of Physical Medicine and Rehabilitation, New York University-Bellevue Medical Center.

15. Licht, S., editor: Massage, manipulation and traction, New Haven, Conn., 1960, E. Licht, Publisher.

16. Licht, S., editor: Therapeutic electricity and ultraviolet radiation, New Haven, Conn., 1959, E. Licht, Publisher.

17. Licht, S., and Johnson, E. W., editors: Therapeutic exercise, New Haven, Conn., 1961, E. Licht, Publisher.

CHAPTER 19

REHABILITATION OF PATIENT WITH PARAPLEGIA OR QUADRIPLEGIA

GENERAL PRINCIPLES

Rehabilitation of the patient with paraplegia or quadriplegia can be accomplished only with the closely knit cooperation of an experienced team. The medical members of the team include the neurosurgeon, urologist, orthopedic surgeon, internist, neurologist, psychiatrist, and physiatrist. The rehabilitation nurse, physical therapist, occupational therapist, social worker, psychologist, and vocational counselor are all invaluable assets in meeting the total needs of these severely disabled patients. The role of the physiatrist is to coordinate a program designed to (1) prevent decubitus ulcers, (2) maintain range of motion, (3) improve the power of those muscles that are intact, and (4) prevent contractures.

At the onset it is important that the patient be kept in a neutral extended position. The therapist must be cautioned that there should be no flexion, hyperextension, or torsion of the vertebral column. As is pointed out later in this chapter, the neurosurgeon makes the final decision as to whether a laminectomy should be performed. Until the exact condition of the spinal cord has been determined, extreme caution to protect the cord must be taken by physicians, nurses, and therapists.

On admission to the hospital the patient should be placed on an ordinary hospital bed with a foam rubber mattress or a Stryker frame. Special nursing care during the early weeks of hospitalization—with special attention to nutrition, skin care, and posture—is mandatory.

INJURIES

Radiologic findings. In injuries to the cervical spine, fractures, subluxations, and dislocations with encroachment on the neural canal may occur and be easily recog-

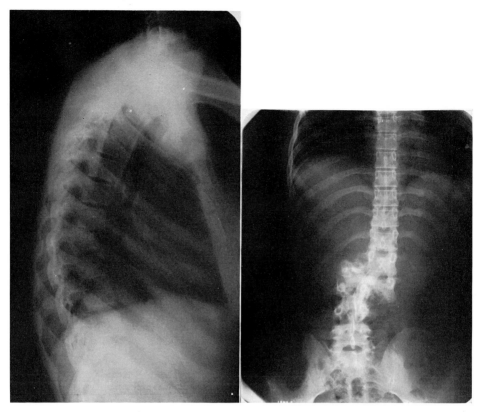

Fig. 19-1. Compression fractures and fracture-dislocations involving the dorsal and lumbar spine.

nized (Fig. 19-1). On the other hand, a transection of the cord can occur without any x-ray evidence of a subluxation or dislocation.

In spinal injuries there may also be damage to the disks. Herniation and/or degeneration may occur, and secondary posttraumatic bone (productive) changes may develop to help fixation. Oblique projections will demonstrate the intervertebral foramina, whether there is any encroachment on or any decrease in the size of any of the intervertebral foramina, and whether there is any subluxation between the articulating facets. Lateral projections will reveal the position of the vertebral segments and the height of the intervertebral spaces. Flexion and extension studies may be indicated to demonstrate whether there is any interference with mobility or the normal gliding movements. Ingenuity and cautious handling, however, need to be exercised in obtaining x-ray studies.

In the dorsal area, fractures are of compression type, as a rule, and these occur most frequently in the lower dorsal area (Fig. 19-1). With the compression there may be an increase in the dorsal kyphosis and encroachment on the neural canal. Subluxations and dislocations seldom occur.

In the lumbar area, fractures may be identified as either compression or com-

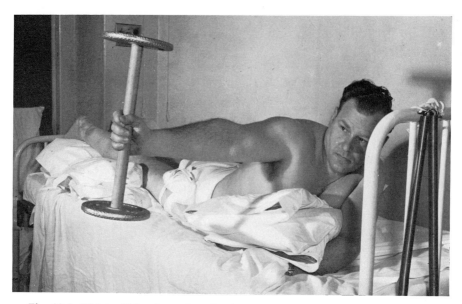

Fig. 19-2. Weight lifting is used to develop the muscle group around the shoulder.

minuted, and associated with a fracture there may be subluxation, encroachment on the neural canal, or even complete fracture-dislocation.

Skin care. The patient must be turned at least every 2 hours, day and night, in order to prevent decubiti. Nurses must be instructed to inspect the skin carefully over the following pressure points: sacrum, trochanters, ischial tuberosities, knees, and heels. As the patient is turned, these areas should be cleansed very carefully with soap and water, dried gently, and massaged lightly; then a bland oil or neutral ointment should be applied. (See Chapter 13.)

Maintenance of range of motion. All joints of the affected extremities should be taken through a complete range of motion twice daily. Patients who have muscle power, but with some weakness, should be encouraged to carry through a range of motion actively, as far as possible, and the therapist should complete the range of motion passively. These patients are prone to develop short heel cords with subsequent foot drop. To prevent this, bedclothes should *not* be pulled down tightly over the foot of the bed. The patient's feet must be kept at right angles to the legs at all times, either by means of a footboard and pillow or sandbag support or, preferably, by a posterior splint.

Therapeutic exercises. Muscle-strengthening exercises should be started early, with particular emphasis on conditioning the shoulder depressors, triceps, and the latissimus dorsi (Figs. 19-2 and 19-3). These are the muscles that the patient with paraplegia must utilize to walk with braces and crutches. Exercises for these muscle groups should be started early and increased in intensity as rapidly as possible. (See Chapter 4.)

Contractures. Prevention of contractures is fundamental. To neglect this preventive procedure may delay rehabilitation by weeks or even months. Special care

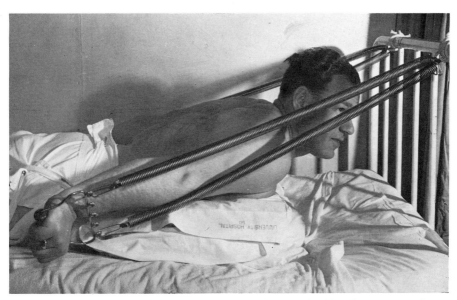

Fig. 19-3. A spring pull exerciser is used to develop the shoulder depressor muscles.

must be given to the hip and knee joints, where flexion contractures are prone to occur but can be prevented by a complete range of motion passively twice daily. Patients who have developed sacral decubitus ulcers present special problems, and modified technics must be devised (in cooperation with the plastic surgeon) that will produce results without jeopardizing the plastic procedures. If range of motion is not carried out, many patients will develop stiffness at the hip and knee joints, in extension, that may take months to bring back to a satisfactory range of motion.

Drugs. Habit-forming opiates should be avoided except when absolutely necessary postoperatively.

Most paraplegic pain can be controlled with a mild analgesic. Codeine may be used sparingly if needed. Persistent pain that is not relieved by these drugs may require neurosurgical or orthopedic intervention. It is practically impossible to rehabilitate a patient with paraplegia who has become addicted to drugs.

Physiologic doses of testosterone (usually 50 mg. daily) will decrease the incidence of tissue breakdown, osteoporosis, weight loss, and decubitus ulcer formation and is indicated in the early weeks. It should be discontinued when the patient is started on an active rehabilitation program.

Feeding. A high-caloric diet (4000 calories) and a high-protein diet (150 grams) are essential. Here it is important to see that the diet is not only prescribed but also ingested.

Early standing. It is important that the paraplegic patient be put in a standing position as soon as possible. Since more and more neurosurgeons are doing early laminectomies, it is possible to determine the exact condition of the cord. The patient whose spinal cord has been irreparably damaged, as determined by a laminectomy, should be put in a standing position by means of a tiltboard as soon as possible.

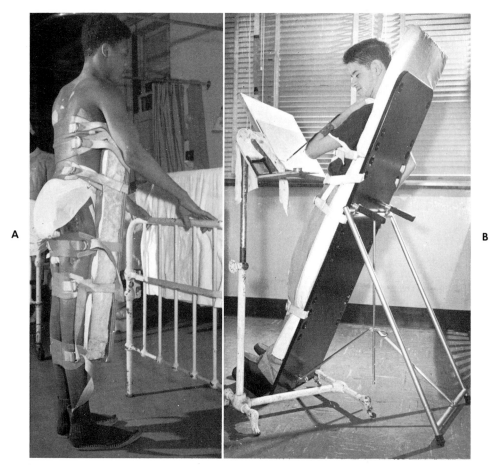

A

B

Fig. 19-4. A, An ordinary board can be used for standing if the back is to be left free because of decubitus ulcers. **B,** Otherwise a standard tiltboard can be used for early standing.

Usually this can be accomplished in from 10 days to 2 weeks following surgery, depending on the clinical condition of the patient. Patients without transection, with a hope of return of muscle function, must be carefully protected orthopedically as long as necessary. Here, it is best to err, if at all, on the side of conservatism. (See Chapter 13.)

The tiltboard is an invaluable therapeutic tool. There are many types of tiltboards and tilt tables available. An ordinary board, 2 feet wide (Fig. 19-4) padded with foam rubber, is quite satisfactory. The patient can be maintained in an erect position by means of straps around the chest, the pelvic region, the thighs, and below the knees. This type of tiltboard is particularly useful for those patients who are having plastic repair of decubitus ulcers over the sacrum, ischial tuberosities, or trochanters.

Another type, as illustrated in Fig. 19-4, *B,* is a board 2 feet wide and 6 feet long with a small second board attached at one end, at a right angle, for the patient to stand on. Again, straps are used to secure the patient. There are also tilt tables on the market that are constructed to bed height. These make it easy for one person to roll the patient over, apply the straps, and elevate him to a standing position.

At the onset, the angle of inclination must be small, usually from 20 to 30 degrees. The patient should be watched closely for signs of insufficient cerebral circulation, evidenced by pallor, sweating, and tachycardia. At any of these signs the tiltboard should be promptly placed in a horizontal position. The angle of inclination and length of time that the patient can stand are increased rapidly, so that at the end of 2 to 3 weeks the patient should have reestablished cerebral circulatory equilibrium and should be able to stand in a vertical position for a minimum of 1 hour daily.

Abramson[1] has pointed out that an hour of standing each day will prevent osteoporosis in the lower extremities and helps to prevent the formation of urinary calculi and genitourinary infection, which is the "killer" of patients with paraplegia and quadriplegia. Circulation and nutrition, as well as morale, are also aided by keeping the patient in the upright position for several hours each day. Only when maximum physiologic response is obtained is it time to consider plastic repair of existing decubitus ulcers. Large decubitus ulcers over the sacrum, trochanters, ischial tuberosities, and heels must be covered by full-thickness flaps to provide adequate skin thickness to properly brace the patient and to allow him to ambulate and learn the Activities of Daily Living.

Upon completion of definitive medical and surgical treatment, the rehabilitation program becomes primary, and a full day's program, usually from 4 to 6 hours, is indicated. General conditioning exercises performed on the mat should be stressed, and, if the physical condition of the patient is satisfactory, he should be given from 2 to 3 hours of this activity every day. Exercises should be designed to hypertrophy the muscle groups of the upper extremities, particularly the shoulder depressors and triceps, and the abdominal and trunk muscle groups. The patient is taught sitting balance. Short sawed-off crutches or wooden blocks may be used to practice push-ups on the mat while sitting.

Braces. Braces are prescribed and fitted in this phase of training (Chapter 12). In prescribing braces, it has been found practical, as a rule of thumb, to use the tenth thoracic vertebra as a basic landmark. Patients with lesions at the level of the tenth thoracic vertebra and above are usually given double-bar long leg braces with a pelvic band and a Knight spinal attachment for the back. (Fig. 19-5.) These braces are fitted with sliding box-type locks at the hips and knee joints. The ankle joints should have stirrup attachments with a 90-degree stop. Patients with lesions from the tenth thoracic to the first lumbar vertebra are given double-bar long leg braces with a pelvic band, and those with lesions below the first lumbar vertebra, depending on the specific functioning muscle groups around the hips, are given an ordinary long leg brace. Adequate bracing is fundamental in the rehabilitation of the patient with paraplegia. It is much better, at first, to err on the side of too much bracing than too little. Braces can always be cut down, but the morale is always traumatized if it is decided that additional bracing is necessary after the initial fitting.

Activities of Daily Living. This is one of the most important facets of the rehabilitation program. Here the patient is taught to take care of his own daily needs,

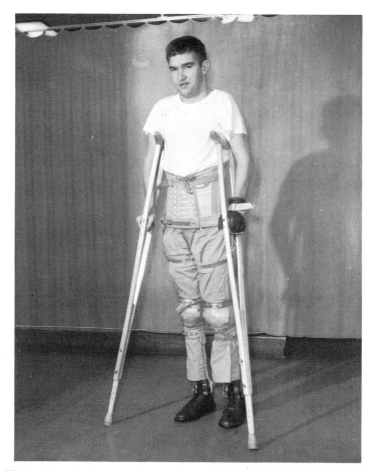

Fig. 19-5. Bilateral long leg braces with a pelvic band and the Knight spinal attachment are used for paraplegic patients with lesions at the level of the tenth thoracic vertebra or above.

such as cleansing and dressing himself. He must be able to put on and take off his braces, to get from the bed to the wheelchair and back again, and to rise to a standing position from his wheelchair, pick up his crutches, start to ambulate, and return to his chair. The program in Activities of Daily Living includes bed, toilet, eating, dressing and undressing, hand, wheelchair, elevation, walking, climbing, and traveling activities. (See Chapter 6.)

Most paraplegic patients have emotional and social problems as well as physical disability. Each patient should be seen as early as possible by a social service worker. It is also important for the members of his family to be interviewed at the beginning and at the end of the rehabilitation program so that they may prepare for home-coming. Emotional problems must be evaluated and, if serious, they often require definitive psychiatric and psychologic help. It is important, particularly in patients with spinal cord injury, that the vocational counselor be brought in early to start the individual thinking in terms of work and ultimate placement. As the patient becomes more proficient in his Activities of Daily Living, more and more emphasis should be placed on vocational testing, counseling, and training, the final plan being based on the patient's desires and aptitudes and the opportunities available in his community.

THE PATIENT WITH QUADRIPLEGIA

Quadriplegia may be caused by trauma or disease. In the majority of patients undergoing rehabilitation it is caused by trauma with a resultant fracture-dislocation of one or more of the cervical vertebras. Poliomyelitis is the leading cause of non-traumatic quadriplegia.

Traumatic quadriplegia is often caused by diving into shallow water, body surfing, falling from high places and landing on the heel and neck, or by water skiing accidents, automobile accidents, airplane crashes, and football injuries.

Standard orthopedic management consists of immedate skeletal traction, with adequate weighting for a period of from 2 to 6 weeks after traction. It may be necessary to provide some type of neck support for 2 to 4 months after traction. If the dislocation is not reduced by traction, cervical laminectomy may be required.

We are far more reluctant to do a cervical laminectomy than a thoracic or lumbar laminectomy. There is always a surgical risk connected with cervical laminectomy, and the percentage of morbidity and mortality increases in proportion to the higher the lesion is in the vertebral column. There is also increased risk of exacerbating the existing condition.

It is important during this period of hospitalization that the patient's extremities be carried through a complete range of motion of all joints twice daily. Extreme care must be taken by the therapist to protect the cervical vertebral fixation. Little can be done at this time toward strengthening the remaining muscles.

After the patient is released from traction, it is important to get him into a standing position on a tiltboard. Again, as in the patient with paraplegia, standing an hour or more a day on the tiltboard helps to prevent urinary infection and calculi.

A great majority of patients with quadriplegia are not braced for ambulation. A few with low lesions (at the seventh cervical vertebra and below) have been given a full Knight spinal brace attached to a pelvic band with bilateral long leg braces and are able to stand and take a few steps in the parallel bars. However, they are not and never will be functional walkers.

The majority of patients with quadriplegia have lesions of the sixth or seventh vertebra. The reason for this is both anthropologic and anatomic. There is more movement between the sixth and seventh cervical vertebras than between the other cervical and thoracic vertebras, which provides a greater opportunity for dislocation.

A typical quadriplegic patient with a lesion of the sixth or seventh vertebra retains certain movements around the shoulder; the rotator cuff is fully innervated and the latissimus and pectoralis major muscles have partial innervation. The biceps are completely innervated, and the extensor muscles of the hand on the wrist are partially innervated. All of these muscles are very important and must be given a thorough program of muscle reeducation and muscle-strengthening exercises. Flexors of the hand are usually not present. However, the extensors of the hand on the

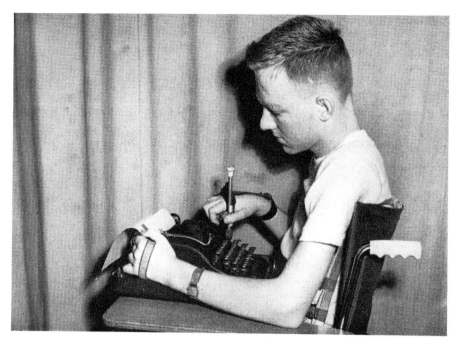

Fig. 19-6. Self-help leather cuff device assists the quadriparetic patient in typing.

wrist may provide enough grasp for the patient to hold large objects in his hand. The majority of the patients, however, need adapted self-help devices (Fig. 19-6). (See Chapter 9.)

These patients will always need some attendant care. Usually they are unable to get from the bed to the wheelchair and back again without assistance (Fig. 19-7). However, a quadriplegic patient with a lesion of the sixth or seventh cervical vertebra should be able to propel his wheelchair if knobs are added to the handrims so that he can use his biceps rather than triceps muscles. This enables him to move around within his home, but he will need help for traveling outside his home.

Patients with lesions above the sixth cervical vertebra need more continuing care. Electric devices are available that can be operated with the pressure of one muscle in the forearm or the elbow, or even the touch of the chin can raise and lower and adjust an electric bed, turn on and off a television set or radio, and even answer a telephone (Fig. 19-8).

Many quadriplegic patients are homebound. However, many have been able to do gainful work at home such as providing a telephone-answering service or selling by telephone, and a few are transported to their offices and operate their business very satisfactorily.

The following case illustrates the possibilities for a quadriplegic patient who has motivation and intelligence.

Case study. J. B., a 17-year-old male, suffered a fracture-dislocation of the sixth and seventh vertebras while diving into shallow water. He was immediately taken to a local hospital, where he was put on skeletal traction with weights. At the end of 4½ weeks, the dislocation had been reduced and the patient was placed in a high Thomas collar. At the

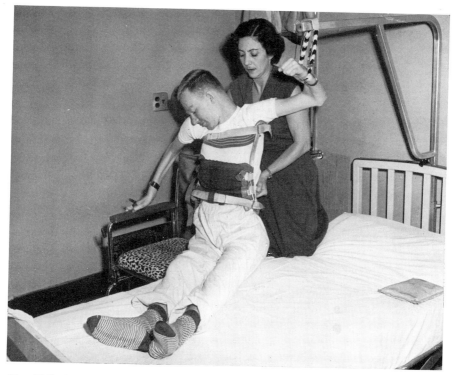

Fig. 19-7. Quadriparetic patient transfers from the bed to a wheelchair with assistance.

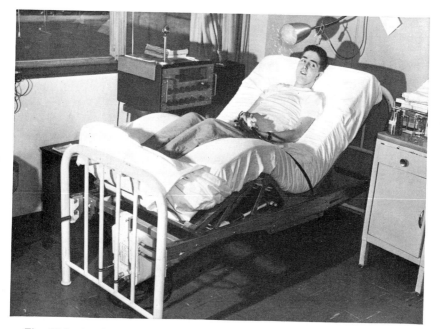

Fig. 19-8. An electric device enables the patient to change his position in bed.

end of 4 months he was brought to the Institute of Rehabilitation Medicine. It was found that the muscles around the shoulder were present but very weak and that he had the use of the biceps and some wrist extensors.

He was started on a tiltboard and could tolerate only 3 to 4 minutes at a 30- to 40-degree angle initially. This was gradually increased, and at the end of 3 weeks he was able to stand in an erect position on the tiltboard for 30 minutes twice daily. He was placed on a program of muscle reeducation for all the existing muscles and occupational therapy, where, again in a standing position, he was given work to do to help develop the muscles of the shoulder, arm, and wrist.

He was seen in the psychosocial unit because there was a family problem. The mother was brought in and counseled by both the social worker and the psychologist. It was found that this boy had better-than-average intelligence. He was motivated to finish his schooling, and he completed his studies while at the Institute. Special self-help devices were developed so that he could button his shirt and feed himself. With assistance he learned to transfer from the bed to the wheelchair and back again. A wheelchair was provided that he could propel on a level surface himself. The family was told he would always have to have attendant care with some of the more difficult activities of daily living.

This boy was determined to become a lawyer and he continued his college career. He was graduated from law school and is now employed by a large firm, doing legal work in his own home at a very substantial salary.

It must be remembered that with lesions above the sixth cervical vertebra a patient will be able to do less for himself. Patients with lesions below the sixth cervical vertebra have increasing musculature; in fact, a patient with a lesion at the level of the first thoracic vertebra will have full innervation of all of the upper extremity muscles, including the essential intrinsic muscles of the hand. This patient will be able to do more than the patient with a lesion at the level of the sixth or seventh cervical vertebra.

The prescription for a wheelchair for a patient with a spinal cord injury should be as carefully thought out by the physiatrist as any prescription for drugs. The condition of the patient's home and community must be discussed very carefully with the patient. There are many attachments that may be added to or subtracted from the standard folding chair and that may make all the difference in the world in a patient's self-sufficiency. It is better to wait until the latter part of the patient's rehabilitation program to prescribe exactly what he needs when he returns home. (See Chapter 7.)

In the past, quadriplegia has been considered to be a hopeless disability. However, this is no longer true. With present rehabilitation technics and patient perseverance and motivation, a majority of the patients can live dignified, productive lives.

NEUROSURGICAL ASPECTS

The neurosurgical consultant should see the patient with spinal cord injury immediately on admission to the hospital. After a detailed neurologic examination and appraisal of the roentgenograms, the first decision that the neurosurgeon is called on to make is whether any immediate neurosurgical procedure is indicated.

Early neurosurgical treatment

In cases of fracture-dislocation of the cervical spine in which reduction must be carried out, skeletal traction is the standard procedure. This is most commonly accomplished by applying Crutchfield tongs to the crown of the skull or by passing wires through trephine openings in the skull. In a great majority of patients, proper reduction with decompression of the spinal cord, which is the initial treatment of choice, can be accomplished by skull traction. In those in whom skeletal traction properly applied has not reduced the dislocation or in whom a subarachnoid block persists despite adequate use of traction, cervical laminectomy with decompression is indicated. Another indication for open operation is progressive increase in neurologic deficit despite satisfactory traction. In some instances of violent injury to the cervical portion of the spinal cord, acute herniation of the intervertebral disk also occurs. Usually clinical and roentgenographic studies will aid in the diagnosis of this lesion, although occasionally myelography is necessary. In such instances laminectomy with decompression of the spinal cord should be performed promptly.

In cases of fracture-dislocation of the thoracic or lumbar spine, the subject of laminectomy versus more conservative management continues to be controversial, although an increasing number of neurosurgeons now feel that laminectomy and decompression is the treatment of choice in almost every case. It is the policy of the neurosurgical service at the Institute of Rehabilitation Medicine, New York University Medical Center, to recommend that laminectomy be performed early in every patient with paraplegia that is caused by injury of the thoracic or lumbar spine, provided his condition permits the procedure. Routine laminectomy in such patients is actually the conservative course of action and can be done safely and without causing any delay in the ultimate rehabilitation of the patient.

There are many reasons why early laminectomy should be performed.

1. In the thoracic and lumbar regions it can be performed quickly, under local anesthesia if necessary, without interfering with the stability of the spinal column and with a mortality rate that is practically nil.

2. In patients demonstrating complete paraplegia there is no way during the immediate posttraumatic period to ascertain, by either neurologic or roentgenographic examination, whether there has been an anatomic transection of the spinal cord. An examiner is not justified by either clinical or roentgenologic examination in concluding that the spinal cord has been irreparably damaged. In patients demonstrating complete loss of motor and sensory function, the only way one can determine early whether the spinal cord is anatomically transected is by direct exploration. Therefore, the patient should not be denied laminectomy because the examiner has the impression that the spinal cord is already hopelessly injured.

3. In some cases it can be considered that the removal of depressed laminae and other posttraumatic fragments of bone, as well as intraspinal hematomas, may have a beneficial effect on future spinal cord function.

4. Wide debridement with intradural exploration may help to prevent late arachnoiditis and other later complications, such as pain and spasticity.

5. If wide decompression of the injured area of the spinal cord is performed shortly after injury, early mobilization and rehabilitation may be instituted more safely than if the patient still has spinal cord compression.

6. Paraplegic patients logically conclude that direct exploration and decompression of the spinal cord should be attempted, if there is not some definite medical contraindication, before they are asked to resign themselves to a lifetime of paraplegia and a long period of rehabilitation for their paralyzed extremities.

7. Laminectomy demonstrating irreparable damage aids the patient in acceptance of and in early determined participation in the rehabilitation training program.

Alleviation of sequelae

During the late phases of care for the patient with spinal cord injury—that is, from 3 months to several years after injury—the neurosurgeon may be able to play a role in alleviating some of the undesirable sequelae of paraplegia. Two of the most intractable and distressing complications are pain and spasticity. Although these phenomena are not both invariably present in the same patient, there is some justification for considering pain and spasticity simultaneously. Both of these phenomena are paradoxical neurologic sequelae of spinal cord injury. In the case of pain, this sensation is usually referred to areas rendered anesthetic and analgesic by the spinal cord lesion, so that heightened pain sensation is present in an area from which the application of painful stimuli cannot be appreciated. Likewise, spasticity in patients with paraplegia reflects exaggerated involuntary motion of muscles that cannot be moved voluntarily. Thus these phenomena—pain referred to analgesic areas and exaggerated activity in paralyzed muscles—are considered to be physiologically related. Moreover, in evaluating the etiology of these distressing sequelae, there is considerable evidence that not only the destructive elements of traumatic lesions but also the irritative effects of such injuries must be taken into account. For these reasons pain and spasticity may be considered together.

Management of pain. Approximately 75% of the patients who incur spinal cord injury will, at some time during their illness, complain of some type of severe pain. Because of the nature of such an injury, which may render a large part of the body numb, pain in these patients is particularly difficult to evaluate. No single classification of pain in paraplegic patients has been suggested that can be considered inclusive. Moreover, an etiologic classification is not feasible at this time, inasmuch as the individual roles of the sympathetic nervous system, spinothalamic tract systems, and unrelated pathways have not yet been elucidated in these patients. Since there are some obvious differences in the mechanisms productive of pain during the early weeks after injury and those productive of pain after this period, it is well to consider both early and late pain after spinal cord trauma.

Early pain. A surprisingly small percentage of patients complain of severe pain over the site of vertebral injury. Many vertebral fracture-dislocations are unattended by local pain. In our experience, local discomfort is more common in patients with cervical fracture-dislocation than in those with fractures in other regions of the vertebral column. The treatment dictated by consideration of spinal cord lesions, as outlined earlier (that is, early skeletal traction in patients with cervical lesions and early laminectomy in patients with thoracic or lumbosacral lesions), will invariably alleviate the pain produced by the bony abnormality. In patients in whom spinal fusion is indicated, this procedure will also lessen the likelihood of prolonged local pain. Consequently, local pain over the site of vertebral fracture seldom becomes intractable and is usually managed without difficulty.

The two common neurologic causes of pain in the nerve root are cartilaginous compression and complete intraspinal subarachnoid block. The presence of a complete subarachnoid block, particularly in the lower thoracic or in the lumbar region, will often produce severe pain in both lower extremities. This pain is often sharp and lancinating in character, as well as constant and excruciating. It may be diffuse throughout the extremities, and it is sometimes accompanied by burning paresthesias. The pain is aggravated by extension of the back, coughing, sneezing, or other activities that tend to increase intra-abdominal and, thus, intraspinal pressure. Roentgenologic examination may provide evidence of narrowing of the intraspinal canal, and the presence of a subarachnoid block is confirmed by spinal puncture with careful manometric studies. Once this block is demonstrated, early laminectomy with decompression of the block is indicated. Surgical decompression will usually result in prompt relief of pain due to this cause. Since the presence of an intraspinal block is in itself indication for laminectomy, further discussion of pain due to such a cause is not necessary.

Pain due to compression of one or more nerve roots even in the absence of an intraspinal block is not infrequent in patients with vertebral fracture-dislocation. Compression of such roots may be due to anterior dislocation of a fractured lamina, posterior dislocation of part or all of the body of a fractured vertebra, protrusion of bony fragments into an intervertebral foramen, or traumatic protrusion of an intervertebral disk. The occurence of the latter phenomenon is more common than had formerly been thought and has been emphasized recently, particularly as a cause of neurologic damage in the cervical region.

Radicular pain due to one of the aforementioned causes of nerve root compression is recognized by those qualities peculiar to root pain: (1) The pain follows the anatomic distribution of one or more nerve roots; (2) paresthesias along the same distribution are not uncommon, particularly in patients with incomplete transection of the spinal cord; (3) the pain is aggravated by coughing, sneezing, or straining; (4) the pain is usually intensified at night after the patient has been in an extended or hyperextended posture for some time; and (5) the pain is usually sharp and lancinating in character.

Careful neurologic evaluation and roentgenologic study will usually indicate the site at which radicular compression is occurring. Roentgenographic demonstration of a fracture extending through one of the intervertebral foramina, an extreme narrowing of an intervertebral disk space, or an obvious bony abnormality may indicate the vertebral site of nerve root compression. In some instances, in the absence of such roentgenologic findings and in the absence of a subarachnoid block, contrast myelography may be utilized as an adjunctive aid in evaluating the site and cause of nerve root compression following spinal injury. However, this latter procedure is to be avoided whenever possible. Many spinal cord injuries are accompanied by subarachnoid or intramedullary hemorrhages, arachnoid lacerations, and other types of lesions conducive to fibroblastic reaction. Superimposed on such a nidus, the introduction of iodized oil (Lipiodol) or iophendylate (Pantopaque) may contribute to the development of an adhesive arachnoiditis. This is a common finding in patients in whom exploration is done one or more years after injury, and such arachnoiditis is undoubtedly a contributing factor in the production of the sequelae of spinal cord trauma.

Surgical exploration and decompression is the treatment of choice in patients with pain due to nerve root compression. Hemilaminectomy, foraminotomy, or removal of a protruded intervertebral disk, depending on whichever proves to be indicated, will often relieve radicular pain. Treatment of this type of injury to the spinal cord or cauda equina will be more successful in those patients in whom decompression is carried out early, before the pain has become constant and before the patient has become dependent on narcotics.

One other cause of pain in paraplegic patients during the early days following spinal cord injury is the presence of concomitant injuries to other body parts, such as a ruptured spleen or traumatic hemothorax. Such a category is too obvious to require further elaboration and is mentioned only for the sake of completeness. Unfortunately, extraspinal injuries have not infrequently been overlooked because of the initial examiner's preoccupation with the major symptom of paraplegia. Vigilance in a complete physical examination obviates such errors.

Late pain. Pain of an intractable nature is more often a complaint of the patient 3 or more months after injury than during the early posttraumatic period. This pain has been classified by several authors. Most writers seem agreed that pain in the lower extremities may be of an intermittent, sharp, lancinating, radicular quality, a diffuse dull aching quality, or a burning causalgia-like quality. The first type is most frequently seen in patients with injury to the cauda equina, whereas the second type is more commonly encountered in patients with injuries of the spinal cord proper. The diffuse burning type of pain may be present in patients with injury to either the spinal cord or the cauda equina. Although the burning pain complained of by persons with paraplegia is similar to that usually described as causalgic, this type of pain in paraplegic patients is seldom, if ever, entirely relieved by sympathectomy. Holmes[21] postulated that such persistent burning pain might be due to changes in the anterolateral columns of the spinal cord. Foerster's[22] investigations demonstrating that the entire pain pathway in the spinal cord is excitable substantiated Holmes' hypothesis. This still offers the best physiologic explanation of the presence of diffuse aching, burning, tingling, or other nonradicular pains of the insensitive lower extremities of the paraplegic patient.

Experience has shown that all the categories of pain described in the preceding paragraph, present 3 months or later after spinal cord injury, are best treated by bilateral high thoracic or cervical spinothalamic chordotomy. The results with the use of spinothalamic chordotomy for pain in paraplegic patients have been disappointing. However, if patients are selected who have not yet become addicted to narcotics and whose pain is not psychologically motivated and if such patients are operated on under local anesthesia to allow instant appraisal of the level of analgesia achieved, the incidence of successful relief of pain will be reasonably satisfactory.

Relief of spasticity. In persons with traumatic paraplegia due to injuries to the cervical and thoracic portions of the spinal cord, spasticity of the paralyzed lower extremities becomes marked enough to impede rehabilitative measures in approximately 40%. The relief of this spasticity will often enable a bedridden paraplegic patient to become ambulatory and self-sufficient. The four types of neurosurgical procedures that will alleviate intractable spasticity in the paraplegic patient are (1) peripheral neurectomy, (2) subarachnoid alcohol block, (3) rhizotomy, and (4) subtotal spinal cordectomy.

Peripheral neurectomy. The resection of peripheral nerves has rather limited usefulness in relief of spasticity in paraplegic patients. It is confined largely to resection of the obturator nerves for relief of adductor, or scissors, spasticity. When the incapacitating spasticity is largely confined to the adductor muscles of the thighs, the procedure of choice is bilateral transabdominal obturator neurectomy. This procedure can be performed easily and provides immediate relief. The nerve is approached retroperitoneally through a McBurney abdominal incision and is readily identified as it emerges from the medial side of the psoas muscle to cross the brim of the pelvis. Stimulation of the nerve to produce adduction of the thighs positively identifies it prior to resection.

Subarachnoid alcohol block. Subarachnoid alcohol block has emerged as the most useful and most applicable single method available for relief of paraplegic spasticity. The indications for subarachnoid alcohol injection are essentially those previously applied to anterior rhizotomy. The chief indication is intractable spasticity of the lower extremities, which have no useful motor or sensory function. Enough time should have elapsed from the time of injury so that the lesion may be known to be stationary; this eliminates the possibility of future neurologic recovery. The procedure has certain advantages over anterior rhizotomy; it is safer and more easily performed and is free of the hazards of major surgical procedure. Its effects are immediate and, in most instances, of long duration. However, it is not selective. Thus, if certain nerve roots, such as those to the detrusor urinae, are to be excluded, this method is contraindicated in favor of selective anterior rhizotomy. Such contraindications, however, are rare, particularly since it has been demonstrated that satisfactory autonomic micturition and sexual potency in the male can develop after subarachnoid alcohol injection.

The technic of subarachnoid injection is as follows: The patient is placed on his side in bed, and the foot of the bed is elevated 18 inches. An 18-gauge spinal puncture needle is inserted at the first lumbar interspace. A syringe containing 15 ml. of dehydrated alcohol is attached to the needle, and the alcohol is injected at the rate of 1 ml. every 15 seconds. An assistant constantly checks the patient's vital signs and the level of anesthesia during the injection. The injection is continued until both legs have become completely flaccid or until 15 ml. of alcohol has been injected. The patient is then turned on his back and kept in this position for 4 to 6 hours. He is allowed out of bed on the following day. If the injection has not produced satisfactory flaccidity, it is repeated the following day, with an additional 10 ml. alcohol being used.

Rhizotomy. Rhizotomy, or intraspinal resection of nerve roots, is the time-honored method of relieving paraplegic spasticity and has been used since the introduction by Foerster of posterior root section for spastic paraplegia. The fact that posterior rhizotomy alone did not provide adequate relief of spasticity led to the development of the technic of anterior dorsolumbar rhizotomy. This method requires laminectomy and division of all anterior nerve roots from the tenth thoracic through the first or second sacral vertebra. The last firm dentate ligament is usually regarded as indicating the level of the first lumbar nerve root, although this relation is not invariable. If one is anxious to avoid sacrificing the motor supply of the detrusor urinae (usually the second and third sacral roots), these roots can be positively identified by stimulating each sacral root individually and, at the same time, recording the intravesical pressure. Unless one is desirous of sparing certain nerve roots, such as those that innervate the urinary bladder, or unless some sensation exists in the lower extremities, subarachnoid alcohol injection is probably to be preferred to anterior rhizotomy.

Subtotal spinal cordectomy. Occasionally the problem of recurrent spasticity after subarachnoid alcohol injection is encountered. In these circumstances the operation of anterior rhizotomy is made very difficult by the dense adhesive arachnoiditis produced by the prior alcohol injection. This arachnoiditis usually makes positive identification of individual roots impossible. In such patients subtotal resection of the spinal cord is indicated. This subtotal cordectomy involves resection of the spinal cord and nerve roots below the level of the ninth or tenth thoracic segment. The operation requires a laminectomy extending from the ninth or tenth thoracic to the first lumbar level and is performed more quickly and easily than is anterior dorsolumbar rhizotomy. MacCarty[23] has demonstrated that removal of the entire cord in man, below the second thoracic level, is not incompatible with life. Removal of that section of the spinal cord below the ninth thoracic level has had no untoward or unexpected sequelae.

•   •   •

In summary, during the early management of the patient with spinal cord injury, the neurosurgeon may contribute to the neurologic evaluation. He may perform skeletal reduction of cervical fracture-dislocation and laminectomy, as well as decompressive procedures when these are indicated. The principal sequelae that often require the attention of neurosurgeons during the later phase of this problem are intractable pain and spasticity. There are various neurosurgical methods that may be utilized in the management of these sequelae.

NEUROGENIC UROPATHY

When one considers the hazards of the paraplegic state and weighs the gravity of each, there remains the impressive fact that urosepsis and its allied complications are highest on the list of mortality statistics for this group of individuals.

The virtual elimination of an early primary mortality from urosepsis after spinal cord injury *has not* been followed by a corresponding reduction in the incidence of later renal changes among the patients.[16]

Briefly speaking, paraplegia patients tread a well-known urologic path to renal destruction. In the beginning there is stasis that is readily followed by infection. Subsequent to infection, calculus formation, renal obstruction, and gradual loss of kidney function ensue in a cycle as repetitious as it is vicious.

If there is to be an eventual urologic rehabilitation of the patient with paraplegia, it must of necessity begin with the elimination of urinary retention. Until this can be accomplished we are obliged to treat the complications of retention conservatively, expectantly, and with utmost vigilance.

Evolution of bladder function in paraplegia

After traumatic lesions of the spinal cord, whether complete or incomplete, the bladder enters a state described as the stage of spinal shock. During this phase the viscus remains a completely nonfunctioning organ, emptying small amounts peri-

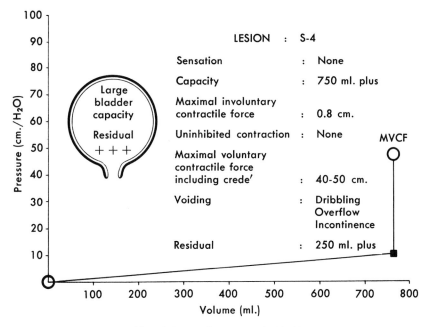

Fig. 19-9. Atonic neurogenic bladder.

odically only because mechanical force overcomes the forces of external resistance. This stage may last for a period of several weeks to 18 months or longer, depending on the level of the spinal lesion; patients with quadriplegia, for instance, recover some automatic function much sooner than those with paraplegia.

Cystometric examination. The actions of the vesical detrusor as a functioning unit may be studied manometrically by changes in hydrostatic pressure within the viscus, brought about by the contractions of the muscle. This study, known as a cystometric examination, affords a graphic picture of the general detrusor activity and fairly accurately reveals significant changes in the progress of bladder recovery.

Atonic neurogenic bladder. In the stage of spinal shock and in lower motor neuron lesions involving the posterior sacral roots of the reflex arc, the detrusor reflex is abolished and sensation of bladder fullness is absent.

Cystometry at this stage shows little or no increase in intravesical pressure on filling until the viscus is overdistended, at which point a rise in pressure comes about by sheer mechanical force overcoming the forces of inflow. Clinically, the atonic bladder empties small amounts periodically only as the mechanical pressure overcomes the forces of external urethral resistance. (See Fig. 19-9.)

Autonomous neurogenic bladder. Lesions affecting both afferent and efferent limbs of the reflex arc leave the bladder a completely autonomous organ that is capable of exerting some expulsive force by reason of the myoneural activity manifested by the vesical ganglionic plexuses within its walls. This bladder is classically demonstrated in clinical fashion following anterior and posterior sacral rhizotomy, spinal anesthesia, or spinal cordectomy. (See Fig. 19-10.)

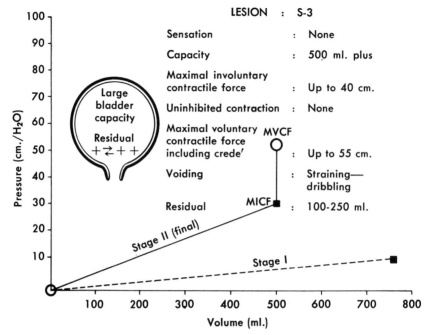

Fig. 19-10. Autonomous neurogenic bladder.

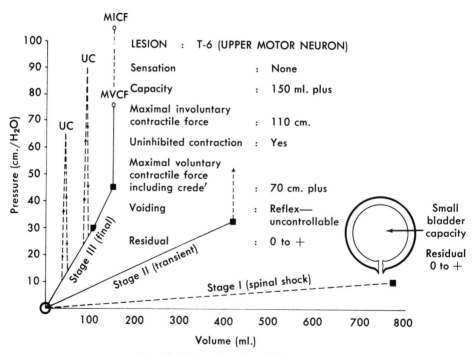

Fig. 19-11. Spastic reflex bladder.

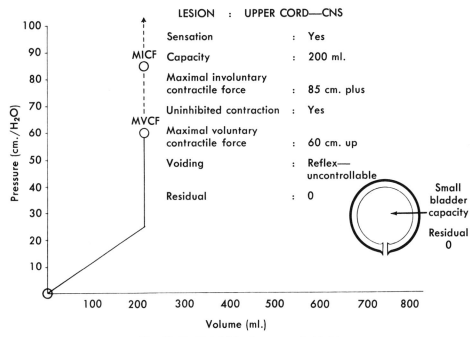

Fig. 19-12. Uninhibited neurogenic bladder.

Spastic reflex bladder. In patients with upper motor neuron lesions, bladder recovery after the stage of spinal shock is usually manifest by precipitate micturition of variable amounts without warning. Occasionally, preliminary erections, fullness in the head, or strange autonomic reactions may precede an involuntary evacuation. Bladder capacity in these patients is usually small, and voiding contractions are not sustained long enough to allow complete emptying. Furthermore, spasticity of the external sphincter is usually present coincident with peripheral spasms, causing interruption of the stream at any point. (See Fig. 19-11.)

Uninhibited neurogenic bladder. This type of neurogenic dysfunction is rarely found after traumatic lesions of the spinal cord unless they are minimal. Acquired lesions of the cerebral cortex due to hemiplegia or tumor or partial destruction of spinal cord pathways by multiple sclerosis commonly result in this type of bladder. Sensation is intact and there is usually great urgency coupled with inability to inhibit bladder contraction, with resultant incontinence. Lack of cerebral inhibition caused by cerebral or cord disease is the fundamental factor underlying this type of bladder dysfunction. (See Fig. 19-12.)

The cystometric examination is a test of detrusor activity alone and must not be considered an index of what we may expect in the way of satisfactory micturition.

Management of the neurogenic bladder

At the time of injury the bladder enters the stage of flaccid inactivity known as spinal shock. During this phase there is no reflex voiding; thus the viscus, unless catheterized, becomes greatly distended, emptying small amounts of urine by reason of overflow incontinence.

Electrode studies of the detrusor during this stage of paralysis have shown that a degree of muscular activity persists through the action of autonomic nerve fibers within the bladder walls and hence the viscus is not truly atonic in a strict sense. This autonomic tonus plays a vital part in the functional evolution of some bladders and must be preserved. Overdistention quite promptly destroys this tonus and may easily convert a partially functioning detrusor into a hopelessly atonic sac. Prevention of overdistention is easily achieved by prompt catheterization.

Another factor of equal importance is the management of this catheter as a substitute for micturition. In the normal individual the bladder is an organ in continuous operation, specializing in filling and emptying itself while fulfilling its role as a temporary reservoir for the excretory products of renal function. When this view of bladder function is kept in mind, the catheter will be managed in such fashion that will enable the organ to perform in a manner closely resembling the normal.

Various mechanical devices have been constructed, based on principles of hydrophysics. One such device distends the bladder to a certain capacity level at which flow is automatically reversed by syphon action and the viscus emptied.

Tidal drainage, once widely used in many centers, has now been abandoned in favor of a more simple procedure that consists in applying a clamp to the urethral catheter and releasing it every 2 or 3 hours—depending on the amount of fluid intake. Straight drainage is permitted throughout the night. By this simple means the bladder is allowed to fill and empty in a normal but passive manner.

In choosing the catheter, one should remember that trophic disturbance from pressure can exist in the urethra as easily as it can over the sacrum. It is important, then, to select the smallest size catheter, one which will exert the least pressure and still be an effective drain. For most purposes a No. 16 or 18 Foley catheter will prove satisfactory.

In general, catheters are foreign bodies and their existence in the urethra and bladder results in some degree of urethritis and cystitis. If the catheter employed is too large, its pressure on the urethral wall with the concomitant urethritis may quickly result in the development of a urethral fistula, usually at the penoscrotal junction.

Although the catheter must be changed periodically, no set rule can be established as to frequency of change, since this is dependent on a number of factors. If the urine remains clear with a minimum of white blood cells, it may not be necessary to change the catheter more often than every two weeks. However, in the presence of persistent alkaline cystitis, with a tendency toward encrustation, the need for change is, of course, more demanding; it may be required as often as every 5 to 7 days.

In changing and irrigating catheters it is well to remember that in both instances, regardless of the aseptic precautions taken, new organisms may be introduced into the viscus, burdened as it is already. The marked tendency for an alkaline, encrusting form of cystitis to develop may be offset by irrigating the bladder twice daily with solution G or 1% acetic acid; 50 ml. of the solution is instilled and allowed to remain within the viscus for 30 minutes. Such irrigations should not be used in the presence of hematuria. Pronounced pyuria may be reduced by irrigation with solutions of benzalkonium chloride (Zephiran) (1:5000) or nitrofurazone (Furacin) (1:10).

Stone formation in the urinary tract of the paraplegic patient is quite common within the first 2 years following injury. A combination of contributory factors may promote such formation, that is, long periods of recumbency, associated fractures, urinary stasis, etc. Hence, it is important to investigate the urinary tract by x-ray examinations at 3-month intervals for the first 2 years of illness.

A low-grade infection within the bladder and urethra in the presence of a catheter may be considered the normal status of a paraplegic patient, and it is not therapeutic surrender to think otherwise. Where infection is concerned, efforts must be directed toward the prevention of the acute type rather than the elimination of the low-grade chronic types.

In choosing prophylactic medication, it is best to begin with the simplest and then to progress to the more complex and specific medications when indications warrant them. Antibiotics, as we know them today, are effective agents against various organisms, but these same organisms have a way of developing resistance to antibiotics proportionate to the frequency and duration of their use. Hence, it is not considered clinically feasible to treat the early paraplegic patient by the antibiotic dictates ascertained from his urine culture and sensitivity tests.

Experience has shown that, in the beginning, simple prophylaxis in the form of chemotherapy is quite effective in the prevention of acute episodes of upper tract infection. Mandelamine (mandelic acid and methenamine) is useful when a benign urinary antiseptic is required. The acidifying action of mandelic acid, fortified by the breakdown of methenamine into formaldehyde, affords a certain measure of defense against the flourish of intravesical organisms. The minimum effective dose is 0.75 Gm. four times a day. Toxicity from the drug is minimal. In most cases, this drug alone will act as a deterrent to the development of acute infection. However, its action in the presence of urea-splitting organisms is appreciably nullified unless preliminary urinary acidification is provided by other means, for example, ammonium chloride, sodium acid phosphate. Where persistent *Proteus* and *Alcaligenes* organisms are present, the use of methenamine mandelate (Mandelamine) is generally not indicated.

In the patient with paraplegia the clinical evaluation of infection is far more important than are laboratory findings. Increasing microscopic pyuria is much more significant than are cultures and sensitivity tests. For the patient with paraplegia, kidney flare-ups mean episodes of fever, chills, and malaise, presumably due to ascending infection in the upper urinary tract. But such episodes should not immediately indict the urinary tract as the source of trouble until all other possible causes have been eliminated.

Throughout the life of the paraplegic patient the urinary bladder becomes a site vulnerable to invasion by bacteria of incredible variety. The commonest pathogens present are (1) coli-aerogenes group, (2) *Streptococcus faecalis,* (3) staphylococci, (4) *Pseudomonas pyocyanea,* (5) *Proteus* and *Alcaligenes* group, and (6) yeasts.

Of all these organisms perhaps the most persistent and most likely to become well entrenched are the urea-splitting bacteria of the *Proteus* and *Alcaligenes* group. Efforts to eradicate these organisms from the urinary tract by chemotherapy are generally unsuccessful in the presence of catheters or other foreign bodies. Further-more, their tendency to foster an alkaline cystitis is notorious and, in such cases, crystalline deposits of calcium and magnesium salts within the bladder occur read-ily—a situation that should double one's vigilance for intravesical calculi.

Complications

Upper urinary tract

Pyelonephritis. In the paraplegic patient, infection is perhaps the most common complication involving the urinary tract and usually the first to be manifest. Episodes of acute pyelonephritis increase in frequency and severity whenever an inefficiently working bladder is left to function on its own. Stasis in the lower urinary tract that harbors residual urine soon invites infection. It is this instance that begins the vicious cycle of hazards that shorten the life of the patient. Infection may reach the kidney by the hematogenous or lymphatic routes or by direct retrograde inoculation through ureteral reflux. Clinically, pyelonephritis is characterized by septic temperatures, chills, general malaise, and pyuria. Pain may or may not be present, dependent, of course, on the level of the cord lesion. Treatment should be instituted at once and should consist of open catheter drainage, use of a broad-spectrum antibiotic while awaiting the results of the urine culture (for predominant organism), and forced oral fluids. Pyelonephritis is a parenchymal disease, and each attack may compro-mise renal function to a greater or lesser extent. Therefore, one must not lose sight of the danger inherent in allowing patients to retain large amounts of residual urine in the hope that these will lessen by themselves or in conducting endoscopic pro-cedures in the presence of gross pyuria.

Calculi. Stone formation is the next most frequent complication in the para-plegic patient, and it is particularly troublesome in the first 2 years of illness. During this period, factors other than infection are at work to enhance the precipitation of urinary salts. Hypercalciuria and hyperphosphaturia, coupled with changes in protein metabolism, usually follow long periods of recumbency and subsequent osteoporosis. Increased urinary excretion of calcium and phosphorus increases the tendency for these minerals to precipitate out in the urinary tract. Intercurrent in-fection provides a further stimulus to calculus formation and is most common in the presence of urea-splitting organisms causing alkaline urine. The most common calculi encountered are those found in infected alkaline urine and consist of various combinations of calcium and magnesium with ammonia, oxalates, and phosphates. Except for the application of certain basic tenets regarding the prevention of calculi, there are as yet no specific therapeutic measures that may be considered phophylactic. A high fluid intake and measures for the prevention of infection, of course, have practical indication.

Hydronephrosis. Hydronephrotic dilatation of the upper urinary tract is usually due to calculus obstruction or to vesicoureteral reflux (Fig. 19-13). In regard to the latter, reflux of urine from the bladder is a common process, especially noted in patients with lesion of the lower motor neuron type. Episodes of recurrent pyelo-nephritis in children with spina bifida are often the first manifestation of bilateral advanced hydroureter and hydronephrosis.

The existence of reflux in the paraplegic patient is not always indicated by intra-

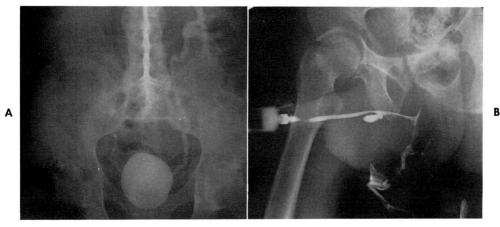

Fig. 19-13. X-ray films showing, A, vesicoureteral reflux and, B, urethral diverticulum.

venous pyelograms. For this reason, a cystogram must be made with an adequate amount of contrast material before reflux can be ruled out. A suitable contrast material is instilled into the bladder via the urethral catheter in amounts up to 400 ml., depending on bladder capacity. Anteroposterior and oblique films are then taken and repeated after manual pressure (Credé maneuvers) over the bladder. The latter procedure often makes reflux manifest when hitherto undetected. Delayed cystograms after ambulation occasionally reveal the complication, especially in children with spina bifida.

It should be stated that the development of reflux may signal the end point in urologic rehabilitation. In most cases, although satisfactory micturition may be accomplished, it will be noted that the patient voids in both directions and by doing so will not only foster the proximal dilatation but also allow the upper tract to be constantly seeded with any bacteria present in the bladder—an event that surely increases both the incidence and severity of renal infection.

Lower urinary tract. In the lower urinary tract, infection once again is the forerunner of all complications. Introduction of the urethral catheter, with attendant peripheral infection, invites the extension of organisms into the prostate and epididymis. Fistulas and diverticula in the urethra (Fig. 19-13, *B*) are common occurrences, and the role of the catheter in their development has been mentioned previously.

Treatment of complications

Infection. Low-grade chronic infection within the bladder is, at best, held in abeyance by such drugs as methenamine mandelate (0.75 Gm. four times a day) fortified, if necessary, by sulfisoxazole (Gantrisin), 1 Gm. four times a day. Acute episodes of infection involving the bladder or kidney call for the employment of antibiotics. Urine cultures consistently reveal a host of pathogens, any of which may be the chief offender. However, the predominant organism isolated is usually the one at fault. The broad-spectrum antibiotics are employed, when possible, in relation to in vitro sensitivity studies. During episodes of acute infection, oral fluids are encouraged and the urethral catheter opened for constant drainage.

Calculi. An alkaline encrusting type of cystitis is prone to form calculi around the bulk of the inlying catheter. The tendency for this occurrence may be diminished by irrigating the bladder twice daily with an acidifying agent such as solution G or 1% acetic acid. An attempt to acidify the urine by administering methenamine mandelate is indicated along with the elimination of milk, dairy products, and the more alkaline-ash foods (fruits, juices, etc.) from the diet. Small calcium phosphate stones, if soft, may be dissolved by solution G, but the harder types will generally require open or transurethral litholapaxy. Calculi in the upper tracts usually demand open surgery for removal, particularly if they are causing obstruction to the kidney or ureter, in which case prompt removal is necessary for the preservation of renal integrity.

Hydronephrosis and vesicoureteral reflux. Nonobstructive hydronephrosis due to ureteral reflux should dictate a cautious approach, since kidney function may be further compromised by injudicious surgery. Reimplantation of the ureters into the bladder does nothing to correct the underlying disturbance. A more recent approach to the problem of vesicoureteral reflux in patients with paraplegia has been the employment of urinary diversion procedures utilizing isolated segments of ileum to serve as substitute bladder reservoirs. This operation, largely perfected by Bricker,[17] consists of the isolation of a distal segment of ileum into which both ureters are implanted. One end of the ileum is closed and the other transplanted to the skin as a stoma. This ileal conduit, by its active peristalsis, functions adequately as a new bladder, diverting the urine to an exterior collection appartus affixed to the skin. Of fifteen such operations performed at Bellevue Hospital, 75% of the patients have demonstrated good results, many showing fairly complete resolution of the hydronephrosis. Similarly, isolated segments of ileum alone—attached to the dome of the bladder—may function more adequately than permanent cystostomy tubes. The success of these procedures is, in a large measure, accounted for by the manner in which they permit free urinary drainage without obstruction or residual urine. Also, since foreign bodies in the form of catheters and cystostomy tubes are not required, infection is eliminated.

Lower tract complications. The patient with acute epididymitis is best managed by rest in bed, elevation of the testes, and antibiotics. In spite of its being the probable etiologic factor, the urethral catheter must remain in situ for adequate bladder drainage, since infected residual urine will contribute as much to the persistence of infection as will the catheter.

Urethral diverticula, if infected, will require surgical excision after preliminary diversion of the urinary stream, either by suprapubic cystostomy or perineal urethrostomy. This is also true for infected fistulas. Occasionally, widemouth diverticula will not require surgery, provided that they may be emptied by manual compression and that no foreign body exists in the urethra.

Correction of disturbances of micturition by surgical intervention

The basic aim of surgery in the urologic rehabilitation of the paraplegic patient is the elimination of residual urine and the foreign body (catheters, cystostomy tubes, etc.). To date, numerous procedures have been developed to enable the paraplegic patient's bladder to perform in a satisfactory fashion. Satisfactory is not meant to imply normal; rather, it refers to the action of the viscus in its ability to empty as completely as possible and at reasonably predictable periods.

The following procedures are currently in use, each of which has its own specific indication: (1) transurethral resection of the bladder neck, (2) pudendal neurectomy, (3) sacral rhizotomy, (4) spinal cordectomy, and (5) intrathecal alcohol injection.

Transurethral resection. This procedure was initially advocated for all types of neurogenic bladder dysfunction regardless of the level or completeness of the lesion. We have learned by experience, however, that it is of benefit in a relatively small percentage of paraplegic patients. The circumferential excision of tissue from the rim of the vesical neck serves to lessen the resistance to urinary outflow, provided that resistance can be shown to exist in some form or other, such as true hypertrophy or neurogenic spasticity. In theory, spasticity of the internal sphincter, or muscular fibers grouped around the vesical neck, may result in hypertrophy if it is of long standing. In practice a true hypertrophy is not often encountered endoscopically, the major factor being an external sphincter obstruction that could be corrected by other means.

The best application of the transurethral procedures is in the patient with a lower motor neuron lesion in whom voiding is accomplished by straining and Credé maneuvers. In these patients the resection of the rim of the vesical neck may remove an element of resistance, thereby enabling a more complete evacuation with less effort.

Pudendal neurectomy. In patients with upper motor neuron lesions, particularly at the cervical and upper thoracic levels, varying degrees of peripheral spasm will be found. Included in the spastic category are the muscles of the perineum. Since we know that a relaxation of the pelvic floor is the first requirement of normal micturition, one can realize to what extent spasms of these muscles could interfere with emptying of the bladder. Spasticity of the external urinary sphincter is usually concurrent with this mass contraction. The combination of these factors causes a gradual or sudden cutoff of the urinary streams. If the cystometric examination in the patient exhibits sufficient and sustained force, section of the pudendal nerves may abolish the perineal muscle spasm that prohibits bladder emptying. A preliminary test by local anesthetic block of the pudendal nerves is first performed and, if *satisfactory voiding* is achieved, both nerves may be surgically excised.

Sacral rhizotomy. The essential anatomic difference between an automatic and a reflex bladder lies in the reflex arc for micturition at the second, third, and fourth sacral segments. A dysfunction of the sacral reflex arc, resulting in urinary retention, is a common cause of failure in the establishment of a reflex bladder. It may be due to the presence of inhibitory impulses within this arc. It is the elimination of the arc itself that serves to establish automaticity, and this is mediated through the action of the myoneural components within the detrusor. Transection of the cauda equina clearly exhibits this type of function, and rhizotomy has been shown to duplicate it.

Anterior and posterior sacral rhizotomy converts an upper motor neuron lesion into a lower motor neuron lesion and, in effect, amounts to the trading of a greater evil for a lesser one. Generally speaking, patients with low cord lesions have better urologic progress than those with high ones, which fact establishes the theoretical basis for this procedure. Rhizotomy is best indicated in paraplegic and quadriplegic patients with severe peripheral spasticity, which in turn denotes a spastic reflex bladder. The bladder is of small capacity and is associated with involuntary sphincter contractions that cut off the urinary stream.

Preliminary sacral block studies, employing a long-acting anesthetic, enable us to evaluate the procedure from the functional standpoint. For practical purposes, determination of bladder capacity, voiding pattern, and residual urine before and after the block will give the examiner the necessary information for the indication of rhizotomy.

It is interesting to note that Hutch[18] has observed the resolution of bilateral ureterohydronephrosis (due to reflux) after anterior and posterior sacral rhizotomies on four patients. This fortuitous result, if observed in further cases, may establish the procedure as being of even greater benefit than originally intended and hence another step forward in urologic rehabilitation.

Spinal cordectomy and intrathecal alcohol injections. A relatively radical procedure that serves to create a more completely autonomous bladder consists of the removal of the spinal cord below the level of the lesion or the intrathecal injection of absolute alcohol.

Cordectomy, first described by Browder and Corradini in 1949, resulted in notable improvement in bladder function in two patients.[19] Recently, Cooper and Sullivan have demonstrated marked improvement in ten of twelve patients subjected to cordectomy or to alcohol wash. Clinically a most outstanding result from these procedures has been a notable increase in bladder capacity and in detrusor activity, supporting the concept of inhibitory impulses in the posterior roots of the sacral reflex arc.

Rhizotomy and cordectomy, of course, have a dual effect on the paraplegic patient. The relief of peripheral skeletal spasm enables more rapid general rehabilitation and also improves bladder function.

Regarding any surgical operation aimed at establishing better bladder function, it must be emphasized that such procedures should be undertaken only when the patient has reached his plateau of recovery and when no further improvement is anticipated, either in bladder function or peripheral mobility.

Bladder-training program

In the urologic rehabilitation of the patient with paraplegia, our efforts are directed mainly toward complete or nearly complete evacuation of the bladder (Chapter 13). If this act can be performed at regular intervals, we may consider the patient as having satisfactory bladder function. Unfortunately, such a situation is all too rare in the paraplegic patient, especially in those individuals with the spastic reflex activity of upper motor neuron lesions.

The surgical conversion of upper into lower motor neuron lesions usually results in more predictable bladder function. Even though sensation of bladder fullness is seldom present, the patient may aid emptying by straining and Credé maneuvers, performed at reasonably regular times. The functionally adequate spastic reflex bladder is a viscus of absolutely unpredictable action, dependent in certain instances on peripheral spasm and in others on the stretch reflexes within the detrusor. The majority of these patients will require some type of external urinary apparatus for the involuntary and unpredictable evacuations characteristic of this type of erratic bladder function. Efforts have been made to regulate fluid intake to correspond with bladder emptying in these patients. Almost all of them have proved fruitless because of the reasons cited previously. The spastic reflex bladder defies training, and it must be emphasized that the ability of the organ to empty variable quantities of urine at stated intervals does not alone signify a trained bladder.

Summary

In the urologic management of the neurogenic bladder, adherence to the fundamental principles of clinical urology is of prime importance.

Certain investigative procedures in patients with paraplegia have as their basic aim the elimination of residual urine and the achievement of satisfactory bladder function. Although a measure of progress has been attained, there is, on the other hand, much to be learned and much to be accomplished. The paraplegic patient is a fragile individual with a modified life expectancy. To subject him to surgery of doubtful merit is unjustly experimental and extremely hazardous. As our knowledge of normal and abnormal physiology increases, so will our success in his urologic rehabilitation.

ORTHOPEDIC MANAGEMENT

Unlike the neurosurgical and urologic members of the team who take immediate charge of the patient with paraplegia, the orthopedic surgeon does not play an early active role. He confines his activities to watchful observation, mindful of preventing complications. Because of a great variety of undesired developments following in the wake of cord transections, the orthopedic management is a highly individualized one.

Spinal fusion. It has been pointed out earlier in this chapter that patients with spinal cord injury in the thoracic and lumbar regions should have an early laminectomy. This brings up the question of fusion of the spine. There can be little doubt that laminectomy, as such, does not of necessity require subsequent fusion. If the vertebral bodies have more or less maintained their rectangular shape and are found in mechanically sound alignment, such a spine needs no further surgical stabilization. Conversely, with sharp wedging of the vertebral body and corresponding gibbus formation, internal fixation by fusion may well be desirable. The same is true if pain exists at the fracture site and, particularly, if it is absent on recumbency but recurs on being upright. Generally, if a fusion makes it possible to eliminate an otherwise needed back brace we believe this to be a sound indication for the procedure.

Spinal fusion is not an emergency measure and, as a rule, should not be done in combination with the initial decompression. Not only would it unnecessarily increase the size of the operation, but more often than not the need for it is not sufficiently apparent. Unlike the fusion in Pott's disease, we are more lenient with postoperative immobilization for these patients. A plaster jacket is contraindicated because of the possibility of decubitus ulcers. It is a well-known fact that patients with paraplegia and quadriplegia are prone to such ulcer formation. Rather, a well-fitted back brace is made preoperatively and is put on a few days postoperatively. This allows observation of the skin and affords as effective an immobilization as the plaster jackets, if not better. With such a brace the patient is allowed to sit up about a week later. If bony fusion fails to occur, it is no great calamity. The dense fibrous scar tissue inside will still hold the spine better than any outward support.

This is not the place to enter into details of operative technic, but one point needs emphasizing—the cancellous bone grafts, best obtained from the iliac crest, are laid down far out laterally on the denuded spine so as not to interfere with the neurosurgeon's decompression of the cord.

Pathologic fracture. Pronounced demineralization of bone may occur rather rapidly in patients with this lesion. Not only are the minerals likely to be redeposited in form of concrements throughout the urinary tract, but also their absence from the long bones facilitates pathologic fractures, often on the slightest provocation. To counteract demineralization, standing the patients up in braces or strapping them to the tiltboard for at least an hour daily has proved extremely helpful. Pathologic fractures, painless as they are, are easily overlooked and are often discovered only because of persistent swelling of the area involved. In our experience these fractures heal under conservative treatment, without great difficulty and usually with no marked delay. Since this type of bone is not well suited for internal fixation by metal gadgets, use of such procedures should be avoided wherever possible. For the same reason, skeletal traction is not well tolerated. Skin traction is hazardous and invites decubitus ulcers. Immobilization in padded plaster splints or bivalved circular casts that permit watching the skin underneath is the technic of choice.

Neurotrophic joint. Occasionally a Charcot type of neurotrophic joint is encountered, often the hip, in which the femoral head seems to be melting away partially or in toto. Actually, it is not surprising that a cord transection should give rise to a neurotrophic joint disorder; it is perhaps more startling that it does not occur more often. Dislocations of the hip are not uncommon and seem to occur on a similar basis in the later stages. When there is no history of trauma or infection, one can assume the capsule of the joint to be extended by effusion, a familiar occurrence in a Charcot joint, thus allowing the femoral head to slide out with ease. It can sometimes be reduced without difficulty, but it is likely to redislocate promptly. Indeed, it is almost impossible to hold the head in place. Since the patients wear braces anyway, the dislocation has little practical significance. If pain develops to a troublesome degree, resection of the head and neck has proved beneficial.

Myositis ossificans. Extensive myositis ossificans, usually occupying large segments of pelvic and upper thigh musculature, may also be a troublesome feature of the later stages (Fig. 19-14). Despite its impressive x-ray appearance, it is usually not productive of grave symptoms. This is fortunate because surgical excision of the bony masses is likely to be followed by recurrence. No effective and dependable treatment, surgical or otherwise, is available.

Contractures and their treatment. Patients with long-existing, neglected paraplegia often present very extreme flexion contractures of the hip, knee, and ankle joints, complicated by large decubiti in the areas of bony protrusions. To avoid osteomyelitis that would vitiate all further attempts at rehabilitation, it is wise to correct the contractures by sectioning soft parts only, avoiding osteotomies and, if possible,

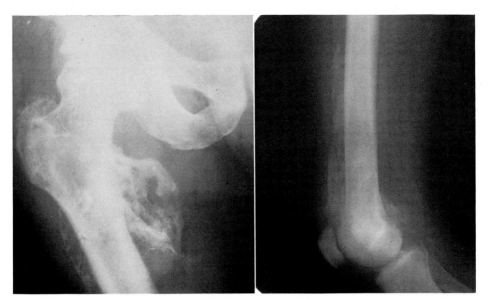

Fig. 19-14. Two common sites for myositis ossificans are about the hip and in the anterior thigh.

capsulotomies also. Such strippings and release operations may not bring about full correction at once, but what little remains of fixed deformity would yield readily to stretchings and manipulation after healing of the incisional wounds. It is true that an osteotomy would effect correction, perhaps quicker and more completely, but aside from shying away from opening the medulla cavity of the bone in the presence of a decubitus nearby, an osteotomy requires rigid immobilization, which we try to avoid and which is not necessary in release operations. With this method it is possible even in the most bizarre cases to make the extremities fit for braces. Fortunately, these extreme deformities due to long neglect are becoming a problem of the past. Indeed, fixed deformities need not occur at all, and they reflect on the initial professional services.

Equally unfortunate are extension contractures of the hip. Patients with large decubiti over the sacrum and ischium who are being treated by plastic repair and who remain on their abdomen for months may end with their hips stiff in extension. This is entirely too high a price for closing a skin ulcer, particularly in a patient who, by the very nature of his condition, is forced to sit most of the time. This type of contracture can be prevented by turning the patient to the side for only a short interval daily and flexing the hip through its full range of motion. In the prone position, contractures of the knees can be avoided by flexion of the leg at the hip and knee, and equinus deformities of the ankle can be avoided by placing the foot between the mattress and footboard.

Teen-age proneness. In spite of meticulous care, some patients develop one or the other of these orthopedic complications. The ones most likely to be afflicted with most severe complications seem to be the teen-agers. The adult paraplegic patients appear to be the more resistant group. This observation is consonant with the observation that the higher the general physical fitness level, the more rapid are the deconditioning phenomena during bed rest. These phenomena are obviously associated with the level of endocrine imbalance.

Amputations. The more common orthopedic conditions accompanying paraplegia have been outlined. One bizarre problem should be mentioned. A suggestion heard occasionally is "Why not amputate the paraplegic's legs as they are no use to him anyway?" Such reasoning violates the most rudimentary understanding of physiologic surgery and human psychology. Who can say that one day the riddle of cord healing with function may not be solved? Stranger things have happened through research. Amputation is only indicated for pathologic necessity. The paraplegic patients with amputations are an unhappy lot.

Hand function in quadriplegia. In the quadriplegic patient, naturally, the principal problem is hand function. If some function is preserved, fixed deformities can easily develop, if not guarded against, by imbalanced muscle pull and the forces of gravity.

An occasional tendon transfer may greatly enhance hand and finger function. Whether such procedures are feasible needs particularly careful individual study and evaluation. In patients with lesions at the level of the sixth or seventh cervical vertebra, the only action of the hand is dorsiflexion of the wrist. By fastening the deep finger flexors and the thumb flexor securely to the radius, with the fingers extended and the wrist flexed, the fingers will automatically bend when the patient actively extends the wrist. Functionally this means much. The obvious advantage surgically lies in the fact that no functioning structure is taken away, since the tendons used are paralyzed.

In patients with lesions between the seventh cervical and first dorsal vertebras, there is loss of flexion of all fingers and loss of all intrinsic muscles. Here the extensor carpi radialis brevis can be transferred to the thumb flexor and the wrist flexors can be attached to finger flexors. As is the case with all tendon transfers, only normally functioning tendons should be used as transplants and also only those that can be spared and are not needed for other essential action.

Comprehensive imaginative orthopedic services are fundamental in a rehabilitation program.

REFERENCES

1. Abramson, S. A.: Bone disturbances in injuries to spinal cord and cauda equina (paraplegia), J. Bone Joint Surg. **30-A**:982, 1948.
2. Berns, H. S., Lowman, E. W., Rusk, H. A., and Covalt, D. A.: Spinal cord injury—rehabilitation costs and results and follow-up in thirty-one cases, J.A.M.A. **164**:1552, 1957.
3. Bors, E.: Spinal cord injuries, Vet. Admin. Tech. Bull. Series 10, **2**:503, Dec., 1948.
4. Heyl, H. L.: Some practical aspects in rehabilitation of paraplegics, J. Neurosurg. **13**:183, 1956.
5. Primer for paraplegics and quadriplegics, Patient Pub. 1, New York, 1957, Institute of Physical Medicine and Rehabilitation, New York University–Bellevue Medical Center.
6. Rusk, H. A.: Meeting needs and life problems of paraplegic patients, Merck Rep. **62**:3, July, 1953.
7. Schlesinger, E. B.: Observations on rehabilitation of neurologically handicapped, Bull. N. Y. Acad. Med. **28**:229, 1952.
8. Hoen, T. I., and Cooper, I. S.: Acute abdominal emergencies in paraplegics, Amer. J. Surg. **75**:19, 1948.

9. Covalt, D. A., Cooper, I. S., Hoen, T. I., and Rusk, H. A.: Early management of patients with spinal cord injuries, J.A.M.A. **151:**89, 1953.
10. McCravey, A.: War wounds of spinal cord: plea for exploration of spinal cord and cauda equina injuries, J.A.M.A. **129:**152, 1945.
11. Scarff, J. E., and Pool, J. I.: Factors causing massive splasm following transection of cord in man, J. Neurosurg. **3:**285, 1946.
12. Crutchfield, W. G., and Schultz, E. C.: Fractures and dislocations of spine, Amer. J. Surg. **75:**219, 19 8.
13. Cooper, I. S.: Relief of spasticity in paraplegia, J. Int. Coll. Surg. **22:**53, July, 1954.
14. Cooper, I. S., and Hoen, T. I.: Intrathecal alcohol in the treatment of spastic paraplegia, J. Neurosurg. **6:**187, May, 1949.
15. Freeman, L. W., and Heimburger, R. F.: The surgical relief of spasticity in paraplegic patients: I. Anterior rhizotomy, J. Neurosurg. **4:**435, 1947.
16. Talbot, H. S., and Lyons, M. K.: Late renal changes in paraplegia, J. Urol. **53:**4, April, 1950.
17. Bricker, E. M., Butcher, H., and MacAfee, C. A.: Late results of bladder substitution with isolated ileal segments, Surg. Gynec. Obstet. **99:**469, 1954.
18. Hutch, J. A.: Treatment of hydronephrosis by sacral rhizotomy in paraplegics, J. Urol. **77:**2, Feb., 1957.
19. Browder, J., and Corradini, C. W.: Report of the proceedings of the seventeenth annual meeting, Harvey Cushing Society, June 2-4, 1949, New Haven, Conn.
20. Bunnell, S.: Surgery of the hand, ed. 3, Philadelphia, 1956, J. B. Lippincott Co.
21. Holmes, G.: The Garrlston lecture on spinal injuries of warfare: the clinical symptoms of gunshot injuries of the spine, Brit. Med. J. **2:**815, 1915.
22. Foerster, O.: Die Lietungsbahnen des Schmerzgefühls und die chirurgische Behandlung der Schmerzzustände, Berlin, 1927, Urban & Schwarzenberg.

CHAPTER 20

REHABILITATION OF PATIENT WITH METABOLIC DISEASES

ARTHRITIS

Of all chronic diseases, with the exception of nervous and mental diseases, the arthritides inflict the greatest toll in morbidity. In the United States arthritis and rheumatism cause more years of disability than do all types of accidents. As a result of a carefully conducted Public Health survey carried out in 1951, it was estimated that more than 10 million persons over 14 years of age, exclusive of those with rheumatic heart disease, considered themselves victims of arthritis or rheumatism. Of this group 75% had consulted physicians because of their symptoms and almost 6½ million had had such a diagnosis confirmed. About 25%, or 2½ million persons, were afflicted to such a degree that they had had to change, modify, or stop their work because of the severity of disease. From other statistic sources it is estimated that in the United States 144,000 persons annually are invalided from rheumatic diseases.

With the medical conquest against fatal infectious diseases almost a complete success, the sharply rising incidence of chronic and degenerative diseases assumes increased significance. Furthermore, since the arthritides inflict a low mortality and high morbidity, the potential reservoir of those afflicted and disabled by this group of chronic diseases is great. The socioeconomic consequences of this situation is readily apparent. It is estimated that more than 97 million man days are lost annually as the result of rheumatic diseases and that the economic toll in terms of wages lost is more than a quarter of a billion dollars. Finally, it must be remembered that arthritis is not primarily a disease of the aged but that it may affect infants and adolescents as well; the two most common and most crippling forms, rheumatoid arthritis and rheumatoid spondylitis, preponderantly affect younger persons, in their third and fourth decades.

346

Classification and causes

Arthritis by definition means inflammation of a joint. It is thus a symptom complex, the etiology of which may be variable. It has been estimated that there are more than a hundred different causes of arthritis, or joint inflammation, the treatments and prognoses of which differ widely. (See Chart 12.) Therefore, in considering the physical, medical, and rehabilitation aspects of arthritis, it is of basic importance that an etiologic, as well as a pathologic, diagnosis be established.

Although the causes of arthritis are numerous, the majority of cases fall within seven major groups: (1) arthritis due to specific infection, (2) arthritis due to rheumatic fever, (3) rheumatoid arthritis and rheumatoid spondylitis, (4) arthritis due to direct trauma, (5) arthritis of gout, (6) degenerative joint disease, and (7) nonarticular rheumatism. In order of frequency, rheumatoid arthritis accounts for 30% to 40% of cases; degenerative joint disease, for 25% to 30%; and nonarticular rheumatism, for 10% to 20%. These three groups alone compose at least two thirds of all cases. Hence, although the etiologic possibilities have great diversity, the problem, statistically and from a treatment standpoint, is largely restricted to a small group of types.

Arthritis due to specific infection

Infection may be introduced into a joint by direct implantation, as from a stab wound, by contiguous spread from a local area of infection, or by transport via lymphogenous or hematogenous routes. Since the discovery and widespread use of antibiotics and other chemotherapeutic agents, the incidence of uncontrolled infection has been greatly reduced, and specific infectious arthritis has become a relatively rare disease. Even when it does occur, the prompt use of agents to eradicate infection usually averts gross destruction of the articular structures, which in the past too frequently resulted from such purulent processes as pneumococcal and gonococcal arthritis.

When the specific infectious process involves a joint, the primary approach to treatment is the institution of therapy for the eradication of the infection. In the active phase of the infection the objectives of physical medicine are analgesia and the maintenance of ranges of motion of the involved joint to prevent deformity. Heat applied locally, constantly or intermittently, is usually indicated, and warm moist heat in the form of compresses, fomentations, or packs is generally best tolerated. In the less active phase, hydrotherapy in the form of whirlpool baths may be used more effectively to permit early active exercising. Even in the active phase of a specific infectious arthritis, the involved joint should be carried passively through normal ranges of motion at least once a day to prevent capsular and tendinous tightening and the development of deformity. As the process subsides, an active exercise program should be initiated to maintain normal ranges of motion and to prevent muscle atrophy.

Following the eradication of the infection, the problem of physical rehabilitation is contingent on the degree of muscle weakness that has resulted, the range of motion that has been lost, and the extent of structural intra-articular damage produced by the infection. The first two conditions should be actively treated with range-of-motion stretching exercises and progressive resistive exercise programs. Intra-articular dam-

CHART 12

NOMENCLATURE AND CLASSIFICATION OF ARTHRITIS AND RHEUMATISM OF THE AMERICAN RHEUMATISM ASSOCIATION*
(Tentative)

I. Polyarthritis of unknown etiology
A. Rheumatoid arthritis
B. Juvenile rheumatoid arthritis (Still's disease)
C. Ankylosing spondylitis
D. Psoriatic arthritis
E. Reiter's syndrome
F. Others
II. "Connective tissue" disorders
A. Systemic lupus erythematosus
B. Polyarteritis nodosa
C. Scleroderma (progressive systemic sclerosis)
D. Polymyositis and dermatomyositis
E. Others
III. Rheumatic fever
IV. Degenerative joint disease (osteoarthritis, osteoarthrosis)
A. Primary
B. Secondary
V. Nonarticular rheumatism
A. Fibrositis
B. Intervertebral disc and low back syndromes
C. Myositis and myalgia
D. Tendinitis and peritendinitis (bursitis)
E. Tenosynovitis
F. Fasciitis
G. Carpal tunnel syndrome
H. Others (See also shoulder-hand syndrome, VIII E)
VI. Diseases with which arthritis is frequently associated
A. Sarcoidosis
B. Relapsing polychondritis
C. Henoch-Schönlein syndrome
D. Ulcerative colitis
E. Regional ileitis
F. Whipple's disease
G. Sjögren's syndrome
H. Familial Mediterranean fever
I. Others (See also psoriatic arthritis, I D)
VII. Associated with known infectious agents
A. Bacterial
1. Brucella
2. Gonococcus
3. Mycobacterium tuberculosis
4. Pneumococcus
5. Salmonella
6. Staphylococcus

7. Streptobacillus moniliformis (Haverhill fever)
8. Treponema pallidum (syphilis)
9. Treponema pertenue (yaws)
10. Others
B. Rickettsial
C. Viral
D. Fungal
E. Parasitic (See also rheumatic fever, III)
VIII. Traumatic and/or neurogenic disorders
A. Traumatic arthritis (viz., the result of direct trauma)
B. Lues (tertiary syphilis)
C. Diabetes
D. Syringomyelia
E. Shoulder-hand syndrome
F. Mechanical derangements of joints
G. Others (See also degenerative joint disease, IV; carpal tunnel syndrome, V G)
IX. Associated with known biochemical or endocrine abnormalities
A. Gout
B. Ochronosis
C. Hemophilia
D. Hemoglobinopathies (e.g., sickle cell disease)
E. Agammaglobulinemia
F. Gaucher's disease
G. Hyperparathyroidism
H. Acromegaly
I. Hypothyroidism
J. Scurvy (hypovitaminosis C)
K. Xanthoma (tuberosum)
L. Others (See also multiple myeloma, X G; Hurler's syndrome, XII C)
X. Tumor and tumorlike conditions
A. Synovioma
B. Pigmented villonodular synovitis
C. Giant cell tumor of tendon sheath
D. Primary juxta-articular bone tumors
E. Metastatic
F. Leukemia
G. Multiple myeloma

*From Blumberg, B. S., and associates: Nomenclature and classification of arthritis and rheumatism (tentative) accepted by the American Rheumatism Association, Bull. Rheum. Dis. **14**:340, 1964.

CHART 12, cont'd

X. Tumor and tumorlike conditions— cont'd
  H. Benign tumors of articular tissue
  I. Others
     (See also hypertrophic osteo-arthropathy, XIII G)
XI. Allergy and drug reactions
  A. Arthritis due to specific aller-gens (e.g., serum sickness)
  B. Arthritis due to drugs (e.g., hydralazine syndrome)
  C. Others
XII. Inherited and congenital disorders
  A. Marfan's syndrome
  B. Ehlers-Danlos syndrome
  C. Hurler's syndrome
  D. Congenital dysplasia
  E. Morquio's disease
  F. Others

XIII. Miscellaneous disorders
  A. Amyloidosis
  B. Aseptic necrosis of bone
  C. Behcet's syndrome
  D. Chondrocalcinosis (pseudo-gout)
  E. Erythema multiforme (Stevens-Johnson syndrome)
  F. Erythema nodosum
  G. Hypertrophic osteoarthrop-athy
  H. Juvenile osteochondritis
  I. Osteochondritis dissecans
  J. Reticulohistiocytosis of joints (lipoid dermato-arthritis)
  K. Tietze's disease
  L. Others

age leaves the joint mechanically impaired; the consequence of this is in direct re-lation to the extent of the damage and the amount of physical activity to which the joint must be subjected. Mechanical derangement in a weight-bearing joint as the result of infection is thus considerably more vital than damage of a similar degree in a nonweight-bearing joint. Principles for the protection against further wearing of structurally damaged joints are the same as those for similar mechanical problems imposed by trauma or degenerative arthritis.

Arthritis due to rheumatic fever

The joint involvement of rheumatic fever, next to that of acute gouty arthritis, is the most dramatically acute and painful of all the arthritides. On the other hand, it is the most benign in its residual effects on joints. Rheumatic fever never leaves residual damage within joints; it has aptly been said that it "licks the joints and bites the heart." Physical therapy, therefore, has no place in the treatment of acute rheu-matic fever except in the application of physical agents locally to joints for ameliora-tion of pain. In the convalescent phase of the disease, a graduated exercise program will expedite the reconditioning of the patient after his prolonged period of bed rest.

Rheumatoid arthritis and rheumatoid spondylitis

Rheumatoid arthritis and rheumatoid spondylitis are highest in incidence among the arthritides and exact the greatest toll in crippling. Although they may have their onset in infancy or in old age, the peak age of onset is in the third and fourth decades of life. Their mortalities are extremely low, and when death occurs it is usually from an intercurrent complicating disease. These then comprise a group of diseases of high morbidity and low mortality with a high potential for crippling disability among a relatively young age group. The causes of rheumatoid arthritis and rheumatoid spondylitis are not known, and all treatment, therefore, is non-specific. Since the introduction of adrenocortical steroids as a nonspecific chemical for the suppression of inflammation, effective control of these types of arthritis has been improved. Despite this therapeutic addition, however, the basic approach to treatment remains the same: additional rest and adequate diet for improving body resistance, salicylates for analgesia, and physical therapy for the prevention of deformity. Physical therapy therefore remains a keystone of treatment, and the more diligent its application, the more effective the prevention of crippling and deformity.

The prominent clinical manifestations in a rheumatoid arthritic joint are the effusion, pain, and heat consequent to the inflammatory synovitis. The intra-articular effects of the disease, however, may extend to produce a chondritis, an osteitis, and a sterile osteomyelitis. In addition, in the neuromuscular system, nodular myositis and nodular neuritis may occur. As a result of this diffuse inflammatory process, pain and muscle spasm in and about the involved joint predispose to functional splinting with resultant tendon and capsular tightening. When this latter occurs, normal ranges of motion become impaired and deformity results. In addition, the functional splinting and the neuromuscular pathology predispose to atrophy of the supporting muscles about the joint. Finally, the intra-articular inflammation may produce structural distintegration of the cartilage and subchondral bone of the joint.

The joint lesions vary greatly, depending on the stage and the severity. Early in the disease there is proliferation of the synovial cells, thickening of the synovial membrane, effusion in the joint, and periarticular soft tissue swelling. Subchondral osteoporosis follows this inflammatory intra-articular process and is the earliest radiologic change. The proximal interphalangeal joints are often the first to be involved with characteristic fusiform swelling. If the disease progresses, the articular cartilage may be destroyed, with subsequent necrosis of subchondral bone, subluxation, dislocation, and deformity (Fig. 20-1). Fibrous ankylosis, and ultimately bony ankylosis, may ensue. One or more joints become swollen and others follow. In severe cases almost every joint in the body may become involved. Remissions are common.

Radiologically, the early findings may be only those of soft tissue swelling, intracapsular effusion, and subchondral osteoporosis. Subsequently, there may ensue narrowing of the joint space, diffuse demineralization, and marginal destructive changes involving the articulating cortex and subchondral bone. Secondary productive and/or sclerotic bone changes may develop. These latter destructive changes may variously resemble the discrete punched-out areas of destruction seen in gout, those typical of a neurotrophic joint, and at times may be so major as to give the impression of either resorption or excision of the entire articulating portion of the joint.

Treatment of rheumatoid arthritis. Physical therapy has no curative effect for arthritis, nor does it alter the course of the disease process. The objectives of therapy are (1) analgesia, (2) maintenance of normal ranges of motion in involved joints, (3) maintenance of normal power in muscles about involved joints, and (4) protection of joints against additional trauma that might result in further structural deterioration and/or deformity. Therapy is thus prophylactic; as such it must be carried out daily and indefinitely as long as the active disease process persists. The more active the arthritic process, the greater is the threat to joint mechanics and the more assiduously must the physical therapy program be performed.

Since physical therapy for rheumatoid arthritis must be carried out daily and indefinitely, it is rarely economically or geographically feasible to have this done by

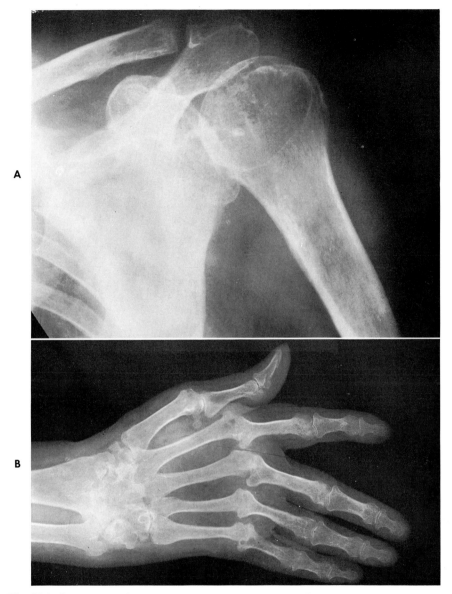

Fig. 20-1. A, Osteoporosis secondary to rheumatoid arthritis. **B,** Marked metacarpal subluxation with ulnar deviation deformity. The carpal articulations also show marked rheumatoid changes.

a professional therapist. It is preferable, therefore, that the patient be instructed in a home program of therapy that may be performed alone or with assistance. Such instruction usually can be accomplished in several treatment-instruction sessions with a physical therapist. The patient should then be rechecked at regular intervals to ascertain progress and to make necessary changes or modifications.

The basic components of the home physical therapy program are (1) heat, (2) therapeutic exercise, and (3) the application of energy-saving and protective measures to involved joints.

Fig. 20-2. Using paraffin heated in a double boiler is a simple and effective method of applying heat to the hands and wrists at home.

Heat. Heat applied locally to an arthritic joint relieves muscle spasm and pain. The effects are the same, regardless of the physical agent used to produce tissue heating, the only difference being one of degree. From the patient's standpoint the analgesic effect of heat is the important one. From a therapeutic standpoint, however, this analgesia provides the means toward a definitive end, in that it makes it tolerable for the patient to carry out a therapeutic exercise program. In general, patients with rheumatoid arthritis prefer damp or wet heat to dry heat and superficial heat to deep heat (hot towels, for example, in preference to a body baker, and infrared in preference to diathermy).

Selection of the type of heat to be used in the home program should be based on simplicity, facility, and economy of application (Fig. 20-2). Methods of heating requiring large financial investments in equipment are unnecessary. Similarly, complex methods are to be avoided. Finally, methods involving considerable expenditure of a patient's daily time render the home program impractical and predestine it to failure. For the patient with involvement of the hands, feet, wrists, and ankles, the contrast bath can be simply, cheaply, and quickly carried out. For those with bilateral knee involvement, Hydrocollator packs might effectively be prescribed. For those with disseminated joint involvement, on the other hand, it is far more practical to heat all involved joints simultaneously in a hot tub or with a body baker than to rotate an infrared lamp from joint to joint over the prolonged period that would be required. Following the application of heat, the patient carries out his therapeutic exercise program.

Exercise. Therapeutic exercises for the arthritic patient are of two types: range of motion and resistive.

Range-of-motion exercises. These are maneuvers to preserve the normal ranges of motion of involved joints. Since the types of motions of each joint vary and since most joints have multiple motions, the patient must be taught the specific motions and the ranges of these motions for each joint to be exercised. For example, shoulder range-of-motion exercises must include flexion, extension, abduction, adduction, and internal and external rotation. It is important that each joint be carried to the extremes of each of its ranges of motion at least once daily, even in the acute or subacute phase of the illness. Loss in range of motion of a joint always begins at the extreme of the range, and intermediate joint excursion is the last to become restricted. This sequence is the reason for the acute importance of forcing a joint through its full ranges of motion daily if the insidious development of deformity is to be averted.

Resistive exercises. These are repetitive exercises done against graduated resistance for the maintenance of power in muscles predisposed to weakness from disuse or from intramuscular rheumatoid pathology. These should be performed in nonweight-bearing positions to avoid additional intra-articular damage from trauma. They must be graded to the weakness of the muscle involved and may vary from nongravity exercises on a powder board to progressive resistive exercises with a barbell, boot, and weights. The extent to which a rheumatoid joint may be subjected to such exercises depends on the acuteness of the inflammatory disease process. The importance of resistive exercises is secondary to range-of-motion exercises, for while the latter are imperative daily procedures, the resistive exercises must be gauged to the patient's tolerance and needs. This latter is accomplished in the frequent checkup sessions of the therapist with the patient. (See Chapter 4.)

Energy conservation and protection of joints. The joint affected by rheumatoid arthritis often sustains structural intra-articular damage as a consequence of the disease; in addition, impairment of its normal physiology predisposes it to traumatic wearing. Such joints, therefore, should be spared as much unnecessary trauma as can be eliminated by the application of energy-saving technics and devices in the performance of essential physical activities. (See Fig. 20-3.) Furthermore, the use of night splints to prevent contractures and hand supports to avoid ulnar deviation deformity should be anticipated before the need demands them. Finally, crutches, canes, and wheelchairs should be regarded as energy-conserving devices and should be used freely as prophylaxis against joint traumas and damage that with wisdom, may be prevented. Too often these are regarded as resorts "in extremis" rather than as aids for the protection that they can temporarily or permanently afford.

Treatment of rheumatoid spondylitis. The objectives of physical medical treatment of rheumatoid spondylitis are, with a few modifications, basically the same as

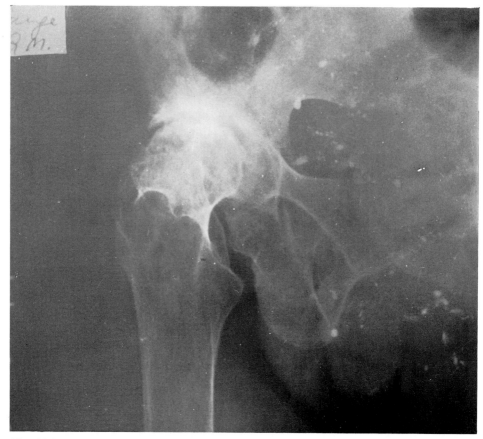

Fig. 20-3. Structural damage as a result of rheumatoid arthritis. For the preservation of function such a joint should be protected against additional unnecessary wearing.

those of rheumatoid arthritis. Similarly, the basic components of the treatment program are the same. There are, however, certain special considerations that must be incorporated into the treatment program of spondylitis.

The patient with rheumatoid spondylitis tends to develop a kyphotic deformity of the spine (Fig. 20-4). This may involve only he lumbar area or the entire spine, depending on the extent of the disease process. Exercise programs, therefore, are directed primarily toward the prevention of such a kyphosis rather than toward maintenance of normal range of motion as is true of peripheral rheumatoid arthritis (Fig. 20-5). The one spinal area in which efforts are directed toward maintenance of range of motion is the cervical spine. The exercise program for the patient with rheumatoid spondylitis therefore consists of (1) upper back extension exercises and deep-breathing exercises, both directed toward maintaining an erect spine, (2) cervical range-of-motion exercises, and (3) stretching exercises for maintaining maximal elasticity in the hip extensors, the fascia lata, and the hamstring muscle groups. Maintenance of maximal hip flexion through such exercises provides flexibility that functionally can compensate for loss of flexibility in the spine.

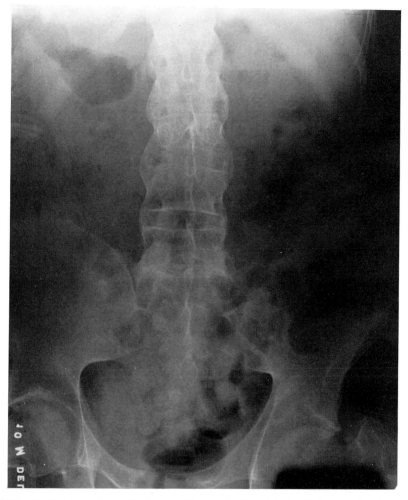

Fig. 20-4. Typical far-advanced rheumatoid spondylitis with fixed "bamboo" spine and obliterated sacroiliac joints.

In the prevention of a kyphotic deformity, the patient with rheumatoid spondylitis should also sleep on a firm bed with a bedboard and without pillows to avoid a prolonged arched position at night.

Development of a kyphotic deformity results in both a loss of erect height and a decrease in respiratory excursion due to forward compression of the thoracic cage. As a simple gauge of progress the spondylitic patient should measure his height and chest expansion regularly and maintain a record of these. Loss of either indicates an inadequate treatment program and a need for revising the program.

Plaster shells and back and neck braces rarely are indicated today for the patient with rheumatoid spondylitis, since phenylbutazone and steroids have proved to be more effective in the control of the disease process.

When, in conjunction with rheumatoid spondylitis, joints peripheral to the spine

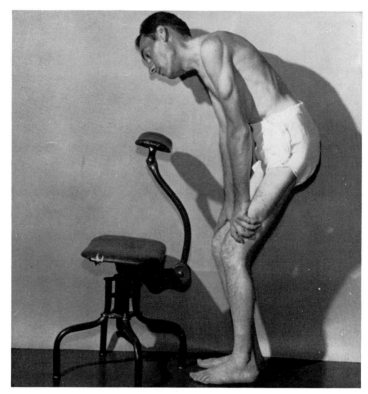

Fig. 20-5. Therapeutic exercise for the patient with spondylitis is directed toward prevention of the deformity illustrated. This patient's spine is fixed in a kyphotic deformity. In addition, there are compensatory hip and knee flexion contractures.

also become involved, the physical medical approach to treatment is the same as that for peripheral rheumatoid arthritis.

Arthritis due to direct trauma

The mechanical integrity of a joint may become impaired from trauma to articular structures; this results in a decrease in the functional efficiency and capacity of the joint. Use of such a damaged joint may mechanically produce intra-articular irritation with consequent joint pain and swelling. Joint damage may result from a single major trauma, such as a fracture extending into the joint space. More commonly it results from an accumulation of repeated microtraumas, such as those incurred from occupational stresses, from relaxed or torn periarticular ligaments, from joint instability as a result of muscle weaknesses, from scoliosis or other postural derangements resulting in abnormal joint stresses, and from recurrent meniscus or joint dislocations (Fig. 20-6). Single joints generally are affected.

Treatment of the traumatic arthritic joint should be directed toward (1) elimination of the etiologic factor, (2) relief of pain, (3) therapeutic exercise, and (4) protection of the joint against further wearing. Relief of pain is best attained by local application of heat and by rest of the involved joint. This, however, is an immediate, not a definitive, measure. The latter is attained only by reducing the amount

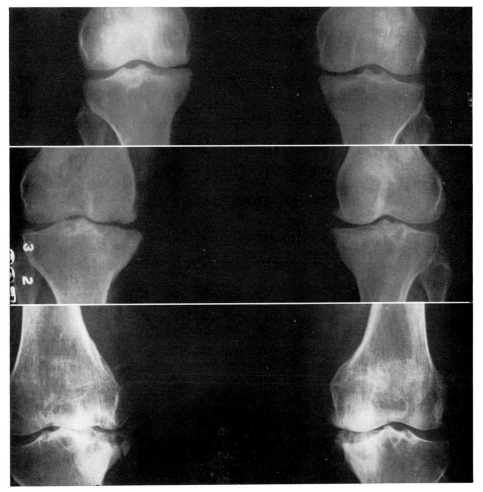

Fig. 20-6. Progressive development of traumatic arthritis within the medial portion of the knee joints of a patient with rickets and relaxed lateral ligaments.

of physical work imposed on the damaged joint structure. This entails curtailment of physical activity, therapeutic exercises to build periarticular muscle power and joint stability to a maximum, and transfer of work to other uninvolved joints through the use of such measures as shoe lifts, stabilizing supports, crutches, canes, braces, and energy-saving devices.

Degenerative joint disease

In the process of aging, tissues of articular structures undergo senescent deterioration. The degree to which this develops varies greatly among individuals but is present to some extent, microscopically or macroscopically, in all persons past the age of 50 years. The degenerative changes in joints are variously termed degenerative joint disease, hypertrophic arthritis, osteoarthritis, and senescent arthritis. While the etiologic factors in the production of joint degeneration are not fully understood, it is

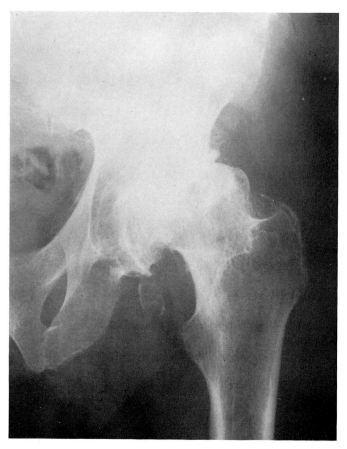

Fig. 20-7. Degenerative joint disease of the hip. There is sclerosis of the acetabulum and of the femoral head. There are subarticular areas of cystic rarefaction and productive spur formation at the superolateral margin of the acetabulum.

generally accepted that the microtraumas inflicted by "wear and tear" contribute directly to, if they are not the major cause of, the degenerative changes. The joints most commonly involved are those subjected to the greatest trauma, that is the knees and hips of the lower extremities and the lumbar and cervical sections of the vertebral column. In addition, for reasons not well explained, the terminal interphalangeal joints of the hands commonly are affected. This involvement, occurring more often in women than in men, results in the knobby enlargements known as Heberden's nodes.

The earliest discernible pathologic change in degenerative joint disease is a loss of the normal glistening character of the articular cartilage. Subsequently there develops an increased friability with a tendency toward fibrillation and pitting. While this is microscopic initially, it may progress to macroscopic proportions and major defects in the articular surfaces. The result then is a mechanical one of friction and irritation, with consequent inflammation and pain when the irregular articular surfaces glide against one another. In a reparative effort, periosteum lays down new bone peripherally; this may progress to macroscopic proportions, in which case these proliferations are known as osteophytes or "spurs." Basically, then, degenerative joint disease is not an arthritis; the arthritis, or the inflammation of the joint, is the consequence of the intra-articular defects that mechanically produce friction on use of the joint.

Radiologically, the fundamental changes that are recognizable are productive changes at the margins of the articulating surfaces or margins of the vertebral bodies, so-called spurs or lipping. The joint spaces may be narrowed because of degeneration and destruction of the articulating cartilages. With destruction of articulating cartilage and cortex, there may develop subchondral bone sclerosis, cystic changes, subluxations, and deformities that result in restricted mobility (Fig. 20-7). Osseous ankylosis is not the rule, but it may follow. Except for the characteristic subchondral eburnation, or sclerosis, bone density is usually normal; this may be decreased, however, in long-standing cases, in the aged, and in patients showing an associated atrophy due to disuse. At times, joint bodies or an effusion may also develop.

In hands and feet these degenerative changes commonly involve distal interphalangeal articulations. When they involve the spine, there is partial destruction of intervertebral discs, with marginal productive changes and spur formations. When the latter are posterolaterally situated, they may project into and encroach on intervertebral foramina. There may also be malalignment between vertebras, loss or reversal of normal curves, and interference with—as well as loss of—normal movements. Smaller articulations as well as sacroiliac joints may be similarly involved.

Not infrequently the above changes are incidental findings on an x-ray examination and may be asymptomatic. This is often so in the aged.

The objectives of physical medicine in the treatment of degenerative joint disease are (1) palliation of pain, (2) maintenance of normal ranges of motion in involved joints, (3) maintenance of maximum muscle power in muscles about the joint, and (4) protection of the joint against further wearing and deterioration.

Palliation of pain. Pain in an osteoarthritic joint is the result of frictional irritation within the joint and is an indication of overuse. The restriction of physical activities to within the tolerance level of the joint should constitute the basic approach to the problem of pain; overuse of a mechanically impaired joint not only is productive of pain as a result of the abuse but also adds further traumatic damage. The amount of physical activity permitted a patient with osteoarthritis is thus dictated by the tolerance of the joint. If there is no discomfort after a particular activity, the tolerance of the joint has not been exceeded. If transient aching or stiffness results from a particular activity, the degree of activity demands future caution. On the other hand, if after physical activity there is resultant pain and stiffness that persist with gradual remission over a 12-hour period or longer, the tolerance of the joint has been exceeded, and activity to such a degree should not be repeated. This simple rule of thumb may be applied to any activity and to any osteoarthritic joint. When pain results as a consequence of overuse of a joint, rest and the application of heat locally are helpful palliative measures. As in rheumatoid arthritis, the type of heating apparatus chosen depends more on simplicity and economy than on any major variance in therapeutic effectiveness among apparatus. This is a particularly important consideration, since the problem is a chronic and frequently a worsening one; measures adaptable to home use and self-application should therefore be prescribed. Ultrasonics, microthermia, diathermy, and other measures for relatively deeper heating of tissues require expensive apparatus for application and, clinically, offer to the osteoarthritic patient little if any more than can be obtained from infrared or luminous heat lamps, bakers, contrast baths, hot tubs, and other simple and economic home treatment measures.

Maintenance of ranges of motion in joints. Pain within an osteoarthritic joint results in muscle spasm as well as in voluntary splinting of the joint by the patient; both of these splinting factors predispose to capsular and tendon tightening with consequent loss of range of motion of the joint. When this occurs, the normal mechanics of the joint become impaired, thus superimposing an additional source of pain. Therefore, the patient should be taught range-of-motion exercises to be done actively or with assistance to prevent such a complication.

Maintenance of maximum muscle power. As in ranges of motion, pain with consequent splinting of a joint predisposes also to disuse atrophy of muscles mechanizing the joint. Muscle weakness in turn impairs the stability of the joint and predisposes to additional joint trauma and pain. To prevent such a cycle a resistive exercise program should be incorporated in the patient's program to build and maintain maximum muscle power about the involved joint.

Protection of the joint against further wearing. In addition to measures already mentioned, other means for reducing the work load on joints can and should be employed. Crutches, canes, wheelchairs, orthopedic shoes with appropriate corrections, braces, self-help devices, and planning for energy conservation may appreciably deter progress of the degenerative process when judiciously included in the total treatment program.

Special considerations

Degenerative arthritis of the hands. The hallmark of degenerative arthritis is Heberden's node, which involves the terminal interphalangeal joints of the hands. These nodes occur most frequently in women and in a hereditary pattern. Acute symptoms usually occur in the early phases of development of the nodes and are more disturbing than disabling. There may be acute pain with redness and local tenderness, but such a flare-up rarely persists for more than a few days and generally involves only one joint at a time. During acute episodes the use of paraffin or contrast baths once daily will provide pain relief and speed remission. With the passage of time, acute episodes usually cease; degenerative intra-articular changes, however, generally follow. These changes may be slight, being limited to the terminal interphalangeal joints, or they may be major, with involvement also of the proximal interphalangeal joints. Involvement of the terminal joints, however, is always greater than that in the proximal ones and the resultant enlargement of the joint is osseous rather than synovial. When these degenerative changes supervene, the problem becomes a totally mechanical one, and symptoms of joint stiffness and soreness are in direct proportion to the extent of joint damage and to the amount of physical activity imposed on these joints. Management at this stage is the same as for degenerative arthritis in any other joint. Since finger function, however, is largely one of dexterity, the need for building muscle power is minimal, and physical treatment is directed

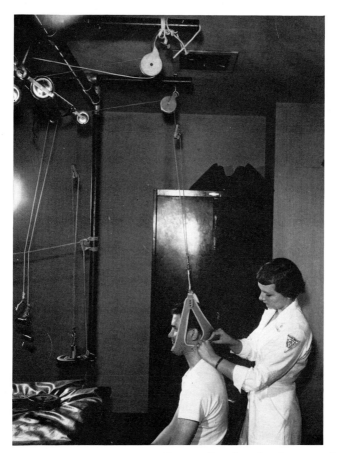

Fig. 20-8. Simple and economical Sayre neck sling readily adapted to home use for the patient with cervical osteoarthritis.

toward relief of pain, maintenance of joint ranges of motion, and elimination of work, through restricted activity and use of energy-saving devices.

Degenerative arthritis of the spine. Degenerative joint changes in the spinal column occur most frequently in the lumbar and cervical segments; presumably this is because of the greater mechanical stresses in these areas. Symptoms generally are the result of the encroachment of osteophytes on the intervertebral foramina and compression of nerve roots. The clinical consequence of this is radicular pain in the dermatome distribution of the involved nerve. This is a consideration of importance in treatment in which measures are directed toward decreasing the foraminal compression and/or immobilizing the area of involvement to reduce irritation induced by motion. In the cervical region of the spine, widening of the foramen with decompression can often be accomplished by traction. Heat is generally applied first for relief of pain and for relaxation of muscle spasm; this is then followed by traction in a Sayre traction apparatus. Since the therapeutic effect is a transient one, the procedure must be repeated once or twice daily; for this reason patients should purchase the equipment necessary for carrying out the procedure at home and should be instructed in its proper use. (See Fig. 20-8.) Although usually unnecessary, it may at times be beneficial for the patient also to sleep in cervical traction and to wear a molded leather collar or neck brace during the day.

Unlike the cervical spine, the lumbar spine cannot be so easily stretched to relieve nerve pressure. In addition to the usual measures of heat, exercises, etc., for the treatment of degenerative arthritis, it may be desirable to stabilize the lumbar spine with a corset or brace to reduce nerve root irritation induced by motion. In such cases, however, it is important to remember the vital role of the back and abdominal musculature as supporting structures and to direct the patient to carry out exercises for these areas even more assiduously because of the potential disuse atrophy that may develop as a result of the bracing.

Nonarticular rheumatism

Among the types of nonarticular rheumatism, psychogenic rheumatism is one of the most often encountered. Among patients seen in general arthritis clinics, incidences of psychogenic rheumatism ranging from 15% to 40% have been reported. Because of the magnitude of the problem and the generally good prognosis as compared to the organic arthritides and because most patients eventually are treated by physiatrists, it is important that the syndrome be clearly appreciated.

Psychogenic rheumatism is a muscular response to the stress of emotional tensions and, physiologically, is probably a manifestation of muscular fatigue. It is, therefore, a functional rather than an organic process. Increased muscle tone as the result of emotional stresses is a normal physiologic response of the vegetative system as long as this vegetative response is mobilized intermittently in fear, anger, and other normally transient emotional reactions. When the muscular system is called on for sustained reaction, the result is a dysfunction of the responding system, leading to psychogenic rheumatism. The person with psychogenic rheumatism is unable to dispense emotional tensions through adult interpersonal relationships and other mature vents and, as a consequence, suffers the cumulative effects of this closed system viscerally. He is thus usually the tense, anxious, nervous individual—with or without dynamism—whose psyche is deeply scarred from previous and persisting traumas.

Psychogenic rheumatism may be acute in onset, or it may develop insidiously as a chronic process. The acute syndrome usually follows in the wake of overwhelming and unusual stress. With the removal of the precipitating emotional stresses, the muscular system is released from tension and the visceral symptoms dissipate. The chronic process constitutes a more complex problem. In these cases it is helpful to give the patient some insight into his difficulty by acquainting him with neuromuscular functional interrelationships and the consequences of emotional tension. The use of such commonly recognized analogies as tension headache, nervous indigestion, and spastic colitis renders this explanation of psychogenic rheumatism more tenable. In accepting this tension relationship, the patient accepts a personal responsibility for his difficulty and cannot transfer hopes for cure to the physician. Supportive assurance that much can be done in a palliative way should then be extended and discussed. This entails adoption of a program of physical therapy, sedatives, salicylates, and extra rest and a plan for avocational outlets.

Palliative physiotherapeutic measures consist of various heat modalities, massage, and hydrotherapy. These are valuable for their relaxing effect. Any source of heat may be used. Choice in the individual case should depend on the desired extent of application, the patient's general physical state, and his response to particular types of therapy. Since physical therapy is totally palliative and is an undertaking of treatment for an indefinite period, the simplest measures possible should be prescribed. The patient with insight may be instructed to carry out these measures at home with a minimum of economic burden.

In a small percentage of cases the immaturity and neurosis will be so pronounced that the patient will be incapacitated by his psychogenic rheumatism. The psychotherapeutic approach then is the only productive one.

Rehabilitation of the arthritic cripple

Rehabilitation of the arthritic cripple is not a simple undertaking. Unlike the patient severely disabled from poliomyelitis, in whom improvement is to be expected, or the patient with transverse myelopathy, in whom the disability is a static one, the patient with arthritis is afflicted with a continuing disease process that may worsen. Also, in contrast to most other patients with severe disabilities, the arthritic patient must contend with pain from his disease. The problem is thus a compounded one.

Although the pathologic damage from arthritis is primarily in the joints, the consequences of the disease, as in other chronic diseases, ramify into every sphere of the patient's living. Its effects are reflected not only physically but also socially, economically, psychologically, vocationally, and avocationally. Therefore, when rehabilitation is considered, the physical aspects of the patient's problem constitute only one facet. Unless the total patient is evaluated in the light of all the complex ramifications and unless treatment is directed toward all angles, the problem will be only partially met and solved. (See Chart 13.)

Medical evaluation. Of first importance in the establishment of rehabilitation goals for the arthritic patient is the accurate assessment of the type and severity of the arthritic process. Since there are more than 100 different causes for arthritis, the prognoses of which differ, a positive categoric diagnosis is imperative before treatment can effectively be undertaken. Not only is it important that the diagnosis be positively established; the disease process also must be assessed, in terms of its activity and potential for progression. In the case of rheumatoid arthritis, for example, the success of physical and vocational rehabilitation to a large extent depends on how adequately this progressive disease process may be converted into a stabilized one through the use of steroids or other antirheumatic medications.

Functional evaluation. In addition to the medical appraisal of the patient's problem, it is important in the projection of a rehabilitation goal to have other information that may modify overall functional objectives. Muscle power must be tested to identify the extent and location of muscle weakness. This evaluation of muscle weakness is an important determination; weakness in muscles about arthritic joints, which makes them unable to provide adequate stability, predisposes to additional intra-articular traumatic damage and resultant additional joint pain.

Next, the ranges of motion of joints must be measured to determine both the total ranges of motion and the phases of the arcs in which there may be limitation

CHART 13

REHABILITATION EVALUATION

of the

ARTHRITIC PATIENT

medical history
physical examination
specialist consultations ▶ DIAGNOSIS and PHYSICAL PROGNOSIS
laboratory examinations

muscle test
joint range of motion
speech and hearing evaluation ▶ FUNCTIONAL CAPACITY and POTENTIAL
activities of daily living

psychologic testing
social survey ▶ ECONOMIC POTENTIAL
vocational testing

REHABILITATION POTENTIAL

the TOTAL PATIENT must be evaluated

SOCIALLY

PSYCHOLOGICALLY

VOCATIONALLY

as well as MEDICALLY

and FUNCTIONALLY

before a realistic goal may be established

for the individual

---

of motion. The latter aspect is of considerably more acute significance functionally than the former. A knee, for example, may have an excellent range of motion of greater than 90 degrees in flexion and still be contracted in flexion by a crucial 15 degrees, thus seriously hampering the mechanics of weight bearing and predisposing to additional traumatic degeneration within the knee joint if ambulation is undertaken.

Finally, the patient is directly tested in the performance of activities. Although determination of muscle power and measurement of joint ranges of motion in themselves are valuable data suggestive of functional capacity and potential, inferences as to function are not always valid, for pain may be an additional limiting factor in patients with arthritis. Patients are therefore tested directly in the performance of activities considered necessary for self-sufficient living. The extent of functional impairment is thus directly measured and scored and, with the muscle power chart and the range-of-motion measurements, correlation of functional with physical deficiencies can be assessed. This sum total gives an indication not only of the deficiencies but also of the physical and functional potentials of the patient toward which treatment may be directed.

Psychosocial evaluation. Since the effects of arthritis ramify beyond the physical stigmas, the disturbances wrought in the psychologic, social, and vocational milieu must be assessed, and in the overall rehabilitation planning, positive action toward solution of these intimately related problems must be undertaken. The identification and solution of minutiae in these areas may prove as acutely decisive toward the overall success of the rehabilitative undertaking as any of the more obvious medical facets of the case.

Treatment programs. All of the medical, functional, and psychosocial data are used collectively to project a rehabilitation goal for the patient and to plan and initiate treatment toward that end. Goals that are established initially often are fluid ones that may be changed repeatedly throughout the course of rehabilitation as dictated by unanticipated changes in the patient's progress.

Initiation of medical therapy for the treatment of the specific arthritic disease should precede initiation of a physical rehabilitation program. Whether this therapy be use of salicylates, chrysotherapy, uricosuric agents, or steroids, maintenance levels for antirheumatic and analgesic effects should be attained before the physical program is undertaken.

Physical treatment programs should be intensive, covering 3 to 5 hours a day, 5 days a week, and must be individually prescribed to meet the needs of the particular patient. Treatment might include all or various combinations of physical therapy modalities, occupational therapy, remedial exercise, functional training in activities of daily living, training in the use of self-help devices, psychologic and psychiatric assistance, vocational testing and counseling, and job retraining. The two major treatment objectives are functional independence and protection of joint structures against further structural damage. For example, manual stretching of a knee contracted in flexion may be incorporated in the treatment program while, at the same time, the patient is being treated with strengthening exercises for quadriceps muscle groups, both treatment procedures being directed toward improving joint mechanics and increasing functional proficiency. Similarly, stretching of back muscles may be prescribed, along with abdominal muscle exercises, to promote the flexibility of the spine and the abdominal muscle power essential for sitting up in bed or bending to dress the lower extremities. Regardless of the area under treatment or the type of treatment employed, the objectives are the same.

Factors influencing rehabilitation goals. The success or failure in the rehabilitation of the severely disabled arthritic patient can be anticipated with some accuracy

CHART 14

FACTORS INFLUENCING REHABILITATION FEASIBILITY

AND GOALS

1. Medical control of disease process

2. Extent of joint damage

3. Psychologic economy of patient

4. Functional training

5. Corrective orthopedic surgery

6. Applicability of self-help devices

7. Vocational and socioeconomic resources

---

in terms of seven major factors (Chart 14). The positive and negative aspects of these factors constitute assets and deficits, not only useful in determining feasibility of rehabilitation for a patient but also helpful in predicting the extent of the rehabilitation goal for a particular patient.

Medical control of the disease process. It is generally agreed that in the case of rheumatoid arthritis, the best candidate for rehabilitation is the patient in whom the disease has becomed burned out or is quiescent. In such patients the disease has reached a static level, and control of the disease process has been accomplished. Disabilities in these instances are of a static nature, and the worrisome problem of a continuing disease process that will progressively worsen the disability is no longer a factor. Unfortunately, rheumatoid arthritis that occurs with sufficient intensity to produce severe crippling infrequently goes into a complete remission or does so only in the very late stages. As a consequence it is usually necessary to face the problem of disease activity and accept the need for combining medical therapy with physical rehabilitation measures. Furthermore, those patients who have reached a crippled stage requiring intensive rehabilitation have usually also been treated unsuccessfully in the course of their disease with the gamut of antirheumatic therapies and therefore require the use of steroids. Not only is this more drastic antirheumatic therapy needed for control of the inflammatory disease process but it is usually positively indicated also as a protective measure to safeguard against any further loss of function, which is already crucially impaired. The effectiveness with which the inflammation of rheumatoid arthritis may be controlled by steroids, therefore, directly modifies the rehabilitation goal. Similarly, effective uricosuric therapy for the patient with gouty arthritis and analgesia for the pain of osteoarthritis directly affect the goal.

In the medical management of the patient with chronic rheumatoid arthritis, at the start it must be accepted that if steroid therapy is initiated, it will probably have to be continued on an indefinite and uninterrupted basis. It is essential, therefore, that steroid dosage be maintained as low as possible so that the danger of troublesome side effects is minimized and so that, through excess dosage, the undesirable rheumatoid state is not traded for the equally undesirable physiologic state of hyperadrenalism. When the intensity of the rheumatoid disease process is such that adequate control cannot be attained by the use of levels of steroid dosage that are small and recognized as safe, then it is necessary to accept the partial control rather than risk the complications inherent in higher dosage levels. In such instances the capacity of the patient to participate in a physical rehabilitation program will be directly modified.

In arriving at maintenance steroid levels for the rheumatoid patient, one pitfall should be stressed. Since rheumatoid patients suffer pain from mechanical damage within joints as well as from the inflammatory disease process, steroid dosage should be regulated solely to alleviate the pain caused by the latter process. When dosage is progressively increased in an attempt to alleviate the joint discomfort that follows use of mechanically impaired joints, dosage is frequently increased to much higher levels than would be necessary if the foregoing criterion were adhered to. Steroids should not be used in an attempt to alleviate pain consequent to mechanical damage within joints; such pain is best managed by restriction of activity and local palliative measures. Careful differentiation must be made clinically in identifying the two sources of symptoms and the drug restricted accordingly if serious hyperadrenal pitfalls are to be avoided and if long-term therapy is to be carried out successfully.

Extent of joint damage. The amount of physical activity that an arthritic joint will tolerate is dependent on the severity of the intra-articular inflammation and on the extent of mechanical joint damage.

It is not always easy to predict the amount of activity that a joint can tolerate. Roentgenograms may be helpful, but at times they may also be grossly misleading indices; they are best relied on only as confirmatory evidence. Trial, therefore, is the most reliable means for determining a patient's tolerance, and physical goals should be projected in conformance with the amount of activity his joint can tolerate. For example, it may be apparent at the very start of rehabilitation treatment that weight-bearing joints will not tolerate such demanding work as walking, and treatment and training accordingly will have to be restricted to self-care, bed, and wheelchair activities.

It should be noted that destructive changes in joints of the lower extremities are considerably more restrictive than comparable changes in the upper extremities. Goals, therefore, in terms of self-care and other self-sufficiency activities may be aimed much higher in the face of upper extremity damage than in the face of lower extremity damage of comparable degree. This is because lower extremities are primarily "work horses," whereas upper extremities are mainly concerned with activities involving dexterity and agility and are not weight bearing.

Both limitation of range of motion of a joint and muscle weakness about the joint may appreciably modify the tolerance of the joint for activity and thus compound the problem already created by destructive articular changes. Capsular and tendinous tightness about a joint, particularly a weight-bearing joint, may impose mechanical stresses that not only produce pain but also predispose to even greater

CHART 15

PSYCHOSOCIAL FACTORS

MOTIVATION of the patient
is essential if
REHABILITATION GOALS
are to be attained

Many rheumatoid arthritic patients
develop a PASSIVE
DEPENDENCY to their disease
and to disability
which restricts REHABILITATION
possibilities

At times this DEPENDENCY
may be irreversible
and it is wise to screen patients
carefully
from this standpoint
before instituting a
PROGRAM

In the psychosocial survey
of our patients
the following factors
frequently
have been the indices pointing
toward MOTIVATION
and possible
SUCCESS IN REHABILITATION

● SPECIFIC AND REALISTIC GOALS IN REHABILITATION

● FAIR OR REASONABLE STANDARD OF LIVING

● SECOND-GENERATION AMERICAN

● GOOD WORK HISTORY

● SUCCESSFUL MARRIAGE

● MODERATE RELIGIOUS PRACTICE

● FINANCIAL INDEPENDENCE AT LEAST 5 YEARS PRIOR TO ILLNESS

● MINIMAL FOCUS ON HYPOCHONDRIACAL SYMPTOMS

● ADEQUATE EGO STRENGTH   ● ADEQUATE BODY IMAGE

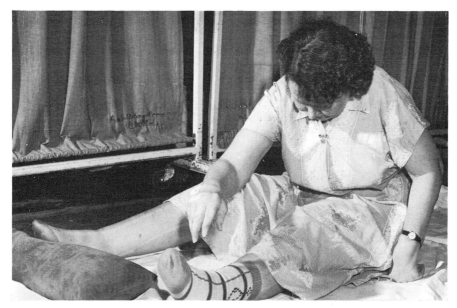

Fig. 20-9. Patient is instructed in exercises to stretch the back and leg muscles, aimed at providing the suppleness necessary for sitting up in bed, getting from a bathtub, etc.

articular destruction. Similarly, weakness in retaining muscles about a joint jeopardizes the stability of the joint, thus inviting additional trauma with consequent further destruction within the joint. The degree to which joint tightness may be stretched, for normal mobility and for mechanically good weight-bearing positions, and the degree to which muscle strength may be restored about the joint are vital determinants of the work tolerance and efficiency of joints.

Psychologic economy of the patient. More than any other factor, the psychologic economy of the arthritic patient is a major force in determining success or failure in attainment of rehabilitation goals (Chart 15). In the course of a progressive disease such as arthritis, accompanied as it is by mounting disability and a fluctuating but constant pain, the psyche suffers from the constancy, the hopelessness, and the frustration to which it is subjected. The result is an exhaustion of psychologic economy and a reversion to passivity and dependency; frequently this develops to a degree unequaled in any other chronic disease. Since success in rehabilitation is largely dependent on the active participation of the patient in his treatment program, motivation of the patient is essential if goals are to be attained. It is easy to evaluate a patient medically and physically and thus determine feasible goals, but unless the patient's goals are in accord with the physician's, the divergence in results will duplicate the difference in objectives.

Functional training. While muscle power, joint range of motion, and pain tolerance are valuable indices of the patient's potential functional capacity, these are useless unless they can be transposed into function. Therefore, therapy directed toward increasing power in weakened muscles and toward increasing ranges of motion in restricted joints must be prescribed, with the projected objectives of putting these physical gains to functional use. Functional training is thus an integral part of the

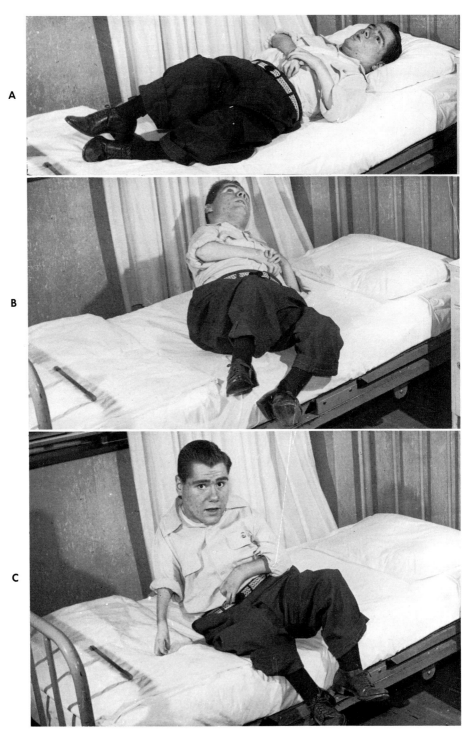

Fig. 20-10. Training in the Activities of Daily Living increases the mechanical efficiency of the body. The patient has learned to swing himself into a sitting position despite severe joint deformities.

rehabilitation program and is directed toward training the patient to become maximally independent in those activities essential to self-sufficient living. Stretching of tight back muscles (Fig. 20-9) may be done in physical therapy, for example, to increase flexibility that may then be utilized in functional training to teach the patient how to bend and dress the lower parts of his body. Similarly, strengthening of the quadriceps muscle groups may be undertaken to provide muscle strength needed for training the patient to climb stairs or curbs or to arise from a toilet seat. In every instance, function is the ultimate objective.

In addition to gross function, efficiency in function is an objective in functional training. The human body as a machine normally functions on a grossly inefficient level, estimated by some to be less than 25%. Even in the face of severe disabilities, therefore, one may be taught to attain a considerably higher degree of efficiency to compensate for irreversible physical deficiencies. (See Fig. 20-10.) This may entail months of tedious training and practice, but such an approach frequently opens the door to functional independence. Such functional training is an important part of a patient's daily treatment program, and as he accomplishes proficiency in one sphere, emphasis is then shifted to the next.

Corrective orthopedic surgery. It is no longer felt that arthritic patients should wait for their disease to reach far-advanced stages before offering themselves for corrective surgical procedures. Furthermore, from a standpoint of protection of joints against additional mechanical wearing, corrective surgical procedures are at times urgently indicated. A patient with knee flexion contracture, for example, should not be permitted to ambulate in this mechanically disastrous position, at the risk of superimposing marked structural damage, when such a deformity can be surgically corrected by a capsulotomy (Fig. 20-11). It should be emphasized that success of corrective orthopedic surgery is not dependent on quality of surgical technic alone; just as important, if not more so, are the diligence of the physician and the cooperation of the patient in the preoperative and postoperative therapeutic program.

Self-help devices. Special apparatus and devices should be avoided unless they are essential for increasing function or for the protecting of already impaired joints. (See Chapter 9.) The great numbers of self-help devices that have been developed for the arthritic patient are too numerous to mention except for those more important ones notable for their simplicity and usefulness (Fig. 20-12).

Devices for dressing. Dressing of the upper or lower parts of the body may at times be impossible because of mechanical restrictions imposed by deformities of joints of the upper or lower extremities. Zippered shoes, long-handled shoehorns, devices for putting on socks and stockings, long-handled combs, elastic shoelaces, etc., are often of advantage in such problems (Fig. 20-13). Recently attention has been focused on the designing of clothes especially adaptable to the severely crippled person.

Devices for feeding and self-care. When deformities exist in joints of the upper extremities, special long-handled devices for feeding, washing, applying cosmetics, and toilet care are helpful. Similarly, when severe hand and wrist involvement exists, special devices are available for writing, shaving, holding feeding utensils, etc. (Fig. 20-14).

Adapted chairs and toilet seats. Chairs and toilet seats of standard height often

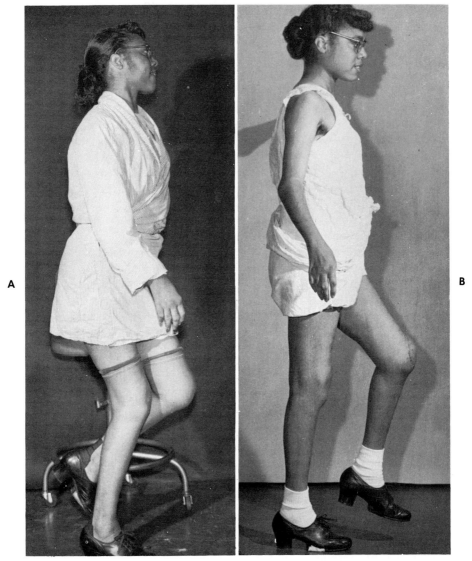

Fig. 20-11. Knee flexion contracture before and after capsulotomies. The restoration of full extension eliminates the mechanical stresses imposed by the contractures.

pose major mechanical problems for the arthritic person with knee or hip involvement (Fig. 20-15). These are usually 17 or 18 inches in height, and the patient either may experience considerable pain in using them or, in cases of ankylosis, may find it mechanically impossible to use them. Chair heights may be increased by structurally adding leg length or by the simple expedient of using sponge rubber seat cushions to give extra elevation. For patients with more disseminated involvement and especially for those with rheumatoid spondylitis with peripheral joint involvement, it may be necessary to individualize modifications. This is especially true when there is ankylosis of one or of both hips, in which cases seats may be molded from fiberglass impreg-

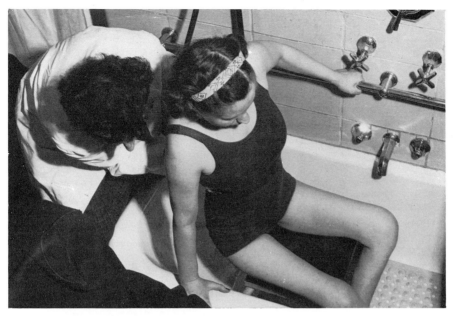

Fig. 20-12. Bathtub stool and wall bars help solve the problem of bathing.

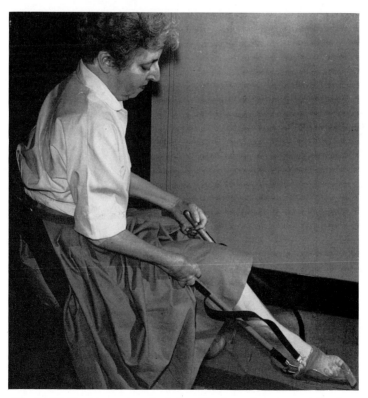

Fig. 20-13. Stocking-put-on device enables the patient to reach her feet.

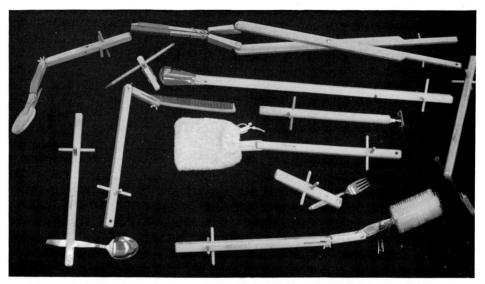

Fig. 20-14. Assorted self-care devices, with added length and pegs, assist in compensating for a loss of reach and a deficiency in grasp.

Fig. 20-15. One of the many types of elevated toilet seats. The ᴸ wall bar provides additional support.

Fig. 20-16. Careful planning of homework areas can eliminate unnecessary energy expenditure by the housewife returning to care for her home.

nated with plastic and mounted on legs at a height suitable to accommodate the specific deformity. Toilet seating may be elevated by using a stationary or removable built-up seat that adds 4, 5, or 6 inches in height.

Wheelchairs. Wheelchairs must be individually prescribed, depending on the needs of the patient. In addition to various models that are commercially available, there are more than twenty-two different attachments that may be prescribed. Recently there has been developed a motorized, collapsible wheelchair that is compact, serviceable, reasonable in price, and easily operated. The control panel for this chair can be placed within operating range of any patient's deformed upper extremities, and the control buttons can be managed easily by even the most distorted and weakened fingers. (See Chapter 7.)

Crutches. Because of the various combinations of joint deformities that may develop in the upper extremities, standard axillary crutches often may not suit the needs of the arthritic patient. Therefore, depending on the residual function in the upper extremity joints, crutches must be prescribed or adapted to meet the special needs. The types of crutches available are numerous, ranging from the standard handgrip type to those permitting elbow weight bearing. (See Chapter 8.)

Energy-saving devices for homework. The return of the housewife to her household duties is as important an objective from both the standpoint of her personal dignity and her economic productivity as is the return of the man to his job (Fig. 20-16). Careful planning and reliance on such energy-saving devices as worktables on casters, adjustable ironing boards, long-handled cleaning utensils, lightweight and easily used kitchen equipment can effectively reduce work that otherwise might be placed on the joints in the course of her job as housekeeper. (See Chapter 10.)

Socioeconomic and vocational factors. The ultimate goal in the rehabilitation of the arthritic patient is the reestablishment of the patient in a social and economic environment in which he may function to the maximum of his capabilities, despite and within the limitations of his disability. The attainment of this goal often requires exhaustive resourcefulness and patience. Success in this sphere is directly proportional to two factors: (1) the resources of the patient that may be worked with and (2) the extent of positive assistance afforded the patient by the social worker, the psychologist, and the vocational counselor, the latter factor probably outweighing the former in productive importance. Since chronic disease and disability ramify permanent effects into all spheres of living, the need for dynamic and positive assistance in readjustment to these effects is acute; this is in marked contrast to the problem of acute illness, which precipitates only a temporary derangement in the socioeconomic and vocational areas of living. Therefore, for the patient with a chronic disability problem, the customary cursory social service is inadequate; it must be expanded to include detailed inquiry into the family structure and an intimate knowledge of the living and working conditions to be faced by the patient after his discharge from the structured and protective environment of the hospital. Apparent minutiae may assume crucial importance in determining the capacity of the patient to function outside the hospital. The inability to maneuver six stairsteps from house to street may make the difference between ability and inability to travel to a job. A doorway of sufficient width to permit passage of a wheelchair or having a bed, chair, or toilet of sufficient height to ensure independence in use by the patient may be the factor determining whether attendant care or nursing home care will be necessary. In most cases the social worker must make one or several home visits in her positive assistance in clarifying and solving such problems. This sort of attention obviously is time-consuming, and social workers concerned with problems of chronically disabled persons cannot be expected to assume large case loads if their results are to be productive. The old concept based on problems of acute illness—that one social worker in a clinic setting could handle the population of several clinics—is now outmoded by the prevalence of chronic disease; success in rehabilitation is largely dependent on awareness of this fact and the establishment of a realistic ratio of social workers to patients with chronic disease. Similarly, the psychologic complexities and vocational problems consequent to chronic disability are not transient ones but residuals that, for their solution, must be energetically pursued by psychologists and vocational counselors. The adequacy, therefore, of all of these services is a major factor in reaching the optimum goal for the patient. (See Chapters 15 and 17.)

Job placement of the disabled arthritic patient remains the most difficult objective to attain. Limited ability to utilize public transportation to and from work, the competitive labor market, and the need for specialized placement, within the limitations of the patient's disabilities pose restrictive factors in employment. These factors are, however, by no means insoluble if approached positively and with understanding, nor do they require sympathy and preferential considerations for solution. If the patient's psychologic and vocational aptitudes are carefully assessed by the psychologist and vocational counselor in collaboration with the social worker, plans for job placement or for suitable job retraining usually can be worked out. (See Chapter 17.)

DIABETES MELLITUS

The importance of rehabilitation of the diabetic patient is apparent when it is realized that there are 2,600,000 known diabetics and an estimated 1,600,000 unknown diabetics in the United States.[18]

Since insulin was discovered in 1922, the life-span of the diabetic person has been greatly prolonged. Diabetic coma and diabetic gangrene remain the most frequent complications reported as causing death. Early detection and proper treatment of the disease are therefore most important in the prevention of these serious complications.

The medical treatment of diabetes consists of diet, diet and insulin, and diet and oral antidiabetic drugs. Oral preparations have been helpful in treating the majority of patients with maturity-onset diabetes.[19]

Management of the patient with diabetes mellitus should include the maintenance of a normal physical, mental, social, and economic existence. The patient should be impressed with the fact that he is no different from other persons except that he must give closer attention to his daily routine. He may eat the same food as the nondiabetic person, but it must be measured. He must test his urine daily, and if he requires insulin, he must administer this every day. He must report to his physician periodically and cooperate in the prevention of the complications of the disease.

The person with diabetes must be made to realize that he can earn a living and support his family and thereby maintain his dignity and self-respect. Although he is unsuitable for certain hazardous jobs, he may be capable of performing in a vast majority of occupations. Although prevention of the complications of diabetes is the best treatment, this is, unfortunately, not always possible. The diabetic person should understand, therefore, that when a complication occurs, prompt treatment will speed his return to his usual place in society.

Rehabilitation of the diabetic patient at times poses difficult problems. For purposes of discussion, these will be considered in three main categories: (1) the emotional response to the disease, (2) the problems of employment, and (3) the rehabilitation treatment of organic complications.

Emotional response to the disease. The knowledge that one has developed diabetes may result in serious emotional trauma. The patient may be overcome with fears that have developed as a result of unscientific gossip, and the responsibilities of testing his urine, taking insulin, and adhering to a diet may at first seem overwhelming. However, with proper guidance, education, and understanding on the part of the physician, these fears can be quickly dissipated. The most important step in the handling of a new diabetic patient is to acquaint him with the problems of his disease. The physician should dispel his fears by explaining in detail the proper diet, the method of testing urine for sugar and acetone, the details of insulin administration, principles of proper foot care, and general hygiene. Furthermore, he should recommend a good diabetic manual so that the patient may learn the carbohydrate, protein, and fat values of food and thus be equipped to adjust his diet easily wherever he may have to eat.

The care of the child who has just developed diabetes may present a difficult problem that requires tactful handling by the physician. The diabetic child resents being "different," yet he cannot take candy and sweets like other children. In addition, he is required to take insulin injections every day and to follow a carefully prescribed routine. With the younger child, the testing of urine, the calculation of diet, and the injection of insulin can be made a game. The importance of proper care of the diabetes and the prevention of insulin reaction must be made clear without causing anxiety or fear. The child should be made to feel that he is allowed to play and compete equally with other children. Teachers and school authorities should cooperate by making no distinction between the diabetic and the nondiabetic child. They must, however, carefully observe the diabetic child and be familiar with the symptoms of insulin reaction and its treatment.

Every diabetic patient requiring insulin should inject the insulin himself. He must be taught to sterilize the needle and syringe, to measure the insulin properly, to administer it, and to overcome any fear of self-injection. The patient, whether child or adult, thus is self-sufficient and independent.

Group therapy can play an important part in the rehabilitation of the young diabetic patient. Children coming to a clinic should meet each other and receive group instruction in self-care. The older children should encourage the younger ones while discussing mutual problems. Group therapy may also be employed in summer camps for diabetic children, such as those sponsored by diabetes associations throughout the country. The camps maintained by Dr. Elliott Joslin and by the New York Diabetes Association are good examples of what can be accomplished. Diabetic children attend these camps for 3 or 4 weeks in the summer, during which time they test their own urine, administer their own insulin, and play games together. Older children are given career guidance to prepare them for future employment, and they meet with numerous alumni who tell the campers how they have progressed. A summer at a camp like this helps resolve the insecurity, anxiety, and fear of the diabetic child.

Occasionally, a child will resent his illness and as a result become aggressive and difficult to treat. Similarly, adults may react with undesirable emotional responses to their disease. The patient may even become hysterical and refuse to administer insulin or adhere to a diet. Usually these reactions depend on the preexisting emotional structure. Conversely, some patients lead too careful an existence and worry unnecessarily when mild glycosuria appears. Thus understanding of the emotional responses of the diabetic patient and prompt treatment of the emotionally unstable patient are of utmost importance in rehabilitation.

Problems of employment. The diabetic person should consider himself as capable of supporting his family as the nondiabetic person. It is unsafe, of course, for him to participate in certain types of work if he requires injections of insulin. He is unsuited for employment as a railroad engineer, airplane pilot, bus operator, window washer, or other hazardous occupation if he requires insulin. The possibility of sudden unconsciousness due to insulin reaction renders him a danger both to himself and to the public.

It is important not only to educate the diabetic person about the work he may expect to do, but also to educate labor and industry to the fact that the diabetic person is just as capable of doing good work as the nondiabetic. Diabetes should not prevent the pursuit or maintenance of a responsible position. Unfortunately, the diabetic person on occasion has had to conceal his illness to obtain employment.

The American Diabetes Association, in a statement of the Committee on Employment published in January, 1953, stated:

> It is the philosophy of the American Diabetes Association that controlled diabetics without complications are good employable risks. They should not be classed as vocationally handicapped, for they are capable of performing a full day's work to satisfaction despite their diabetes, for most positions.
>
> Diabetics seeking employment naturally fall into several groups. One is composed of the mild diabetics who require no insulin for the control of their disease. The employment of these people presents no special difficulty. There are also many selected diabetics who are controlled with the aid of tolbutamide (Orinase) only.
>
> Another group includes more severe diabetics who require insulin and a carefully regulated diet. This group itself is divided into two very important subgroups: the well-controlled and well-regulated patient, and the uncooperative and/or poorly regulated patient.
>
> The latter subgroup, composed of poorly-controlled and/or uncooperative diabetics, should be refused employment solely on the basis of their diabetic condition. This is not only because they are poor employment risks in their present condition, but also because they may develop complications which would result in increased absenteeism on their part. A prejudice against all diabetics would thus be created in the mind of the employer. There are a small group of unstable diabetics in whom it is difficult to maintain normal standards of control in spite of adequate medical supervision. These workers should not be denied employment, but should be individually appraised.
>
> When it is said that a diabetic is a good employment risk, reference is made to the well-controlled, well-regulated diabetic who keeps himself constantly under medical supervision. These people should prove to be quite acceptable employees, as the experience of hundreds of companies employing such diabetics indicates. Indeed, because they are self-disciplined and well-balanced individuals, they often become outstanding workers in whatever positions they occupy.*

Data concerning the diabetic person in industry were collected from sixty-three companies, representing 780,823 employees.[14] The known incidence of diabetes among 286,662 employees of thirty-nine reporting companies was 0.5%. Forty-three of the sixty-three companies employed persons with known diabetes and twenty did not. Thirty-three companies restricted the diabetic persons to nonhazardous work, twenty-four had no restrictions, and six had no set policy. Thirty-eight companies permitted an employee who developed diabetes to continue in his job, twenty-four restricted the type of work after diabetes was discovered, and one reported having no set policy.

---

*From American Diabetes Association: Employment of diabetics, a statement of the Committee on Employment, June 7, 1952. Revised Jan., 1953, May, 1957, and June, 1958.

The New York Diabetes Association has established a vocational and counseling service to assist diabetics in obtaining employment. This counseling service has developed a work classification sheet to evaluate the diabetic and determine the candidate's job potential. Counselors maintain a relationship with other vocational service agencies and they also prepare educational material for personnel directors and employers. This service advises the diabetic concerning suitable employment.

In industry the importance of close cooperation between the private and industrial physicians cannot be overemphasized. Together they can alleviate the fears of the diabetic person concerning his employment while ensuring his proper care. In addition, proper job placement depends on cooperation between the industrial physician and the personnel department.

Studies have shown that the work record of the diabetic person in industry is generally satisfactory as compared with that of the nondiabetic.[14] In a study of ninety-two known cases of diabetes, Dublin and Marks[15] showed excellent work attendance records in 40%, average in 19%, fair or poor in 26%, and unsatisfactory in only 15%.

Because most diabetic workers are over 40 years of age and are experienced employees, retaining them in their jobs for which they are trained is good management.

Workers who become diabetic seldom have to change their jobs, although in exceptional cases a shift may be desirable for physical reasons. Hurwitz[17] believes that the degenerative complications of the disease, especially peripheral vascular disease, make the diabetic person unusually susceptible to trauma. It is of interest, however, to note that absentee records show that the time off for industrial accidents is no greater for the diabetic than for the nondiabetic person.[14]

Industry must be educated to a reasonable attitude toward the diabetic person. Wise treatment of employees with diabetes leads to better industrial relations. Conversely, an unfavorable attitude is detrimental in that it leads to concealment of the disease, discouragement and discontent among diabetic persons, and a higher rate of unemployment.

The diabetic employee must also be taught that his disease, if properly cared for, need not pose a problem in industry. He must be educated in the necessity for obtaining good medical care and the importance of close cooperation between his private physician and his company medical department. He should treat his disease without embarrassment, and his fellow employees should be notified of his disease so that emergency care may be rendered if necessary.

In summary, the problems of employment that concern the diabetic person are best resolved by proper job placement. The diabetic patient on insulin therapy must avoid certain jobs as noted previously. But for these exceptions, he is able to compete with the nondiabetic person if he is under regular medical care and his diabetes is controlled.

Rehabilitation treatment of organic complications. The most serious complications of diabetes are ketosis, arteriosclerosis, gangrene, and retinopathy. In most instances these complications can be prevented by early and proper care.

Diabetic ketosis is best prevented by avoiding infection and by maintaining proper diet and insulin dosage. Once ketosis develops, prompt and rigorous therapy should be instituted.

Arteriosclerosis is an inevitable complication in the older diabetic patient. Proper diabetic care is the best method of delaying the appearance of arteriosclerotic changes.

Foley[16] suggested that the medical profession reconsider the long-accepted practice of bed rest for patients with gangrene of the lower extremities from arterial insufficiency. He believes, conversely, that the exercise from walking stimulates collateral blood flow and improves the metabolic state of the involved tissues. As a result of his successful treatment of 21 such patients, he advocates that walking be added to the medical regimen of oscillating bed, reflex heat, abstinence from tobacco, antimicrobial ointments, bandages, release of vasospasm, and treatment of associated conditions, such as diabetes and cardiac decompensation.

Gangrene and lesions of the lower extremities may be prevented by proper foot care. Instructions in foot care should be given to every new diabetic patient and should include instructions for daily foot soaks, massage with alcohol and lanolin, and Buerger's exercises. The patient should visit the foot room in a clinic or receive foot care from his physician or a trained podiatrist for the treatment of corns, calluses, and toenails. When necessary, corrective shoes and foot appliances should be prescribed. Ulcers of the lower extremities should be treated promptly and with great care. This, in most instances, will prevent the necessity of amputation; however, in spite of good preventive care, surgery is frequently necessary. When it becomes necessary, it should be performed in such a way that rehabilitation can be started early and easily.

New surgical procedures have been developed to increase the circulation in the lower extremities. These procedures have increased the circulation to such a degree that amputation may be avoided in many instances.[11, 12]

Every patient with an amputation of a lower extremity should be thoroughly evaluated by a rehabilitation team prior to prosthetic training. Before the expenditure of time, effort, and money, the evaluation team should be sure that the patient will be able to make use of the artificial extremity. Frequently, after many weeks of effort, a patient who has occupied a valuable hospital bed is discharged and never makes use of his prosthesis. Improper evaluation may have failed to consider the patient's physical capacity or the possible presence of another disease such as congestive heart failure that prevented the utilization of his prosthesis. It has been found that after prosthetic training a patient with a unilateral amputation may be prevented from using his prosthesis because his remaining extremity becomes ulcerated or traumatized as a result of the extra burden. Occasionally a patient with bilateral amputations learns to ambulate better than the patient with unilateral amputation. It has also been observed that an older arteriosclerotic diabetic patient with an exceptional attitude may have years of usefulness after prosthetic rehabilitation. However, expert and careful evaluation is necessary before training the patient for prosthetic rehabilitation.

The diabetic patient with a cerebrovascular accident due to arteriosclerosis should be treated similarly to any other patient with a cerebrovascular accident. The diabetes should be properly controlled with diet, tolbutamide (Orinase), or diet and insulin as required.

Loss of vision due to diabetic retinopathy is progressive and irreversible. It usually prevents the patient both from earning a living and from administering insulin to himself. These visual changes, however, usually occur in the older diabetic patient who is already limited by other complications and already requires custodial care.

The rehabilitation of the diabetic patient follows the same basic principles that are applied to any other disability, plus intelligent, scientific management of the diabetes as the primary problem.

HEMATOLOGIC DISEASES

Involvement of the neuromuscular, skeletal, and other organ systems in the course of various hematologic diseases and syndromes may lead to the development of long-term or permanent disability. Such disabilities pose problems of great importance in the total management of patients thus afflicted.

Specific treatment of the hematologic abnormality, with few exceptions, offers little in the way of improvement of already established disability.

The armamentarium of total rehabilitation is necessary to ensure the patient maximal function within the limits established by the disability and to offer him the best life possible.

The mechanism and nature of these disabilities will be discussed in this chapter, and appropriate rehabilitation technics, procedures, and goals will be described. Description of the underlying disease and its treatment will be limited to those aspects directly concerned with and related to the complicating factor of long-term disability.

Pernicious anemia and related deficiency states

Involvement of the nervous system is so common in addisonian pernicious anemia that it is the rule rather than the exception in any given case. Involvement of the spinal cord is estimated to occur in from 70% to 90% of the cases. In the advanced state of the disease the picture is that of subacute combined degeneration that involves primarily the posterior and lateral columns of the spinal cord but may involve other portions of the central nervous system as well. The lesions of the nervous system reflect the same deficiency of vitamin B_{12} responsible for the abnormal blood picture, with cobalt containing vitamin B_{12} being necessary for the maintenance of a normal nervous system. There is no real correlation between the severity of the combined system disease, the degree of anemia, and the blood picture. In related conditions (severe liver disease and subtotal gastrectomy) an identical lesion of the central nervous system may develop due to impaired absorption or storage of vitamin B_{12}. The use of B_{12} can reverse the early symptoms of combined system disease, and in advanced cases some improvement or arrest of progress occurs. The state of the hematopoietic system is no indication of the dose of vitamin B_{12} necessary in cases of neurologic involvement. Much larger doses are required in the management of the central nervous lesion than in the control of the hematologic abnormality. In a comparison of spinal cords of treated and untreated patients with combined system disease, treatment showed a marked effect on the histologic picture. There was much less involvement and more gliosis in the spinal cords of the treated patients as compared to the untreated.

A patient with moderate or advanced combined degeneration of the spinal cord presents a major rehabilitation problem and requires a coordinated and long period of rehabilitation treatment. Reversal of symptoms is not to be expected, and the goal of treatment is to gain maximum function despite the disability. The rehabilitation of patients with combined system disease is described in detail in Chapter 23, but it should be noted here that experience has shown a rehabilitation program to be more successful when combined with adequate vitamin B_{12} therapy. Also, in early cases, in which reversal of symptoms can be expected, better results are obtained when appropriate exercises are combined with vitamin B_{12} therapy than when the latter is used alone. The specific rehabilitation of subacute combined degeneration is considered in Chapter 23.

The occurrence of peripheral neuropathy in persons with pernicious anemia is a recognized phenomenon. There is disagreement, however, as to its incidence and severity. Some investigators[31] believe that peripheral neuropathy is common and is responsible for many of the early sensory symptoms associated with pernicious anemia. Peripheral neuropathy and combined system disease can coexist. Van der Scheer and Koek[31] described a series of well-documented cases of polyneuropathy complicating pernicious anemia. Pathologic study of the involved nerves confirmed their clinical impressions of a primary peripheral nerve lesion. The polyneuropathy appeared to be an integral part of the clinical picture, and no extraneous cause for this complication could be found in their patients. Peripheral neuropathy poses a serious problem in treatment and rehabilitation and can, if ignored, result in severe and irreversible disability.

Polycythemia vera

The importance of ploycythemia vera as the cause of severe disability stems from the high incidence of vascular complications. Such complications are due to a greatly increased tendency toward intravascular thrombosis resulting primarily from significant slowing of the circulation. Thrombosis and vascular accidents involving any organ system can occur. Of particular concern, from the rehabilitation point of view, are those involving the central nervous system and the extremities. Cerebrovascular accidents are fairly common and result in variable degrees of hemiplegia, with or without aphasia. This catastrophe superimposes a major disability on the patient with polycythemia—a disability in which total and early rehabilitation is of the greatest importance. (For the rehabilitation of the hemiplegic patient see Chapter 29.) The incidence of cerebrovascular complications is much higher in patients with polycythemia vera than in a comparable control group without this disease.

Vascular occlusion of the extremities may result in situations requiring amputation, and the incidence of this complication is higher in persons with polycythemia vera than in the comparable general population. The management of the stump may be complicated by the sluggish circulation and its sequelae. The problem is further complicated by the possibility of subsequent vascular catastrophe to the remaining extremities. Psychologic as well as physical preparation for amputation is of great importance, and a frank but hopefully oriented discussion of the implications of amputation and of what can be offered to counteract them should be entered into at the earliest opportunity, preferably prior to the actual amputation. The possibility of similar future complications should be minimized, and, indeed, the vascular complications can to a large degree be alleviated by the specific treatment of the underlying hematologic condition.

Sickle cell disease

Sickle cell disease, an abnormality of hemoglobin formation, produces chronic disability primarily in the vascular, musculoskeletal, and cardiovascular systems.

The neurologic disabilities result from vascular accidents with resultant hemiplegia and other signs of cerebral involvement. In 31 cases of sickle cell disease with central nervous system symptoms reported by Hughes and co-workers,[26] hemiplegia was noted in 14 and aphasia in 9 patients.

There is some difference of opinion as to the nature of the vascular lesion responsible for the central nervous system involvement. The most frequently offered explanation is that the abnormal, sickled red blood cell is responsible for capillary stasis, with capillary thrombosis, infarction, and perivascular hemorrhage. Involvement of larger blood vessels is thought to result from a retrograde phenomenon. Other observers feel that the basic vascular lesion is an obliterative endarteritis of a special kind and that thrombosis, although seen at times, is not an essential component of this process. Occlusion in vessels may occur without actual thrombosis, through proliferative changes alone.[20] Another concept of the vascular lesion suggests that its primary cause is vascular spasm.[27]

Hemiplegia and other disabilities due to involvement of the cerebrovascular system may be the first clinical evidence of sickle cell disease. Since the majority of people so afflicted are children or young adults, the importance of a broad rehabilitation program as an integral and important part of the treatment of such patients cannot be overstated. The detailed management of hemiplegia is described elsewhere, as is the rehabilitation of patients with aphasia (Chapter 14). The psychologic impact of disability on a young age group must be borne constantly in mind and every measure taken to ensure the disabled person the fullest life possible. Problems of education and vocational rehabilitation are of particular importance.

One of the clinical features of sickle cell disease is recurrent arthralgia. Joint involvement may be quite severe, with much pain. Primary permanent changes in the joints are not found, but complications in the form of contractures around the joints and muscle atrophy may cause significant disability. These complications are due to fixation of the joints because of the immobility caused by pain and disuse of the muscles. Such disabilities may be avoided. Carrying a joint through its range of motion as soon as feasible will often prevent the development of contractures. Such range-of-motion activities must be performed methodically and as often and completely as is feasible in view of the patient's tolerance. Treatment of the involved joint with a form of heat will help decrease the pain and spasm. If contractures have already developed, the carrying out of systematic range-of-motion exercises is essential. The joint is manipulated to the point of tolerable discomfort. The goal is the ultimate stretching out of the tight and fibrotic joint capsule. Pretreatment with heat enhances the procedure. Rarely such procedures are unsuccessful, and surgical relief (capsulotomy) may have to be resorted to.

Exercises are valuable in preventing disuse atrophy of the muscles surrounding involved joints. When atrophy is already present, special exercise programs, depending on the degree of weakness, are required. (See Chapter 4.)

Gradual collapse of the vertebras is described in sickle cell disease. This is thought to be due to the replacement of cancellous bone by connective tissue through an intense erythropoietic reaction in the marrow. The collapse is related to the severity of the disease, its duration, and the body weight of the patient. For this reason it is likely to occur in adults. Back pain may result. If sufficiently severe, the use of special supports may be indicated to obtain relief. Hormonal therapy, sometimes helpful in osteoporosis, is not indicated in collapse due to sickle cell disease.

Symptoms of cardiovascular involvement may be present in sickle cell disease. Murmurs and enlargement of the heart may develop. Involvement of the pulmonary vascular bed may add the factor of cor pulmonale. Congestive heart failure in some degree is usually present in patients with cardiac involvement. Suitable programs for cardiac rehabilitation may be of value to such patients in teaching them to function maximally. (See Chapter 27.)

Hemophilia

The abnormal bleeding characteristic of hemophilia can produce severe disability as a result of hemorrhage into a number of structures. Important among these structures are the striated muscles, the joints, and the central nervous system.

Hemorrhage into striated muscle may result in contracture, with secondary involvement of the peripheral nerves and fixation of the joints as a result of disuse. Early physical therapeutic measures designed to hasten the absorption of blood and to preserve the integrity and function of the involved muscles can do much to minimize this development. When the disability is already established, it may be possible to reduce it through rehabilitation procedures aimed at increasing the range of motion of the involved joints and decreasing the amount of muscle contracture. Procedures useful in the management of peripheral neuropathy are applicable when this complication is present. When the residual disability interferes significantly with "normal activity," training in the Activities of Daily Living is important in achieving maximum independence.

Repeated hemarthrosis is common in hemophilia and can result in chronic arthritis, with complete destruction of the joint as a not infrequent end result. The disability resulting from such joint involvement poses serious problems in the patient's ability to function independently.

A single hemorrhage into the joint may be absorbed and leave no aftereffects. The x-ray appearance is similar to that of ordinary synovitis. Repeated hemorrhages into the joint may lead to fibrosis of the synovial membrane and contraction of the joint capsule. Hemorrhages may occur also in the bone and subchondral areas, resulting in erosion or destruction of cartilage and bone. The joints commonly affected are the knees, elbows, and ankles. The shoulders, hips, wrists, and the smaller articulations of the hands and feet are less often involved. The radiologic appearance of the joint in chronic conditions is one of destructive changes involving cartilage and bone. The joint space is narrowed and there is trabecular destruction, with cystlike areas in the subchondral regions. Ankylosis may result.

The use of appropriate physical agents and rehabilitative procedures during the stage of acute joint involvement may do much to minimize the residual involvement. Rehabilitation problems resulting from residual hemophilic joint involvement are similar to those resulting from other disabling joint diseases, and their management has been described earlier in this chapter. The value of self-help devices in achieving maximal independence in those patients with severely involved joints should be emphasized. Return to a nonhospital environment may be made possible through such devices even when little or nothing can be done to increase joint function per se.

Paralysis of the lower extremity can occur as a result of retroperitoneal bleeding, producing pressure on the femoral nerve. Complete or partial dysfunction of that nerve may be seen. The involvement of the lower extremity is of the type seen in peripheral neuropathy, which indeed it is. The pattern of the disability depends on the degree and severity of the femoral nerve involvement. Active measures must be taken to avoid contractures around joints, muscle atrophy, and other complications of peripheral nerve dysfunction. Appropriate rehabilitation measures to ensure maximal function should be started as early as feasible. The latter include appropriate bracing and training in the Activities of Daily Living. (See Chapters 6 and 8.)

Bleeding into the brain and spinal cord produces serious and catastrophic disability in the hemophilic patient. Hemiplegia, paraplegia, or quadriplegia can result from hemorrhage into vital areas of the central nervous system. Seemingly negligible trauma to the head or spine may result in sufficient bleeding to cause a major neurovascular disability. The management of the catastrophic disability becomes the primary problem in the patient who has suffered one. It is in the rehabilitation of paraplegic or quadriplegic patients that an integrated team approach is most essential if life is to be preserved and a maximal reintegration into active living achieved. (See Chapter 19.)

An approach to the rehabilitation of the hemophilic patient must include significant attention to the psychologic and sociologic aspects of the illness. The behavior of the individual, particularly the young individual with hemophilia, is at least as much conditioned by the behavioral response and the social and emotional responses of others as it is to the disease itself. Indeed, the successful management of the disease and the minimization of untoward complications may in large measure be a function of the emotional state and the self-concept and self-value held by the individual patient. The counseling of the family over an adequate period of time is essential to the maximum rehabilitation obtained in the patient and cannot be too strongly stressed.

A 3-year demonstration project sponsored by the Southern California Chapter of the National Hemophilia Foundation vigorously confirmed the necessity of including psychosocial and rehabilitation aspects in the total treatment program, particularly with the increased availability of more potent antihemophilic factor concentrates. Multidisciplinary comprehensive medical, psychologic, sociologic, and the rehabilitation services were offered to the patients in the project. It is interesting to note that the project demonstrated the value of intensive exercise programs for every patient and confirmed the impression that a decrease in bleeding episodes is related to increased activity. Such dynamic exercise programs and subsequent maximum use of muscle strength were considered a preventive measure not duplicated by splinting or bracing. The study confirmed the tremendous importance of appropriate counseling and psychologic aid for the family if the hemophilic patient was to achieve his maximum functional and social capacity. The guilt and anxiety present in parents and the need for reasonable and active close relationship between father and son, and mother and daughter, are examples of the kinds of problems that must be dealt with. The study noted that the more positive the adolescent hemophiliac valued himself as a person, the greater acceptance he had of his medical condition and the more positively he accepted the need for normal risk taking. "It should be constantly kept in mind the frustrations faced by the hemophiliac, particularly the younger age groups, and the need for both social and psychologic mechanisms to minimize these frustrations and mobilize energy into the most productive channels possible"*

Purpura

Hemorrhage of the brain and spinal cord occurs in purpura. The most common residuals are those of hemiplegia, but quadriplegia and paraplegia may also occur. These complications were described as early as 1886 and were extensively described by William Osler. With present-day treatment a greater number of patients with such disabilities may be expected to survive and present the difficult rehabilitation problems inherent in these severe disabilities. The management of such patients is described in the discussions dealing with hemiplegia, quadriplegia, and paraplegia.

Bleeding into the muscles and joints is seen also in patients with purpura. The problems are the same as those described in the discussion of hemophilia.

Xanthomatosis

Xanthomatosis, a disturbance of the reticuloendothelial system, gives rise to prolonged disability of various types. Of particular interest to rehabilitation medicine are those disabilities arising from involvement of the central nervous and osseous systems.

Diffuse involvement of the central nervous system can take place. The essential finding is that of plaques of foam cells and fibrous tissue. The clinical problems arise primarily from involvement of the pyramidal tracts, cerebellum, and peripheral nerves.

Problems requiring management are primarily those of spasticity, paralysis, and ataxia in varying combinations and peripheral neuropathic syndromes. The specific management of these conditions is dealt with in Chapter 23.

Marked bony involvement may occur in long-standing chronic cases of xanthomatosis. The long bones are commonly involved, but involvement of the vertebras, with compression fractures, is also not unusual. Gibbus formation can result from vertebral involvement, and syndromes resulting from compression of the spinal cord can be seen. Compression fractures of one or both femoral heads can result from osseous involvement with xanthomatosis tissue. Patients with the resulting disabilities require a program of rehabilitation geared toward maximum function within limits of the disability, since little can be done to improve the underlying condition. The specific measures depend on the nature and extent of the functional loss and can be found in appropriate discussions elsewhere in this book.

---

*From Hemophilia: A total approach to treatment and rehabilitation, Los Angeles, 1968, privately printed by Orthopaedic Hospital of Los Angeles, Calif.

POISONS AND NOXIOUS SUBSTANCES

Long-term disability may be a serious sequela of poisons and certain other substances when used in toxic amounts. Such a disability poses rehabilitation problems of the greatest magnitude and becomes the primary concern of both the patient and those concerned with his management. The management of such disabilities from the rehabilitation viewpoint will be considered in this chapter.

Ethyl alcohol poisoning

Although ethyl alcohol, when used properly and socially, can hardly be considered a poison or a noxious substance, its abuse by the chronic alcoholic has, essentially, a parallel effect.

Chronic alcoholism is a total disease, involving the whole patient. Regardless of the presenting problem in an individual patient, unless a total rehabilitative approach is made, little can be expected in the way of improvement, either in the specific disability or in the patient generally. As is well known, relapse is the rule. Only a broad program that considers all of the ramifications of personality, physical state, home and vocational environment, etc., can attain any degree of success. The integrated services of a psychiatrist, a clinician, a vocational counselor, and a social worker are necessary. Total abstinence from alcohol of all types is, of course, basic in rehabilitation in any disability problem in the chronic alcoholic.

Peripheral neuropathy is the principal motor disability in chronic alcoholism. Current opinion is that this is principally due to a deficiency of vitamin B_1 (thiamine chloride) concomitant with long-standing alcoholism. However, in addition, there may be a specific deleterious effect of alcohol on nerve tissue. The lower extremities are affected far more frequently than the upper, and the condition is more likely to occur bilaterally. Pathologically, there is extensive degeneration of the myelin sheaths, and the axis cylinders may be considerably involved in severe cases. Brain and spinal cord changes also have been observed in patients with peripheral neuropathy with involvement of the anterior horn cells and posterior ganglion cells.

Leg fatigue and night cramps are early clinical signs of alcoholic peripheral neuropathy. The reflexes are hyperactive at first and then become hypoactive and absent. In established cases, calf pressure, either applied or incident to walking, results in severe pain due to the tenderness of the inflamed nerves. Foot drop may be noted early. Ataxia, marked weakness of the legs, and, finally, paralysis may occur. Recovery is slow, even with complete abstinence and massive doses of the B vitamins. Permanent residuals frequently result. Unless an active program of physical medicine and rehabilitation is undertaken, complicating problems such as joint contractures (particularly in the ankles) and marked muscular atrophy may impose greater disability than the polyneuropathy. The specific rehabilitation for peripheral neuropathy is described in detail elsewhere. The value of massive doses of the B vitamins, particularly thiamine chloride, has been clearly established, and their use for long periods of time is of primary importance in the management of this condition.

Rarely, atrophy of the cerebellum or the anterior cerebrum complicates chronic alcoholism. The physical management is similar to that described for cerebellar ataxia. With cerebral involvement, signs of dorsal column disease predominate in the physical findings, but a mental picture very similar to that seen in a patient with paresis may complicate the problem.

Lead poisoning

Lead poisoning can result in a number of chronic conditions, the most important of which, from a rehabilitation point of view, is involvement of the neuromuscular system. Poisoning, resulting in so-called "lead palsy," is due to the chronic absorption of lead. Most cases occur among industrial workers. Instances are seen in children as a result of the ingestion of paint that contains lead; however, these cases are becoming rare with the increased use of lead-free paint for children's furniture and fixtures. Lead palsies are ordinarily late manifestations of chronic plumbism and are usually preceded by gastrointestinal symptoms of lead colic.

The muscle groups characteristically affected by plumbism are those most frequently and vigorously used by the individual. Palsy also affects functionally related muscle groups rather than those with a common nerve supply. Muscle fatigue seems to be the important factor in localization. Paralyzed muscles show a reaction of degeneration and atrophy, but fibrillation and sensory changes are absent. Although definite involvement of the nerves supplying affected muscles is noted, experimental evidence suggests that lead is not primarily toxic to the peripheral nerve. The nerve lesions, usually peripheral in location, are of an atrophic degenerative nature and could represent an ascending process. Certain investigators, notably Reznikoff and Aub,[38] hold that the toxic process occurs in the muscle and that the weakness and subsequent paralysis are due to the formation in the muscle tissue of insoluble lead phosphate that interferes with muscle metabolism.

Since lead palsy follows certain typical patterns of involvement, it has been classified into a number of "types." Following is a discussion of the most common of these types.

1. *Antibrachial type.* The extensors of the wrist and fingers are involved but not the flexors, resulting in wristdrop. The supinator longus is not involved, although it has the same nerve supply as do the affected muscles (a diagnostic point). The process begins in the extensor communis digitorum and spreads to involve the other muscles. The interossei are usually involved quite late.

2. *Brachial type.* The deltoid muscle is primarily affected with or without involvement of the biceps, brachialis anticus, and supinator longus.

3. *Arau-Duchenne type.* The thenar and hypothenar muscles are primarily affected; the interossei may or may not be involved. Atrophy is extensive, and wristdrop is common.

4. *Peroneal type.* The peroneal muscles and the extensors of the foot and toes are involved. The tibialis anticus usually escapes, and the gastrocnemius is rarely involved. This type is more common in children.

Lead palsy, regardless of "type," has certain distinctive characteristics. It is painless, essentially motor, and usually affects extensor muscles. Those muscles most frequently used are most commonly involved.

The management of lead palsy presumes the use of vigorous measures to prevent the development of contractures and atrophy. These measures are the same as those used in the peripheral neuropathies and are described in the discussion of peripheral neuropathies. Complete recovery is generally possible but may take one or more years.

Specific treatment has been revolutionized recently by the introduction of chelating agents, and a much more rapid recovery from lead palsy is possible. Disodium calcium ethylenediaminetetra-acetate (Ca EDTA) is the agent used. It forms, with lead, a chelate that is stable, water-soluble, and nonionizable and that is excreted by the kidneys. The oral or the intravenous route can be used; they are equally satisfactory. As an oral dose, 30 mg./kg. of body weight with liberal amounts of water, taken before breakfast and supper, seems to be adequate. For intravenous administration, 0.4 Gm. in 5 to 10 ml. of saline solution is given once or twice daily, according to Sidbury.[40]

In addition to palsy, a form of progressive muscular atrophy is sometimes observed in chronic lead poisoning. Lead is toxic to ganglion cells in the brain and spinal cord; involvement of anterior horn cells seems to be responsible for the progressive muscular atrophy. A localized form of muscular atrophy involving the small muscles of the hand is often seen in association with the antibrachial (wristdrop) type of lead palsy. Rarely is progressive muscular atrophy generalized; involvement of the pyramidal tracts has been described in association with it. The rehabilitation of patients with progressive muscular atrophy and associated conditions is described in Chapter 22.

Arsenic poisoning

Peripheral neuritis is a fairly common finding in cases of chronic poisoning from solid arsenic. Neural involvement is generally symmetric, with the lower extremities more frequently involved than the upper. Paresthesia and intense neuritic pains are prominent. A severe sensory ataxia can result, with marked impairment of position and other sensory modalities, closely simulating tabes. Although motor disability is ordinarily slight, severe palsy can develop in the course of arsenic peripheral neuropathy. The disuse and malpositioning of the involved extremities because of the pain may result in secondary contractures around the joints and lead to prolonged disability, even though significant palsy may never develop. Measures to avoid such disabilities must be instituted and carried on as effectively as possible, despite the problem posed by the pain. Motor involvement, if it appears, is managed as in other forms of peripheral neuropathy. (See Chapters 4, 6, and 23.)

In acute poisoning with arsenic trioxide (As_2O_3) very severe general motor paralysis may be a late manifestation. The problems of management in this condition are those of severe paralysis in general. Care of the patient to prevent the occurrence of decubiti is of great importance. The complications introduced by decubiti as far more difficult than the measures required for their prevention. Adequate attention to bowel and bladder function is essential. Prevention of contractures and significant atrophy requires specific measures. When permanent residual paralysis is apparent, a program in rehabilitation designed to make available to the patient maximal function within the limits of the disability is instituted. The combined efforts of a rehabilitation team are necessary when the residual disability is severe.

Carbon monoxide poisoning

Severe poisoning with carbon monoxide that does not end fatally can cause widespread damage to the central nervous system, with serious residual disability. Removal from the carbon monoxide does not always halt progression of central nervous system damage. The exact nature of the central nervous system lesions is not clear, but anoxia and vascular disturbances seem to be of great importance. The central nervous lesions exhibit a predilection for the cerebral cortex and the corpus striatum. In adults the globus pallidus seems to be most affected, whereas in children the caudate nucleus and putamen seem to be more prone to involvement.

Clinical manifestations include aphasia, apraxia, choreoathetosis, and parkinsonism. Pyramidal tract syndromes, hemiplegia, peripheral neuropathy, and spinal cord involvement also have been observed as sequelae of carbon monoxide poisoning.

Improvement of severe neurologic disability may occur after very long periods of time, and prognosis as to the extent of permanent residual disability should be extremely guarded.

The rehabilitation program depends on the nature of the disability. Specific measures are described in those discussions on the particular disabilities. General measures to avoid superimposed problems are of the greatest importance. In patients in whom prolonged coma is present, range-of-motion exercising and proper positioning will do much to prevent the development of disabling contractures. Evaluation of bladder function and proper measures to avoid urinary tract infection are important. The prevention and treatment of decubiti are essential.

Tilt tables (Chapter 13) and rocking beds will do much to prevent disuse osteoporosis in patients when prolonged immobilization is unavoidable. A cheerful and optimistic attitude in dealing with the patient can do much to foster cooperation and stimulate motivation.

Since disabilities resulting from carbon monoxide poisoning are likely to persist for long periods of time, much can be accomplished by a long-range total rehabilitation program, with maximum advantage taken of late return of function.

METABOLIC BONE DISEASE

Metabolic bone disease includes several generalized diseases of the skeletal system in which a metabolic or endocrine factor, or a combination of both factors, seems to be of fundamental importance. The pathologic chemistry and physiology involved and the specific endocrine and metabolic treatment are thoroughly considered in works on these subjects. This discussion will be limited to problems of disability incurred through such diseases and to measures directed toward the prevention and rehabilitation of such disability.

Osteoporosis

Osteoporosis is a disorder in which there is a decrease in the formation of bony matrix, apparently due to abnormality of protein metabolism of diverse origin. As a result there is a progressive decrease in bone mass with increasing weakness of the skeletal structures. The condition is a fairly common one.

Osteoporosis may be a complication of many disease states. Disuse atrophy of bone produces osteoporosis as a result of prolonged inactivity and the absence of normal stresses and strains on the skeletal system. This is of great significance, since osteoporosis can seriously complicate severe disabilities in which prolonged immobility is a necessary corollary.

Osteoporosis is encountered in many endocrine disorders, among which are hypogonadism, Cushing's syndrome, hyperthyroidism, acromegaly, and diabetes mellitus. It is probable that senile osteoporosis has an endocrine basis. Osteoporosis can complicate treatment with the adrenal steroid hormones and may be of significance in the rehabilitation of arthritic patients receiving such treatment.

Osteoporosis, although generally distributed, tends to involve the vertebras and pelvis more severely than the remainder of the skeleton. Because of this, symptoms stemming from compression of the vertebral bodies are most common. Back pain can be of the most severe sort and refractory to ordinary treatment. The use of back braces to immobilize the spinal column may be of value, but frequently disabling pain persists. Heat and light massage may prove valuable through their effects of reducing inflammation around the spinal nerve roots. Ultrasonic therapy may be used for the same purpose. This latter permits much more accurate localization of treatment, and its effects appear more marked. Therapeutic trials with ultrasound seem to indicate that benefit results from a reduction of perineural inflammation and the relief of the muscle spasm associated with the changes in the vertebras. Treatment directed against the underlying cause of the osteoporosis is, of course, an essential part of the total management of the condition.

Pathologic fractures are not uncommon. These may occur when a disabled patient begins active rehabilitation, placing new strains on his bony skeleton, or as the result of seemingly insignificant trauma.

In the management of patients with severe disability such as quadriplegia or paraplegia, specific measures must be taken to avoid osteoporosis or to correct it if already present. High-protein diets and treatment with anabolic hormones should be instituted. Early mobility is of the greatest importance. Tiltboards and rocking beds should be used as early as is feasible and mobility begun as soon as the patient can tolerate it. (See Chapter 13.)

It should be emphasized that a decrease in bone density is not radiologically detectable early (Fig. 20-17). A loss of approximately 25% or more of the mineral content of bone is necessary before the naked eye can detect a decrease. In order to better quantitate changes in bone density, Barnett and Nordin[37] have described a new approach. In the long bones they measure in millimeters the thickness of the compact cortex (medial and lateral) of a femur, at its thickest part, divide this figure by the entire thickness of the bone, and multiply the result by 100. The same process is repeated, measuring the mid-part of the second metacarpal of either hand. A third measurement is made, that of the best-centered lumbar vertebra. The vertical height of that body, usually L-3, is measured in the central or narrowed portion; this is divided by the greatest vertical height, which is usually anteriorly, and the result is multiplied by 100.

Femoral measurements range between 32 and 71 with most above 46, the second metacarpals between 33 and 75 with most above 44, and the vertebras between 74 and 97 with most above 81. When the femoral and the hand scores are added together, most total above 89. A spinal score of 80 or less represents spinal osteoporosis,

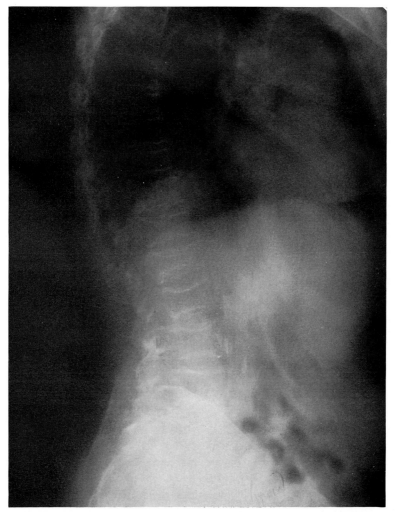

Fig. 20-17. X-ray study of the spine showing diffuse osteoporosis, with invagination of the intervertebral disks into the vertebral bodies producing a typical "fishmouth spine" appearance.

and a total peripheral score (femur plus hand) of 88 or less represents peripheral osteoporosis. A total score of 168 or less (spinal plus peripheral) is regarded as abnormal. A total figure above 168, however, may still conceal spinal or peripheral osteoporosis.

Other methods for determining bone density have also been used, for example, densitometry, bone and aluminum wedges or ladders, and bone density–computing machines.

Osteomalacia

Osteomalacia results from inadequate calcium or phosphorus for deposition in the bone matrix. It follows either decreased intestinal absorption of calcium of vitamin D, as in the steatorrhea of chronic pancreatitis and the sprue syndrome, or increased urinary excretion of calcium, as seen in chronic renal disease and in essential hypercalciuria.

Osteomalacia due to marked dietary restriction of calcium and vitamin D is common in parts of India and China but is rarely, if ever, seen in the United States.

As the result of the failure of bony matrix to calcify, bones soften, and in advanced cases marked bony deformity may result. Marked bone pain and tenderness, particularly of the back, and profound muscular weakness are characteristic of the condition. The pain and weakness respond rapidly to metabolic treatment. The bony deformities noted in severe cases are permanent and may cause considerable disability.

Radiologically one may see focal areas of demineralization in the cortices of the diaphyseal portions of the long bones, which in time extend across the shaft of the bones as linear translucencies (Looser's zones). These changes also occur in the flat bones. They are usually multiple, often symmetric, and present the appearance of fractures, but there is no malalignment of fragments and no history of trauma. The syndrome described first in 1934 by Milkman and since referred to as Milkman's syndrome is considered a mild type of osteomalacia. As full-blown osteomalacia develops, fractures, malalignment, and deformities may occur, and the entire skeleton may show considerable loss of calcium content. With skeletal softening, deformities occur as the result of weight bearing. The skull may show an invagination of the base into the cranial cavity. The spine develops curvatures, with the vertebral bodies becoming biconcave and the intervertebral disks becoming enlarged or swollen. The normal curves of the long bones may be accentuated. The bones of the pelvic girdle may encroach on the pelvic cavity. The lumbosacral angle may be increased and the sacrum assume a horizontal position. As the pelvis becomes deformed, the directions of stress are altered and degenerative changes ensue. Arthritic changes, fractures, and deformities are painful; these result in physical inactivity and secondary disuse atrophy as an additional source of difficulty.

Careful disability evaluation should be carried out, and a survey of deficiencies in the Activities of Daily Living should be made. (See Chapters 2 and 6.) Rehabilitation procedures designed to minimize these deficiences can greatly increase the functional ability and comfort of the patient.

Self-help devices and specially designed equipment may permit the patient a considerable degree of independent living. (See Chapters 9 and 12.)

Osteitis deformans (Paget's disease)

Osteitis deformans is a bone disease of adults characterized by areas of bone destruction and of new bone formation. Although it may be relatively asymptomatic, in advanced progressive cases severe deformity and disability can result. Repeated pathologic fractures contribute much to the development of disability. The use of rehabilitation measures designed specifically to overcome deficits in function can do much to increase the independence and functional ability of patients in such cases.

REFERENCES
Arthritis

1. Acker, M.: Vocational rehabilitation of patients with rheumatic diseases, Arch. Phys. Med. **37:**743, 1956.

2. American Rheumatism Association: Primer on rheumatic diseases, J.A.M.A. **190:**127, 425, 509, 741, 1964.
3. Hollander, J. A., editor: Arthritis and allied conditions, ed. 7, Philadelphia, 1966, Lea & Febiger.
4. Lowman, E. W., editor: Arthritis, general principles, physical medicine and rehabilitation, Boston, 1959, Little, Brown & Co.
5. Lowman, E. W., et al.: Rehabilitation of the rheumatoid arthritic patient (audio tape No. 27), New York, 1963, Publication Unit, Institute of Rehabilitation Medicine, New York University Medical Center.
6. Lowman, E. W., and Klinger, J. L.: Aids to independent living, New York, 1969, McGraw-Hill Book Co.
7. Lowman, E. W., Lee, P. R., and Rusk, H. A.: Total rehabilitation of the rheumatoid arthritic cripple, J.A.M.A. **158:**1335, 1955.
8. Manheimer, R. H., and Acker, M.: "Back to work" program for physically handicapped arthritics, J. Chronic Dis. **5:**770, 1957.
9. Woolsey, T. D.: Prevalence of arthritis and rheumatism in the United States, Public Health Rep. **67:**505, 1952.

Diabetes

10. American Diabetes Association: Employment of diabetics, a statement of the Committee on Employment, Jan., 1953.
11. Baron, H. C.: Arteriosclerotic lesions affecting the lower extremity, J. Med. Soc. New Jersey **65:**107, March, 1968.
12. Baron, H. C., Schwarz, A. W., Cabaluna, W., and Rodrigues, R. J.: Gas endarterectomy in the treatment of the ischemic lower extremity, Arch. Surg. **98:**754, 1969.
13. Brandaleone, H.: Diabetic worker in industry, Geriatrics **17:**691, 1962.
14. Brandaleone, H., and Friedman, G. J.: Diabetes in industry, Diabetes **2:**448, 1953.
15. Dublin, L. I., and Marks, H. H.: The diabetic in industry and his employer, Industr. Med. Surg. **19:**279, 1950.
16. Foley, W. T.: Treatment of gangrene of the feet and legs by walking, Circulation **15:**689, 1957.
17. Hurwitz, D.: The diabetic in industry, National Safety News **62:**94, Oct., 1950.
18. Public Health Service, Pub. 1000, Series 10, No. 40, Washington, D. C., Aug., 1967, U. S. Government Printing Office.
19. Levine, R., and Duncan, G. G.: Special editor's symposium on clinical and experimental effects of sulfonylureas in diabetes mellitus, Metabolism **5:**2, Nov., 1956.

Hematologic diseases

20. Bridges, W. S.: Cerebral vascular disease accompanying sickle cell anemia, Amer. J. Path. **15:**353, 1939.
21. Davison, C.: Xanthomatosis and the central nervous system (Schüller-Christian syndrome), Arch. Neurol. Psychiat. **30:**75, 1933.
22. Davison, C.: Effect of liver therapy on pathways of spinal cord in subacute combined degeneration, Arch. Intern. Med. **67:**473, 1941.
23. Gordon, G. L.: Osseous Gaucher's disease; report of two cases in siblings, Amer. J. Med. **8:**332, 1950.
24. Hall, B. E., Krusen, F. H., and Woltman, H. W.: Vitamin B_{12} and coordination exercises for combined degeneration of the spinal cord in pernicious anemia, J.A.M.A. **141:**257, 1949.
25. Henkin, W. A.: Collapse of the vertebral bodies in sickle cell anemia, Amer. J. Roentgen. **62:**395, 1949.
26. Hughes, J. G., Diggs, L. W., and Gillespie, C. E.: The involvement of the nervous system in sickle-cell anemia, J. Pediat. **17:**166, 1940.

27. Kimmelstiel, P.: Vascular occlusions and ischemic infarction in sickle cell anemia, Amer. J. Med. Sci. **216:**11, 1948.
28. Longscope, W. T.: Cerebral and spinal manifestations of purpura hemorrhagica, Med. Clin. N. Amer. **3:**279, 1919.
29. Norman, I. L., and Allen, E. V.: The vascular complications of polycythemia, Amer. Heart J. **13:**257, 1937.
30. Hemophilia: a total approach to treatment and rehabilitation, Los Angeles, 1968, privately printed by Orthopaedic Hospital of Los Angeles, California.
31. Van der Scheer, W. M., and Koek, H. C.: Peripheral nerve lesions in case of pernicious anemia, Acta Psychiat. Neurol. **13:**61, 1938.
32. Woltman, H. W., and Heck, F. J.: Funicular degeneration of the spinal cord without pernicious anemia: neurologic aspects of sprue, nontropical sprue and idiopathic steatorrhea, Arch. Intern. Med. **60:**272, 1937.

Poisons and noxious substances

33. Brain, R.: Diseases of the nervous system, ed. 5, New York, 1955, Oxford University Press.
34. Browne, R. C.: Metallic poisons and the nervous system, Lancet **1:**775, 1955.
35. Hamilton, A., and Hardy, H. L.: Industrial toxicology, ed. 2, New York, 1949, Paul B. Hoeber, Inc.
36. Mayers, M. R.: Lead poisoning. In Christian, H. A., and others: Oxford loose-leaf medicine, vol. 4 (part 3), New York, 1951, Oxford University Press.
37. Barnett, E., and Nordin, B. E.: The radiological diagnosis of osteoporosis: a new approach, Clin. Radiol. (London) **11:**166, July, 1960.
38. Reznikoff, P., and Aub, J. C.: Lead studies; experimental studies of lead palsy, Arch. Neurol. Psychiat. **17:**444, 1927.
39. Sanger, E. B., and Gilliland, W. L.: Severe carbon monoxide poisoning with prolonged coma, J.A.M.A. **114:**324, 1940.
40. Sidbury, J. B.: Lead poisoning; treatment with disodium calcium ethylenediamine-tetraacetate, Amer. J. Med. **18:**932, 1955.
41. Thompson, G. N., editor: Alcoholism, Springfield, Ill., 1956, Charles C Thomas, Publisher.

Metabolic bone disease

42. Albright, F., and Reifenstein, E. C., Jr.: Parathyroid glands and metabolic bone disease, Baltimore, 1948, The Williams & Wilkins Co.
43. Albright, F., Smith, P. H., and Richardson, A. M.: Post-menopausal osteoporosis: its clinical features, J.A.M.A. **116:**2465, 1941.
44. Black, J. R., Ghormley, R. K., and Camp, J. D.: Senile osteoporosis of spinal column, J.A.M.A. **117:**2144, 1941.
45. Gutman, A. B., and Kasabach, H. H.: Paget's disease (osteitis deformans): analysis of 116 cases, Amer. J. Med. Sci. **191:**361, 1936.
46. Joliffe, N., and Goodheart, R.: Nutritional polyneuropathy, Med. Clin. N. Amer. **32:**727, 1948.
47. Snapper, I.: Medical clinics on bone disease, New York, 1949, Interscience Publishers, Inc.

CHAPTER 21

REHABILITATION OF PATIENT WITH MUSCULOSKELETAL PROBLEMS

MUSCULOSKELETAL PROBLEMS DUE TO TRAUMA

Efficient function is dependent on good biomechanics: painless and free motion, good alignment and posture, and adequate coordinated muscular control. The musculoskeletal system is responsible for static and dynamic stability of the extremities and trunk in the performance of essential functions. When stability is disturbed, function is lost. There are a great many traumatic states that are accompanied or followed by temporary or permanent disability. Most often the trauma occurs as a single incident of violence, inflicting its maximal damage all at once so that it can be recognized. In such cases there is usually a conventional form of treatment that can be selected by the traumatic or orthopedic surgeon. Traumatic injuries include fractures, dislocations, strains, sprains, lacerations of every type of soft tissue, and thermal injuries. Frequently, the source of trauma is more subtle and may result from faulty posture, overweight, malalignment, muscular imbalance, reflex irritation, or muscular fatigue. In such cases the tissue damage is insidiously and cumulatively developed to the clinical level over a period of time, sometimes years. The pathologic condition is usually masked by tissue reactions in the form of muscle spasm, fibrositis, myositis, and mechanical inflammations (bursitis, synovitis, tenosynovitis, arthritis, etc.) and by the various "syndromes" (scapulocostal, cervicodorsal, scalenus anticus, sciatic, etc.). Some of the mechanisms of these static forces are well understood, whereas others are diagnosed symptomatically and are very difficult to treat.

Those injuries incurred by direct violence have a tendency to improve from the beginning, whereas those which develop slowly often become chronic and recurrent or progressively worse. The application of good principles of rehabilitation medicine

397

is necessary in both groups, but it is from traumatic sources that a great number of severe physical disabilities arise. The following principles may be set forth as "ten commandments" for good rehabilitation medicine in acute injuries.

"Ten commandments" for good rehabilitation medicine in injuries. Irrespective of the cause, type, or extent of the injury, there are certain basic principles of treatment that should be followed:

1. Evaluate each case and plan a program consistent with the definitive care based on specific indications.
   a. Medical rehabilitation technics are not a substitute for conventional medical and surgical management of acute injuries.
   b. Medical rehabilitation technics are indispensable adjuncts to good definitive treatment.
   c. Anticipate the scope of therapy needed based on the nature of the injury and the possible unavoidable restrictions that may be superimposed on it by the treatment method.
   d. Write all prescriptions as a matter of record and change them to keep the treatment current with indications.
   e. Discuss the progress of each patient with the treating therapist.
2. Prevent deformities if possible.
   a. Immobilize the affected part or parts in the optimal position.
   b. Supervise bed posture and the relationship of injured parts to the rest of the body if the patient is ambulatory.
   c. Avoid unnecessary or excessive fixation of joints.
   d. Maintain function of uninjured parts.
   e. Stimulate muscles that the patient cannot move.
3. Mobilize as early as possible without jeopardizing healing.
   a. Early motion retards atrophy, prevents permanent scar formation, preserves joint range, disperses edema, reduces the painful period, stimulates reciprocal muscle functions, and reduces the extent and period of disability.
   b. Fractures into joints often heal better without rigid fixation and loss of the joint.
   c. Establish function through occupational therapy and other purposeful use rather than by abstract motions.
4. Relieve pain before doing therapeutic exercises.
   a. Use the various forms of heat, massage, and electric stimulation to depress pain and muscle spasm to make exercises more easily tolerated.
   b. Patients do not engage in painful activities and usually resent them.
   c. Seldom exceed any patient's tolerance for pain.
5. Protect weakness.
   a. Avoid activities that the patient does not have a reasonable chance to accomplish.
   b. Splint and brace to aid weak muscles and prevent or correct malalignments.
6. Treat weak muscles specifically.
   a. Selective muscle reeducation and exercises are a proved method for regaining power, coordination, length, reciprocation, and relaxation.
   b. Avoid repeated muscle fatigue.
7. Treat joints gently.
   a. Avoid forced motions. Encourage and educate the patient to initiate essential active motions—and force the patient. Muscle stretching is a form of extension, not leverage.
   b. Do not place resistance against swollen joints.
   c. Combine complete rest with short periods of motion for painful joints.
8. Begin weight bearing as early as possible.
   a. Make sure that weight bearing does not exceed the structural capacity of the bones, capsules, and ligaments.
   b. Limited weight bearing is best achieved in deep water in a therapeutic pool. It is also accomplished in parallel bars and with crutches.
   c. Crutch walking must be taught and supervised; it is not an automatic skill.

9. There is an end point in every case.
   a. Do not treat forever.
   b. Recognize that many injuries result in irreversible damage to tissues and that residual limitations are frequently inevitable.
   c. Treatment stops when all functioning elements approach high efficiency in relationship to the available physical capacity.
   d. Do not delay surgery by persistence with ineffectual treatment if surgery is indicated.
   e. The patient should be made aware of his status when discharged. Tell him what his capacity is and what his limitations are. Give detailed instructions in maintenance therapy at home. Do not fail to proceed with his remaining rehabilitation needs when his medical requirements have been met.
10. Do not be deluded by the healing powers of machines, manipulation, and magic to achieve a miracle.
   a. Effective therapy must be directed toward the reestablishment of voluntary function.
   b. Efficient physical function, once lost, must be trained for, just as an athlete trains for maximal efficiency. This is a graded process of daily superimposition and summation of improvement—slow, steady, and sure. The development of the patient's own strength and function is primary; it cannot be baked, rubbed, soaked, or vibrated into the muscles and joints.

Chronic states of disability due to trauma may follow acute injuries or develop without apparent external force. When a specific diagnosis can be made, the treatment is clear, but the majority of patients require symptomatic treatment. It requires considerable skill and understanding to make the type of examination necessary to identify the physical manifestations of the presence of pain. This type of competence will be demonstrated by the recognition of restricted movement in joints and of spasm, loss of elasticity, weakness, shortening, imbalance, or other abnormalities of muscle function. The recognition places one in a position to formulate a reasonable plan of treatment. Such treatment usually takes the form of the application of superficial heat and massage to the area to relieve pain and provide muscular relaxation. This is followed by stretching of tight or fibrotic muscles, regaining part of lost motion at a tight joint by gentle passive exercise, instituting appropriate exercises for the strengthening of weak muscles, or giving coordination exercises, occupational therapy, or ambulation training.

Chronic trauma. Chronic trauma may result from postural or occupational habits that cause persistent muscular fatigue; for example, the stenographer who leans forward a great part of the day and the taxi driver or the dentist who performs unilateral movements are subject to such trauma. In these instances the external force is gravity. Many people work, sleep, stand, or play in persistently poor mechanical positions. Muscular fatigue results from lack of opportunity for muscles to relax. If one carries a heavy suitcase along the street, it appears to become heavier and heavier and the arm muscles begin to fail in holding it up. If the suitcase is then placed down on the sidewalk and the muscles relaxed for a few seconds, it can be picked up and carried again until the muscles become sore. Individuals who lean forward are "hanging on the back muscles" a large part of the day, and the muscles do not have an opportunity to relax. The fatigue mechanism is interpreted in terms of the accumulation of metabolites within the muscle as a result of persistent contraction. Because of the tension within the muscle, the circulation is diminished to the extent that a state of tissue anoxia can ensue. This emphasizes the importance of a detailed history in traumatic conditions in which external trauma apparently is not involved.

In addition to treatment for the specific manifestations accompanying the pain, the key to the situation is the relief of persistent postural pull on the muscles. This is accomplished by changing the type or height of the seat or workbench for a bench worker, prescribing a support for improved posture for the patient with a protuberant abdomen, or a low back support for a woman with exaggerated curves of the lumbar and dorsal spines. Attention to individual requirements is the basis for success. All patients should be made conscious of their posture in standing, walking, working, and sleeping. While at work most persons require a change in position at 1- or 2-hour intervals to get out of the bent-over attitude and to stretch and then relax. Many postural syndromes can be prevented by good habits.

When stress phenomena are accompanied by inflammatory manifestations in the vicinity of joints, such as lateral epicondylitis or bursitis in the region of the shoulder, local infiltration with procaine hydrochloride (Novocain) and one of the injectable corticosteroids is often helpful as an anti-inflammatory agent. The principles of restoration of normal motion then can be applied.

Certain types of trauma are notorious because of the severe physical handicaps that follow. Among these are amputations, severe injuries to the hands, and injuries causing low back pain. Because such injuries create a large pool of disabled workers, they will be discussed in further detail.

AMPUTATIONS

Amputations are musculoskeletal problems of a special nature because the disability results not from a pathologic condition but from treatment that has eliminated the pathologic condition.

It is estimated that 1 out of every 300 persons in the United States has had a major amputation. Each year there are 35,000 amputations as a result of congenital defects or surgery, with a ratio of 3 in the upper extremity to each 10 in the lower extremity.

During World War II considerable attention was drawn to the 21,000 amputee war casualties, whereas six times that number of amputations were performed among civilians in the United States in the same 4 year period without it being recognized as a national medical problem.

The causes for amputation are (1) accidental violence to extremities, (2) death of tissues from peripheral vascular insufficiency of arteriosclerosis and diabetes, (3) death of tissues due to peripheral vasospastic diseases such as Buerger's disease and Raynaud's disease, (4) malignant tumors, (5) long-standing infections of bone and other tissues that leave no chance of restoration of function, (6) thermal injuries—both from heat and from cold, (7) uselessness of a deformed limb that is objectionable to the patient, (8) other conditions that may endanger the life of the patient, such as vascular accidents and snakebites, and (9) congenital absence.

In peacetime in the United States and other industrialized and western countries, the majority of amputations in the lower extremity are performed on patients over 60 years of age. These are primarily for peripheral vascular diseases and obstructions. More than half of this group has diabetes. Up until recent years, amputation above

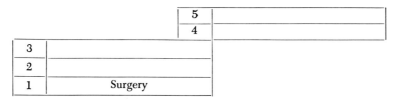

Fig. 21-1. First stage in rehabilitation of the amputee—surgery.

the knee was essentially an automatic indication for the diabetic with gangrene of the toes. This concept is now considered to be obsolete.

Since 1964 intensive research into the comprehensive rehabilitation of the amputee and improved interpretation of procedures for evaluating viability of the limb have resulted in salvaging more of the extremity. In centers in which surgeons are familiar with the modern concept of amputee management and newer surgical technics, amputations below the knee far outnumber those above the knee in vascular cases. This is the by-product of research investigation of immediate postsurgical fitting and studies of early ambulation. Details of the technics of these procedures have been published in a manual by the Veterans Administration[11] and in a monograph by the Institute of Rehabilitation Medicine, New York University Medical Center.[12]

Rehabilitation of the patient who has had an amputation is an achievement reached through successful management of the patient in a number of stages and is a step-by-step procedure that must be carefully adhered to. Teamwork among doctors, nurses, therapists, and the limb fitter is essential.[12]

Surgery. The first stage is surgery (Fig. 21-1). The objective here is to eliminate the pathologic condition and leave a healed stump of adequate length, shape, comfort, and function. Amputation that was once considered mutilating surgery must now be regarded as a reconstructive procedure resulting in a stump that is a functioning organ, capable of tolerating and controlling available sophisticated prostheses. To accomplish this the surgeon must treat each tissue in such a way that it will make its best contribution to a functioning stump.

Skin. In above-knee amputations the anterior and posterior flaps may be of equal length but longer than half of the diameter of the limb at that level. This will allow for much looser skin closure than previously was considered adequate. The most common fitting problem and one that produces pain is skin that is too tight. Few problems arise from skin flaps that are closed loosely. In amputations below the knee equal flaps may also be made. Recently it has been considered more advantageous to close the below-knee amputation stump with a single long posterior flap. This is desirable in vascular cases, since the blood supply is better in the posterior portion of the leg than it is in the anterior. The resulting scar is subjected to much less shear force in a total contact socket. The skin flaps of an amputation stump should be approximated edge to edge loosely with single loop sutures, preferably of plastic material rather than silk or wire. Mattress sutures and tension sutures should be avoided, since they can produce islands of necrosis by strangulation, particularly in vascular cases, when postoperative edema places them under great tension.

Subcutaneous tissue and fascia. This tissue and fascia should be preserved in the course of the operation without developing unnecessary lines of cleavage between them so that they may be interposed between the skin and the underlying tissues. The fascia should be closed carefully and completely under very slight tension so that it forms a closed envelope within which the tensions and pressures of the amputation stump will be equally distributed and will stabilize the stump. The subcutaneous layer should be firmly approximated to reinforce the effect of the closed envelope produced by the fascial layer; this will then allow for edge-to-edge approximation of the skin flaps without tension. There should be no angulations or depressions.

Muscles. Muscles are to be cut cleanly and shaped to allow gentle tapering of the stump. They must not be allowed to retract, but should be anchored to tissues close to the bone or to the bone itself. Muscles that are well shaped and tapered may be sewn to each other over the end of the bone and anchored down to the bone so that they will contract isometrically within the socket of a prosthesis. All slack should be taken out of muscles, and anchorage of the muscles should take place under slight tension. This is called tension myoplasty. There are some instances in which drill holes are made in the bone to anchor the muscles firmly to insure that they will not retract. Muscles that have a positive point of insertion so that they contract isometrically provide a degree of proprioception for the patient in the use of his prosthesis that he could not otherwise obtain. It was previously thought that muscle over the end of the bone would become fibrotic and interfere with fitting. Fibrosis does not occur if the muscles are tapered and anchored to each other over the end of the bone when a total contact socket is used. The hardening and fibrosis observed in years past were due to the use of open-end sockets, which left a condition of chronic edema at the end of the stump that eventually produced fibrosis.

Blood vessels. Adequate hemostasis is essential. If this is not accomplished, there may be a hematoma at the end of the stump, which causes it to expand in bulbous formation, making it very difficult to fit with a prosthesis and eventuating in a hard, brawny area that can be a persistent source of discomfort. The use of the rigid dressing is extremely helpful in the prevention of the formation of hematomas. When the rigid dressing is used, it is not necessary to place a drain in the stump. When the rigid dressing is not used, every major amputation stump should be drained for 48 hours to avoid the formation of hematoma.

Nerves. These are treated best by pulling them down and cutting them cleanly with a sharp knife. They should be allowed to retract upward away from the scar. It is not necessary to inject nerves. It is often necessary to place a ligature around the sciatic nerve to contain the bleeding of the blood vessel incorporated in it. This suture should be tied tightly enough to stop the bleeding, but not so tightly that it will cut the sheath and fibers of the nerve itself.

Bone. Before severing bone, one or two sharp knife cuts through the periosteum should be made and the periosteum stripped distally. The femur and the tibia should be cut parallel to the ground without scuffing of the periosteum or piling it up proximally. Exostoses from traumatized and raised periosteum are very common and very troublesome. Below the knee the fibula should be cut from $\frac{1}{4}$ inch to 1 inch shorter, depending on the length of the amputation stump. In amputations above the knee, unless end weight bearing is anticipated, the femur should not be divided

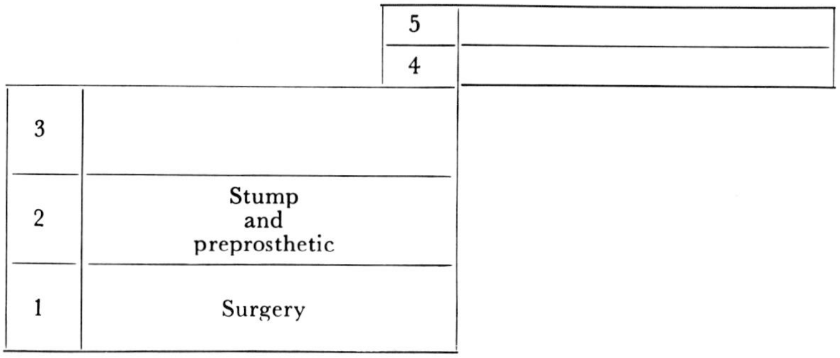

		5	
		4	
3			
2	Stump and preprosthetic		
1	Surgery		

Fig. 21-2. Second stage in rehabilitation of the amputee—stump and preprosthetic stage.

nearer than 4 inches from the knee joint. This is the amount of space necessary to incorporate an automatic knee in the prosthesis. If the femur is severed closer to the knee, the thigh of the prosthesis will be longer than that of the opposite side.

Stump and preprosthetic stage. The second stage may be called the stump and preprosthetic stage (Fig. 21-2). Here attention is drawn to the physical characteristics of the patient and the stump.

Diabetes or other systemic disease must be brought under control and also careful attention must be given to the cardiac status of the patient. Special attention is given to the condition of the remaining limb. The muscles must be strengthened, and it is necessary to make sure that all of the joints can be carried and controlled through their full range of motion. Note should be taken of the status of the peripheral vascular supply of the remaining limb and whether the patient has independent balance on it. Standing, balance, and ambulation in parallel bars and ultimately with crutches are to be developed as soon as the patient can tolerate it. The patient's posture on the remaining leg and the stability of his pelvis are very important items to control and develop. The upper extremities must be strengthened with progressive resistance exercises, since in the early phases of learning to walk the patient is practically holding his entire body weight on the upper extremities.

As far as the stump is concerned, it requires conditioning, shaping, a full range of motion at the hip or knee, and activities to encourage shrinking. When immediate fitting or early ambulation is employed, the entire preprosthetic stage with regard to the stump is bypassed, since nothing is more effective in placing the stump in satisfactory condition than actual use of the tissues themselves. This is accomplished primarily by fitting as early as possible with a temporary prosthesis properly shaped and properly aligned. Makeshift devices do not serve the same purpose and are very inefficient. If early ambulation with the use of a temporary prosthesis is not available, conditioning of the stump is achieved by a graded program of exercises—at first gentle active and assistive exercises and finally resistive exercises—to make sure that there is no flexion contracture of any of the joints and that there is no abduction contracture of the hip. The gluteal muscles, hip flexors and extensors, and adductors and abductors must be made strong and competent. Shaping of the stump is achieved concomitant with shrinkage by means of elastic compression. The elastic bandage must be applied precisely and efficiently; otherwise it should not be used at

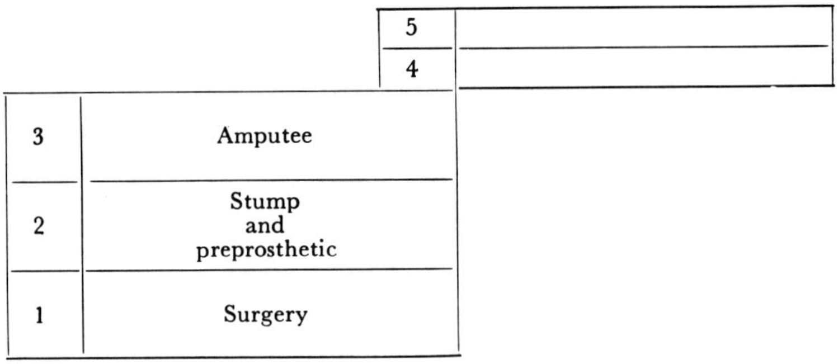

Fig. 21-3. Third stage in rehabilitation of the amputee—that of being an amputee.

all. Most patients are unable to apply the bandage properly. The repeated encircling of the limb by several turns of a tight elastic bandage can actually constitute a tourniquet. It is preferred that elastic stump-shrinking socks be used. These have a predetermined degree of compression necessary for shaping and shrinkage of the stump and are far superior to bandaging by inexperienced patients or attendants.

Being an amputee. When the second stage is well under way, the patient enters the third stage, that of being an "amputee" (Fig. 21-3). In this stage every personal aspect of the individual's life must be considered.

Age. Persons over 55 years of age are much less adaptable for using an artificial limb than are persons below that age.

Sex. There are certain factors in sex that will influence the components of an artificial limb, particularly those relating to housewives and the kind of work they are required to do. Among these are body build, weight, and cosmesis.

Occupation. It is obvious that a person who intends to work while wearing an artificial limb will require a limb that is equal to the work demands. It must be quite different from the type of limb that makes it possible only to get out of one chair and move to another.

Need and desire. Does the patient really need and want a prosthesis? This question should be specifically asked of the patient because many people do not want a prosthesis.

Cooperation and understanding. Has the patient cooperated up to this point and does he understand what is being done? The amputee should be made fully aware of what is taking place to gain his full understanding and participation in the program.

Motivation and acceptance. This is important, because patients who do not accept their disabilities, who resist all attempts to help, and who have no real drive to become mobile cannot be worked with successfully.

Expectations. What does the patient expect of a prosthesis? This question is important too, because there are people who have the feeling that if they simply had an artificial limb they would be restored to normal and be able to walk about freely. They have no knowledge of what it takes to achieve such activity.

Realism. Is the patient realistic and active or is he passive and full of wishful thoughts? The individual who verbally expresses a desire to become active again

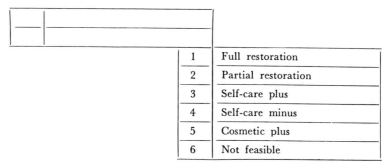

	1	Full restoration
	2	Partial restoration
	3	Self-care plus
	4	Self-care minus
	5	Cosmetic plus
	6	Not feasible

Fig. 21-4. Classification of function of amputee with an artificial limb.

but who does not reveal much cooperation in his program must be suspected either of not accepting his disability or of expecting too much of the prosthesis.

Past experience. What kind of life has the patient led? Those people who have had athletic experience, either in sports or in their work, will find such experience extremely useful in their training period with a prosthesis.

Classification of function. Knowing the physical and emotional state of the patient as well as his personality and the condition of his amputation stump, one may be in a position to prescribe an artificial limb. Before this is done, however, the patient's potential for the use of the artificial limb and his objective must be determined. In formulating an objective, it must be emphasized that any patient's performance with an artificial limb must be compared to what would be considered "normal." The patient's performance must be measured against his potential rather than the normal. When a patient achieves his potential by way of performance, he has been successfully rehabilitated, even though his performance is low and his potential is low. It is therefore necessary to know the classifications of function for the purpose of establishing an objective for each individual patient. Toward this end, objectives and functional performance are divided into six classes (Fig. 21-4).

Class 1—full restoration. When what one would consider to be normal is discounted, the patient who achieves the full restoration level of function has no disability, although he has a handicap. He is able to work, participate in limited sports, engage in all and any business and recreational activity, and maintain his stature as the head of a family without help from another person. This level of performance is rare, but there are many instances of the achievement of such a high level of function by amputees.

Class 2—partial restoration. This means that with an artificial limb the individual will be completely functional. He will be able to work. He may possibly have to make changes in his job. He can still dance, but he does not dance as much. He does not walk as far as he used to, and there are some activities such as tennis and golf in which he no longer participates, but he is a functioning individual who can live an acceptable and full life. He will not need to make frequent visits to the limb fitter.

Class 3—self-care plus. Although people in this class are disabled and have physical limitations, they are capable of doing everything for themselves on a personal basis and more. Many are able to work if they can obtain the kind of job that does not require too much standing or walking. They no longer dance. They no longer carry weights or walk distances, but they are functioning individuals who can live in the home and discharge their responsibilities to their families. These amputees require frequent adjustments of their sockets.

Class 4—self-care minus. People in this category are better off with an artificial limb than they are without one. Their best achievement falls somewhat short of taking care of all of their needs. They require help from others. They are severely disabled people. They cannot go up and down stairs without the assistance of another person or crutches. They do not go out alone because they may not be safe. Many can never be fitted comfortably and they may change artificial limbs frequently. They can, however, be taken into a car. Many businessmen who have assistance in getting to their place of business are quite efficient in their desk chairs and live an acceptable life that is made possible by the use of a prosthesis, whereas they could not do so without it. Also in this category are people who may be able to live at home with their families if they have a prosthesis because they can take care of themselves without too much help from other people, whereas they might require institutionalization if they did not have an artificial limb. Many people do not achieve even this type of objective.

Class 5—cosmetic plus. The amputee in this class is better off with a prosthesis than without one, but the difference is quite small. Many people want a prosthesis for personal reasons because they value their appearance or they wish to meet the public even though they require considerable help. They fall far short of taking care of their personal needs, but they do better with the prosthesis for the small gains that they may achieve, even from the point of view of appearance alone. They do little more than they would without the prosthesis.

Class 6—not feasible. The amputee in this class does not have a prosthesis, and prostheses should not be prescribed for them. However, they should be trained to do as much as they can from a wheelchair in the way of self-care activities.

• • •

These classifications that determine objectives based on the summation of those defects or factors that interfere with function apply specifically to the unilateral above-knee amputee. The unilateral below-knee amputee performs one level higher than the above-knee amputee, all other things being equal. Objectives for amputees with other than standard above- or below-knee stumps must be individually developed.

How does one select the proper objective for a patient? This is not difficult if the step-by-step process indicated on p. 407 has been adhered to. In the first three stages, for every defect that the patient demonstrates, he will drop down one class. Therefore, a patient with two defects has a class 3 objective; for example, a young man with a painful stump and flexion contracture has two defects, and his best achievement will be class 3.

Age per se is a deficiency if the patient is more than 55 years old; for example, a 60-year-old obese man with impaired circulation in the remaining leg has three defects and belongs in class 4.

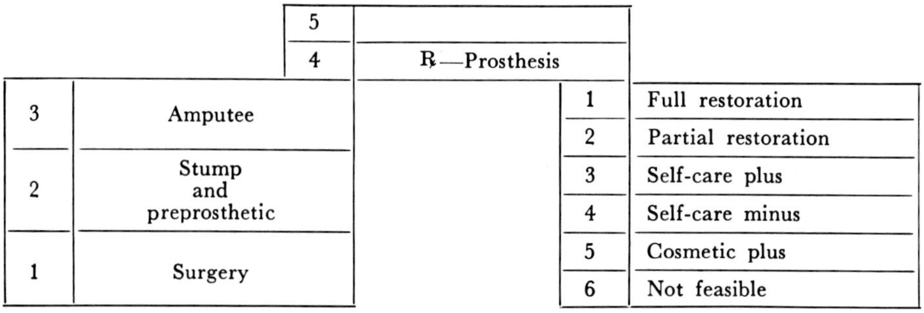

	5				
	4	℞—Prosthesis			
3	Amputee		1	Full restoration	
			2	Partial restoration	
2	Stump and preprosthetic		3	Self-care plus	
			4	Self-care minus	
1	Surgery		5	Cosmetic plus	
			6	Not feasible	

Fig. 21-5. Factors on which the prescription of an artificial limb is developed.

Any patient with five defects should not have a prosthesis ordered; for example, a 65-year-old man with a painful stump and with a flexion contracture at the hip, who is not motivated and is unable to walk with crutches, has five defects and will not get anywhere with an artificial limb, since he has—at best—a class 6 objective.

It becomes clearer at this point why such careful and detailed step-by-step control and supervision constitute the foundation for successful rehabilitation.

Classifying the amputee is important also from the point of view of the length of time he will need training with his prosthesis. Any patient who has an objective in class 3 and who wants to be fully restored will never achieve it. If the physician, therapist, and patient do not know what his best achievement is in class 3, it is not possible to determine when he has been physically rehabilitated and so his training may go on indefinitely. This patient will be unhappy and so will the therapist and the physician, even though he may have reached his best level of function. By classifying the patients, it is possible to know when they may be discharged from their training programs. Paradoxically, the patients with the lower levels of achievement require more training to reach their objectives.

Prescription stage. Fig. 21-5 shows the factors upon which the prescription of an artificial limb is developed. The components of an artificial limb consists of a socket, suspension, knee joint, a shin, an ankle joint, and a foot that simulates the functions of the gastrocnemius and the tibialis anticus muscles. The artificial limb is a piece of precision equipment made on prescription. The type of socket will depend on individual requirements. The type of suspension whereby the limb is held to the body is determined by all of the information that has been accumulated up to this point, plus the objective. A suction socket should not be prescribed for the older patient with instability and numerous defects. Instead, a so-called conventional socket should be prescribed, and the suspension should be in the form of a hip joint and pelvic band. Suction is merely a form of suspension to keep the limb from falling off. It does not influence the circulation, since there is positive pressure rather than negative pressure when the limb is placed on the ground. The suction applies only when the limb is off the ground and it has no detrimental effects on the circulation; in fact, it will probably help circulation because it simulates intermittent suction and pressure.

The alignment of the limb that is prescribed should be related to the weight and

		Rehabilitation		
	5	Ambulation training		
	4	R —Prosthesis		
3	Amputee		1	Full restoration
			2	Partial restoration
2	Stump and preprosthetic		3	Self-care plus
			4	Self-care minus
1	Surgery		5	Cosmetic plus
			6	Not feasible

Fig. 21-6. Rehabilitation schedule.

balance, muscular status, and stability of the patient. Here all of the preprosthetic training comes into play in the prescription, because when the physician knows what his patient can do, he knows what kind of artificial limb he will be able to use to best advantage.

Ambulation training. Having prescribed the prosthesis, it is now possible to place the patient on training (Fig. 21-6). Here again, the step-by-step process becomes important, since it is now known what the patient is being trained for and what kind of training is necessary to achieve that objective. Training is the process of having the amputee develop independent static and dynamic stability on the amputated side. The technic is described in detail in Chapter 4. It is a slow process of educating the patient and developing movement and balance patterns to the point where the patient becomes stable. It does not consist of watching the patient to see what he can do. The preprosthetic strength and control that have been placed into the stump now begin to function in the prosthesis.

•   •   •

In the case of patients with amputations below the knee, all the facets of the step-by-step procedure apply, with the exception of the classification of the objective. Since the below-knee amputee possesses a knee, he requires less effort for rehabilitation and is usually more successful than the above-knee amputee, provided he has a satisfactory and comfortable fit. Whereas the above-knee amputee expends from two and one half to three times as much energy as his nonamputee counterpart in walking, this is not the case with the below-knee amputee. It has been determined that young men with above-knee amputations who walk well require about 250 calories of energy to walk a mile at the rate of 2 miles per hour. The normal young person can walk a mile on 75 calories. This implication achieves importance when one considers that about 75% of all amputations above the knee are done on people over the age of 55 years. The investment of energy, therefore, increases with age in attempting to walk with an artificial limb. A good number of these persons have peripheral vascular disease in the remaining leg, coronary insufficiency, and other cardiovascular problems. Physical fitness, therefore, must always be considered in the prescription for an artificial limb, and the objective must be established in advance.

THE HAND

Rehabilitation of the patient with a poorly functioning hand requires the closest of collaboration between the disciplines of surgery and rehabilitation. The benefits of active and passive exercise, heat, massage, and related technics may not be optimal unless superficial scars, densely adherent tendons, and nerve injuries are repaired. Nor will maximum benefit to the patient accrue in many instances through surgery unless rehabilitation measures are also applied. Frequently the full benefits of reconstructive surgery may be gained only if the guidance of a physiatrist is available during the postoperative period.

The hand is a closely knit anatomic and physiologic composite in which skin, tendons, nerves, bones, and joints must all work harmoniously. Disease of or injury to any of these components may seriously affect the function of any or all of the others. After a preliminary effort of several weeks to months to obtain maximum function by means of physiotherapeutic measures, attention in reconstruction of the hand is usually directed first to the reconstitution of the superficial covering, the skin and subcutaneous tissues. When these have been surgically improved, attention may be focused on the reconstitution of the deeper structures. Such an approach is dictated by the need for protection by the skin for the delicate healing process involved in repair of tendons, nerves, bones, and joints and for the useful function of these structures.

Skin. In the reconstruction of cutaneous deficiencies, three major technics of plastic surgery may be needed, individually or in conjunction with each other. These technics are the use of the Z-plasty, the application of split-thickness grafts, and the insertion of pedicled flaps of skin and subcutaneous tissue. The Z-plasty is especially useful for the eradication of relatively narrow bands of scar tissue that, by their position, overlying joints, limit the range of motion either in extension or flexion. Simple excision of the scar, followed by resuture of the wound margins, would be followed by the redevelopment of a narrow scar that could be expected to contract again, with a return of the original contracture. The Z-plasty, by the interdigitation of small triangular flaps from one side of the wound with similar flaps from the other side, leads to the development of a zigzag or pleated scar, the total length of which is far greater than the original scar. Although the individual segments of the pleated scar will undergo normal shortening during the postoperative period, the effect of this upon limitation of joint function will have been eliminated. The principle is very much the same as that employed in the construction of the bellows of a collapsible camera.

The application of split-thickness grafts following the excision of a broad, flat scar is, in general, useful only when sufficient subcutaneous tissue remains to provide a cushion over bony protuberances or a bed for the movement of tendons. Although scar may be removed and replaced with split-thickness skin, it is to be expected that a line of scar tissue will form all about the margins of the graft with the surrounding normal skin. Lest such scar develop into a contracted band overlying any joint, it is frequently necessary to make the line of junction between the graft and the surrounding skin in the form of a series of indentations and projections to produce a marginal scar in the shape of the scar of a Z-plasty.

When skin as well as subcutaneous tissue is lacking, both must be supplied, and one of a variety of pedicled flaps may be employed for this purpose. A pedicled flap of skin and subcutaneous tissue may be elevated from the abdomen and sutured into place on the hand after the scar tissue has been removed. Then when the flap has healed into position, the pedicle is severed from the abdomen. The abdominal skin and fat now surviving on the hand not only will provide durable cover and a suitable bed for the repair of tendons, nerves, and bones but also may, in some instances, aid in the restoration of contour.

Tendons and tendon transfers. Freshly lacerated tendons may frequently be repaired within the early hours after injury. Should the fear of infection or the loss of superficial tissue dictate delay in repair, such repair may well be postponed for several months, during which time physiotherapeutic efforts designed to maintain the mobility of joints and the integrity of muscles may be instituted. Usually, whenever the repair of the involved tendons is carried out, whether early or late, immobilization for several weeks is mandatory for the development of firm union at the suture lines. The position of the hand during immobilization will depend in great part on the nature of the surgical repair, but in general the principle to be followed is to avoid tension at the suture line while maintaining the joints, as nearly as possible, in the neutral position of function. This principle often calls for a delicate decision on the part of the surgeon to achieve the combined purpose of strong, primary healing to be followed by restitution of function. Three weeks is usually adequate for immobilization, after which progressively more intensive efforts must be directed toward the reestablishment of function.

Any one of a number of physical medicine measures such as heat, passive and active exercise, and massage may be utilized at this time. However, the best results are usually obtained and return of function hastened through the more interesting and purposeful activities provided in occupational therapy. (See Chapter 5.) The patient's own motivation will be a factor in the prescription of activities designed to promote restoration of function.

In addition to the early or late repair of lacerated tendons, surgery for the improvement of tendon function may include the following types of procedures: (1) tendon grafts for the replacement of missing portions, (2) substitution of one tendon with a useful muscle for another whose muscle is paralyzed (tendon transfer), (3) construction of pulleys for the improvement of direction of tendon pull, (4) stabilization of joints by arthrodesis, (5) rotation of fingers by osteotomy, (6) phalangization, and (7) pollicization.

In general, the period of rehabilitation following tendon transfer is longer than that following tendon repair or tendon grafting. This is because of (1) the problem of the reeducation of the muscles to perform unaccustomed tasks, (2) the frequent occurrence of weakness in the muscle of the transferred tendon, and (3) the generally more massive disability in the patient requiring the tendon transfer. Just as it is unwise to rush into tendon transfer before it is absolutely certain that the return of function of the hand has been maximal, it may also be unwise to prolong tendon transfer beyond this time in the vain hope that a miracle may happen. Undue prolongation of the time for tendon transfer may mitigate against the effectiveness of surgery. The physiatrist should be mindful that delay beyond a certain time for each patient may not be in his best interest. What is called for is a course of action combining the benefits of surgery, when indicated, with the benefits of rehabilitative efforts.

TRAUMA TO THE BACK AND BACK PAIN

Pain in the back may result from single or mutiple incidents of external violence, from excessive effort, and from cumulative stress and strain; or it may develop without apparent cause. Many cases of back pain have been traced to structural, congenital, or developmental defects of the spine and its supporting structures. Injuries to the back account for 12½% of all injuries incurred at work in New York State.[8] Workmen with back injuries, however, lose 25% more time than do those on the average with all other disabilities, even though other injuries are ten times as likely to leave a permanent residual as is an injury to the back.

Etiology. Chronic back pain as a symptom has the distinction of having divided the medical profession, with regard to etiology, to a greater extent than any other medical symptom. Schools of thought have variously ascribed chronic pain in the back to (1) pathology of the intervertebral disks, (2) muscular imbalance or muscular weakness, (3) stresses and strains, or (4) emotional instability. There is ample evidence in the literature that all of these etiologic factors have been responsible for pain in the back in large groups of patients. Obviously, it is necessary to determine the etiology before appropriate treatment can be given, since the treatment of back pain arising from one of these causes is different from that which may be necessary for each of the others. Much of the disagreement with regard to the diagnosis and treatment of chronic back pain can be attributed to the fact that the back is a complex structure. The integrity of the back depends on alignment of many bones separated by intervertebral disks. These articulations are compounded by small interarticular joints supported by muscles, tendons, and ligaments, all of which must be maintained in a state of balance permitted by an accurate reciprocal elasticity of all of these structures. When the trunk is moved out of the vertical position, it becomes a long lever at the points of motion at the hips and lumbosacral joint. This results in the requirement of considerable forces passing through a small area on the opposite side of the fulcrum to restore the trunk to normal alignment.

Strait[9] has estimated that the tensile force in the erector spinae muscles necessary to support a 180-pound man standing with his trunk flexed 60 degrees from the vertical line will be approximately 450 pounds (Fig. 21-7, *A*). The tensile force necessary to hold 50 pounds in this position will approach 750 pounds, and the compression force on the fifth lumbar vertebra will be 850 pounds. As trunk flexion occurs, most of the motion in the spine is noted in the cervical and lower lumbar vertebras. The gravitational pull of the trunk requires considerable force to restore the upright posture, placing torque and stress against the lumbosacral joint. When we assume that equal symmetry exists in the bony structures of the spine, pelvis, and lower extremities, the symmetric stresses in the flexors, extensors, lateral flexors, and rotators maintain postural control (Fig. 21-7, *B*).

There are some 140 muscles attached to the spine that perform this work. Strain of any one or a group of these muscles will produce damage to the muscle, its tendon, or its point of insertion, resulting in local hemorrhage and exudate and a tendency for organization of this exudate to produce scar tissue. The elasticity of the remain-

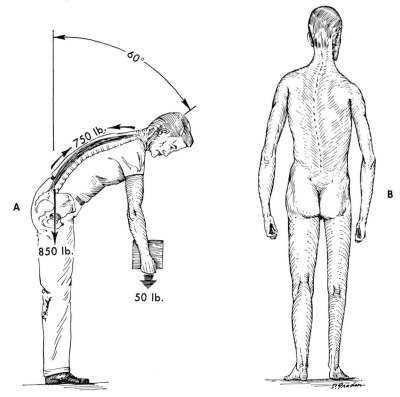

Fig. 21-7. A, Forces passing through the lumbosacral structures. **B,** Reciprocal muscle balance controls the center of gravity.

ing muscles, particularly the erector spinae, will be temporarily reduced by the protective mechanism of spasm.

In the majority of cases of acute strain or sprain of a ligament, the pathology will spontaneously resolve when the local tissue damage has healed. In some instances there will occur a significant alteration in the total elasticity of the posterior muscles, with painful restriction of motions of the trunk. If this persists for a long time, fibrosis of the muscles occurs. This is a prominent cause for back pain in patients who have been subject to prolonged postural defects. The lumbosacral joint, which is less stable than the sacroiliac joint, bears four fifths of the mechanical stress passing through the lower back region.

Sprains of the various ligaments in the vicinity of the lumbosacral joints are more likely to occur than true sprains of the sacroiliac ligaments. The stress that produces lumbosacral sprain will very likely be accompanied by strain of the sacrospinalis muscles. The sustained restriction of motions of the spine due to increase in tone of muscles as a protective mechanism will ultimately produce changes within the joints. Fibrosis of the muscles also occurs, and if osteoarthritic changes are already present, a vicious cycle of pain, restriction of motion, loss of elasticity, and further pain may result. About one fourth of the length of the spinal column is contributed by the intervertebral disks. These cushion the vertebras and contribute to the normal elastic quality of spinal motion. In the newborn infant these disks contain approximately 80% water. They receive no blood supply after a person is 20 years old, and slow dehydration is noted in these structures with aging. Their capacity for repair when injured is limited. Normally disks are insensitive, but, as the changes of aging occur, they become sensitive and painful. For this reason, relatively trivial violence to the osteoarthritic spine may produce severe symptoms and restriction of motion in older people, whereas in younger people the disks withstand considerably more violence to the spine. In addition, the dehydration that takes place with aging makes less likely the herniation of disks through the annulus fibrosus as a result of the stresses of activity or injury. This accounts for a greater incidence of disk pathology in young or middle-aged adults as compared with that in the older age groups, although older people complain more of back pain, which is produced by the wear and tear of time and cumulative damage to the spine as a whole. The treatment for back pain depends on determining the cause.

Acute back pain. Acute back pain is not difficult to treat, particularly when there has been an injury. X-ray examinations will determine whether bony injury has occurred. When such injury is present, standard orthopedic procedures, which have stood the test of time, may be applied. From a rehabilitation standpoint, however, the physician must make every effort to avoid the superimposition of residual disability that may be the result of the treatment itself or the deteriorating effects of prolonged immobilization and disuse atrophy of parts not involved in the injury. Acute injuries of the back without damage to the bony structures usually heal spontaneously under protected conditions. The patient should be placed on complete rest for a period consistent with the degree of violence and symptoms.

During this period of rest, supportive therapy for the release of muscle tensions, for the relief of pain, and for reassurance of the patient constitutes complete treatment. Physical therapy in the form of superficial heat and gentle motions without weight bearing are of great value during the acute stage. When a patient with an injury to the back fails to respond to such therapy and pain becomes prolonged, it becomes necessary to make a detailed examination for purposes of identification of the specific nature of the pain.

At this point, an all-out effort should be made to prevent the condition from falling into the category of the refractory chronic back case so familiar to all. A detailed review of the history to correlate the nature and circumstances and extent of injury should once again be made. A systematic evaluation of the patient's posture, gait, balance, and spinal movements and the elasticity, length, strength, and balance of the supporting musculature of the trunk, pelvis, abdominal walls, and hips should then be recorded. All of the orthopedic leg tests should be evaluated and a neurologic examination made or requested on consultation. Consultation with other specialists to rule out malignancy or back pain as a manifestation of some other disease should not be neglected. (See Figs. 21-8 to 21-11.)

Herniation of the intervertebral disk can almost be diagnosed clinically. The history of sharp pain occurring after sudden stress, followed by steady sciatic distribution, eased by lateral tilt away from the affected side and aggravated by hyperextension is almost pathognomonic. Local tenderness at the fourth and fifth lumbar

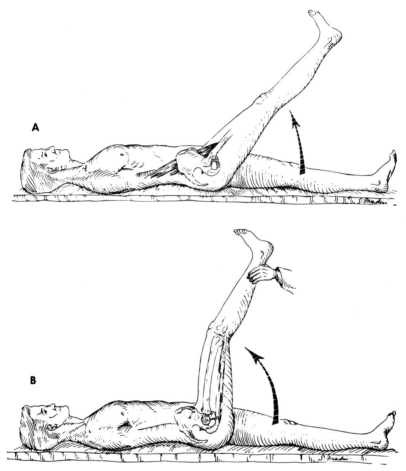

Fig. 21-8. Straight leg-raising test. **A,** Active. **B,** Passive.

interspaces, calf atrophy, loss of ankle jerk reflex, and diminished-to-absent sensation over the lateral malleolus and fourth and fifth toes may later confirm the impression. Finally, myelogram and fibrillation potentials in the electromyogram will conclusively establish the diagnosis. Herniations of the cervical and thoracic disks are being reported.

The x-ray examination will reveal any congenital abnormality, such as spina bifida, or perhaps some inflammatory or granulomatous lesions of the vertebras. Spondylolisthesis, fracture, particularly of the transverse process, osteoporosis, compression fracture, and epiphysitis all can be noted on x-ray examination.

Chronic low back pain with marked splinting of the back and with reduced chest expansion should raise suspicions of rheumatoid spondylitis. If the patient is a 20- to 25-year-old male presenting chronic symptoms and an elevated erythrocyte sedimentation rate and if x-ray films of the sacroiliac joints demonstrate blurring, the diagnosis is strongly suspected.

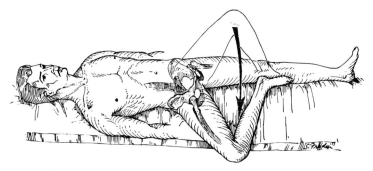

Fig. 21-9. Patrick's test, used to rule out hip joint pain.

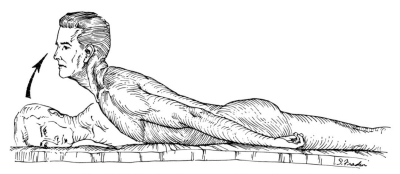

Fig. 21-10. Testing the power of the back muscles.

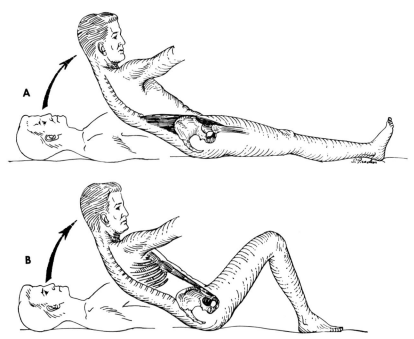

Fig. 21-11. A, Testing combined abdominal and hip flexor power. **B,** Testing abdominal muscles with work of hip muscles eliminated.

The remaining diagnostic possibilities should be considered as suggested by the results of the x-ray examination, blood count, and urinalysis. These vary from renal disease, aortic aneurysm, and mycotic infections to vascular accidents of the vertebral vessels.

When the diagnosis of a protruded intervertebral disk has been positively made, the treatment should be conservative at first, since the majority of cases can be brought under satisfactory control. If recurrences are frequent or insufficient relief follows conservative treatment, surgery should be considered. Most authorities agree that the best results after surgery are obtained in the patients operated on relatively early. Because of the secondary changes in muscles, posture, habits, and emotional reactions with time, it may be too late for surgery to relieve more than a small part of the complex problem after a case has been chronic for years.

In time the chronic back case becomes less an orthopedic or neurologic problem and more a rehabilitation problem. A study of 400 long-standing cases of back pain was made at the Institute of Rehabilitation Medicine. The duration of symptoms was from 6 months to 25 years, averaging over 1 year. In this group 15% of the patients had one or more operations, and all were unable to work. Using the diagnostic, medical, social, psychologic, and vocational resources of a rehabilitation center, it was possible to return only 2% of these patients to their former occupations and 60% to other jobs demanding physical capacities that they could meet. This study revealed the following:

1. In patients with long-standing cases, although those on a noncompensation basis responded better to treatment, the general pattern was very similar to that in patients on a compensation basis.
2. The patients operated on late had successful results from a surgical viewpoint, but symptoms persisted and the psychologic problems increased.
3. At least 80% presented such serious emotional and psychologic obstructions to successful rehabilitation that it was obvious that physical treatment alone was inadequate.
4. For 10%, intensive psychotherapy was required before any other treatment could be considered. Cooperation for such treatment was difficult to obtain.
5. Another 10% presented purely psychiatric problems that were considered nonfeasible even for psychotherapy at the stage seen.

The urgency for resolving back cases as early as possible is pointed up by these data.

Principles of treatment. The treatment should be directed toward reversing the pathology and the objective manifestations of disturbed function. A patient who has suffered recent back trauma should be placed at rest, muscular spasm released, range of motion of the trunk restored, weight bearing gradually introduced, activities tolerance developed according to each patient's needs, and continuous reassurance given to prevent excessive concern. The relief of pain during the treatment period must be achieved to permit the patient to cooperate. The specific technics are discussed under physical therapy in Chapter 4, and a written prescription for the complete regimen should be made following the examination.

When back pain is chronic, it may be impossible to make a pathologic diagnosis, so that treatment can be directed only to the manifestations of altered function. It is well to stress that measures for the relief of pain alone are not likely to bring many back cases under control.

Even if a specific diagnosis cannot be made, a plan for therapy should be established. If the symptoms are still relatively acute, the patient should be kept at rest on a firm mattress or on one under which a bedboard has been placed. Sedative and analgesic drugs should be used to control pain. As soon as they can be tolerated, active setting exercises for the gluteus muscles, the abdominal muscles, and the back muscles should be begun. Nonweight-bearing stretching exercises in the bed are sometimes effective in relieving tension from spasm. The use of muscle relaxant drugs during the acute or chronic stage has not proved effectual. When symptoms begin to subside, the use of local heat in the form of infrared, diathermy, tub bath or shower, and histamine ion transfer for the sedative effect should follow. As the symptoms permit, passive stretching of tight muscles should be attempted. Exercises in a therapeutic tank with the water at 100° F. are frequently followed by rapid improvement in range of motion. Occasionally, walking in the deep end of a therapeutic pool restores coordination and elasticity of muscles.

The use of leg traction in the acute and chronic phases of the disability is of questionable value. Rothenberg, Mendelsohn, and Putnam[10] failed to observe any effect on the disk or appreciable distraction of the adjacent vertebras that were observed during laminectomy with 25 pounds on one or both legs. Lumbar puncture needles inserted into the fourth and fifth lumbar and first sacral vertebras failed to demonstrate any appreciable distraction when traction was applied to a conscious patient. There is no convincing evidence that leg traction is superior to complete bed rest with adequate control of pain and sedation.

Manipulation of the spine should be employed only when the diagnosis of mild sprain is made and rapid reduction in spasm can be achieved. When the patient is capable of resuming activities on a progressive basis, it may be useful to protect the back from the stresses of movement and support of body weight by the use of reinforced belts and braces.

It is necessary to keep the patient on a sustained program of exercises of increasing intensity, designed to rebalance the supporting musculature of the trunk and pelvis, while he is wearing these supports, with the ultimate purpose of removing them later when he has a muscular corset sufficient to handle the stresses. When the patient's progress permits, he should be placed in graded activities in an occupational therapy program designed to simulate his work requirements. If his symptoms were severe and his work is strenuous, it may be advisable to discuss a change in occupation to avoid recurrence under working conditions. If the patient fails to respond to these methods and chronic back pain persists, he may develop numerous psychogenic superimposed symptoms that complicate the picture. Such patients should obtain the supportive therapy of professional personnel such as psychologists, social workers, and the psychiatrist when necessary.

Prevention of backache. The early recognition of predisposing factors of backache should receive prompt attention. All postural defects, differences in leg length, asymmetry, ankle pronation, and foot deformities should be evaluated and treated. Lateral chest films should be made for all children with "round backs" to note any vertebral epiphysitis (Scheuermann's disease). Corrective posture, restriction of flexion activities, and splinting may aid in reducing eventual deformity. Lifts, wedges, corrective posture exercises, and stretching may all serve to avoid later disablement.

Every child who has had poliomyelitis followed by muscle weakness anywhere should be seen regularly to observe any spine malalignment and should have periodic x-ray examinations of the spine. Educational programs that will emphasize the need for physical fitness and proper use of the back should reduce adult problems in this area.

Preventive hygiene instruction in correct sitting posture, bending, lifting, pulling, pushing, and reaching is beneficial. Work areas laid out to minimize excessive stretching, bending, and twisting will reduce backache problems. One firm has a pre-employment examination, including x-ray films of the back, to rule out any congenital or other structural defects, since they may cause backache. Another firm would not employ any man over 5 feet 10 inches tall for a job that was conducive to excessive backaches. An understanding of the role of the employee's personality, proper placement, and a consideration of the working environment should help reduce this problem.

Disability evaluation. To evaluate the disability of backache, the personality and social considerations must not be overlooked. It would be wise to refer any patient with back pain of more than 6 weeks' duration to a rehabilitation service where the physical, social, and emotional factors will be clearly defined. The frequency with which many patients are reported as evidencing no objective abnormalities to corroborate their subjective complaints should direct our attention in such a case to the total patient and not to his back. The earlier this is done and the sooner a direct attack is made on the neurotic manifestations, the more likelihood there is of success. Repeated examination by many specialists, with many diagnostic studies, only aggravates the neurosis. It is well to repeat that, second to headache, backache is one of the most frequent symptoms of psychologic and emotional disturbances.

The patient should be observed in various spine and trunk activities if any real evaluation of his disability is to be made. The following list of movements should be considered: standing, walking, climbing, sitting, twisting, bending (flexion, extension, lateral), crouching, lifting (reaction to various weights), pulling, pushing, raising (observe for weight and height), rocking (flexion-extension as in sawing), and striking (hammering).

Occupational therapists will be able to evaluate the progress of the patient. Additional activities and loads in the various movements should be attempted. Only by these means can an accurate appraisal of the patient's progress and capacity to return to work be made. If return to former employment cannot be accomplished, the need for vocational counseling is apparent. This service is described in Chapter 17. Retraining, a new job, or modifications of the existing job—all can serve to restore the patient to work.

Conclusion. The primary objective in the management of backache is to relieve symptoms and avoid chronicity. The physician must be alert to the possibilities of this disability being a manifestation of neurotic behavior.

Through cooperation and collaboration with the attending physician and by the services of the rehabilitation team, this problem can be hopefully reduced.

REFERENCES

1. Bowers, W. F., editor: Surgery of trauma, Philadelphia, 1953, J. B. Lippincott Co.
2. Peterson, L. T.: Amputations. In Bowers, W. F., editor: Surgery of trauma, Philadelphia, 1953, J. B. Lippincott Co.

3. Klopsteg, P. E., and Wilson, P. D.: Human limbs and their substitutes, New York, 1954, McGraw-Hill Book Co., Inc.
4. American Academy of Orthopaedic Surgeons, Inc.: Orthopaedic appliances atlas, Ann Arbor, 1952, J. W. Edwards.
5. Bunnell, S.: Surgery of the hand, ed. 3, Philadelphia, 1956, J. B. Lippincott Co.
6. Hampton, O.: Wounds of the extremities in military surgery, St. Louis, 1951, The C. V. Mosby Co.
7. Furlong, R.: Injuries of the hand, Boston, 1957, Little, Brown & Co.
8. Russek, A. S.: Medical and economic factors relating to compensable back injury, Arch. Phys. Med. **36:**316, 1955.
9. Strait, L. A., Inman, V. T., and Ralston, J. J.: Sample illustrations of physical principles selected from physiology and medicine, Amer. J. Physiol. **15:**375, 1947.
10. Rothenberg, S. F., Mendelsohn, H. A., and Putnam, T. J.: Effect of leg traction on ruptured intervertebral discs, Surg. Gynec. Obstet. **96:**564, 1953.
11. Russek, A. S.: Follow-up study of amputees served by a prosthetic team—Research Report. Sponsored in part by a grant from the Office of Vocational Rehabilitation, Washington, D. C., 1961, Department of Health, Education, and Welfare.
12. Russek, A. S., and others: Investigation of immediate fitting and early ambulation following amputation in the lower extremity, Rehabilitation monograph 41. Sponsored in part by a grant from the Social Rehabilitation Service, Department of Health, Education, and Welfare, Washington, D. C., 1969, Institute of Rehabilitation Medicine, New York University Medical Center.

CHAPTER 22

REHABILITATION OF PATIENTS WITH DISEASES OF MUSCULAR AND NEUROMUSCULAR SYSTEMS

There are few diseases of the muscular and neuromuscular systems about which there is greater confusion and misunderstanding than the triumvirate of progressive muscular dystrophy, progressive muscular atrophy, and arthrogryposis. One cannot discuss the rehabilitation aspects of these illnesses effectively without consideration of the natural history of the illnesses in terms of prognosis. For example, there is a widespread belief that progressive muscular dystrophy is almost universally a rapidly progressive, fatal condition, and thus an attitude of defeatism exists with reference to rehabilitation of patients with this disease. Likewise, it is commonly believed that one of the varieties of progressive muscular atrophy—amyotonia congenita—is a temporary condition from which the infant or the child gradually recovers. On the other hand, there are those who believe that there is no such clinical condition as amyotonia congenita and that the disease represents an example of the earlier described progressive muscular atrophy of Werdnig-Hoffmann, which has very dismal prognosis. With respect to arthrogryposis, it is not clearly established whether this condition is a congenital disease of the connective tissue, the muscles, or the nervous system. It is therefore necessary to discuss our concept of the diseases themselves prior to discussion of the principles and practices of the rehabilitation of patients with these diseases. This discussion is essential because the problems of rehabilitation and the attitudes toward the rehabilitation of these patients are intimately related to the prognosis of the illness. Principles and practices of rehabilitation will be discussed collectively for the muscular dystrophies and atrophies.

PROGRESSIVE MUSCULAR DYSTROPHY

In 1891 Erb clearly delineated a primary progressive degenerative disease of striated muscle.[1] Although cases had been described earlier by Meryon[2] and Duchenne,[3] these authors had confused muscular dystrophy with a neurologic disease in which progressive muscular weakness is secondary to degeneration of the motor nerves that innervate the muscles. This latter syndrome is called progressive muscular atrophy. Progressive muscular dystrophy and progressive muscular atrophy are quite different diseases; nevertheless, even today they are frequently confused. We shall, therefore, discuss the salient features of the several clinical types of progressive muscular dystrophy and then mention the progressive muscular atrophies and several other types of myopathy sometimes confused with progressive muscular dystrophy.

Definition and classification. Progressive muscular dystrophy is a primary progressive degeneration of the striated musculature of the body, of unknown etiology. Although numerous so-called clinical types have been described, most of the cases can be placed in one of five broad categories. Classifications of progressive muscular dystrophies are frequently designated according to the appearance of the muscles (pseudohypertrophic), the group of muscles primarily involved (facioscapulohumeral, distal, ophthalmoplegic), the age of onset (childhood, adolescence, adulthood), or the changes in the muscle tone and reactivity (myotonic). There are, therefore, numerous classifications of this disease[4] and no general agreement in this regard. We shall adopt a commonly used clinical grouping without discussing the merits or the deficits of the classification.

We wish to emphasize, however, that classification of the clinical types of progressive muscular dystrophy is not just an academic exercise. Accurate classification is essential if we are to learn the natural history of the several types of the disease. Accurate genetic and medical counseling of patients and families depends on this information. The diverse clinical and genetic aspects of progressive muscular dystrophy suggest that we may be dealing with a group of genetically determined biochemical aberrations in which degeneration of striated skeletal muscle is the common feature.

Incidence and prognosis. The number of persons afflicted with progressive muscular dystrophy is not accurately known; however, on the basis of results of Chung and Morton's studies, converted into table form (Table 22-1) it would appear that the incidence of progressive muscular dystrophy is less than is generally believed, but from the prognostic viewpoint it is one of the most serious disabling diseases of childhood. More than half of those affected are children between the ages of 4 and 15 years. The prognosis varies considerably with the type of dystrophy. For example, in the childhood type, which has its onset between the ages of 2 and 10 years, the disease progresses rapidly and the great majority of the patients die before the age of 20 years. On the other hand, the facioscapulohumeral type, which has its onset between 10 and 18 years of age, and the late form, which may make its appearance in adult life, often progress so slowly that the patient may well live throughout a virtually normal life expectancy and succumb to some unrelated illness. The clinical, genetic, and anatomic characteristics of five types of progressive muscular dystrophy and other conditions that must be considered in differential diagnosis are outlined in Table 22-2.

Table 22-1. Prevalence, incidence, and mutation rate of three clinical types of progressive muscular dystrophy*

Type of dystrophy	Prevalence	Incidence	Mutation rate
Duchenne	66 living cases/million males	279 males/million male births	89/million gametes
Limb-girdle			
Autosomal recessive cases	12 living cases/million	38/million births	31/million gametes
Sporadic cases	8 living cases/million	27/million births	31/million gametes
Facioscapulohumeral	2 living cases/million	4/million births	5/10 million gametes

*Based on computations of Chung and Morton: Amer. J. Hum. Genet. **11:**339 and 360, 1959. From Swinyard, C. A.: Progressive muscular dystrophy and atrophy and related conditions, Pediat. Clin. N. Amer. **7:**703, 1960.

Etiology. Although it is generally agreed that there is a hereditary basis in all types of progressive muscular dystrophy, little is known about the modus operandi of the defective gene in production of the disease. Most investigators believe it to be a biochemical defect, but as yet the nature of the defect is unknown.

It has long been known that in progressive muscular dystrophy there is increased urinary creatine and decreased secretion of creatinine. Similar alterations, however, are also found in other diseases characterized by muscle wasting. The decreased ouput of creatinine may be of greater significance.[5] The observation that there is a diminished creatine tolerance in muscular dystrophy[5] has not been confirmed.

In 1928 Evans and Burr[8] observed that laboratory rats deprived of dietary vitamin E develop degeneration of striated muscle and profound weakness. Subsequent work has shown that this so-called "nutritional dystrophy" can be experimentally produced in over twenty different species of animals. This work naturally led to a wave of hope that human muscular dystrophy might be caused by vitamin E deficiency. However, Mason and co-workers[9] found no deficit of vitamin E in dystrophic muscles. Administration of vitamin E to human beings with muscular dystrophy has no effect on the muscle degeneration and weakness.

Although it has been postulated, there is no evidence that the etiology resides in a disturbed carbohydrate metabolism.

One of the types of muscular dystrophy is commonly associated with testicular atrophy, baldness, and cataracts (myotonic dystrophy).[10] This has led to speculation that the etiology may be due to a hormonal defect or a hypothalamic lesion; however, there is no objective evidence for this theory.[6]

Denny-Brown[11] believes that when the nature of the primary defect becomes known, it may be a defective function of the muscle nucleus that interferes with the production of myofibrils rather than an inadequacy of the contractile properties of muscle.[12] It has recently been postulated that the connective tissue of dystrophic muscles synthesizes 5-nucleotidase, an enzyme that destroys muscle adenylic acid. Thus the etiology of the disease is a prime objective of future research.

Drug treatment of progressive muscular dystrophy. The discovery of the vitamin E nutritional dystrophy focused attention on the possibility of alleviating human hereditary muscular dystrophy with vitamin E. Many forms of vitamin E and related compounds have been administered for years without beneficial effect. Claims made for protein hydrolysates[13] and more recently[14, 15] for an anabolic steroid compound combined with a small amount of digitalis have not been substantiated. Despite the numerous drugs, hormones, and dietary regimens that have been tried over the past 25 years, we must still admit that there is, at this time, no drug available that will prevent, arrest, or modify the inexorable course of the disease.

Pathologic histology. Although there are minor differences in the histologic picture in the several varieties of dystrophy, only the general features of the degeneration will be briefly mentioned.

As the degenerative process advances, the striations become faint and the fibers become smaller, fragment, and disappear. There is a tremendous variation in size of the muscle fibers, with the large, small, and atrophic fibers scattered throughout the muscle. The muscle nuclei are increased in number and are more varied in shape. Frequently they appear to line up in chains in the center of the degenerating muscle fibers. Severely atrophic fibers show only chains of darkly staining nuclei.

When the muscle fibers disappear, quantities of fat and connective tissue replace them. In one of the types of dystrophy (pseudohypertrophic) the large fat deposits in the calf and shoulder muscles are characteristic and give the false appearance of extremely well-developed musculature, although the muscles are actually quite weak. There are records of cardiac muscle involvement in several types of dystrophy[16, 17] and resultant problems of cardiac insufficiency. The central and peripheral nervous systems are normal.

Hereditary aspects. The classic genetic studies of Bell,[18] Tyler and Stephens,[19-21] Bradburne,[22] and Fleischer[23] have firmly established that all types of muscular dystrophy are hereditary or heritable. The mechanism of hereditary transmission is different in the several types of dystrophy and will be mentioned in connection with the type under discussion.

Although clinical expression of the dystrophies may not occur until after years of apparently normal existence, recent studies have shown that there may be enzymatic and histologic changes in the muscles years before clinical weakness occurs.

The mechanism of inheritance of the clinical types of progressive muscular dystrophy is indicated in Table 22-2. One of the most important considerations of a rehabilitation program is patient and parental counseling with reference to prognosis and inheritance. Space precludes extensive discussion of these problems, but some of the factual circumstances that are related to the genetic transmission of the three most frequent clinical types of progressive muscular dystrophy are summarized below.

Duchenne or childhood type of dystrophy. This is a sex-linked recessive trait that is transmitted through females to males, and the vast majority of affected individuals are males. (See biochemical discussion for possible explanation for affected females and detection of female carriers.) A female sibling of an affected male has a 50% chance of being a carrier and will pass the defective gene to 50% of her sons. Affected males rarely marry.

Limb-girdle dystrophy. An autosomal recessive inheritance, this occurs equally in both sexes. It appears when two unaffected carriers of the recessive gene marry. One fourth of their children, regardless of sex, will possess the two recessive genes

Text continued on p. 428.

Table 22-2. Conditions that require consideration with reference to insidious onset of

Disease	Age at onset	Initial area of muscle weakness	Prognosis
Progressive muscular dystrophy			
Duchenne	2-10 yr.	Glutei, abdom., ant. tibial erector spinae	Poor, die before 25
Limb-girdle	10-40 yr.	Pectobrachial or pelvi-femoral groups	Poor, death middle age
Facioscapulohumeral	10-18 yr.	Shoulder, face, pectoralis maj., low trapezius	Fair, slow progress
Myotonic	15-80 yr.	Hand, forearms, sterno-cleidomastoid, masseters	Poor
Ophthalmoplegic	1-40 yr.	Extrinsic eye muscles	Good—only disabling
Progressive muscular atrophy			
Werdnig-Hoffmann	6-12 mo.	Trunk, shoulder, proximal limb	Poor, 5 yr.
Peroneal muscular atrophy	12-50 yr.	Extensor, evertor foot muscles	Fair, slow
Amyotonia congenita	Congenital	Universal	Fair, depends on severity
Benign congenital hypotonia			
Walton's	Congenital	Universal	Fair-good
Ataxia			
Friedreich's	6-15 yr.	Distal lower extremity	Poor
Marie's	After 20 yr.		Fair
Inflammatory disease of muscle			
Polymyositis			
Acute	Children	Shoulder and proximal ext.	Poor
Chronic	Adults	Lower extremity, upper	Poor
End-plate disease			
Myasthenia gravis	10-70 yr.	Ocular, facial, deglutition	

*From Swinyard, C. A.: Progressive muscular dystrophy and atrophy and related conditions, Pediat.

muscle weakness*

Associated defects or diagnostic points	Inheritance	Neurologic	
		Motor	Sensory and/or mentation
EMG rapid, low voltage; cardiac involvement	Sex-linked recessive	No abnormality	No abnormality
	Autosomal recessive	No abnormality	No abnormality
	Somatic dominant	No abnormality	No abnormality
Cataracts, gonadal atrophy, myotonia	Somatic dominant	Myotonic EMG pattern	No abnormality
Myopathic EMG pattern	Somatic dominant	Myopathic EMG	No abnormality
Neurogenic EMG	Recessive	Lower motor neuron	None
Neurogenic EMG	Dominant or sex-linked recessive	Lower motor neuron	Peripheral neuropathy sensory deficit common
30% tongue fibrillation, neurogenic EMG	Occ. dominant	Lower motor neuron	None
Myopathic EMG		No abnormality	No abnormality
Pes cavus, scoliosis, nystagmus	Recessive or dominant	Dim. reflexes in low. ext.	Proprioceptive loss, vibratory loss
Involves olive, pons, and cerebellum	Dominant or recessive	Hyperactive	
Fever and pain frequent		No abnormality	No abnormality
Skin lesions; muscle tender		No abnormality	No abnormality
Tensilon test, EMG		Muscle weakness only	No abnormality

Continued.

Table 22-2. Conditions that require consideration with reference to insidious onset of

Disease	Age at onset	Initial area of muscle weakness	Prognosis
Demyelinating diseases Schilder's disease	Birth-senil.	Neck and extremities	Poor—var. wks.-yr.
Krabbe's disease	4-6 mo.	Dysphagia, then generalized	Poor—death 1 yr.
Multiple sclerosis	15-50 yr.	Variable	Poor
Amyotrophic lateral sclerosis	20-50 yr.	Variable	Poor—5 yr.
Metabolic disorders of nervous system Lipoidoses Cerebromacular degeneration (Tay-Sachs)	3-6 mo.	Neck, trunk, extremities; becomes universal	Poor—2-4 yr.
Cerebromacular degeneration (Spielmeyer-Vogt)	4-18 yr.	Universal	Slow—10-15 yr.
Niemann-Pick disease	Same as Tay-Sachs except for hepatosplenomegaly and skin discoloration; defect of sphingomyelin metabolism		
Hurler's disease (gargoylism)	Defect of phospholipid metabolism; impaired vision; dwarfed; hepatosplenomegaly; mental retardation; generalized weakness		
Gaucher's disease	Defect of keratin metabolism; hepatosplenomegaly; rapid paralysis when nervous system is involved; dysphagia and laryngospasm		
Carbohydrate Galactosemia	Absent P Gal transferase activity; generalized weakness		
Glycogen storage disease	Deposit of glycogen in muscles and nerve cells; generalized weakness		
Maple sugar disease	Defective metabolism of valine, leucine, and isoleucine; retarded motor development		
Protein Phenylketonuria	Deficient phenylalanine hydroxylase; mental retardation; delayed motor development		
Amino-aciduria	Hypoamino-acidemia and hyperamino-aciduria; delayed motor development		
Hormonal disorders Thyrotoxic dystrophy	Pectoral and pelvic girdle muscle weakness; disappears in euthyroid state		
Menopausal dystrophy	Pectoral and pelvic girdle muscle weakness; responds to cortisone administration		

muscle weakness—cont'd

Associated defects or diagnostic points	Inheritance	Neurologic	
		Motor	*Sensory and/or mentation*
	Familial	Spasticity, athetosis, ataxia	Cortical blindness, hearing loss, dementia
	Familial	Rigidity/flaccidity	
Paresthesia, diplopia, vertigo	Not of significance	Spasticity variable	Impaired
Fasciculation	Not hereditary	Upper and lower motor neuron	No abnormality
Prog. blindness, dementia, paralysis, retinal cherry-red spot	Recessive	Flaccidity, spasticity	Retinal blindness
Macular brown spot	Dominant or recessive	Flaccidity, spasticity	Retinal blindness

necessary for the disease and will have the disease. Consanguinity increases the chances of occurrence of the disease.

Facioscapulohumeral dystrophy. This autosomal dominant inheritance occurs equally in both sexes. An affected child must have an affected parent (unless it is a new mutation). The affected parent and the affected child will each pass the disease to one half of his children of either sex.

Review of biochemical abnormalities in progressive muscular dystrophy. Although the early hope for diagnostic and prognostic assistance from study of creatine and creatinine excretion did not reach initial expectations, studies of muscle enzymes that appear in the blood serum of patients have contributed significantly to our understanding of the natural history, preclinical status of the patient, and genetic aspects of progressive muscular dystrophy. In recent years much attention has been devoted to altered serum levels of a large variety of intramuscular fiber enzymes that are liberated into the serum from abnormally permeable sarcolemmal membranes or from degenerating muscle fibers. A few of the significant studies of this subject will be reviewed briefly.

Muscle tissue enzymes in blood serum. In 1949 Sibly and Lehninger[24] directed attention to the occurrence of serum aldolase in serum. Elevation of serum aldolase in progressive muscular dystrophy was described by Shapira and co-workers[25] and later by Pearson.[26] Their observations precipitated a large number of investigations on the serum levels of other muscle enzymes in progressive muscular dystrophy. These studies have been reviewed by Dreyfus and Shapira[27] and by Bourne.[28] Pearson[26] found transaminase activity always elevated in pseudohypertrophic dystrophy and suggested that this enzyme evaluation might have merit in evaluating therapeutic procedures. On the other hand, Conrad[29] found transaminase elevated in 92% of a series of patients with myocardial infarction and 50% of the patients with acute rheumatic carditis. He felt that transaminase had no diagnostic value. Ratner and Sacks[30] found transaminase elevated in muscle-splitting operations.

In 1959 Ebashi and associates[31] called attention to elevated creatine phosphokinase (CPK) activity in the serum of patients with progressive muscular dystrophy. This was confirmed by Dreyfus and co-workers,[32] by Okinaka and associates,[33] and by others. Among all the muscle enzymes found in the serum of patients with progressive muscular dystrophy, it appears that aldolase and CPK are the most reliable and sensitive indicators of muscle degeneration.[34] The degree of activity of these enzymes in the serum is proportional to the rapidity and amount of muscle fiber breakdown. The enzyme studies are useful as an adjunct to clinical findings and help to determine the rate of muscle degeneration. These enzymes are elevated in the infantile atrophies also, but the level of activity is generally less. It is likely that the principal value of the enzyme determinations resides in the possibility of detection of the preclinical and carrier states of progressive muscular dystrophy.

Biochemical and histologic aspects of preclinical status of dystrophy. Dreyfus and associates[35] and Pearson[36] were the first to note that some healthy appearing siblings of children with Duchenne dystrophy had elevated serum aldolase and CPK levels. In one of the kindred studied by Pearson, some of the unaffected boys with elevated aldolase and CPK levels developed clinical signs of Duchenne dystrophy in 3 to 6 years. This significant observation suggests that the preclinical stage of dystrophy may last several years. For another kindred with benign form of Duchenne dystrophy studied by Pearson, siblings with elevated enzymes were found who did not develop clinical symptoms. These observations lend support to the concept of a benign type of Duchenne dystrophy. In a more recent investigation, Pearson[37] made histologic studies of muscle biopsy specimens from the clinically unaffected siblings of patients with Duchenne dystrophy and found histopathologic changes characteristic of very early dystrophy. Thus the biochemical and histologic studies indicate that there may be muscle degeneration for years before the muscle changes are reflected in evidence of clinical weakness.

Biochemical detection of carrier state in Duchenne dystrophy. In 1961 Okinaka and co-workers[33] observed elevated serum CPK activity in sisters of patients with Duchenne dystrophy. This observation was confirmed by Chung and colleagues[38] and by Shapira and co-workers.[39] Pearson[34] studied creatine phosphokinase activity in the sera of many mothers and sisters of Duchenne dystrophy males. He found CPK levels to be the most sensitive indicator of the carrier state and found elevated values in approximately 50% of the mothers and female siblings of affected boys.

These studies have important bearing on the existence of a significant number of females we have seen who appear to have Duchenne dystrophy as described by Dubowitz.[40] Until very recently, our understanding of the genetic transmission of sex-linked, male-affected diseases was that the females should not manifest any evidence of the disease because the influence of the affected female X chromosome was believed to be suppressed by the other normal X chromosome. However, recent studies by Pearson[37] indicate that, in addition to the elevated CPK enzyme, the mothers and carrier sisters of Duchenne dystrophic males also have focal histopathologic changes of early dystrophy in their muscles.

It is possible that these apparent conflicts with established genetic concepts can be explained by the recent suggestions of the British geneticist, Mary Lyon.[41, 42] Lyon believes that the "Barr body" which exists in one half of the somatic cells of females is in fact a hyperpyknotic X chromosome that has been inactivated. She believes that at each cell division one of the X chromosomes of the female is inactivated, in a random manner. The Lyon hypothesis, therefore, postulates that, with respect to the X chromosome in somatic cells, females are genetic mosaics. In other words, one half of the female's somatic cells have an X chromosome derived from the mother and one half from the father. If the female is a carrier of Duchenne dystrophy, one half of her muscle fibers may contain active X chromosomes that carry the defective gene. Pearson[34] and his associates[43] have suggested that if, in the random inactivation of chromosomes, the female possesses a larger number of active X chromosomes with the abnormal gene, she will have biochemical, histologic, and clinical evidence of Duchenne dystrophy. If, on the other hand, she receives a disproportionate number of active X chromosomes without the abnormal gene, she will be a carrier without even biochemical or histologic suggestions of dystrophy. Pearson's application of the Lyon hypothesis to the explanation of female examples of Duchenne type dystrophy has also been advanced by Emery,[44] with supporting biochemical and histologic data.

Isoenzyme studies in progressive muscular dystrophy. One of the most recent advances in biochemical research in human progressive muscular dystrophy has been derived from the observation that enzymes may exist in multiple forms in the same tissue (isoenzymes). One of the enzymes most widely studied in connection with its isoenzyme pattern in progressive muscular dystrophy is lactate dehydrogenase (LDH). It has been shown that LDH exists in five different molecular varieties in man.[45] By use of an electrophoretic mobility technic, Weime[46] described five types of LDH (LD1E to LD5E). In some preliminary studies Pearson[34] found an increase in serum in total LDH activity with an increase of the LD1 fraction and a loss of the LD5 fraction. Kaplan and Cahn[47] found a similar change in the isoenzyme pattern of LDH in hereditary muscular dystrophy of chickens.

These most intricate biochemical studies are being pursued by an increasing number of competent investigators and strengthen the hope that the nature of the fundamental biochemical abnormality in dystrophy will eventually be found.

Following a brief discussion of the progressive muscular atrophies, the problems of rehabilitation of patients with progressive muscular dystrophy and the several types of progressive muscular atrophy will be considered.

PROGRESSIVE MUSCULAR ATROPHY

The progressive muscular atrophies are neurologic diseases in which there is failure of development or progressive degeneration of the lower motor neuron cell body and resultant degeneration of the motor peripheral nerve fibers with neurogenic atrophy and paresis of the muscles denervated. The etiology of all the muscular atrophies is as much a mystery as is that of the primary muscular dystrophies previously discussed. There are hereditary aspects in these diseases, and the onset may occur at almost any age. However, unlike muscular dystrophy, in many of these cases individuals show clinical evidence of muscle weakness at or shortly after birth.

Lower motor neuron involvement

Werdnig-Hoffmann disease. This syndrome was described in 1891[48] and 1893[49] as a progressive muscular atrophy whose symptoms do not appear until the sixth to twelfth month of postnatal life, at which time the paralysis rapidly progresses from trunk involvement to shoulder and pelvic girdle and then to the proximal limb muscles. The distal limb and facial muscles are spared, but death usually occurs within 5 years, from weakness of the muscles of respiration and intercurrent infection. The biopsy, autopsy, and electrodiagnostic studies clearly indicate this disease to be a lower motor neuron denervation of the affected muscles.[50] Brandt[51, 52] found the disease to be inherited as a recessive trait or as a dominant trait with low penetration. There is no sex linkage.

Amyotonia congenita (Oppenheim). There is so much uncertainty concerning the identity of this syndrome that one scarcely knows how to designate the condition. For this reason it is essential to consider the nature and status of this condition as a clinical entity. In the United States the appellation amyotonia congenita is commonly applied to a congenital syndrome consisting of almost universal muscle weakness and hypotonia, coupled with diminished to absent deep tendon reflexes and hypermobility of joints, presumably due to lower motor neuron degeneration. The condition is generally believed either to be static or to be subject to gradual improvement. The favorable prognosis was emphasized by Oppenheim[53] in 1900 in his original description of 3 cases of this syndrome.

Shortly after its description, speculation developed regarding the relationship of this condition to progressive infantile atrophy that had been previously described by Werdnig-Hoffmann. The latter condition is a lower motor neuron disease of unknown etiology. The former was described as manifesting itself either at birth or during the first half of the initial postnatal years and as improving gradually; the latter was described as appearing during the latter half of the first postnatal year and as progressing rapidly to a fatal termination.

Numerous etiologic factors will produce hypotonia and hypermobility of joints. Turner[54, 55] concluded that the clinical picture of amyotonia congenita can be given in the early stages of a primary myopathy. We have found that in a muscular dystrophy clinic a number of cases of amyotonia congenita are labeled progressive muscular dystrophy. Ordinarily a careful history will indicate a later onset and a definite anatomic localization and sequence of progression of the muscular weakness in dystrophy, in contrast to the universal muscle weakness of amyotonia congenita.[53] The dystrophies are also characterized electromyographically by frequent low-voltage motor units, in contrast to the fibrillation potentials, prolonged insertion potentials, and positive sharp waves of lower motor neuron disease.[56] Muscle biopsy is also a valuable adjunct in separating the dystrophies from the atrophies.[50] The great rarity of progressive muscular dystrophy of the childhood type in female patients and lack of sex linkage in the hereditary cases of progressive muscular atrophy and amyotonia congenita are also worthy of consideration.[19, 21, 52] Other clinical syndromes characterized by universal muscular hypotonia, such as glycogen storage diseases,[57, 58] cerebellar disease, cerebral dysgenesis, and other conditions, must be considered.[59] Hypotonia with normal muscle power and electric activity of the muscles has been described.[60] The feature of hypermobility of joints without muscle weakness has been described as a sex-linked hereditary characteristic,[61] and similar hypermobility of joints due to overlengthened capsular and ligamentous tissues was described by Sutro.[62] Universal hypotonia has also been reported in association with congenital heart disease,[63] without commitment as to the relationship, if any, of the two conditions.

The relationship between amyotonia congenita and Werdnig-Hoffmann disease is important and confused. In 1927 Greenfield and Stern[64] concluded from necropsy findings that these two syndromes are but variants of a single pathologic process. Brandt[51] studied the hereditary aspects of seventy families exhibiting 112 cases of infantile progressive muscular atrophy and found a recessive type of transmission with perhaps two or more coacting genes modifying the severity and rate of progression of the disease. Brandt[65] later followed the prognosis of the same 112 patients and found that 80% died before the age of 4 years. Brandt finally encompassed his years of study of these syndromes in a masterful monograph[52] and pointed out that most of the cases that have been designated as amyotonia congenita eventually progress to a fatal termination. He observed that even the 3 cases initially described by Oppenheim,[53] which gave rise to the original concepts of this syndrome as being of benign nature, finally terminated fatally, and necropsy study of the cases by Oppenheim's students revealed extensive lower motor neuron degeneration.

Walton[66, 67] made a follow-up study on 109 patients in whom the diagnosis of amyotonia congenita had been made. In 42 of the 109, either the condition was static or the patients experienced variable degrees of recovery (8 complete). In 67 of the 109 cases the disease showed evidence of progression, and 55 patients died between the fifth week and the twelfth year. Twelve patients were surviving at the ages of 4 to 20 years, with severe disability, at the time of the survey (1956). Both Brandt and Walton have suggested that the term amyotonia congenita be discarded; however, the term is still so widely used in this country that we have elected to use it. It is our belief that there are cases of this syndrome that are benign in the sense that they may be static rather than progressive.

Buchthal and Clemmesen[68, 69] made one of the early electromyographic studies of myogenic and neurogenic atrophy. They found maximal contraction to be accompanied by interfering activity of many motor units. In slight muscle contraction, interference from neighboring units also appeared and single oscillations were difficult to produce. Neurogenic atrophy was characterized by synchronous activity of different motor units. Brandt,[65] using the technic of Buchthal and Clemmesen, electromyographically studied 19 patients with a diagnosis of progressive infantile atrophy. In 10 out of 11 patients the data indicated neurogenic atrophy. A normal electromyogram was found in 1 patient whose biopsy indicated a myopathic lesion. In 6 patients the diagnosis of spinal atrophy could not be confirmed by electromyography.

In 17 of Walton's[66, 67] patients who were designated as having benign congenital hypotonia with *complete recovery,* electric stimulation was normal in 5, although 2 of these had shown excessive polyphasic potentials with volition during infancy. Short-duration potentials with volition were found in 3 patients. Although the electromyographs were "myopathic," the muscle biopsies were normal. On the other hand, of Walton's 67 patients with progressive spinal atrophy, 23 were studied with electric stimulation, and loss of faradic response was found. In 5 patients in whom strength-duration curves were obtained, incomplete to complete denervation was indicated. Four out of 10 patients studied electromyographically showed spontaneous fibrillation. The interference pattern of voluntary maximal effort was reduced in all patients, but the surviving motor units were either normal or above average in amplitude and duration. Biopsy specimens in 10 cases revealed denervation atrophy, and spinal cord study in 5 necropsy cases showed a significant reduction in the number of anterior horn cells.

In cases that were considered under this syndrome, all were characterized by either congenital symmetric muscle hypotonia and weakness or the detection of these phenomena during the first 6 months of life. The deep reflexes were either greatly diminished or absent. In the great majority of patients careful study of the tongue revealed visible fibrillation. Although tongue fibrillation has been previously observed in this syndrome,[52] we believe this clinical sign has not been studied as carefully as it should be. It is positive evidence of hypoglossal nuclear involvement and lends strong suspicion to the belief that the marked weakness of the trunk and extremities is on the same basis. Greenfield and Stern[64] observed diminution of hypoglossal nuclear cells in all of the necropsy cases they studied. The cases we have considered as amyotonia congenita have marked disability and apparently static neurologic state. In virtually all patients the tongue revealed visible fibrillation, and fibrillation potentials were recorded from skeletal muscle in approximately 25% of the patients.

Peroneal muscular atrophy (Charcot-Marie-Tooth disease). This syndrome was simultaneously described in 1886 by Charcot and Marie[70] and Tooth.[71] It is a progressive muscular atrophy that has its onset anytime from 12 to 50 years of age. It is ordinarily initiated with symmetric atrophy and weakness of the extensor and evertor muscles of the foot and results in equinovarus foot deformity. In time most of the leg and foot muscles become involved, and weakness of the intrinsic muscles of the hand and forearm follows. In some instances the upper extremity may be involved before the lower extremity. The face, trunk, and proximal limb muscles are usually spared. There may be paresthesias and sensory loss.[72]

Postmortem study of the spinal cord clearly reveals loss of anterior horn cells and degeneration of peripheral nerve fibers. There may also be loss of cells in dorsal root ganglia and posterior columns. Study of the muscle pathology has been in some dispute. Wohlfahrt and Wohlfahrt[73] described cases in which the muscle changes were dystrophic. Adams and co-workers[12] believe they are characteristically atrophic. There appears to be little evidence that this syndrome is in any way a dystrophy. It is inherited in several patterns, as a dominant, as a sex-linked recessive, or as a simple recessive.[72, 74]

Combination of upper and lower motor neuron disease

Amyotrophic lateral sclerosis. This disease is so protean in its variable manifestations that it frequently becomes a consideration in the diagnosis of later-appearing types of muscular dystrophy. The onset varies from 20 to 50 years of age. It is not hereditary.

The lower motor neuron atrophy of either spinal or cranial musculature may precede, follow, or occur simultaneously with neurologic signs of upper motor neuron involvement. Sensation remains intact. The disease is progressive, and death results after about a 5-year course of progressive atrophy and paresis of the muscles of deglutition and respiration.

Affected muscles fasciculate (involuntary twitch) markedly. The neurologic lesions are in the anterior horns of the spinal cord, the nuclei of motor cranial nerves, and the cells of the motor cerebral cortex. The muscle pathology is characteristically that of denervated striated muscle.[75]

REHABILITATION OF PATIENTS WITH PROGRESSIVE MUSCULAR DYSTROPHY AND ATROPHY

Unfortunately there is a fairly widespread opinion in the medical profession that a diagnosis of progressive muscular dystrophy is a death sentence and that it is futile to attempt any rehabilitation procedures. This defeatist attitude stems largely from a knowledge of the rapid progression of the childhood type of progressive muscular dystrophy and the Werdnig-Hoffmann type of progressive muscular atrophy. In the former the majority of patients die before the age of 20 years, and in the latter death usually occurs from 5 to 10 years after onset. On the other hand, it should be appreciated that in facioscapulohumeral progressive muscular dystrophy and amyotonia congenita there may be severe disability but virtually a normal longevity. These patients should keep muscular efficiency at as high a level of efficiency as the progressive weakness will permit, and both they and their families require constant psychosocial assistance. During the changing phases of the illness and as the muscle weakness involves successive groups of muscles, the rehabilitation requirements change drastically. For example, in facioscapulohumeral dystrophy, with the increasing weakness initially in the face, shoulders, and arms, problems of self-care are early problems, and ambulatory difficulties arise later. On the contrary, in childhood dystrophy, which has inital muscular weakness in the trunk and lower extremities, deficiencies in ambulation are early serious problems. The universal weakness of amyotonia congenita presents combinations of both problems. In myotonic dystrophy the weakness in muscles of deglutition, respiration, and mastication may lead to death from intercurrent infection while the patient is still able to ambulate and is moderately efficient in his self-care needs.

These variations in prognosis are illustrated by a series of eight functional stages of ability, based on the chronologic change in the type and method of ambulation.[76, 77] The criteria used for separating the stages of disability are listed as follows:

> *Stage 1*—ambulates with mild waddling gait and lordosis; elevation activities adequate (climbs stairs and curbs without assistance)
> *Stage 2*—ambulates with moderate waddling gait and lordosis; elevation activities deficient (needs support for curbs and stairs)
> *Stage 3*—ambulates with severe waddling gait and marked lordosis; cannot negotiate curbs or stairs but can achieve erect posture from standard height chair
> *Stage 4*—ambulates with very severe waddling gait and marked lordosis; unable to rise from standard height chair
> *Stage 5*—wheelchair independence: good posture in chair; can perform all Activities of Daily Living from chair
> *Stage 6*—Wheelchair with dependence: can roll chair but needs assistance in bed and wheelchair activities
> *Stage 7*—wheelchair with dependence and back support: can roll chair only short distance; needs back support for good chair position
> *Stage 8*—bed patient: can do no Activities of Daily Living without maximum assistance

In the first three stages of progressive disability the patient ambulates independently with disturbed pattern primarily because of weak abdominal and gluteal muscles (lordosis and bilateral Trendelenburg gait).

In the fourth stage a critical point is reached in which the patient can no longer push himself erect from the sitting posture but can ambulate if assisted in getting into an erect position. Such a patient can be handled in any one of the following ways: (1) someone lifts him to a standing position and then he ambulates independently; (2) he is supplied with a wheelchair with a hydraulic lift and, by elevating the chair seat, he can become sufficiently erect to begin ambulating independently; or (3) he is placed in a wheelchair and is taught all Activities of Daily Living, except ambulation, from the wheelchair.

There is an unfortunate tendency to allow patients in this (fourth) stage to become bed patients, whereas if they are properly managed they can learn to transfer from the wheelchair to the bed and toilet and reverse, and thus be independent for a significantly longer period of time. If the case is one of childhood dystrophy with rapid progression, this independent wheelchair stage may last only from 2 to 3 years; on the other hand, this stage may last from 10 to 15 years or more in a patient with facioscapulohumeral dystrophy or amyotonia congenita.

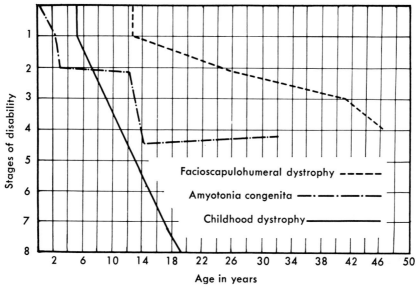

Fig. 22-1. Changes in functional ability as related to age in patients with childhood dystrophy, facioscapulohumeral progressive muscular dystrophy, and infantile progressive muscular atrophy (amyotonia congenita).

The patient in the wheelchair (stages 5 to 7) becomes gradually less efficient in performing Activities of Daily Living as the illness progresses. In stage 7 the weakness has involved the trunk to such a degree that the patient cannot maintain a satisfactory wheelchair position without the assistance of a back brace. Often the improved chair posture provided by the brace will enable the patient to handle the chair with much greater efficiency. It is essential that the wheelchair patient (stages 5 to 7) and the bed patient (stage 8) be given the physiologic and psychologic value of daily erect posture by use either of appropriate braces or of a tilt table.

The use of these criteria to visualize prognosis with reference to ambulation is illustrated in Fig. 22-1, in which the stage of ambulatory ability is plotted on the ordinate and the age at which the patient reached this stage is represented on the abscissa, for two types of progressive muscular dystrophy and amyotonia congenita.

It is apparent from examination of the longitudinal perspective of the two types of progressive muscular dystrophy and a case of amyotonia congenita (Fig. 22-1) that there is a striking difference in the rehabilitation potential of patients with childhood dystrophy and those with facioscapulohumeral dystrophy and amyotonia congenita.

These data emphasize the necessity of making an accurate diagnosis of the clinical type of progressive muscular dystrophy when undertaking an objective study of the effectiveness of rehabilitation procedures.[78, 79] The ambulatory status of the patient with amyotonia congenita (Fig. 22-1) illustrates the great difficulty of overcoming the profound disuse atrophy that accompanies immobilization of such a patient. This patient was ambulating at the age of 14 years, when the back was surgically fused to correct an increasing scoliosis. This procedure required nearly

14 months of immobilization with a body cast. After the cast was removed, the patient could not ambulate, and prolonged physical therapy was required to restore enough power for ambulation.

Weight control is essential to the success of a rehabilitation program for patients with progressive muscular dystrophy and atrophy. Many of the patients with amyotonia have a history of a short period of ambulation, which is lost after a few months. Likewise, some of these patients are quite independent in wheelchair activities when weight is normal and lose these functions as obesity appears. This is because the universal muscle weakness of these patients is at a critical level; that is, there is just enough power to perform a certain activity. The additional weight, due to obesity of the body part concerned, makes the activity impossible with the constant level of muscle power. We believe that in many instances the progressive inability to perform certain activities is sometimes erroneously interpreted as neurologic progression of the disease, whereas it actually is related to the increasing work load in the presence of static marginal muscle power.

It is tempting to extend the curves of Fig. 22-1 beyond the present age by assuming that the current rate of change will continue, and thus to use points beyond for predictive purposes. This should not be done because a patient with myotonic progressive muscular dystrophy might be ambulatory in stage 3, but an acceleration of degeneration in the muscles of respiration and deglutition might result in death from intercurrent infection.

The effect of a physical therapy program in progressive muscular dystrophy has been studied by Abramson and Rogoff[78] and by Hoberman.[79] Abramson treated 27 patients for a period of 7 months and found improvement in muscle strength, reduction of contractures, and increased performance of the Activities of Daily Living. Hoberman,[79] on the contrary, concluded that physical therapy did not increase actual muscle strength; however, most of the patients improved their performance in Activities of Daily Living.

In our experience a physical therapy program can keep muscle power at an optimal level. However, usually the exercise derived from ambulation is adequate, and prolonged programs of physical therapy are not warranted. Without active exercise the patient becomes dependent on a wheelchair sooner than is necessary, and disuse weakness contributes significantly to the overall disability. The objective is to keep the patient functioning as efficiently as possible in the psychosocial, physical, and vocational areas during the changing phases of his illness.

The constantly changing status of the physical disability requires the utmost attention of the rehabilitation team to provide appropriate wheelchairs, back supports, braces, and various types of adaptive equipment to meet the patient's individual needs. A description of this equipment and indications for its use are discussed in Chapters 7, 8, and 12.

There are tremendous psychosocial problems with the patients, parents, and families when there is progressive muscular dystrophy and atrophy. The problems of social and family adjustment have been discussed by Truitt[80] and by Austin.[81] Wexler[82] and Bryan[83] have discussed the importance of the psychologic problems in progressive muscular dystrophy.

One of the most trying rehabilitation problems revolves around the patient with the rapidly progressive childhood type of muscular dystrophy. One is faced not only with the ever-changing requirements of therapy and mechanical assistive devices but also with the anxieties of the patient and parents as they try to fight back against a hereditary disease, of unknown etiology, without a medical treatment program that will, in any way, alleviate or alter the irrevocable course of the disease. Despite this dismal outlook we believe that if appropriate effort is directed to all aspects of the rehabilitation program, there can be brought to these children opportunities and attitudes toward life such as have been so aptly expressed by Wexler.[82]

> This little girl lived her life graciously, fully, meaningfully. She was busy in her own way tasting deeply of each passing moment, profoundly secure in the love of those around her; rich beyond belief with an inner richness that few of us will ever get to know. Before she died she lived.*

Thus it is possible by meticulous attention to the many facets of the rehabilitation program to enable patients with progressive muscular dystrophy and atrophy to live happily and meaningfully within the changing limitations of their disease.

ARTHROGRYPOSIS

In 1841 Otto[84] described a 7-month stillborn that came to his attention in the following terms, "monstrum humanum trunco nemes brevi et extremitalibris incurvatus." The fetus had bilateral flexion contractures of the joints of upper extremities. He believed that this condition was a muscle dystrophy and applied the designation congenital myodystrophy.

This relatively rare syndrome of unknown etiology has subsequently been described under at least ten different names. Following is a list of the authors, dates, and the terms used to designate the syndrome:

1841 Otto[84]	Congenital myodystrophy
1897 Schanz[85]	Multiple congenital contractures
1903 Magnus[86]	Multiple congenital contractures with muscle defects
1908 Howard[88]	Dystrophia muscularis congenita
1912 Nové-Josserand and Rendu[87]	Multiple congenital articular rigidity
1923 Stern[89]	Arthrogryposis multiplex congenita
1932 Sheldon[90]	Amyoplasia congenita
1934 Middleton[91]	Myodystrophia foetalis deformans
1947 Dalmain[92]	Myophagism
1947 Rossi[93]	Congenital arthromyodysplasia

There are also numerous minor permutations of terms used by other authors; however, these will suffice for our purposes. It will be noted from the names applied that the designations in some cases represent effort to indicate the pathogenesis (amyoplasia congenita), others to indicate the visible orthopedic deformity (multiple congenital contractures), and one (arthrogryposis), a peculiar etymologic combination of Greek *(gryposis)* and Latin *(arthro),* to refer to flexion contractures.

---

*From Wexler, M.: Mental hygiene and the muscular dystrophy patient, Proceedings of the First Medical Conference of the Muscular Dystrophy Associations of America, Inc., New York, 1951, pp. 58-67.

Although the syndrome was described in 1841 and many cases were described in the subsequent 82 years, the syndrome was rediscovered in 1923[89] and the inappropriate term arthrogryposis was applied. Most of the reports in the American literature have used the term arthrogryposis; however, because many of the patients do not have bent (flexed) extremities but have straight extremities (contractures in extension), because the term is etymologically inadequate, and, finally, because this term is occasionally used in the pediatric literature of England and the Continent to refer to tetany (carpopedal spasm), we suggest that it be abandoned. James[94] has expressed a preference for the term of Nové-Josserand and Rendu,[87] multiple congenital articular rigiditus; however to avoid any confusion with the neurologic entity of rigidity, the term contracture is used here.

Since the functional disability is obviously of orthopedic nature, most of the earlier case reports appeared in orthopedic journals; the surgeons were impressed with the thickening of the articular capsule of the joints and have therefore felt that the primary pathology resides in the connective tissue.[95] This view has been emphasized by McKusick.[96] On the other hand, a number of the English authors have been impressed with the large amount of connective tissue and fat that occupies the site of the muscle and think of the disease as primarily a failure of myoblasts to differentiate muscle—hence, the term amyoplasia congenita.[90, 91] Finally, the suggestion has been made that the fibrous connective tissue and fat are replacements of muscle fibers that have degenerated secondary to lower motor neuron degeneration.[97-99]

This syndrome is therefore characterized by congenital, painless, static multiple contractures. The skeletal system is without significant defect. In the case originally described there were flexion contractures of all extremities, and numerous subsequent cases have been described in which there were almost any combinations of upper and/or lower extremity involvement.

Some cases have been described in which there was only lower extremity involvement. One might raise the question regarding the extent of the involvement required to fit into this ill-defined syndrome. If the cases are properly labeled as multiple congenital contractures, what is the dividing line between this and bilateral congenital clubfoot, for bilateral clubfoot is in essence a multiple congenital contracture. This relationship has been discussed by Rosenkranz.[102] There are now probably 500 cases of the syndrome mentioned in the literature. Most of the reports described 1 or 2 cases; the more extensive ones describe 8 to 10 cases.

Our studies of this problem[99] have convinced us that this condition is not a clinical entity but a syndrome residual to a variety of prenatal etiologic factors, most of which are but incompletely known. We have studied 11 patients electromyographically and found potentials characteristic for neurogenic muscle weakness in the majority of the patients.[99, 100]

It has been said that arthrogryposis is not hereditary[94]; however, Hillman and Johnson[103] observed the syndrome in one member of each of two pairs of identical male twins. These cases were interpreted as evidence that the condition is not hereditary. On the other hand, Lipton and Morgenstern[104] reported involvement of both members of a pair of identical twins and consider this as evidence of genetic transmission. We have reported evidence of hereditary transmission in three kindred. In one kindred, 8 cases occurred in three generations of the family.[101] We have had the opportunity of examining over 70 patients with this syndrome and have been impressed with clinical evidence for diverse etiologic factors operative in the prenatal period that may alter the range of motion of the joints at birth. Some of these factors have been mentioned in the discussion of amyotonia congenita and have been presented in detail elsewhere[99]; some of the clinical conditions associated with congenital altered range of motion that present problems in rehabilitation are summarized as follows:

A. Restricted range of motion
    1. Multiple congenital contractures (arthrogryposis)
        a. Neurogenic: evidence derived from autopsy and electromyography
            Lower motor neuron: encephalitis and myelitis of variable etiology[97, 98, 100]
        b. Myogenic:
            (1) Amyoplasia congenita—failure of myoblast differentiation[91]
            (2) Congenital progressive muscular dystrophy[106]
        c. Collagenic: congenitally short capsular ligaments; Hurler's syndrome[105]
    2. Contractures limited to one or both ankle and foot joints (congenital clubfoot)
        a. Neurogenic: localized lower motor neuron defect
        b. Myogenic: embryonic malformation of muscle
        c. Collagenic: fibrosis and tight joint ligaments
        d. Postural: postural defects in utero
        e. Mechanical: malpositioned tendon attachment
        f. Miscellaneous: evolutionary variant; embryologic defect; nutritional
B. Excessive range of motion
    1. Neurogenic
        a. Benign congenital hypotonia (amyotonia congenita [may be replaced by contracture later])
        b. Progressive muscular atrophy (Werdnig-Hoffmann): range of motion may be replaced with contracture
    2. Myogenic: congenital progressive muscular dystrophy
    3. Collagenic: hereditary congenitally long joint ligaments; so-called double jointed,[62] Ehlers-Danlos syndrome, Marfan syndrome[105]

Brandt[97] made a careful postmortem study of the spinal cord of a case of congenital multiple contractures (arthrogryposis) and found a diminished number of ventral horn cells. Our kindred exhibiting hereditary arthrogryposis revealed electromyographic evidence of muscle denervation.[99] On the other hand, other autopsy studies have found the nervous system to be intact and many of the muscles to be virtually absent.[91] Thus some cases may be neurogenic and others myogenic. We have made a detailed study of the available pathoanatomic data in the literature and believe that the majority of the cases of arthrogryposis are neurogenic in origin. The evidence for a collagenic basis is equivocal. The thickened joint capsule may be secondary to joint mobilizatiomn. Hereditary overlengthened joint ligaments have been described,[61, 62] and the reverse condition is quite possible. Rare cases of congenital contracture incident to progressive muscular dystrophy have been described,[106] and we have observed identical twins with one member presenting a syndrome of amyotonia congenita and the other member arthrogryposis. As growth and development go on, many children born with the universal muscle weakness and hypermobility of joints develop contractures in the absence of proper care.

The differentiation of multiple congenital contracture (arthrogryposis) from congenital idiopathic clubfoot is in many cases not marked, and there may actually be similar etiology and only slightly variable degrees of severity. The same etiologic factors have been postulated,[107, 108] and Stewart[109] has made an eloquent plea for additional objective anatomic data with reference to clubfoot. It is our opinion that the same diverse, and perhaps additional, factors that cause contracture in postnatal life are operative during intrauterine life and result in variable degrees of contracture in a variety of joints in positions dictated by mechanical imbalance resulting from the operative factor.[99] We have been impressed by the frequency of associated congenital defects in other organ systems and the relatively high stillbirth rate in this condition.

The surviving children pose some of the most difficult problems in rehabilitation. Simple therapeutic exercises rarely result in effective muscle power, and manual stretching is generally futile.

Correction of the deformity should be done by conservative means if it can be done effectively and efficiently by this means. As in idiopathic congenital clubfoot, conservative measures should be applied early, for the contractures become more firmly fixed with the passage of time. However, the urgency scarcely justifies application of devices to the foot of a breech presentation before delivery of the aftercoming head, as has somewhat facetiously been advocated. If correction is achieved by casting, it is frequently necessary, later, to hold the position with the use of braces. The type of brace is indicated by the site and number of joints whose motion must be controlled. (See Chapter 12.)

Usually multiple congenital contracture does not respond to conservative measures alone, and surgical procedures are essential. The clubfoot deformity, for example, responds slower and less completely than does the so-called conventional clubfoot, and surgical intervention is more often necessary. Plaster wedging is often followed by tendo calcaneus lengthening, capsulotomy, medial stripping, and tendon transplantation.

Contractures of the knees are at first approached with plaster cast wedging. Osteotomy is frequently necessary, since any joint may be involved in virtually any degree of flexion or extension contracture. The surgical operations practically run the gamut of orthopedic procedures and will not be discussed here. One of our severely involved patients was hospitalized seventeen times for surgical intervention.

Even after all that can be expected from conservative and surgical procedures has been achieved, the muscle weakness is frequently so profound that if ambulation is possible at all, braces and crutches are essential. Some patients will remain dependent on their wheelchairs, and it is essential for maximal efficiency that the wheelchair be properly prescribed to meet the needs of the patient. The goal of maximal efficiency in Activities of Daily Living requires ingenuity in construction of adaptive equipment for assistance in these activities. No universal rules can be established regarding management, for each patient is an individual problem and the procedure is dependent on the number of contractures, the location, nature, and degree of contracture, and distribution of muscle weakness.

MYASTHENIA GRAVIS

Although the precise pathogenesis of myasthenia gravis is still not clearly understood,[12] there appears to be excessive destruction of acetylcholine at the myoneural junction.[110] The frequent occurrence of thymus gland tumors in this disease remains enigmatic. The onset is usually between 10 and 70 years of age, although congenital cases have been described.[111] There is also a familial incidence.[112, 113]

The disease is characterized by muscle weakness, primarily in the ocular and facial muscles and in the muscles of mastication and deglutition. The muscle power weakens rapidly after repetitive contractions, and the electromyogram shows a characteristic response[114-116] and has diagnostic value.[117] Prompt recovery occurs with proper dosage of cholinergic drugs such as neostigmine,[110] edrophonium chloride (Tensilon),[116] or certain other newer synthetic agents.

There is no objective evidence regarding the effect of muscular exercise on the drug requirements. However, drug requirements do change in different phases of the disease,[118] and patients have been known to become somewhat refractory to anticholinesterase therapy. For these reasons the mortality is quite high (30%).

These brief references to the problems of medical management indicate that proper drug therapy and constant medical surveillance are essential to all rehabilitation objectives. If a patient has experienced near death from a myasthenic crisis from drug lack or a cholinergic crisis from anticholinesterase overdosage, severe anxieties may develop that will withdraw the patient from his vocational potential. With proper medical management the rehabilitation problems revolve around the psychosocial difficulties. Vocational changes must sometimes be made so that the necessary medical facilities are more accessible to the patient. Vocational activity should be restricted or redirected if it appears that it is in any way contributory to the disability.

DYSTONIA MUSCULORUM DEFORMANS (ZIEHEN-OPPENHEIM DISEASE)

Dystonia musculorum deformans is a degenerative disease of the striopallidum of unknown etiology.[119, 120] The disease has interesting similarities to Hallervorden-Spatz disease.[123] It is characterized by twisting movements of the trunk and/or neck and extremities. There is a tendency for the torsion movements to shift from one group of muscles to another.

Although infantile and late-developing cases have been described, most cases are of the so-called juvenile type, with onset between 5 and 15 years of age.

Medical management has been directed toward drug therapy that encompasses the same group of drugs used in other diseases of the basal ganglia.

Surgical alleviation of the torsion movements has been attempted with a variety of procedures that have been reviewed by Meyers[121] and by Hamby.[122] Cooper and associates[124] have treated a number of cases quite successfully with chemopallidectomy.

The nature and magnitude of the rehabilitation problem are related to the distribution and severity of the dystonic movements. In dystonic movements of the upper extremities, difficulties in Activities of Daily Living are primary problems, whereas lower extremity involvement precipitates problems in ambulation. One of our patients at one time could ambulate only on all four extremities in backward progression. In younger patients with mild disease, braces appropriate to control dystonic movements are often very helpful. The dystonic movements of older patients are sometimes so powerful that braces cannot be maintained in alignment.

Involuntary movements of all types are significantly affected by emotional peaks. We have found that dystonic patients exhibit most marked degrees of increase in movements during anxiety states. Psychosocial evaluation and supportive therapy have proved to be helpful.

SUMMARY

These muscular and neuromuscular diseases vary in their rate of progression. To provide proper rehabilitation milieu, it is essential to appreciate the clinical course of the several types of these diseases. An appropriate physical and occupational therapy program will prevent disuse atrophy and keep muscular power at maximal level during the changing phases of these illnesses. As the weakness progresses from one group of muscles to another, the requirements for assistive devices constantly change, and new problems arise in management of activities of daily living. The hereditary aspects of these diseases of unknown etiology, for which there are no medical or surgical means of prevention or cure, initiate feelings of helplessness, and the resultant psychosocial problems are of great importance in the rehabilitation program.

REFERENCES

1. Erb, W. H.: Dystrophia muscularis progressiva, Deutsch. Z. Nervenheilk. 1:13, 173, 1891.
2. Meryon, E.: Practical and pathological researches on the various forms of paralysis, London, 1864, John Churchill & Sons.
3. Duchenne, G. B.: Recherches sur la paralysie musculaire pseudohypertrophique ou paralysie myosclérosique, Arch. Gén. Med. 10(series 6): 5, 179, 305, 421, 552, 1868.
4. Tyler, F. H., and Wintrobe, M. M.: Studies in disorders of muscle: problem of progressive muscular dystrophy, Ann. Intern. Med. 32:72, 1950.
5. Shank, R. E., Gilder, H., and Hongband, L.: Studies on diseases of muscle; progressive muscular dystrophy; clinical review of 40 cases, Arch. Neurol. Psychiat. 52:431, 1944.
6. Tyler, F. H., and Perkoff, G. T.: Studies in disorders of muscle: is progressive muscular dystrophy endocrine or metabolic disorder? Arch. Intern. Med. 88:175, 1951.
7. Milhorat, A. T., and Wolff, H. G.: Studies in disease of muscle; metabolism of creatine and creatinine in progressive muscular dystrophy, Arch. Neurol. Psychiat. 38:992, 1937.
8. Evans, H. M., and Burr, G. O.: Development of paralysis in suckling young of mothers deprived of vitamin E, J. Biol. Chem. 76:273, 1928.
9. Mason, K. E., Dju, M. Y., and Chapin, S. J.: Vitamin E content of tissues in progressive muscular dystrophy, Proceedings of the Second Medical Conference of the Muscular Dystrophy Association, Inc., New York, 1952.
10. Thomasen, E.: Myotonia; Thomsen's disease (myotonia congenita), paramyotonia, and dystrophia myotonica: a clinical and heredobiologic investigation. Translated by Finn Brink Carlsen, Copenhagen, 1948, Universitets-Forlaget i Aarhus.
11. Denny-Brown, D.: Nature of muscular diseases, Canad. Med. Ass. J. 67:1, 1952.
12. Adams, R. D., Denny-Brown, D., and Pearson, C. M.: Diseases of muscle. A study in pathology, New York, 1953, Paul B. Hoeber, Inc.
13. Van Meter, J. R.: Progressive muscular dystrophy. A preliminary report on treatment with amino acids, folic acid and vitamins, Calif. Med. 79:297, Oct., 1953.
14. Dowben, R. M., and Perlstein, M. A.: Muscular dystrophy treated with norethandroline, Arch. Intern. Med. 107:245, 1961.
15. Dowben, R. M.: Treatment of muscular dystrophy with steroids, New Eng. J. Med. 268:912, 1963.
16. Fisch, C.: Heart in dystrophia myotonica, Amer. Heart J. 41:525, 1951.
17. Zatuchni, J., Aegerter, E. E., Molthan, L., and Shuman, C. R.: Heart in progressive muscular dystrophy, Circulation 3:846, 1951.

18. Bell, J.: On pseudohypertrophic and allied types of progressive muscular dystrophy. In Treasury of human inheritance, vol. 4, London, 1943, Cambridge University Press.

19. Tyler, F. H., and Stephens, F. E.: Studies in disorders of muscle; clinical manifestations and inheritance of facioscapulohumeral dystrophy in large family, Ann. Intern. Med. **32:** 640, 1950.

20. Tyler, F. H., and Stephens, F. E.: Studies in disorders of muscle; clinical manifestations and inheritance of childhood progressive muscular dystrophy, Ann. Intern. Med. **35:**169, 1951.

21. Stephens, F. E., and Tyler, F. H.: Studies in disorders of muscle; inheritance of childhood progressive muscular dystrophy in thirty-three kindreds, Amer. J. Hum. Genet. **3:** 111, June, 1951.

22. Bradburne, A. A.: Hereditary ophthalmoplegia in five generations, Trans. Ophthal. Soc. U. K. **32:**142, 1912.

23. Fleischer, B.: Ueber myotonische Dystrophie mit Katarakt: eine hereditäre familiäre Degeneration, Arch. Ophthal. **96:**91, 1918.

24. Sibly, J. A., and Lehninger, A. L.: Aldolase in the serum and tissues of tumor bearing animals, J. Nat. Cancer Inst. **9:**303, 1949.

25. Shapira, G., Dreyfus, J. C., and Shapira, F.: L'elevation du taux de l'aldolase serique, test biochemique des myopathies, Sem. Hop. Paris **29:**1917, 1953.

26. Pearson, C. M.: Serum enzymes in muscular dystrophy and certain other muscular and neuromuscular diseases. I. Serum glutamic oralacetic transaminase, New Eng. J. Med. **256:**1069, 1957.

27. Dreyfus, J. C., and Shapira, G.: Biochemistry of hereditary myopathies, Springfield, Ill., 1962, Charles C Thomas, Publisher.

28. Bourne, G. H., editor: Structure and function of muscle. III. Pharmacology and disease, New York, 1960, Academic Press, Inc.

29. Conrad, F. G.: Transaminase, New Eng. J. Med. **256:**602, 1957.

30. Ratner, J. T., and Sacks, H. J.: Transaminase in coronary artery disease, Canad. Med. Ass. J. **76:**720, 1957.

31. Ebashi, S., Toyokura, Y., Momoi, H., and Sugeta, H.: High creatine phosphokinase activity of sera of progressive muscular dystrophy, J. Biochem. (Tokyo) **46:**103, 1959.

32. Dreyfus, J. C., Shapira, G., and Demos, J.: Étude de la creatine-kinase serique chez les myopathies et leurs familles, Rev. Franc. Etud. Clin. Biol. **5:**384, 1960.

33. Okinaka, S., Kumagai, H., Ebashi, S., Sugeta, H., Momoi, H., Toyokura, Y., and Fuji, Y.: Serum creatine phosphokinase, Arch. Neurol. **4:**520, 1961.

34. Pearson, C. M.: Muscular dystrophy. Review and recent observations, Amer. J. Med. **35:** 632, 1963.

35. Dreyfus, J. C., Shapira, G., Shapira, F., and Demos, J.: Activites enzymatiques du muscle humain, Clin. Chim. Acta **1:**434, 1956.

36. Pearson, C. M.: Histopathological features of muscle in the preclinical stages of muscular dystrophy, Brain **85:**109, 1962.

37. Pearson, C. M.: Biochemical and histological features of early muscular dystrophy, Rev. Canad. Biol. **21:**533, 1962.

38. Chung, C. S., Morton, N. E., and Peters, H. A.: Serum enzymes and genetic carriers in muscular dystrophy, Amer. J. Hum. Genet. **12:**52, March, 1960.

39. Shapira, F., Dreyfus, J. C., Shapira, G., and Demos, J.: Étude de l'aldolase et de creatine kinase du serum chez les meres de myopathies, Rev. Franc. Etud. Clin. Biol. **9:**990, 1964.

40. Dubowitz, V.: Progressive muscular dystrophy of the Duchenne type in females and its mode of inheritance, Brain **83:**432, 1960.

41. Lyon, M. F.: Gene action in the X chromosome of the mouse, Nature (London) **190:** 372, 1961.

42. Lyon, M. F.: Sex chromatin and gene action in the mammalian X-chromosome, Amer. J. Hum. Genet. **14:**135, June, 1962.

43. Pearson, C. M., Fowler, W. M., and Wright, S. W.: X-chromosome mosaicism in females with muscular dystrophy, Proc. Nat. Acad. Sci. U.S.A. **50:**24, 1963.

44. Emery, A. E. H.: Clinical manifestations in two carriers of Duchenne muscular dystrophy, Lancet 1:1126, 1963.
45. Weime, R. J., and Herpol, J. E.: Origin of the lactate dehydrogenase isoenzyme pattern found in the serum of patients having primary muscular dystrophy, Nature (London) 194:287, 1962.
46. Weime, R. J.: Nomenclature of so-called isoenzymes, Lancet 1:270, 1962.
47. Kaplan, N. O., and Cahn, R. D.: Lactic dehydrogenase and muscular dystrophy in the chicken, Proc. Nat. Acad. Sci. U. S. A. 48:2123, 1962.
48. Werdnig, G.: Zwei frühinfantile hereditäre Fälle von progressiver Muskelatrophie unter dem Bilde der Dystrophie, aber auf neurotischer Grundlage, Arch. Psychiat. 22:437, 1891.
49. Hoffmann, J.: Ueber chronische spinale Muskelatrophie im Kindesalter, auf familiärer Basis, Deutsch. Z. Nervenheilk. 3:427, 1893.
50. Rosenberg, H. S., and McAdams, A. J.: Infantile progressive muscular atrophy; value of muscle biopsy in diagnosis of and differentiation from muscular dystrophy, Arch. Path. 58:604, 1954.
51. Brandt, S.: Hereditary factors in infantile progressive muscular atrophy; study of 112 cases in 70 families, Amer. J. Dis. Child. 78:226, 1949.
52. Brandt, S.: Werdnig-Hoffmann's infantile progressive muscular atrophy, Copenhagen, 1950, Einar Munksgaard Forlag.
53. Oppenheim, H.: Ueber allgemeine und localisierte Atonie der Muskulatur (Myotonic) im frühen Kindesalter, Mschr. Psychiat. Neurol. 8:232, 1900.
54. Turner, J. W. A.: Relationship between amyotonia congenita and congenital myopathy, Brain 63:163, 1940.
55. Turner, J. W. A.: On amyotonia congenita, Brain 72:25, 1949.
56. Marinacci, A. A.: Clinical electromyography, Los Angeles, 1955, San Lucas Press.
57. Clement, D. H., and Godman, G. C.: Glycogen disease resembling mongolism, cretinism and amyotonia congenita; case report and review of literature, J. Pediat. 36:11, 1950.
58. Krivit, W., Polglase, W. J., Gunn, F. D., and Tyler, F. H.: Studies in disorders of muscle; glycogen storage disease primarily affecting skeletal muscle and clinically resembling amyotonia congenita, Pediatrics 12:165, 1953.
59. Binney, C., II: The problem of amyotonia congenita, Texas J. Med. 52:12, 1956.
60. Lidge, R. T.: Hypotonia, J. Pediat. 45:474, 1954.
61. Key, J. A.: Hypermobility of joints as a sex-linked hereditary characteristic, J.A.M.A. 88:1710, 1927.
62. Sutro, C. J.: Hypermobility of bones due to "over lengthened" capsular and ligamentous tissues; cause for recurrent intraarticular effusions, Surgery 21:67, 1947.
63. Davison, C., and Weiss, M. M.: Muscular hypotonia associated with congenital heart disease, Amer. J. Dis. Child. 37:359, 1929.
64. Greenfield, J. G., and Stern, R. O.: The anatomical identity of the Werdnig-Hoffmann and Oppenheim forms of infantile muscular atrophy, Brain 50:652, 1927.
65. Brandt, S.: Course and symptoms of progressive infantile muscular atrophy: follow-up study of 112 cases in Denmark, Arch. Neurol. Psychiat. 63:218, 1950.
66. Walton, J. N., and Nattrass, F. J.: On classification, natural history and treatment of myopathies, Brain 77:169, 1954.
67. Walton, J. N.: Amyotonia congenita; a follow-up study, Lancet 1:1023, 1956.
68. Buchthal, F., and Clemmesen, S.: The electromyogram of atrophic muscles in cases of intramedullar affections, Acta Psychiat. Neurol. 18:377, 1943.
69. Buchthal, F., and Clemmesen, S.: On differentiation of muscle atrophy by electromyography, Acta Psychiat. Neurol. 16:143, 1941.
70. Charcot, J. M., and Marie, P.: Sur une forme particulière d'atrophie musculaire progressive; souvent familiae débutant par les pieds et les ambes et atteignant plus tard les mains, Rev. Med. 6:97, 1886.
71. Tooth, H. H.: The peroneal type of progressive muscular atrophy, London, 1886, H. K. Lewis.

72. England, A. C., and Denny-Brown, D.: Severe sensory changes and trophic disorders in peroneal muscular atrophy (Charcot-Marie-Tooth type), Arch. Neurol. Psychiat. **67**:1, 1952.
73. Wohlfahrt, S., and Wohlfahrt, G.: Mikroskopische Untersuchungen an progressiven Muskelatrophien unter besonderer Rückischtsnahme auf Rückermarks—und Muskelbefunde, Acta Med. Scand., supp. 63, p. 1, 1935.
74. Allan, W.: Relation of hereditary pattern to clinical severity as illustrated by peroneal atrophy, Arch. Intern. Med. **63**:1123, 1939.
75. Hassin, G. B., and Dublin, W. B.: Histopathology of muscles in spinal type of progressive muscular atrophy, J. Neuropath. Exp. Neurol. **4**:240, 1945.
76. Deaver, G. G., Greenspan, L., and Swinyard, C. A.: Progressive muscular dystrophy diagnosis and problems of rehabilitation, New York, 1956, Muscular Dystrophy Associations of America, Inc.
77. Swinyard, C. A., Deaver, G. G., and Greenspan, L.: Gradients of functional ability of importance in rehabilitation of patients with progressive muscular and neuromuscular diseases, Arch. Phys. Med. **38**:574, 1957.
78. Abramson, A. S., and Rogoff, J.: Physical treatment in muscular dystrophy: abstract of study, Proceedings of the Second Medical Conference of the Muscular Dystrophy Associations of America, Inc., New York, 1952.
79. Hoberman, M.: Physical medicine and rehabilitation: its value and limitations in progressive muscular dystrophy, Proceedings of the Third Medical Conference of the Muscular Dystrophy Associations of America, Inc., New York, 1954.
80. Truitt, C. J.: Personal and social adjustments of children with muscular dystrophy, Proceedings of the Third Medical Conference of the Muscular Dystrophy Associations of America, Inc., New York, 1954.
81. Austin, E.: Participation of the family and patient in a program of rehabilitation, Proceedings of the Third Medical Conference of the Muscular Dystrophy Associations of America, Inc., New York, 1954.
82. Wexler, M.: Mental hygiene and the muscular dystrophy patient, Proceedings of the First Medical Conference of the Muscular Dystrophy Associations of America, Inc., New York, 1951.
83. Bryan, G. E.: Psychological characteristics of adolescents in families with facioscapulohumeral muscular dystrophy (Ph.D. thesis), 1958, University of Utah.
84. Otto, A. G.: Monstrorum sexcentorum descriptio anatomica. In Vratislaviae, Museum Anatomico-Pathologicum Vratislaviense, 1841.
85. Schanz, A.: Ein Fall von multiplen congenitalen Contracturen, Z. Orthop. Chir. Stuttg. **5**:9, 1897.
86. Magnus, F.: Ein Fall von multiphen congenitalen Contracturen mit Muskeldefecten, Z. Orthop. Chir. Stuttg. **11**:424, 1902-1903.
87. Nové-Josserand, G., and Rendu, A.: Résultats, éloignés et valeur de la méthode de Finck dans le traitement précoce des pieds bots congénitaux, Lyon Chir. **8**:121, 1912.
88. Howard, R.: A case of congenital defect of the muscular system (dystrophia muscularis congenita) and its association with congenital talipes equino-varus, Proc. Roy. Soc. Med. **1**(part 3):157, 1907-1908.
89. Stern, W. G.: Arthrogryposis multiplex congenita, J.A.M.A. **81**:1507, 1923.
90. Sheldon, W.: Amyoplasia congenita (multiple congenital articular rigidity: arthrogryposis multiplex congenita), Arch. Dis. Child. **7**:117, 1932.
91. Middleton, D. S.: Studies on prenatal lesions of striated muscle as cause of congenital deformity (congenital tibial kyphosis; congenital high shoulder; mysdystrophia foetalis deformans), Edinburgh Med. J. **41**:401, 1934.
92. Dalmain, W. A.: Myophagism congenita, Amer. J. Surg. **73**:494, 1947.
93. Rossi, E.: Le syndrome arthromyodysplasique congenital (contribution à l'étude de l'arthrogryposis multiplex congenita), Helv. Paediat. Acta **2**:82, 1947.
94. James, T.: Multiple congenital articular rigidities; a review of the literature with reports of two cases, Edinburgh Med. J. **58**:565, 1951.

95. Steindler, A.: Arthrogryposis, J. Int. Coll. Surgeons **12:**21, 1949.
96. McKusick, V. A.: Heritable disorders of connective tissue; the Hurler syndrome, J. Chronic Dis. **3:**360, 1956.
97. Brandt, S.: A case of arthrogryposis multiplex congenita anatomically appearing as a foetal spinal muscular atrophy, Acta Paediat. **34:**365, 1947.
98. Price, D. S.: Case of amyoplasia congenita with pathological report, Arch. Dis. Child. **8:**343, 1933.
99. Swinyard, C. A., and Mayer, V.: Multiple congenital contractures: public health considerations of arthrogryposis multiplex congenita, J.A.M.A. **183:**23, 1963.
100. Swinyard, C. A., and Magora, A.: Multiple congenital contractures (arthrogryposis): an electromyographic study, Arch. Phys. Med. **43:**36, 1962.
101. Swinyard, C. A.: Multiple congenital contractures (arthrogryposis): nature of the syndrome and hereditary considerations, Second International Conference of Human Genetics, Rome, 1961.
102. Rosenkranz, E.: Ueber kongenitale Kontrakturen der oberen Extremitäten im Anschluss an die Mitteilung eines einschlägigen Galles, Z. Orthop. Chir. Stuttg. **14:**52, 1905.
103. Hillman, J. W., and Johnson, J. T. H.: Arthrogryposis mutiplex congenita in twins, J. Bone Joint Surg. **34-A:**211, 1951.
104. Lipton, E. L., and Morgenstern, S. H.: Arthrogryposis multiplex congenita in identical twins, Amer. J. Dis. Child. **89:**233, 1955.
105. McKusick, V. A.: Heritable disorders of connective tissue: the clinical behavior of hereditary syndromes, J. Chronic Dis. **2:**491, 1955.
106. Banker, B. Q., and Victor, M.: Arthrogryposis multiplex due to congenital muscular dystrophy, J. Neuropath. Exp. Neurol. **16:**119, 1957.
107. Flinchum, D.: Pathological anatomy in talipes equino-varus, J. Bone Joint Surg. **35-A:**111, 1953.
108. Bechtol, C. O., and Mossman, H. W.: Club-foot: embryological study of associated muscular abnormalities, J. Bone Joint Surg. **32-A:**827, 1950.
109. Stewart, S. F.: Club-foot: its incidence, cause and treatment, J. Bone Joint Surg. **33-A:**577, 1951.
110. Osserman, K. E.: Studies in myasthenia gravis. I. Physiology, pathology, diagnosis and treatment, New York J. Med. **56:**2512, 1956.
111. Osserman, K. E., and Kaplan, L. I.: Rapid diagnostic test for myasthenia gravis; increased muscle strength without fasciculations after intravenous administration of edrophonium (Tensilon) chloride, J.A.M.A. **150:**265, 1952.
112. McKeever, G. E.: Myasthenia gravis in mother and her newborn son, J.A.M.A. **147:**320, 1951.
113. Hart, H. H.: Myasthenia gravis with ophthalmoplegia and constitutional anomalies in sisters, Arch. Neurol. Psychiat. **18:**439, 1927.
114. Rothbart, H. B.: Myasthenia gravis in children; its familial incidence, J.A.M.A. **108:**715, 1937.
115. Harvey, A. M., and Masland, R. L.: Electromyogram in myasthenia gravis, Bull. Johns Hopkins Hosp. **69:**1, 1941.
116. Lindsley, D. B.: Myographic and electromyographic studies of myasthenia gravis, Brain **58:**470, 1935.
117. Botelho, S. Y., Deaterly, C. F., Austin, S., and Comroe, J. H., Jr.: Evaluation of electromyogram of patients with myasthenia gravis, Arch. Neurol. Psychiat. **67:**441, 1952.
118. Osserman, K. E., and Kaplan, L. I.: Studies in myasthenia gravis; use of endrophonium chloride (Tensilon) in differentiating myasthenic from cholinergic weakness, Arch. Neurol. Psychiat. **70:**385, 1953.
119. Benda, C. E.: Chronic rheumatic encephalitis, torsion dystonia and Hallervorden-Spatz disease, Arch. Neurol. Psychiat. **61:**137, 1949.
120. Herz, E.: Dystonia; clinical classification, Arch. Neurol. Psychiat. **51:**319, 1944.
121. Meyers, R.: Present status of surgical procedures directed against extrapyramidal diseases, New York J. Med. **42:**535, 1942.

122. Hamby, W. B.: Surgical treatment of dystonia musculorum deformans, J. Neurosurg. **10**:490, Sept., 1953.
123. Netsky, M. G., Spiro, D., and Zimmerman, H. M.: Hallervorden-Spatz disease and dystonia, J. Neuropath. Exp. Neurol. **10**:125, 1951.
124. Cooper, I. S., Poloukhine, H., and Hoen, T. I.: Chemopallidectomy for dystonia musculorum deformans, J. Amer. Geriat. Soc. **4**:40, Dec., 1956.

CHAPTER 23

REHABILITATION OF
PATIENT WITH
NEUROLOGIC DISORDERS

The central nervous system has a limited capacity to respond pathologically and functionally to disease, regardless of its etiology. The symptoms and signs of dysfunction and disability are dependent more on the localization and extent of the pathology than on its nature. An important factor, too, is the extremely limited potential, if any, for regeneration of neuronal elements within the brain and spinal cord.

Abnormalities in neurologic function can be considered to result in irritative, paralytic, or release phenomena, alone or in combinations, and the presenting symptoms and findings are best understood from this viewpoint. Consequently, many disease entities, although of diverse etiology, produce comparable neurologic deficits and resultant physical disability. The rehabilitation management of these problems, once they have reached the subacute or chronic state, will usually utilize the same therapeutic tenets and employ the same basic technics, irrespective of the underlying etiology. An individualized program should be prescribed for the patient with a disability as a consequence of neurologic dysfunction. The program is directed toward correction of the disability if possible. If this is not feasible, the patient is oriented and trained to make maximum utilization of his residual potentials. To determine the extent of the neurologic disability, a comprehensive neurologic examination is necessary. This requires a review of the cranial nerves, the motor status, including coordination performances, muscle tone and power, the reflexes, the sensory state, and the mental and speech facilities. This survey, coupled with information obtained from muscle testing, joint range-of-motion studies, Activities of Daily Living, and psychologic testing, gives us an appraisal of the extent of the functional disability.

Before the specific rehabilitation regimen can be prescribed and realistic goals tentatively outlined, there must be an evaluation of the nature of the underlying pathologic process. Is it progressive, static, or regressive in nature? In patients with the latter two types of disease processes, if there are no significant concomitant medical complications militating against an active rehabilitation program (for example, heart failure or marked mental deterioration), there need be no hesitancy in instituting a program. Unfortunately the neurologic disorders include many that are progressive pathologic processes. In some of these the rate of progression is relatively slow enough to warrant the trial of a comprehensive rehabilitation program. In others, however, the rate of progression of the underlying pathology and its resultant symptoms is so rapid as to make the usual rehabilitation regimen unrealistic. Many of the patients can still frequently benefit from the fund of information available through the rehabilitation worker's use of specific, although perhaps limited, technics. The rehabilitation goals must be completely realistic and at times may be limited to such simple measures as training the patient in the proper use of a wheelchair or teaching his family the technics of proper nursing care and the prevention of complications. Each patient requires a careful individual assessment, and his progression or lack of it should be ascertained by a careful review of his antecedent medical history. Frequently a trial of therapy will be necessary before a final decision as to the value of an intensive rehabilitation program is possible. For many patients minimal goals must be defined initially and then gradually raised as the patient attains increased proficiency and strength. Each patient's medical status as well as his medical and rehabilitation programs must be periodically reviewed if the rehabilitation program is to attain its maximal effectiveness.

A comprehensive regional and differential neurophysiologic and diagnostic survey is beyond the scope of this presentation; nevertheless, a brief review of the general principles is necessary if we are to adequately evaluate how dysfunction at various levels within the central nervous system influences the rehabilitation regimen and goals.

Although the concept of rehabilitation implies a comprehensive consideration and awareness of the patient's capabilities as well as his deficiencies, the primary emphasis is usually directed toward a restitution of or a substitution in the patient's motor deficiencies.

CLASSIFICATION OF THE MOTOR SYSTEM

The motor system can be considered as composed of four main components: (1) the upper motor neuron, (2) the extrapyramidal system, (3) the cerebellar system, and (4) the lower motor neuron.

The upper motor neuron consists of the pyramidal tract and its cells of origin in the cerebral cortex. The fibers funnel through the posterior limb of the internal capsule, pass through the brain stem, decussate in the medulla, become positioned in the lateral columns of the spinal column, and terminate in the ventral motor cell areas. Lesions of the upper motor neuron commonly produce a loss of voluntary movement, with increased muscle tonicity (for example, spasticity and accentuated deep tendon reflexes); pathologic reflexes (Babinski, tonic neck, etc.) appear.

The rehabilitation management of patients with such lesions is primarily and extensively reviewed in the description of the treatment of hemiplegia in Chapter 29.

The extrapyramidal system is a complex system whose function is imperfectly understood. It includes the basal ganglia and nuclei in the reticular formation. These are interconnected and have a diffuse projection system that includes descending pathways to the ventral horn cells. Diseases involving this system produce disorders of muscle tone and involuntary muscular movements. They manifest themselves typically in a picture of (1) parkinsonism, with a combination of tremor and rigidity, (2) athetosis, or (3) choreiform movements. The management of patients with parkinsonism will be briefly described in this chapter.

The cerebellar system consists of the cerebellum and its afferent and efferent projection paths. Its function appears to be that of coordinating motor activities, and its dysfunction results in ataxia, which is brought out best in performing movement patterns. Atonia, too, may frequently be found. The impairment of coordination frequently makes intensive motor retraining necessary. Improvement, if it occurs, requires a prolonged period of time.

The lower motor neuron consists of the ventral horn cell and its projection to the effector unit, that is, gland or muscle fiber. Complete lesions of the lower motor neuron produce atrophy and paralysis of the musculature innervated by it, flaccidity, and absent tendon reflexes. In addition, there are frequently vasomotor and trophic changes. During the active period of degeneration, fasciculations may be noted in the involved muscles. The electric studies and their significant changes are described in the discussion on electrodiagnosis.

Impairments in the sensory portions of the nervous system significantly influence the nature and extent of the patient's neurologic dysfunction. Sensory involvement frequently complicates the rehabilitation program. The usual retraining technics may require modification or special considerations. In managing the hemiplegic patient, for example, it is essential to know whether a visual field deficit is present. Should a deficit be present, the visual stimuli usually employed as part of the rehabilitation program will necessarily have to be presented in the intact field of vision if they are to be effective. Similarly, adequate motor coordination performances require the preservation of position sense recognition. Often it is this defect that impairs the patient's motor performance capacities despite what appears to be an otherwise excellent return of motor power as measured by muscle testing.

COMMONLY ENCOUNTERED DISEASES AND DISORDERS
OF THE NERVOUS SYSTEM
Diseases of the cranial nerves

Facial paralysis (Bell's palsy). Although most commonly facial paralysis is of unknown etiology, it must be borne in mind that occasionally it is the result of a mass lesion (for example, tumor, abscess, cyst). Each patient's systemic medical and neurologic status must be carefully and comprehensively reviewed if some of the rarer causes are not to be overlooked. Fortunately in most instances the process is a relatively benign one, and the majority of patients exhibit a marked degree of spontaneous improvement within 6 weeks after the onset of paralysis.

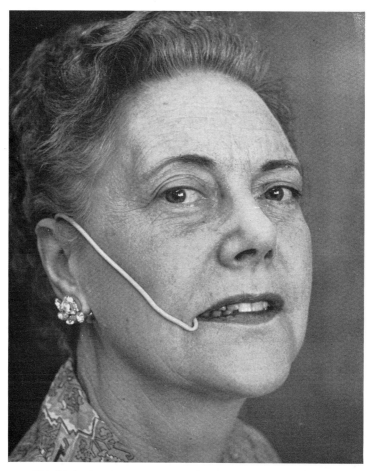

Fig. 23-1. Simplified hook type of splint useful for facial palsy.

The initial treatment is primarily symptomatic and includes the local application of warmth to the involved muscles. Attention to eye hygiene is essential; protective glasses or a patch and the daily use of mild, antiseptic eyewashes are recommended. Gentle massage of the face several times daily in front of a mirror is prescribed. The patient is instructed in the technic of perioral and periorbital massage. He should begin the massage at the medial and inferior margins, gradually stroking upward with a circular motion in the direction of the muscle fibers. Attempts at volitional movement should be made and practiced regularly with the aid of a mirror. The therapeutic use of electric stimulation is controversial and is usually not recommended beyond the acute stage. It is usually considered to be of little value when volitional movement has returned. Facial splints and supports of many types have been designed to prevent stretching and contractures; a simple one extends over the ear on the affected side and hooks into the angle of the mouth (Fig. 23-1). Vitamin therapy and steroid therapy are commonly used as adjuvant medical therapy, but their value is uncertain.

If there is no improvement within 2 weeks after the onset of the paralysis, electro-diagnostic studies, including reaction of degeneration, chronaxie, and electromyo-graphic studies, may prove invaluable in prognosticating recovery or the lack of it.

With recovery there frequently will be transient synkinetic movement of the platysma or other facial musculature that may distress the patient. Mild sedation and reassurance are commonly all that is required in managing the patient since this condition is often a transient phenomenon.

Diseases and disorders of the brain

Hemiplegia. See Chapter 29.

Aphasia. See Chapter 14.

Parkinsonism. Although of diverse etiology, parkinsonism is usually a progressive disorder. The therapeutic rehabilitation program is primarily a symptomatic one directed at amelioration and alleviation of the symptoms of rigidity and tremor and their concomitant resultant functional disabilities. The belladonna group of drugs, especially hyoscine hydrobromide, includes the medications most commonly utilized for this purpose. Many of the newer synthetic solanaceous analog compounds appear also to be of significant value in the management of this entity. These include tri-hexyphenidyl (Artane), caramiphen hydrochloride (Panparnit), ethopropazine (Parsidol), cycrimine hydrochloride (Pagitane), procyclidine hydrochloride (Kema-drin), and benztropine methanesulfonate (Cogentin). These medications are fre-quently used in combination with d-amphetamine. A combination, however, of a well-regulated and periodically reviewed program of drug therapy and an activity program is recommended if the patient is to achieve his maximum potential.

The patient is encouraged to keep as physically active as his general medical condition will permit. In the early stages, gait training is recommended. The patient is encouraged to walk with a broader base; his feet should be held farther apart than is normal, and he is instructed to deliberately lift his feet in walking. Attention is called to the reciprocal pattern of arm swing, and, with volitional and deliberate arm swinging, the gait pattern can be made to resemble a more normal one. Be-cause of the generalized rigidity and loss of associated movements, there is often great difficulty in arising from a sitting position. The patient is therefore instructed not to sit in low or soft upholstered chairs but rather to choose firm-backed chairs. He will find it easier to arise from the sitting position if he first shifts his feet slightly backward beneath the chair or to each side and backward before he leans forward to stand up. A program of general conditioning is often of value. Massage and passive motion of the rigid musculature often prove helpful. Warm baths and pool therapy are frequently excellent aids in relaxation. Occupational therapy is valuable for both its motivating and its exercise activities. The tendency for most patients with parkinsonism to withdraw from social situations and also the psychologic prob-lems associated with social inadequacies must be guarded against. In some cases supportive psychotherapy may be of value. When the patient becomes more inactive, an attempt should be made to improve his proficiency in the Activities of Daily Living, and in very severe cases the program may be designed to teach the family more efficient nursing procedures.

The development of many of the newer neurosurgical procedures directed at amelioration of the symptoms of parkinsonism has evoked a great deal of interest and hope. Although it is too premature to judge its ultimate role in treatment of these cases, the results appear encouraging. The rehabilitation potential of patients with parkinsonism who have had surgery performed must be explored after the operation if maximal benefit is to be obtained from the procedure.

Brainstem and cerebellar disorders

A number of neurologic entities result from the focalization of pathology within the brainstem. Commonly, the disability encountered is not primarily one of weakness, but rather it is the result of ataxia and incoordination as the result of involvement of the cerebellar system. The treatment of such patients is protracted but not hopeless. Attempts are made to strengthen any weak muscles by a gradual program of muscle reeducation exercises, and then a series of coordination exercises appropriate for the involved muscles is prescribed. Intensive gait training is frequently necessary. Often this is initiated in the parallel bars; then as the patient shows progress he is allowed to walk with canes or crutches as necessary. A broad-based gait is deliberately cultivated in many instances. Improvement is commonly painfully slow. Often it is incomplete. With patience and intensive training in Activities of Daily Living, many patients can be made at least self-sufficient. In occasional instances there is such a degree of truncal ataxia that the patient is unable to walk, and it may be necessary for him to rely on and be trained in the proper use of the wheelchair. A trial of speech therapy, too, is usually recommended for the dysarthria commonly encountered.

Degenerative disorders

There is a wide spectrum of neurologic diseases that chronically, and often insidiously, involve the anterior horn cells and the pyramidal tracts in varying degrees. On one hand, these include syndromes in which there is almost exclusive involvement of the anterior horn cells, for example, *primary spinal muscular atrophy;* on the other hand, they include conditions in which the signs of involvement are of only the pyramidal tract, for example, *primary lateral sclerosis.* It is more common, however, to find a mixture of both entities, as in the more fulminating syndrome of *amyotrophic lateral sclerosis.*

The rehabilitation management of each entity is directed toward the underlying disability. Muscle reeducation and progressive resistive exercises are frequently of value when there is some atrophy of disuse superimposed on that which results from the primary anterior horn cell involvement. The primary pathologic process, however, is not influenced by exercises. In many instances it becomes necessary to exploit the patient's Activities of Daily Living potential fully, using a battery of self-help devices to substitute for the motor functions.

With primary pyramidal tract involvement, the treatment is directed toward alleviation of the spasticity and gait abnormality. The patient is coached to walk with a broader base and voluntarily resist any scissoring tendency. Exercises to strengthen any weak musculature are recommended. Relaxation exercises are of value for some patients but require prolonged practice. Stretching the involved extremities is of limited value except for the prevention or correction of contractures. The use of the warm-water pool or tank as a muscle relaxant can be tried. Proper bracing is indicated in selected relatively stationary cases, and in occasional cases obturator neurectomy or block has been beneficial.

Because of the rapidly progressing nature of amyotrophic lateral sclerosis, an intensive rehabilitation regimen is rarely indicated. Treatment is usually palliative in an attempt to prevent further complications. Passive exercise to prevent joint contractures and the daily use of the tiltboard may be warranted. Often the greatest benefit will be obtained from training the members of the family or a nurse in the most efficient way of assisting the patient to carry out his Activities of Daily Living as long as he possibly can.

Demyelinizing disease

Multiple sclerosis. Since multiple sclerosis is extremely variable in its course and symptomatology, each program must be a highly individualized one, depending on the patient's medical status, needs, and problems. The physician's goals must be based on a realistic appraisal of the patient's medical course. When the disease appears to be relatively stationary or retrogressive, a dynamic rehabilitation program is indicated; however, if the disease appears to be rapidly progressive, a careful appraisal must be made as to whether the immediate goals are realistic and warrant expenditures of the required time, effort, and money. In view of the frequency of exacerbations and remissions, regular and frequent reevaluations of the patient's status and the treatment are essential.

Various medical opinions exist as to the value of an exercise program for the patient with multiple sclerosis. Many physicians feel that it is contraindicated, and some even advocate complete bed rest. Others, however, feel that such physical limitation does not positively influence the course of the illness but rather has a harmful effect, in that the patient so treated appears to have a higher incidence of secondary complications. There is no evidence that the rehabilitation regimen, with normal use of physical medicine modalities and aids, has any influence, either positive or negative, on the primary pathologic processes causing the symptoms. Because of this and because of our belief that the secondary complications are less frequent, activity programs are recommended with a realization of the patient's limitations.

One of the characteristic features in multiple sclerosis is the patient's ready fatigability; this must be carefully considered in carrying out the prescribed program. Patients are encouraged to work to the level of fatigue and beyond, but not to the point of exhaustion. A clear distinction is often impossible, and in most instances we must rely on the patient's judgment. During an acute exacerbation, intensive exercise programs should be deferred and the energy directed toward prevention of contractures.

The treatment program is directed primarily toward the acquisition, if possible, of those Activities of Daily Living necessary for proficient function at a specific level in his own environment—bed, wheelchair, or ambulatory status. Whenever possible the specific neuromuscular disability responsible for the deficit in function is corrected by appropriate treatment of abnormal factors, such as correction or improvement of weak muscles or, "coordination exercises." The intelligent use of physical medicine's aids and modalities assumes a significant role in bringing this about. The therapeutic program is directed at functional improvement. An increase in muscle strength is of little value unless it is functionally significant, as evidenced by an improvement in the patient's ability to perform his Activities of Daily Living. If therapy is ineffectual or impracticable, attempts must be made to teach the patient to function in as capable a manner as possible with and despite his disability. In certain patients substituted movements can be adequately utilized. Frequently it will be necessary to employ various compensatory prosthetic or self-help devices, including canes, crutches, braces, and wheelchairs. By the intelligent exploitation of the patient's residual capabilities and by the use of many of these devices, several patients have been made physically independent, and some have been capable of returning to work. Complete self-care is a minimal but significant goal.

The nature and extent of the disabilities encountered in the patient with multiple sclerosis are extremely variable. Occasionally the clinical picture is one of hemiplegia. More commonly spastic paraplegia is encountered. The management of these two disabilities is extensively described in Chapters 19 and 29. The basic rehabilitation principles used in their treatment, however, are applicable for the patient with multiple sclerosis.

Ataxia is frequently encountered and may reflect an involvement of the cerebellar system and its projections or an impairment of proprioception. Frenkel's coordination exercises are commonly employed, and vision is more fully utilized as a sensory medium. The cane tip and shoe tip can be painted white to promote better visualization in dim light. When there is muscle weakness, intensive muscle reeducation and progressive resistive exercises are often necessary. Occupational therapy may be an extremely valuable phase of this treatment. The results may be of marked value in the properly selected patient.

Spasticity is often one of the most distressing problems. Stretching to prevent spasticity is of little sustained benefit. Proper bracing, however, frequently helps in its management. In occasional instances obturator neurectomy or block has resulted in significant improvement in spasticity and correction of "scissoring."

For gait retraining a precise evaluation of the gait abnormalities is necessary. This includes an appreciation of motor power, the particular muscle groups involved, the degree of incoordination due to ataxia and spasticity, the sensory status, etc., and whether canes, crutches, or braces are necessary. The decision as to whether the patient should be braced must be carefully considered in view of the additional weight and the patient's proneness to fatigue. The degree of ataxia and/or spasticity may be severe enough to preclude walking. If so, standing at least 20 minutes a day, either in braces or on a tiltboard, is still advocated to prevent the secondary metabolic complications of inactivity. In retraining, balancing exercises are important before walking is begun. Relaxation exercises are used, and in reeducation, reinforcement technics with residual movement patterns (for example, Strümpell's phenomenon) are used. A broad-based gait is often deliberately cultivated. Walking in a pool between parallel bars may be helpful in the early stages.

Passive exercises are of limited value, except for the prevention and correction of contractures. Only a few repetitions of the joint range of motion appear necessary in the patient's routine management. Progressive resistive exercises have their primary use in strengthening weak muscles and muscles in which there has been a supervening atrophy of disuse; however, in view of the patient's proneness to fatigue, these exercises must be carried out judiciously.

The physical medicine modalities mentioned previously have a limited usefulness. Heat, whirlpool, Hubbard tank, and Moistaire cabinets are essentially preheating technics and useful preliminaries to an exercise program or to stretching for the correction of contracture. Electric stimulation seems to be useful only as an adjuvant in muscle reeducation technics. (See Chapter 3.)

Occupational therapy, in addition to its usefulness in complementing and augmenting muscle reeducation and coordination, psychologically benefits the patient by stimulating interest and motivation. It can also serve for prevocational exploration and prevocational and vocational training. (See Chapter 5.)

Bowel and bladder control are important to the patient. Often it means the difference between social acceptance and rejection, with all its psychologic and economic implications. (See Chapter 13.) Bladder symptoms can often be controlled by medicaments, such as atropine, methantheline (Banthine), caramiphen hydrochloride (Panparnit), and bethanechol chloride (Urecholine). If incontinence persists, the patient must be trained in the proper use of aids, including urinals, urinary pads, and waterproof pants. When bowel control has been lost, strenuous efforts are made to retrain the patient. Most of the patients can accomplish control readily. In only an occasional instance has this proved ineffectual and have the garments and sanitary devices developed for the paraplegic been necessary.

Infectious diseases of the central nervous system

Meningitis and other bacterial infections. During the acute stages the medical management, including the adequate use of the appropriate antibiotic agents, if available, should receive primary consideration. The rehabilitation efforts during this stage are directed toward the prevention of complications—contractures, decubiti, etc. This involves careful nursing technics, frequent turning of the patient, meticulous attention to skin hygiene, proper positioning of the extremities, and careful passive movements at all joints to prevent contractures. In occasional instances the patient may sustain a hemiplegia, with or without aphasia. Once the acute phase has resolved itself, the usual hemiplegia and aphasia rehabilitation regimens are instituted.

Viral disorders. No portion of the central nervous system is immune to the ravages of the viral disorders. Fortunately many of these seem to be self-limited entities that resolve without significant desirability; however, others produce significant neurologic residua. The patient so involved frequently requires intensive rehabilitation.

In the early stages, as for the patient with bacterial infections, the rehabilitation goals are primarily preventive. The program for the patient who has passed through this stage obviously depends on the residual disability that is in turn related to the localization and extent of the damage within the central nervous system. Thus there may remain a hemiplegia, cerebellar system dysfunction, paraplegia or paraparesis, quadriparesis, or peripheral nerve disorder. Frequently the pathology produces residual combinations that include several of these component dysfunctions. The rehabilitation program is directed toward the specific disability, but the combination of several factors makes the rehabilitation program a formidable one.

The late sequela of a parkinsonian syndrome after encephalitis is well known. The rehabilitation program for these patients differs only slightly from that described for parkinsonism, for although the symptoms may appear at an earlier age, the rate of progression is frequently much more rapid.

Transverse myelitis. Transverse myelitis is relatively frequent. The rehabilitation program is essentially the one outlined for the paraplegic patient.

Guillain-Barré syndrome. The etiology is obscure but generally considered to be a "neuronitis" secondary to a virus infection. The typical picture is that of a diffuse peripheral neuropathy, with or without additional central nervous system involvement. The rehabilitation management is essentially that as outlined for the patient with poliomyelitis. However, not infrequently patients with Guillain-Barré syndrome may have sphincter disturbances, and for this aspect the bowel and bladder management outlined for the paraplegic patient may be invaluable. In addition, in some cases residual sensory deficits complicate the problem. An awareness of the affected areas—their localization, extent, and the specific modalities involved—is important, for this information may dictate modifications in the usual "polio program."

Other spinal cord diseases

Tabes. Although tabes is apparently becoming a rarity, an occasional patient is referred for rehabilitation. The symptoms seem to reflect pathology within the dorsal column and roots and are multiplex. A rehabilitation program is often combined with appropriate antisyphilitic treatment. The ataxia may be improved somewhat by intensive training with the reeducational exercises described by Frenkel. In addition, the commonly encountered arthropathies may require treatment despite the fact that they are painless. All appropriate physical modalities can be employed, but adequate bracing, in addition, at the hip, knee, and/or ankle joint is frequently necessary when these joints are involved. A long leg brace with pelvic band may relieve the strain at the hip. For the knee alone, a long leg brace is adequate. A short leg brace may be recommended for the ankle. A shoehorn or spring brace, designed primarily for a dropped foot, will frequently prove inadequate because of its less supportive nature. In occasional cases a spinal brace will prove necessary for the spinal arthropathy.

Posterior lateral sclerosis (combined system disease, subacute combined degeneration). Despite the common concept that posterior lateral sclerosis is usually associated with pernicious anemia, the pathogenesis for the majority of cases remains obscure. The clinical picture usually reflects varied combinations of pyramidal tract, posterior column, and peripheral nerve involvement. The rehabilitation program is directed toward a correction of the specific disability. This will include Frenkel's and coordination exercises for the posterior column involvement and a modified paraplegic regimen to counteract the spastic paraparesis and gait abnormality. The serious problems of spasticity with scissoring occasionally requires obturator neurectomy. Gait retraining (see the discussion multiple sclerosis) is of value for many patients.

PERIPHERAL NEURITIS

Peripheral neuritis is broadly defined as a disturbance in the motor and sensory function of peripheral nerves. It does not necessarily imply an inflammatory lesion, and perhaps the term neuropathy, as suggested by Wechsler, is more appropriately applied to noninflammatory lesions. Either a single nerve or many nerves are involved. The presence of peripheral neuritis with significant motor loss poses a serious problem and can be considered a long-term, and possibly a permanent, disability. It is the cause of significant disability complicating many systemic diseases and can in itself be a primary cause of disability.

Causes. Some of the causes of single or multiple peripheral neuritis (polyneuritis) include (1) trauma—complete division (neurotmesis), crushing, stretching, and pressure; (2) poisons—metals and organic chemical poisons (and drugs); (3) deficiency states and metabolic disorders—vitamin deficiency (especially thiamine chloride), addisonism, pernicious anemia, chronic gastrointestinal conditions, chronic alcoholism, porphyria, diabetes mellitus, and periarteritis nodosa; (4) infectious conditions—specific inflammatory disease of peripheral nerves (acute infective polyneuritis) and complications of other infectious diseases (for example, influenza, exanthemas, mumps, sepsis, diphtheria); and (5) peripheral neuritis of obscure origin—chronic progressive polyneuritis and chronic hypertrophic interstitial neuronitis of Déjerine and Sottas.

Diagnosis and treatment. Although the disability is primarily due to the resultant muscle paralysis in the peripheral neuropathies, secondary changes can complicate the progress and rehabilitation potential significantly. Superimposed contractures and trophic changes are examples of such complications. Proper rehabilitation procedures instituted *early* can do much to prevent their occurrence and, if already present, to correct them.

It must not be overlooked that severe sensory loss may result in disability. Ataxic states may result from the loss of function of fibers carrying proprioceptive stimuli. In addition, the anesthetic area is usually more prone to injury. Healing is frequently delayed in the involved area, and ulcerations with secondary infections are not uncommon.

The proper treatment of patients with peripheral nerve lesions requires an accurate diagnosis of the nerve involved and knowledge of its motor ramifications and the extent of the pathology. A careful clinical examination, when supplemented by special examination technics, will usually supply this information. The electrodiagnostic testing technics are invaluable for this purpose, but examination must be deferred until degeneration has occurred. This will usually require waiting at least 2 weeks if the electric studies are to be considered valuable. (See the discussion on electrodiagnosis.) Sweating tests, too, are frequently carried out to establish the extent and localization of nerve lesions. In occasional instances nerve blocks are of diagnostic value, and in specific cases surgical exploration may be necessary.

Periodic retesting is necessary in many cases for an accurate prognosis. Return of function may be measured by the development and progression of Tinel's sign, alteration in the extent and character of the sensory deficit, retrogression of the vasomotor signs, and gradual return of motor capacities. Serial changes in the electric studies frequently presage obvious clinical improvement.

The rehabilitation program for the patient with peripheral nerve involvement must frequently be of long duration, in view of the fact that regeneration occurs from the proximal section of the nerve and is considered to occur at an average of 1.5 mm./day. In many instances following trauma, surgical nerve suture is essential if there is evidence of anatomic interruption. In this chapter the treatment to be described is the nonsurgical one and is applicable for partial lesions, for lesions with the nerve in continuity, and in the postoperative management.

The rehabilitation program in its initial stages is essentially preventive. Passive motions are utilized to prevent joint contractures and the development of adhesions between tendons and their sheaths. These exercises and massage may frequently retard, but cannot prevent, the reduction in muscle volume and tone. Electric stimulation is of use for the same purpose, but it must be remembered that this in no way appears to influence the rate of neuron regeneration. Special splints, casts, or braces are frequently recommended to prevent stretching of the muscle by the intact antagonist or by gravity. If these are used, they should be so designed to allow daily passive movements of the involved joints. Attempts to stimulate the circulation of the involved part must be carefully carried out. For this purpose, gentle massage, the cautious administration of heat, ultraviolet radiation, and warm baths, with or without whirlpool, can be judiciously utilized. When voluntary motion begins to return, the program is supplemented by intensive muscle reeducation procedures, at first with gravity eliminated but gradually increased until progressive resistive exercises are employed. Causalgia is a commonly occurring, distressing symptom. Its treatment is frequently very difficult. The limb may have to be carefully protected from inclement weather and from almost all usually innocuous trauma to the limb, by adequate padding and support. Cautious exploration as to the value of moist heat and gentle massage is warranted. Despite the discomfort, the importance of passive motion must be emphasized to the patient, and it should be carried out regularly if it is at all possible.

Special problems in the management of specific peripheral neuropathies. The *long thoracic nerve* may be involved by many pathologic agents, resulting in paralysis of the serratus anterior muscle; occasionally this is bilateral. With the lack of scapular fixation the patient finds that it is impossible for him to raise his arm above the horizontal level. In selected patients with intact deltoid muscular function, it has been possible to stabilize the scapula and restore a great deal of arm function by the use of a special brace designed with bilateral firm scapular plates with an incurved medial margin and a firmly affixed breastplate (Fig. 23-2).

The *circumflex nerve* may be involved by injuries or lesions in the region of the neck of the humerus. The arm is splinted in a shoulder-abducted position, and the usual program for peripheral nerve lesions is instituted.

The *musculocutaneous nerve* is seldom involved alone, although it may be injured by direct trauma or dislocation of the head of the humerus. The forearm should be supported by a sling.

The *radial nerve* is frequently involved. It is commonly damaged by injuries to the humerus, dislocations in the axilla, crutch pressure, and various pressure paralyses (malpositioning, Saturday night palsy, etc.). A splint providing for extension of the wrist and of the fingers is essential, but to ensure metacarpophalangeal flexion, elastic

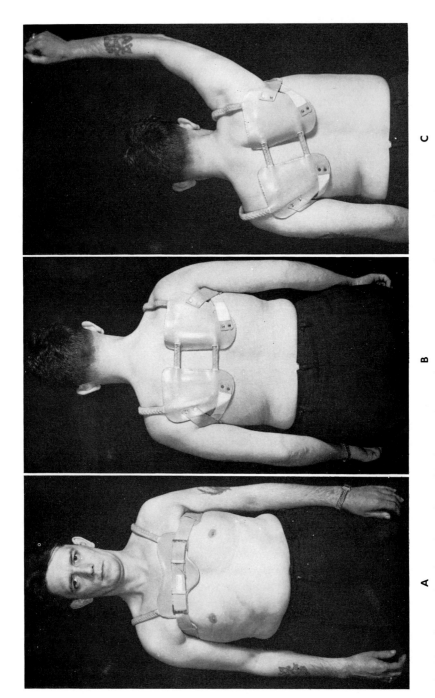

Fig. 23-2. A, Anterior view of a patient with the scapular fixation brace in position. **B,** Posterior view of a patient with the brace in position. Note the turned-in vertebral border on the right side and the shoulder contour. **C,** Posterior view of a patient with elevated arm. Note the shift in position of the brace.

extensions from the fingers, extending over the dorsum of the hand and fixed to the rigid wrist support, are recommended. Splints in which Neg'ator springs are used have been designed for this purpose. (See Chapter 12.)

The *median nerve* is most commonly injured in civilian life by lacerations in the wrist region; in such injuries the ulnar nerve is also frequently involved. A splint to keep the thumb in a position of abduction and opposition is recommended. Causalgia may be a serious problem and hamper rehabilitation.

The *ulnar nerve* may be involved by many pathologic processes, but most commonly the involvement is the result of injuries in the region of the elbow. One commonly overlooked condition is the syndrome of tardy ulnar palsy, in which the paralysis follows, by many years, fracture or injury of the head of the radius that occurred during the patient's early life. With involvement of the radial epiphysis there may be a disproportionate growth of the radius and ulna, gradually producing a stretching of the nerve and symptoms. Involvement has been reported as becoming manifest as late as 35 years after the injury, long after the patient had forgotten the injury. Nerve transplantation is frequently necessary.

A splint designed to posture the hand in a position of flexion at the metacarpophalangeal joints with extension of the interphalangeal joints is of value.

The *femoral nerve,* although vulnerable to many pathologic processes, is most commonly involved in fractures of the pelvis or as a complication of lumbar osteoarthropathies, including herniated lumbar intervertebral disks. The weakness of the quadriceps can be compensated for in many individuals by the use of elastic devices running from the waist or upper thigh to the anterior margin of the lower leg, by special knee braces with spring action to retard knee flexion, or by a long leg brace. (See Chapters 7, 8, and 12.)

The *sciatic nerve* terminates as the *common peroneal* and *tibial nerves.* It and its branches are very susceptible to injury and other pathologic processes, and one of its complicating features is its associated causalgia.

The *common peroneal nerve* is frequently injured by trauma involving the upper end of the fibula. Compression bandages are also a common cause of injury. Usually the peroneal muscles are more involved than the anterior tibial group. In addition to the usual therapeutic procedures, it is necessary to prevent and compensate for foot drop. If the patient is bed bound, the foot may be supported by the use of a posterior molded splint or a footboard. In ambulatory patients the use of a high-topped shoe is frequently all that is necessary to correct a slight foot drop. In others the use of a posterior spring brace, inserted into the heel, or a "shoehorn" brace is valuable. Only occasionally is it necessary to utilize a short leg brace with either a spring or 90-degree stop at the ankle. The latter is usually avoided because of the additional weight. (See Chapters 7, 8, and 12.)

The *tibial nerve* is less commonly involved. Care must be taken to prevent the common tendency toward the formation of trophic ulcers.

CONCLUSION

The patient with a neurologic disease presents a serious problem in general medical practice. Despite encouraging advances in definitive therapy, there frequently are residual disabilities that cannot be overcome by medical or physical modalities.

An accurate appraisal of the nature, course, and extent of the underlying pathologic process, as well as a comprehensive evaluation of the extent of the patient's disability, is essential.

The rehabilitation program is planned to offer the patient the opportunity of functioning at his maximal potential. This requires an appreciation and utilization of the appropriate rehabilitation technics, that is, the employment of specific exercise as required and an intelligent exploration of the various self-help devices currently available.

Each patient's program must be a highly individualized one. Nevertheless, the overall goals should be completely realistic for both the patient and his physician if the program is to be successful. Although the primary goal is usually self-sufficiency, in most instances an attempt should be made to increase the patient's functional capacities to the point at which vocational exploration is warranted.

A carefully planned rehabilitation program, faithfully executed and carefully followed, frequently proves to be of significant value in decreasing the extent of the patient's disability and handicap.

REFERENCES

1. Brain, W. R.: Diseases of the nervous system, ed. 5, New York, 1955, Oxford University Press.
2. Cooper, I. S.: Neurosurgical alleviation of parkinsonism, Springfield, Ill., 1956, Charles C Thomas, Publisher.
3. Fulton, J. F.: Physiology of the nervous system, ed. 3, New York, 1949, Oxford University Press.
4. Gordon, E. E.: Multiple sclerosis: application of rehabilitation techniques, New York, 1952, National Multiple Sclerosis Society.
5. Haymaker, W., and Woodhall, B.: Peripheral nerve injuries, ed. 2, Philadelphia, 1953, W. B. Saunders Co.
6. Head, H., and Rivers, W. H. R.: Studies in neurology, vols. 1, and 2, London, 1920, H. Frowde & Co.
7. Highet, W. B.: Splintage of peripheral nerve injuries, Lancet 1:555, 1942.
8. Marks, M.: Amyotrophic lateral sclerosis, J. Mount Sinai Hosp. 8:113, 1949.
9. Marks, M., and Goodgold, J.: Rehabilitation of the multiple sclerosis patient, J.A.M.A. 156:755, 1954.
10. Marks, M., and Rusk, H. A.: Rehabilitation of the aphasia patient: a survey of three years' experience in a rehabilitation setting, Arch. Phys. Med. 38:219, 1957.
11. Marks, M., Rusk, H. A., Hass, W., and Chase, N.: Management of the patient with cerebrovascular disease, J. St. Barnabas Med. Cent. 2:51, April, 1964.
12. Rusk, H. A., and Marks, M.: Rehabilitation following the cerebrovascular accident, Southern Med. J. 46:1043, 1953.
13. Rusk, H. A., Deaver, G. G., Covalt, D. A., Marks, M., Benton, J., and Turnbloom, M.: Hemiplegia and rehabilitation, West Point, Pa., 1952, Sharp & Dohme.
14. Russek, A. A., and Marks, M.: Scapular fixation by bracing in serratus anterior palsy, Arch. Phys. Med. 34:633, 1953.
15. Taylor, M. L., and Marks, M.: Aphasia rehabilitation manual and therapy kit, New York, 1959, McGraw-Hill Book Co., Inc.
16. Wechsler, I. S.: Clinical neurology, ed. 9, Philadelphia, 1963, W. B. Saunders Co.

CHAPTER 24

REHABILITATION
PROBLEMS OF CHILDREN

A child may be considered physically or orthopedically handicapped if his disabilities prevent him from participating in an acceptable manner in childhood activities of a physical, recreational, or educational nature. It is estimated that of the 28 million handicapped and ill persons in the United States 5,870,000 are under 21 years of age. Approximately one fifth of this number are children disabled because of neuromuscular conditions such as cerebral palsy, spina bifida, muscular dystrophy, and poliomyelitis. It is estimated that in 1 of every 200 live births, the infant will have brain damage that will produce a neuromuscular disability. There is no evidence at this time that this number will be lessened in the near future. In fact, with the increasing birth rate and the saving of the lives of a larger proportion of the premature infants, we may have to contend with an increasing number.

In past years children with orthopedic disability secondary to acute anterior poliomyelitis composed the largest diagnostic group of children requiring rehabilitation services. With the advent and widespread use of the Salk and Sabin types of poliomyelitis vaccine, the prevalence of this type of disability has been markedly lowered.

In recent years there has developed a worldwide interest in birth defects. This interest was given great impetus by the tragic occurrence of congenital limb deformities as a result of ingestion of thalidomide compounds during pregnancy. Since the withdrawal of thalidomide from use, the incidence of congenital limb deformities has dropped to its former level.

This unfortunate episode, however, appears to have helped parents in their understanding and interest in birth defects and has resulted in parents seeking rehabilitation assistance for their children with birth defects. For these reasons children with congenital malformations constitute a larger segment of the population seen in units that are devoted to rehabilitation of children.

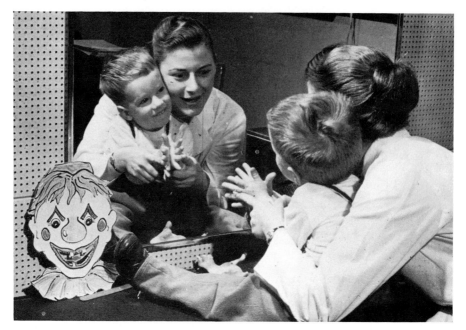

Fig. 24-1. Communication with others is important. The speech therapist helps achieve this goal.

The children with congenital defects such as spina bifida, limb deformities, muscular dystrophy, and amyotonia and those disabled by accidents can all benefit from a rehabilitation program.

The child with a fatal prognosis should not be refused rehabilitation service merely because he may die in 5 or 10 years. Death is the eventual fate of all humanity, but we nevertheless wish to make the most effective use of our remaining time. The list of curable diseases is expanding so rapidly today that it is essential, for example, to help the child with muscular dystrophy retain his skills and prevent deformities, if only because there may be a cure tomorrow. At one time a diagnosis of tubercular meningitis was placed in doubt if a patient survived. Today through modern treatment many survivors are being helped by rehabilitation.

PHILOSOPHY OF REHABILITATION

The rehabilitation program is based on the philosophy that the child with a physical handicap is first a child, whose basic needs are those of a child. In addition to these basic needs, the child with a handicap requires services that will overcome or alleviate the handicap and that will help him to attain the most satisfactory psychosocial and educational adjustment in the environment in which he must function for the remainder of his life.

Fundamental to this concept of the "total child" is the acceptance of the premise that handicapped children are part of a family unit whose other members must be participants in the plan and process of the child's total rehabilitation and whose own needs must be recognized and planned for if the ultimate goal of rehabilitation for the handicapped child is to be reached (Fig. 24-1).

The objectives of a rehabilitation program for children with physical disabilities is directed toward helping them to achieve the maximum physical, social, emotional, educational, and vocational possibilities within the limits of their disabilities. If we are to assist the child to live as full a life as possible so that his social, emotional, and educational development will provide the greatest satisfaction for him, then we must formulate a program that will, as far as possible, overcome the original handicapping and/or associated conditions and teach him ways of adjusting to his irrevocable physical limitations.

CONDITIONS PRODUCING HANDICAPS

There are six conditions that produce the great majority of handicapping conditions: (1) blindness and partial blindness, (2) deafness and partial deafness, (3) cardiac disorders, (4) tuberculosis, (5) mental retardation and socioemotional disturbances, and (6) orthopedic disabilities, which are chiefly on a neuromuscular basis. Children with the so-called orthopedic disabilities are the most difficult to evaluate because of their inability to perform the normal activities of daily living, which is due to the secondary deformities produced by contractures, and—most importantly of all—because of the sensory and perceptual deficits and the communication and mental retardation that limit education, social adjustment, and vocational potential (Chapter 25).

When we study these children from the orthopedic viewpoint, the two following disabilities produce fundamental problems:

1. *Limitation or absence of movements at the joints.* Reduced range of motion results from intrinsic disease of the joint, such as arthritis, from bony blocks resulting from fractures, or from contractures due to muscle imbalance, as seen in patients with poliomyelitis.

2. *Muscle dysfunction.* This may be manifest in weakness from disease, as in patients with amyotonia congenita, or from disuse. The potential motor capacity of long-immobilized children is often surprisingly great, as we have found in children with amyotonia congenita and muscular dystrophy who have been allowed to be in bed and have been discouraged from any activity. Cerebral palsy may produce a variety of defects in movement and posture.

In thinking of a rehabilitation program, it is necessary to evaluate the disability, the extent of the disability and the parts of the body involved, and not to emphasize the diagnosis. Thus hemiplegic patients have the same disability whether the hemiplegia is congenital, caused by cerebral palsy, or is secondary to tubercular meningitis, trauma, or brain tumor. Paraplegic patients present similar problems whether their paraplegia is caused by cerebral palsy, spina bifida, spinal cord injury or tumor, transverse myelitis, or poliomyelitis. The disabilities that handicap the individuals are the same, and the rehabilitation programs are somewhat similar, although sensory changes and bladder and bowel involvement introduce certain nonorthopedic differences.

Rehabilitation should begin as soon as the diagnosis is established. The human body can compensate in one area for deficiencies in another, and a child with paralyzed legs can, with experience and good training, develop powerful arms. In this regard it may be pointed out that the traditional use of the overhead trapeze for long-term bed patients does not constitute adequate training. It is undoubtedly of use in moving about in bed, and it is a good exercise for the elbow flexors, but push-ups are sufficiently practical for the moment and serve as well to develop the elbow extensors that will later be needed for walking with crutches or for propelling a wheelchair.

OBJECTIVES OF REHABILITATION

With slight modification of the definition of the National Council on Rehabilitation, our definition of rehabilitation is as follows: *the restoration of the handicapped child to the fullest physical, emotional, social, vocational, and economic usefulness of which he is capable.* (In congenital syndromes the term *habilitation* is more accurate.) The goal, in short, is to help the child live at maximum capacity within the limits of his disability in dressing, eating, ambulation, school, and other important activities of daily living. *The emphasis is placed on the remaining ability rather than on the disability.*

Early case finding is a prerequisite to success, but it is meaningless without prompt and adequate treatment. If we allow deformities, emotional maladjustments, and other complications to arise, we shall have been negligent.

We can expect that the great majority of children on a rehabilitation program will have a permanent physical disability. Therefore, the objectives of the physical program are to prevent and correct deformities and to build up muscle strength and endurance, so that the child may be taught to perform, insofar as it is possible, all the activities of daily living of which he is capable. Adequate records of progress are essential (Charts 16 to 18).

Five basic objectives that a program of this nature should attempt to achieve are as follows:

1. *Self-care in bed and wheelchair* (Chart 16). Rehabilitation starts at the bed. Should the parents of a child who relies on their help in getting out of bed, attending to toilet needs, and putting on braces become ill, the child must remain in bed. If, however, the child is taught to get out of bed and into a wheelchair, to roll his wheelchair, and to perform the necessary toilet activities, he will never become too much of a burden to others. The evaluation of the child's potential ability and program of training is the responsibility of the rehabilitation nurse.

2. *Maximum use of the hands* (Chart 17). The occupational therapist evaluates the child's ability to perform fundamental movements and necessary activities of daily living. The results of her testing indicate the areas in which the child needs training.

3. *Ambulation and elevation* (Chart 18). Evaluation of the child's potential ability in these areas is the function of the physical therapist. The ability to travel and to get up and down curbs and steps is a necessary function of daily living. Far too much emphasis has been placed on ambulation as the primary goal in rehabilitation. Our studies indicate that even when a person becomes efficient in ambulating with braces and crutches, he frequently finds that it is too time-consuming and requires too much energy to lock his wheelchair and braces, push himself to erect position, reach for crutches, place them under the arms, and back away from the chair before he can ambulate.

CHART 16

INSTITUTE OF REHABILITATION MEDICINE
New York University Medical Center
Children's Unit

Name_____ Age_____ Sex_____ Date_____
Address_____ Appliances_____
Cause_____ Diagnosis_____ Disability_____

Method of reporting test: Fill in spaces with black pencil.

1. Leave blocks blank if the activity cannot be performed independently.
2. Fill in first block when activity is attempted.
3. Fill in two blocks when activity can be performed but with assistance.
4. Fill in three blocks when activity can be performed independently and adequately.
5. Draw diagonal parallel lines if the activity is unnecessary or beyond the age level of the child.

Method of recording progress: Fill in space with red pencil as child fulfills requirements listed above in 2, 3, 4.

WARD ACTIVITIES—REHABILITATION NURSE

I. BED ACTIVITIES WITHOUT BRACES

1. Roll to right side
2. Roll to left side
3. Sit up
4. Sitting balance
5. Put on braces   WITH BRACES
6. Remove braces
7. Bed to wheelchair
8. Wheelchair to bed
9. Bed to standing
10. Standing to bed

III. HYGIENIC ACTIVITIES WITHOUT BRACES

1. Comb hair
2. Brush teeth
3. Wash, dry, hands and face
4. To tub from wheel chair or standing
5. Take bath
6. Dry body
7. From tub to wheel chair or standing   WITH BRACES
8. Wheelchair to toilet
9. Adjust clothing
10. Toilet to wheelchair
11. Standing to toilet
12. Adjust clothing
13. Toilet to standing

II. DRESSING ACTIVITIES WITHOUT BRACES CLOTHING OFF ON

1. Undershirt or bra
2. Hat
3. Socks
4. Slipover garment   WITH BRACES CLOTHING OFF ON
5. Cardigan garment
6. Underpants
7. Trousers or skirt
.8. Shoes

CHART 17

HAND ACTIVITIES

OCCUPATIONAL THERAPIST

Date _____   Patient _____

C.A. at Testing____yr.____mo.

Handedness: Right (___)   Left (___)   Undetermined (___)

I. PREPARATORY ACTIVITIES   RIGHT   LEFT

1. Reach object on table
2. Reach object above head
3. Prehend 1-inch cube
4. Grasp cylindric object (1-inch diameter)
5. Release object into container
6. Prehend coin
7. Drop coin into slot
8. Place hand behind back
9. Pick up object from floor

II. MANIPULATORY SKILLS

10. Pile 1-inch cubes
11. String wooden beads
12. Hammer pegs or nails in board
13. Hold book and turn pages
14. Screw, unscrew jar cap (2-inch diameter)
15. Hold paper, cut with scissors
16. Open and close drawers
17. Turn doorknob
18. Operate faucet
19. Operate light switch
20. Operate doorbell
21. Open and close lock with key
22. Remove money from coin purse
23. Use pencil sharpener
24. Use telephone

Garment fastenings   OPEN   CLOSE

25. Large buttons
26. Small buttons
27. Snaps
28. Zipper
29. Buckle
30. Hook and eye
31. Shoelaces
32. Bow

III. FEEDING ACTIVITIES   WITHOUT APPLIANCE   WITH APPLIANCE

33. Drink through straw
34. Drink from cup with handle
35. Drink from glass (4-ounce) paper or plastic
36. Drink from glass (8-ounce) glass
37. Eat with fingers
38. Eat with spoon
39. Eat with fork
40. Pour liquid into glass
41. Spread butter with knife
42. Cut food with knife

IV. WRITING ACTIVITIES   WITHOUT APPLIANCE   WITH APPLIANCE

43. Initiate stroke with crayon
44. Color in single outline picture
45. Copy single lines and forms, pencil
46. Write with pencil
47. Use typewriter

CHART 18

AMBULATORY ACTIVITIES
PHYSICAL THERAPIST

Date_____   Patient_____

BRACES

	OFF	ON

I. WHEELCHAIR ACTIVITIES

1. Roll wheelchair forward 30 feet
2. Roll wheelchair back 10 feet
3. Lock and unlock brakes
4. Raise and lower footrests
5. Remove wheelchair arms
6. Lock knees of brace
7. Unlock knee of brace
8. Wheelchair to standing
9. Lock hips
10. Unlock hips
11. Standing to wheelchair

II. AMBULATION

Gaits

	IN PARALLEL BARS BRACES		WITH CRUTCHES BRACES	
	OFF	ON	OFF	ON
1. Four-point				
2. Three-point				
3. Two-point alternate				
4. Tripod alternate				
5. Tripod simultaneous				
6. Swing-to				
7. Swing-through				

III. CLIMBING ACTIVITIES

	WITHOUT BRACES	WITH BRACES	CRUTCHES AND/OR BRACES
1. Up the ramp			
2. Down the ramp			
3. Up 4-inch step—rail			
4. Down 4-inch step—rail			
5. Up 8-inch step—rail			
6. Down 8-inch step—rail			
7. Up 8-inch step—no rail			
8. Down 8-inch step—no rail			
9. Up 2-inch curb			
10. Down 2-inch curb			
11. Up 6-inch curb			
12. Down 6-inch curb			
13. Up 15-inch bus step—rail			
14. Down 15-inch bus step—rail			

4. *Speech and hearing.* Approximately 70% of the children with cerebral palsy need speech training. About 5% of the athetoid patients have a hearing loss that requires the use of a hearing aid. The ability to speak and hear is essential for adequate socialization, education, and a vocation.

5. *The appearance of being normal.* The ability to ambulate is important, but we must endeavor to have the child perform this activity in a fashion as nearly as possible like the normal walking pattern. A child's one objective is to go from place to place, in any way possible, but serious emotional problems are likely to arise in adolescence when he realizes his awkward manner of traveling differs from that of his companions. The proper patterns of performing activities must be established during childhood, especially in those who have the potentials for higher education and vocational possibilities.

Many disciplines working together as a group are required to meet the total needs of the handicapped child.

DIAGNOSTIC EVALUATION

The diverse physical, psychological, and educational problems presented by most of the handicapped children are so complex that no one person could possibly provide the requisite skill in more than one area. The handicapped child presents, literally, constellations of interdigitating and complex disabilities that in totality may tax the training and skill of a large variety of professionally trained people. Detailed diagnostic assessment is an essential prelude to formulation of an adequate rehabilitation program. The type of professional person involved in the evaluation will vary with the child's handicap. For example, the urologist's evaluation and recommendations are of paramount significance in planning a program for the child with spina bifida manifesta, whereas the role of the orthopedic surgeon or the ophthalmologist may be the dominant one with another type of condition. We shall not detail the specific responsibilities of the members of the evaluation and therapeutic group but hope that we have made clear that it is through the collective judgment of the group that a program for meeting the tentative and ultimate rehabilitation goals is formulated.

DISCUSSION

In reviewing methods for the rehabilitation of children with physical disabilities, there are certain fundamental factors that must be considered if we are to meet the total needs of the child and his parents.

1. It is just as important to know what child has the disability as it is to know what disability the child has. This necessitates a thorough diagnostic evaluation by the essential specialists.

2. After the evaluation has been completed, a very careful interpretation and understanding counseling must be given to both the child and his parents. Sometimes the child or his parents, or both, have expectations from the program of rehabilitation that are in excess of the child's potential.

3. One of the most important concepts to keep before the parent and the child is that the more independent a child becomes within the expected performance for his age, the more he will be able to benefit from the experiences offered to him

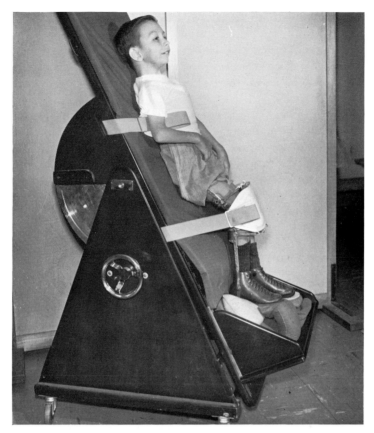

Fig. 24-2. By the use of a stationary tiltboard, even the child who cannot stand independently may view the world from the upright position.

Fig. 24-3. Group activities help a child to develop in intellectual, social, and emotional areas— an important part of his rehabilitation. Socialization while learning hand control is shown here.

through education, recreation, special activities, and vocational counsel and training (Figs. 24-2 and 24-3).

4. A rehabilitation program is based on the motivation level of the child. Many parents think that by daily intensive work a child will attain physical independence more rapidly. This pressure on the child often causes him to rebel and retards his progress.

5. Although evaluation must of necessity be a group process, the recommendations should be transmitted to the parent by one physician who will work closely with the family. It is essential, however, that all members of the therapeutic group know what the counseling program will be. Parents frequently question everyone who is related to the child's rehabilitation program. It is difficult for the therapist or a physician to make a response to parental questions in borderline areas, but if all concerned appreciate the need for a focused counseling program, it is easier to avoid creating confusion in the parent's mind regarding rehabilitation programs and objectives.

PROBLEMS AHEAD

Prevention remains the most gratifying approach. The vaccine for the prevention of poliomyelitis has now greatly reduced the number of children disabled by this condition. The accident prevention program also should reduce the number of disabled children needing rehabilitation.

Education of the handicapped child is a problem. Our schools should admit selected handicapped children to the regular classes so that they may benefit from the socialization with normal children. The normal child in the class will learn to accept the handicapped child, just as industry is learning to accept the handicapped worker. On the other hand, carefully selected special classrooms staffed with skilled educators who understand the child's learning disabilities are essential for many of these children. The educational program must be flexible and geared to meet the child's particular needs. Handicapped persons are an integral part of society today and do not exist as a group apart with separate lives. Their needs and rights are the same as those of any other person; their problems are the problems of all people and should be considered as a part of the whole society.

The end results of a program of rehabilitation for children are as follows:

1. There is a group of children who, because of their serious disabilities, will never be able to function independently and will need assistance the rest of their lives. When their parents can no longer care for them, it will be necessary to place them in an institution for residential care. Every state has institutions for the mentally retarded, but we know of no state that provides residential care for the physically disabled child of normal intelligence. There is a great need for institutions that will provide residential care for the normally intelligent child who has a severe physical disability.

2. As a result of the rehabilitation program, a number of the children will become adequate to meet all the physical demands of daily living but in a slow and awkward manner. They will never have the necessary speed and skill in the use of their hands to be acceptable in industry. Every community has a number of these handicapped young people, of normal intelligence and with good educational back-

grounds, who are unable to find employment. We cannot expect industry to employ handicapped persons if they cannot produce. To meet the needs of these persons, we must provide workshops that will give them an opportunity for employment at work best suited to their abilities and disabilities. If we do not provide these workshops, the time and money spent on the physical rehabilitation and education of these patients have been wasted.

3. As a result of our rehabilitation program there will be those who, if given proper vocational guidance and training, can meet all the requirements for full employment.

Rehabilitation has been shown to be both possible and rewarding. The results of the team approach to the rehabilitation of adults has far surpassed results of any other methods. While objective analysis of pediatric rehabilitation is not yet available, it is axiomatic that the results must be equally commendable.

REFERENCES

1. Bluestone, S. S., and Deaver, G. G.: Rehabilitation of the handicapped child, Pediatrics **15:** 631, 1955.
2. Deaver, G. G.: Manual of procedures of the children's division, Rehabilitation monograph 4, New York, 1952, Institute of Physical Medicine and Rehabilitation, New York University-Bellevue Medical Center.
3. Deaver, G. G.: Cerebral palsy: methods of treating the neuromuscular disabilities, Arch. Phys. Med. **27:**363, 1956.
4. Services for Physically Handicapped Children: The physician and child health services, Pediatrics **12**(part 2): 1, 1953.
5. Switzer, M. E., and Rusk, H. A.: Doing something for the disabled, Public Affairs Pamphlet no. 197, New York Public Affairs Committee, 1953.

CHAPTER 25

REHABILITATION OF PATIENT WITH CEREBRAL PALSY

Although cerebral palsy has existed through the ages, it was not clearly delineated in the medical literature until 1862 when W. J. Little, an English orthopedic surgeon, related birth difficulties to neurologic defects seen in 63 children.[1] During the last 20 years, there has been a tremendous increase in public and professional interest in this condition. There are so many facets important in diagnostic evaluation and rehabilitation of children with this neurologic condition that we cannot discuss all of them adequately in the space available here. However, effort will be made to discuss briefly the more important considerations and to refer the reader to appropriate sources for details.

DEFINITION

Broadly considered, cerebral palsy is a neurologic deficit resulting from cerebral dysfunction and presents a mosaic of neurologic abnormalities. No two children are identical, and the difficulties created by abnormal brain functioning range through the entire gamut of neurologic, orthopedic, and psychologic problems. They also present a variety of learning handicaps and problems of social and vocational adjustment.

In a more conventional and restricted concept, cerebral palsy is a nonprogressive brain disorder occurring during gestation, parturition, or the neonatal period with resultant abnormality of movement and/or posture. Since Phelps[2] first used this term with reference to children with disorders of movement and posture, it has become customary to restrict application of the term to patients with neuromotor deficit. During the last two decades the appellation brain damage has been applied to those children with evidence of organically based brain dysfunction,[3-5] particularly if the neuromotor deficit is minimal. Restriction of the term cerebral palsy to those with neuromtor dysfunction resulted in a tendency to place primary emphasis on the motor deficit. In recent years, Paine[6] and Prechtl and Stemmer[7] have used the term minimal chronic brain syndrome for those children with very mild motor deficits. It is now thoroughly appreciated that these children also have sensory, perceptual, visual, auditory, speech, and cognitive defects in varying combinations and degrees. There have been a number of discussions of definition over the years, which indicate the difficulty in coming to an agreement about a single uniformly acceptable definition.[8-10]

INCIDENCE

It is almost traditional to begin a discussion of incidence of cerebral palsy by quotation of Phelps' figure of a total of 7 new cases per 100,000 population (7 cases per 1500 births).[11] This is quite comparable to the incidence study in Schenectady[12] (5.9 cases per 1000 births) but is higher than incidence figures from England. Asher and Schonell[13] reported a school-age incidence of 1 per 1000. Woods[14] reported 2.5 per 1000 live birth in Bristol, England, in the period from 1943 to 1948 and a reduction to 0.9 per 1000 live births in the period from 1958 to 1962. She attributed the lower incidence to improved antenatal and obstetric care. However, this report has provoked a number of comments that disagreed with the concept of diminished prevalence. It is likely that there may be nearly 500,000 persons with cerebral palsy in the United States.

ETIOLOGY

Many causative factors produce cerebral palsy; in fact, it is the multiplicity of causative factors that led Denhoff and Robinault[15] to refer to cerebral palsy as a group of diseases. The factor common to the group is time of occurrence of the brain dysfunction—in the prenatal or natal period. The question of postnatal occurrence will be mentioned later.

Most of the diverse etiologic factors that produce cerebral palsy can be reduced to the following categories:
1. Developmental defects
2. Hypoxia and hemorrhage
3. Infections, toxins, and poisons
4. Trauma
5. Isoimmunization reactions
6. Defects of biochemical maturation
7. Defects in the hereditary material (genetic)

There are so many specific mechanisms under each of the above major categories that we will not attempt to list them but will briefly discuss some problems of etiologic identification. Cerebral palsy incidence is directly proportional to the degree of prematurity, and one sometimes finds prematurity listed as an etiologic factor. We prefer to think of prematurity as a reflection of many factors including one or more of the etiologic factors mentioned.

The frequency of occurrence of developmental defects of the brain is probably not adequately appreciated. Polani[16] believes that approximately one third of the cases belong in this category. Some of the above categories overlap. For example, a particular developmental defect of the brain may result from defective biochemical maturation of the brain, which in turn may be produced by a gene defect. In similar vein, in past years about 10% of the children with cerebral palsy acquired the brain lesion as a result of jaundice due to immunization mechanisms related to mixing of the fetal Rh-positive blood with maternal Rh-negative blood (isoimmunization). The indirect bilirubin gained access to the brain and stained certain areas of the brain yellow (kernicterus). We now know the genetic mechanisms and much of the biochemical background that produces the jaundice. We also know how to prevent the brain damage (exchange transfusion), and consequently we see children with this etiologic causation much less frequently. Some would now prefer to take this condition out of the heterogeneous grouping of cerebral palsy and simply call it the "kernicterus syndrome." These two circumstances are mentioned under etiologic considerations primarily to illustrate the diversity of affective conditions and agents and to indicate some of the difficulties involved in arriving at an accurate designation of etiology.

With reference to time of initial effect of the etiologic factor, it is customary to use a schema somewhat like the following:

$$\left.\begin{array}{l} \text{Congenital} \\[2mm] \text{Acquired} \end{array}\right\{ \begin{array}{l} \text{1. Prenatal} \\ \text{2. Natal (including 2 weeks neonatal)} \\ \text{3. Postnatal} \end{array}$$

Brain damage in the fetus from maternal rubella in the first trimester is clearly within the category of a prenatal factor. On the other hand, a causation in the prenatal period may not be reflected by noticeable maternal clinical signs and symptoms. In such a case the fetus may be born after a difficult delivery and be quite cyanotic. The finger then would be placed on hypoxia (cyanosis) as the causative agent, whereas that in fact may have had little or nothing to do with the resultant neurologic deficit. The same problems arise with respect to categorizing the natal and postnatal times of occurrence. It is very difficult to know when the brain injury occurs. We have been quite arbitrary about this matter and consider the cerebral palsy to be congenital if the etiologic factor occurred during gestation or the neonatal period (first 2 postnatal weeks) and to be acquired if it appears that onset of the cerebral palsy was clearly after the neonatal period. There are also many occasions wherein this delineation is exceedingly difficult. We found[17] that 11% (143) of a total population of 1283 children appeared to have experienced the disabling factor after the neonatal period. In order of frequency, the etiologic factors in the children with acquired cerebral palsy were infection (47%), trauma (17%), cerebrovascular accident (12%), neoplasms (8%), postseizure complications (7%), surgical complications (6%), and a miscellaneous group (3%). There are some who do not use the term cerebral palsy for these children. However, we prefer to use the term acquired cerebral palsy for such children.

CLASSIFICATION OF CEREBRAL PALSY

The diverse motor- and nonmotor-associated defects of cerebral palsy would suggest the difficulties in categorizing these children. A number of attempts at classification have been made.[10, 18-22] There are significant differences in all of these suggestions for classification. They all violate some of the basic principles of classification and etymologic syntax. For example, the classification of Crothers and Paine[21] makes unwarranted assumptions about the site of the pathoanatomic lesion (extrapyramidal).

No one appears to be completely satisfied with any of the suggested classifications. However, if we are to communicate effectively with each other, a categorization of the children, understood by all, is essential. There is a widespread feeling that the classification developed several years ago by the American Academy for Cerebral Palsy[20] is unnecessarily complex. Several authors have complained that we have borrowed terms from adult neurology and that such terms are inappropriate to describe the disorders of movement and posture observed in neurologically impaired infants and children. Some are disturbed because the type and degree of intensity of the involuntary movements or muscle reactivity change with time even though the brain lesion may be static.

In our clinic we categorize the children with cerebral palsy on the basis of the type of neuromotor deficit and topographic distribution of the deficit according to the following schema, which uses only the major groupings of the classification of the American Academy for Cerebral Palsy.[20]

Neuromotor classification:

1. *Spastic.* This type is characterized by exaggeration of the stretch reflex and increased deep tendon reflexes in the affected parts.
2. *Athetotic.* The chief characteristics of this type are the slow, wormlike, involuntary, uncontrollable, unpredictable, and purposeless motions at rest. Twelve different types of athetosis have been described.
3. *Rigidity.* When the part is moved, there is a continuous resistance in agonist and antagonist muscles, simulating the sensation of bending a lead pipe. The degree of rigidity may, from time to time, be referred to as "lead-pipe," "cogwheel," or "intermittent" rigidity. The principal clinical finding is the hypertonicity of the muscles, which in some patients is so great that no motion is present. The tendon reflexes are normal.
4. *Ataxic.* The principal sign noted in the ataxic patient is disturbance of balance and equilibrium. The ambulation pattern has been described as a reeling, drunken-type gait.
5. *Tremor.* The chief characteristics of this type are the involuntary, uncontrollable motions that are reciprocal and regular in rhythm. This type is very uncommon.
6. *Mixed.* Not all children with cerebral palsy can be diagnosed as true spastics, athetotics, or ataxics. About 1% of the total may be mixed cases in which there is more than one type of the above-described characteristics.

Topographic distribution of neuromotor involvement:

1. *Paraplegia*—involvement of the lower extremities. The patients with paraplegia are practically always of the spastic type.
2. *Hemiplegia*—involvement of an upper and lower extremity on the same side of the body. Persons with hemiplegia are almost always spastic, but occasionally an athetotic hemiplegic patient may be seen.
3. *Triplegia*—involvement of three extremities, usually both lower extremities and one arm. The disability is usually of the spastic type.
4. *Quadriplegia or tetraplegia*—involvement of all four extremities. The term diplegia is sometimes used to indicate that the lower extremities are more involved than the upper extremities. Almost all patients with athetosis have all four extremities involved.

We believe that much of the confusion relating to classification of types of neuro-motor deficit arises by considering the dysmetric voluntary movements of the spastic patient to be evidence of athetosis. True athetotic movements are involuntary and occur at rest but are exaggerated during voluntary effort.

With respect to terminology applicable to topographic distribution of the neuro-motor deficit, we find it less confusing to use the term quadriparesis, with greater involvement of the right or the left side, rather than to use the term double hemi-plegia. Likewise, quadriplegia with greater involvement of the lower extremities avoids the uncertain meaning of the word "diplegia." Many of these areas of con-fusion in classification will eventually be clarified and, perhaps, simplified through forthcoming international efforts to reach accord.

Classification based on severity

One could organize several categories of severity classification, such as severity of mental retardation, severity of speech deficit, or severity of neuromotor involvement. Generally speaking, when a classification of this type is used without qualification, it is based on severity of neuromotor involvement that limits the patient's ability to per-form Activities of Daily Living. On the basis of severity the following categories are useful:

1. *Mild.* The patient needs no treatment, since he has no speech problems, is able to care for his daily needs, and ambulates without the aid of any appliances.
2. *Moderate.* The patient needs treatment, since he is inadequate in self-care, ambulation, and/or speech. Braces and self-help appliances are needed.
3. *Severe.* The patient needs treatment, but the degree of involvement is so severe that the prognosis for self-care, ambulation, and speech is poor.

CLINICAL AND PATHOANATOMIC CORRELATES

Except for kernicterus cerebral palsy, there is very little in the nature of com-prehensive clinical data that has been correlated with complete morbid anatomic studies. There is a distressing tendency to think of spasticity in terms of the pyra-midal tracts and athetotic cerebral palsy as a reflection of extrapyramidal lesions. This is a gross oversimplification of the neurophysiologic mechanisms of movement and posture and is not borne out by postmortem studies. To take an extreme example, there are reports of neurologic rigidity in which the lesion was not even in the brain but in the intrafusal muscle fibers in skeletal muscle spindles. Our knowledge of neurophysiologic mechanisms and reactions of the newborn human brain is too fragmentary to enable us to determine accurately the anatomic localization of the lesion on the basis of clinical examinations of children with cerebral palsy. However, Courville[23] and Towbin[24] have provided good reviews of existing information on findings in the brains of a number of children with cerebral palsy. The neuropatho-logic changes incident to hypoxia are well established, but detailed clinical data accompanied with complete morbid anatomic study are very scanty.

ASSOCIATED DEFECTS IN CEREBRAL PALSY

Seizures. There is great variation in the reported incidence of seizures in cerebral palsy. Woods[25] reported an incidence of 37%. Perlstein and associates[26] an incidence of 50%, and Yannett[27] an incidence of 68%. Melin[28] found the reported incidence to vary from 14% to 75% and concluded that at least one third of the children have a seizure at some time in their lives. Illingworth[29] found an incidence of convulsions of 37.5% in mentally retarded children with cerebral palsy and 31.3% in mentally retarded children without cerebral palsy. About 80% of the children have abnormal electroencephalograms. However, the diagnosis of a convulsive disorder is not precluded by finding a normal electroencephalogram, nor is it established by an abnormal electroencephalogram. Accurate diagnosis is essential to effective anticonvulsive therapy. The electroencephalogram is a useful diagnostic aid that is used routinely in our clinic.

The incidence of seizures is about three times higher in spastic cerebral palsy than it is in athetotic cerebral palsy. It is axiomatic that all seizures should be prevented. Frequent seizures interfere with learning and present hazards of falling. There is also the ever-present danger of additional brain damage as a result of increased intracranial vascular pressure.

There are many clinical types of seizures. Some types respond better than others to certain anticonvulsant compounds. A wide range of effective anticonvulsant drugs exists for the various types of seizures. The drugs effective for the several types of seizures, the dosage required, and the side effects are very well presented by Livingston.[30] By proper selection of anticonvulsant agents in an appropriate dosage schedule, it is possible to keep the great majority of the children free of seizures. Nevertheless, a small percentage of the children appear to have seizures quite intractable to pharmacologic control. Many of these children have been relieved of their seizures by hemispherectomy, and their level of intellectual functioning has improved.[31] These children verify Dr. Wilder Penfield's dictum: "Bad brain is worse than no brain."

Variability of motor capacity and behavior. There is a tendency to think of the cerebral palsied child as one whose motor ability is moderately to severely limited by muscle stiffness, incoordination, involuntary movement, or other neuromotor qualities that provide motor disability. Children of this type present the obvious and classic neurologic signs of neuromotor disability. In recent years, there has been increasing awareness that the continuum of motor involvement extends almost insensibly to the area of normal neuromotor function. Children with very mild neuromotor deficit may exhibit what are frequently considered to be soft or subtle signs of organic neuromotor deficit. These children are the children described by Paine[6] and by Prechtl and Stemmer[7] as children with minimal brain damage. Frequently the so-called brain-damaged child who may present a host of behavioral and learning disabilities without cerebral palsy is described as being awkward or clumsy in motor performance. This awkward and clumsy motor performance is the subtle expression of central nervous system deficit reflected in this minimal degree of motor deficit, which is in fact minimal cerebral palsy.

Bender[32] noted lack of concentration, hyperexcitability, and perseveration as behavioral disorders in brain-damaged children. Strauss and Lehtinen[33] focused attention on the hyperkinetic behavior, and Laufer and Denhoff[34] referred to the excessive motor activity as a "hyperkinetic impulse disorder." Birch[35] has pointed out that hyperkinesis represents but one of a variety of behavioral consequences of brain damage. These children may also have abnormal distractibility, conceptual rigidity, emotional lability, and other abnormalities that suggest that one of the primary and major reflections of organic brain damage is in the area of behavior.

The hyperkinetic child may not necessarily reflect a markedly increased motor energy output. There are indications that his motor behavior is impulsively related to his stream of thought and is released on inappropriate occasions.

In this very brief discussion of motor behavior, we may have given the impression of equating hypokinetic behavior with frank cerebral palsy and hyperkinetic behavior with minimal brain damage. However, the problem is more complex than this because many brain-damaged children without even soft signs of neuromotor involvement do not show abnormal motor activity.

Hearing defects. There is great variation in the reported incidence of hearing loss in children with cerebral palsy. Hopkins and co-workers[36] found 5% with definite defect and 8% with questionable loss among 1293 children tested. At the other extreme, Gerrard[37] found loss in 80% of the children with kernicterus. It is likely that the variable findings reported are influenced to some extent by the number of children with kernicterus included in the population studied. Rosen[38] carefully studied 33 children with kernicterus who had previously been labeled deaf or hard of hearing and found one third of them to have normal hearing. He felt that their distractibility and fluctuating attention led to inaccurate diagnoses of auditory deficit. Hannigan[39] has described aphasic-like language difficulties in 20 cases of kernicterus. Although it is quite possible that errors are made in acoustic evaluation of many children with kernicterus because of the difficulties in testing, it is our opinion that there is a high incidence of hearing loss at the higher frequencies in children with kernicterus. This does not preclude the possibility of there being speech and language problems other than those of dysarthria secondary to athetotic movement.

Involvement of brain components of the auditory system in human kernicterus is apparently quite variable. Gerrard[37] and Dublin[40] found the cochlear nuclei quite heavily stained. On the other hand, Haymaker and associates,[41] in pathoanatomic studies of 87 brains, found the medial geniculate body but rarely stained, and the area of the lateral lemniscus occasionally involved. Windle and co-workers[42] found staining of the inferior callliculus of the monkey brain, in experimental hyperbilirubinemia associated with asphyxia, so constant that he studies this nucleus as an indicator of brain damage in hypoxia. The pathoanatomic studies support the clinical evidence of auditory deficit. All children with delayed speech development should have very careful and, if necessary, repeated audiologic evaluation.

Visual deficits. Gesell and colleagues[43] pointed out that strabismus may be one of the earliest signs of cerebral palsy. Subsequently, Guibor[44] and Breakey[45] found that over 50% of the children have strabismus. Douglas, as he reported in a chapter in Henderson,[46] found a 36% incidence of squint. Esotropia is the most frequent type of strabismus found. Crothers and Paine[21] found 25% of the spastic hemiplegics to have homonymous hemianopsia. Douglas, who carefully examined 166 children with cerebral palsy, found 14 blind and 10 partially blind (visual acuity 4/60 to 6/24). The incidence and degree of visual loss were directly related to the degree of mental retardation. Douglas also found optic nerve atrophy in 9.9% of the children he studied. This is about three times higher than the figures reported by Guibor[44] and Breakey[45] and may reflect the more serious involvement of Douglas' patients. Douglas, Guibor, and Breakey appear to be in agreement that 55% to 60% of all the children have visual defects of some type. The significance of squint and loss of binocular vision in terms of eye-hand coordination, inefficiencies of vergence movements, pattern discrimination, figure background detection, and depth perception is very important and is discussed interestingly by Abercrombie[47] and will be mentioned again later.

General sensory impairment. Phelps[48] appears to have been the first to mention general sensory deficit in children with cerebral palsy. Tizard and co-workers[49] found the general sensory loss to be in stereognosis, two-point discrimination, position and/or vibration, pain, and temperature in that order of decreasing frequency of loss. Loss of touch was about 10% as frequent as stereognosis and occurred nearly twice as frequently in postnatally acquired spastic hemiplegia as in the patients with prenatal or natal onset. Hohman and associates[50] studied the same general sensory modalities in 47 children and found loss in 80% of the spastic children and in 40% of the athetoid children. They referred to light touch, pain, and temperature modalities as noncortical losses. These occurred only in children with spasticity. Tizard and colleagues[49] categorized the patients they studied as having (1) spastic hemiplegia or (2) other cerebral palsies, but appear not to have listed the other cerebral palsy types except as other distributions of spastic involvement. In view of the importance of general sensory input in perception and learning, there appears to be great need for more extensive accurate data on general sensory deficit in children with cerebral palsy.

Perceptual problems. We have briefly mentioned the general sensory modalities and the special visual and auditory input in children with cerebral palsy. We must not, however, allow ourselves to think of these multiple input streams of impulses as something unrelated to the motor performance of the child with cerebral palsy. These flows of impulses into the spinal cord and brain are inextricably interwoven with and are part and parcel of the motor act. In a sense the totality of the sensory input is not only an activating force for motor performance but also a modulating and guidance mechanism in motor performance.

Another aspect of both general and special sensory input, of importance in understanding the problems of children with cerebral palsy, is the influence that these inputs have on the child's consciousness and how they are integrated with each other. For example, the eye may be normal from the anatomic and optical viewpoint: the points of light may be accurately focused on the retina and the nervous pathway to the visual center in the occipital area of the cerebral hemisphere may be intact. The figure that is observed may be seen clearly, but how is the object perceived? The complex process by which these specific sensory impressions are given meaning is known as perception. The cerebral palsied child may thus look at a diamond figure on a complex background. There will be these questions in the area of perception: (1) Does he see all of the lines at the precise angle to each other as they are in the figure? (2) How is the figure oriented in space? (3) How is it perceived in relationship to the background material? Perception also requires the ability to give appropriate relevance to foreground and background visual and auditory input and flexibility in shifting from visual to auditory to kinesthetic perception.

Strauss and Lehtinen[4] have described perception as the mental process that mediates between sensation and thought. Thus perception is intimately related to abstract cognitive functions on the one hand and to the use of symbols in communication (speech, language, reading, and writing) on the other. Mecham and co-workers[51] have described the relationship of perception to speech problems in children very well. Birch and Lefford[52] have shown that there are age-specific characteristics in the development of these intersensory relationships and have outlined strategies for studying perception in brain-damaged children.[35] The psychologic implications of perceptual deficits have been recently reviewed by Birch[35] and extensively discussed by Vernon.[53]

The perceptual deficits that the child with cerebral palsy may have must be carefully and fully analyzed. Psychologic and communicative evaluation tests must be devised and used not only to recognize the deficits but also to take them into consideration with reference to test performance. In view of the frequency of primary general and special sensory loss plus the frequency of brain injury in the cortical-perceptual integration centers, it is likely that these deficits occur far more often than is generally appreciated.

Communication problems. Departures from the broad range of normality in communication may have their origin in the area of emotionality, psychoneurosis, mental retardation, hearing loss, upper or lower motor neuron loss, localized or generalized brain dysfunction, or local defects of the articular apparatus. The communication defects may include speech defects and/or reading and writing disabilities. These capacities of communication require neural integration that is most complex and represents the highest level of neural function. A given cerebral palsied child may present a varied combination of these communication disorders.

We shall provide a brief overall view of a few of the aspects of speech defect and allude to the interrelated problems of dyslexia and dysgraphia.

The reported frequency of speech impairment in children with cerebral palsy varies from 30% to 70%.[54] The range of variation in the incidence figures probably reflects the detail of the examination and the criteria of diagnosis. It would probably be safe to say that two thirds of these children have a speech deficit of some type.[46, 51, 54] The severity of the speech disorders varies from the minor one, which may represent more an annoyance than a significant functional handicap, to unintelligible or absent speech. In general, there is a direct correlation between the severity of the speech handicap and the motor disability. The retardation of speech development is also proportional to the degree of mental impairment.[51] Ingram[58] found dysarthria most common in quadriparetic children and more common among the diplegic than the hemiplegic children. Although the incidence of speech defect in ataxia appears to be very high, the intelligibility of the speech is generally good. On the other hand, the incidence of speech absence is highest in the severely spastic child, and dysarthria in severely athetotic children may make these children almost unintelligible. There have been attempts to describe the characteristics of the speech of the cerebral palsied child,[54, 55] but Gratke[56] feels that it is difficult to generalize about it because of the variability that exists. This is supported by Hedges' study,[57] in which professional judges could not distinguish between the speech of spastic and that of athetotic children.

We have referred elsewhere to the difficulties incident to borrowing terms from adult neurology and applying them to children. It is in the area of speech diagnosis and pathology that this difficulty becomes critical. There are many reports of "expressive," "receptive," or "global aphasia" in children with congenital cerebral palsy in which the verbal capacity and behavior of the child in no way resembles those of the adult who has lost a component of speech after many years of competency in speech and language. An indication of this confusion can be obtained from the discussion in the Proceedings of the Institute of Childhood Aphasia.[59] This problem is more than a matter of semantics and etymology. It has to do with fundamental concepts of the neural mechanisms essential to acquisition of speech. It is intimately related to postnatal brain maturation and problems of the delayed maturation of damaged brains. The range of anatomic aspects of the speech disorders in children with cerebral palsy can be appreciated by reviewing the following classification of common speech disorders, which was developed by Ingram[58] and is discussed by him with reference to cerebral palsy.[60]

1. Disorders of voicing (dysphonia)
   a. With demonstrable disease of the larynx
   b. Without demonstrable disease of the larynx
2. Disorders of rhythm (dysrhythmia)
   a. Clutter
   b. Stammer or hesitation
3. Disorders of articulation with demonstrable dysfunction of articulatory apparatus (dysarthria)
   a. Due to neurologic abnormalities
      Cerebral palsy
      Suprabulbar palsy
      Lower motor neuron lesions
   b. Due to local abnormalities

Jaws and teeth	Hypomandibulosis
	Other malocclusion
Tongue	Tie
	Tongue thrust
Lips	Cleft lip (only)
	Other
Palate	Cleft (with or without cleft lip)
	Other
Pharynx	Large pharynx (palatal disproportion)
	Acquired disease
Mixed	

4. Disorders of articulation without demonstrable dysfunction of articulatory apparatus (secondary speech disorders)
   a. Secondary to hearing defect
   b. Secondary to mental retardation
   c. Secondary to psychogenic disorders
   d. Secondary to dysphasia due to brain damage

5. The developmental speech disorder syndrome (specific developmental speech disorders)
   a. Involving language development and articulation
   b. Involving articulation only
6. Mixed cases
7. Unclassified and other*

It is apparent that contributions to the speech disorder seen in cerebral palsy can be derived from most of the categories listed. The delayed development of speech in children with cerebral palsy is designated under his classification as 4d, secondary to dysphasia due to brain damage.

A number of different approaches to speech therapy are described by Mecham and associates[51] and by Hawk and Young.[61]

Brain damage in children and adults, with resultant communication disorders, is also related to the problem of cerebral dominance and interhemispheric relationships. The fact of cerebral dominance has been known and extensively written about for more than 100 years. It is essential to discuss briefly the problem of cerebral dominance because it is related to some of the communication and learning problems that we see in children with cerebral palsy. There are also theories of cerebral palsy treatment that are based on concepts of cerebral dominance and interhemispheric relationships.

Left hemispheric dominance in speech function in adults is well established. However, the establishment of this dominance appears to occur over a period of several years during postnatal brain maturation. The term cerebral dominance means that function is asymmetrically represented in the two halves of the brain. For this reason, symmetric unilateral lesions do not produce equivalent efforts.[62] This does not mean that one hemisphere exerts control over the opposite one. Section of the corpus callosum in man[63] does not support the idea that the dominant hemisphere controls the minor one. Although there are some exceptions, one can say that in a right-handed adult the probability of right cerebral dominance is practically nil as far as speech is concerned.

In infants extensive damage to the left hemisphere does not prevent speech development but may delay it. These observations suggest that cerebral dominance in the left human hemisphere gradually develops over a number of postnatal years. The hemispheres probably have equipotentiality at birth, but this may be lost—at least as far as speech is concerned—as the acquisition of speech occurs.

The problems of delayed speech development compose one of the most frustrating difficulties with which the parent must contend and also require much skilled parental counseling. Palmer,[64] Van Riper,[65] and Lewis[66] have provided useful manuals to assist parental understanding of speech and language problems of children with cerebral palsy.

Mental retardation. The pendulum of professional opinion regarding the incidence and degree of mental retardation in children with cerebral palsy has taken

---

*From Ingram, T. T. S.: A description and classification of the common disorders of speech in children, Arch. Dis. Child. **34:**444, 1959.

wide swings during the past 100 years. Both Little and Freud referred to the problem, and initially it was thought that virtually all of the children were significantly retarded. On the other hand, undue optimism appeared to result from the statement of Phelps[67] that 70% of the children were normal.

When tests were first used, the test instruments had been standardized on homogeneous populations without handicap, and the test instruments were not modified to make essential allowances for the handicap. Therefore, there was a tendency to underrate the child with cerebral palsy. This led to the impression that if adequate batteries of tests were designed for children with cerebral palsy, little retardation would be found.

In recent years, adaptations of test instruments have been made, and the results suggest that 50% of the children have an IQ below 70 and that 25% test below an IQ of 50 (Cruickshank and Raus[3]). It is likely that too much attention has been given to the intelligence quotient and too little consideration to the child's mental age, in terms of his capacity to acquire certain functions. There is evidence of a close correlation between mental capacity and communication skills.[68] The degree of retardation is greater in spastic children than in those with athetosis. Taylor[69] has discussed the psychologic problems extensively.

For the past 25 years the approach to psychologic evaluation of children with cerebral palsy has been dominated by psychometric principles and technics.[70] The child's intellectual functioning has been evaluated against a background of average age-level achievements. In recent years, coincident with the English translations of the monumental works of the French psychologist, Jean Piaget,[71-73] psychologists have been examining his revolutionary concepts of childhood intellectual achievement. Piaget believes that the child develops his concepts of, and orientation in, space through the avenues of the sensory modalities and that this sensomotor phase of development requires about 18 months. Piaget associates the child's development of spatial representation to his capacity to copy figures, and he separates the concept of numbers from the ability to count. It is apparent that the piagetian concepts of intellectual development are stimulating new approaches to studying problems of intellectual development of children.

Emotional disturbances. Emotional problems observed in children with cerebral palsy vary greatly in frequency and severity. In addition, they appear to wax and wane during different times of life. The direction that the emotionality takes may also vary with age and with sex. Furthermore, the emotional problems are related to the mental capacity of the child and to his awareness and sensitivity to his handicap.

Hersov[74] has emphasized the need for more accurate data concerning the prevalence of emotional disturbance and its relationship to age, sex, intellectual capacity, and other factors. Floyer[75] found that 42% of the affected children displayed "excessive emotionality." Phelps[2] described different types of emotional problems and personality characteristics in relationship to the neurologic type of motor involvement (spastic and athetotic), but Cruickshank and Raus[3] found the emotional and personality problems to be more importantly related to the environment than to intrinsic factors.

There is little doubt that parental and sibling attitudes toward the child contribute very significantly to the child's reaction to his problem. In fact, every significant social contact the child has makes an imprint on the sum total of his emotional attitudes and responses. On the other hand, one cannot entirely discount organic contributions to this problem. There are abundant experimental and clinical data indicating effects of lesions in the hypothalamus and the limbic system on emotional behavior. There is no reason to believe that these vital and primitive brain structures are exempt from organic involvement in many of these patients. In fact, the pathoanatomic data clearly show imvolvement of the limbic system in kernicterus (hippocampus).[41] Emotion is a way of feeling and a way of acting. It may be shown in a positive way toward a person or object or may take the form of negativism, from a person or object. It may be accompanied by motivation and impulse to act or by depression of movement. We should also appreciate that emotional experience and expression may be quite distinct. Ironside[76] pointed out that patients with certain types of neurologic deficit may have bouts of crying or laughing without experiencing emotion and that the parkinsonian patient may remain passive and yet experience a high degree of emotion. We do see emotional lability in many of our children with cerebral palsy.

The nature of the emotionality runs the gamut from temper tantrums and gastrointestinal upsets in younger children to severe depression in the adolescent as he begins to appreciate that normal sexual drives, social contacts, and marriage are beyond his reach. We have seen anxieties intensified in adults with cerebral palsy when they realize that the help and protection provided by their parents may be lost.

The emotional problems we see in cerebral palsy very likely represent contributions from both intrinsic and extrinsic psychosocial influences. These complex problems require attention for many years.

Learning disability. Consideration of the previously mentioned neurologic deficits, frequently observed in children with cerebral palsy, should make it clear that any one of them might, in isolated form, comprise a learning disability. The defects, however, rarely occur singly but are constellations of deficits, with no two patients being precisely alike. Cockburn carefully studied the educational handicaps of 225 cerebral palsied children and found that, when compared with normal children, 85% of the cerebral palsied children over 7 years of age were below age level in reading and 93% were below age level in arithmetic.[46] Only 25% of his sample of cerebral palsied children were of average intelligence or above. The children with the spastic type of neurologic involvement averaged significantly lower in intellectual capacity than those with athetotic movements.

Backwardness in reading (dyslexia) is one of the most common learning disabilities. We do not imply that all children with dyslexia have cerebral palsy or brain damage, for there are problems of instructional methods, motivation, emotional factors, and—perhaps—genetic factors that may play a role in this disability. Visual and auditory dysfunctions also have facets that relate to dyslexia. These problems, including cerebral dominance and altered or delayed concepts of body image, impinge on the learning problems that confront the special educator working with cerebral palsied children. The teacher must give recognition to these problems and be alert to technics for circumventing or minimizing the effect of these disabilities in learning. Finally, it should be acknowledged that although numerous studies indicate poor educational performance of children with cerebral palsy, there are extrinsic factors that contribute significantly to this poor performance, such as the frequent and prolonged periods of absence from school that are related to medical care, the transportation difficulties, and the inadequate numbers of schools providing the special educational opportunities that these children need.

EVALUATION OF THE CHILD WITH CEREBRAL PALSY

It is hoped that constant reference to development and maturation will make clear that evaluation of the child with cerebral palsy is done with reference to the child's level of development as compared to a child of similar age without handicap. The word development is used here in a collective sense. The child's total functioning is seen as representing his status in physical, mental, emotional, psychologic, and social growth and development, and he must be evaluated from all of these viewpoints. It is obvious that no one person could possibly have the requisite information and skills to make the necessary assessment in all these areas. The evaluation of the child with cerebral palsy requires careful study by a group of professionally trained persons.

The treatment or management program is best planned when those who have evaluated the child can meet and discuss the interdigitating problems that must be taken into account. The plan of management must also recognize the social and geographic aspects of the child's environment in order that the recommended treatment plan be acceptable to the parents and compatible with the child's environmental conditions.

MANAGEMENT OF THE CHILD WITH CEREBRAL PALSY

Evaluation goals and parental counseling. We have deliberately avoided the use of the word treatment because the multiple facets of disability in cerebral palsy suggest that one child may need intensive speech therapy, another may require a program with emphasis on occupational therapy, and a third patient may have his principal difficulty related to some special educational technic designed, hopefully, to circumvent a perceptual problem. Despite these diversities, it is possible to phrase the general objectives of a management program that is applicable to all children. This objective is to make the child as physically independent and as socially and vocationally competent as possible within the limits of his irrevocable disability. This means that as far as his neurologic impairment will permit, he must be as independent as possible in all Activities of Daily Living and must be provided with appropriate education and, ultimately, vocational training to achieve maximal social acceptance and vocational independence.

It is necessary that we keep in mind constantly the following items that are literally touchstones to achieve these goals:

1. Continuous parental educational program with emphasis on realistic parental goals
2. Prevention of deformity
3. Motivation of the child

There are some who feel strongly that we should have added a fourth item to the above avenues by which one would attempt to achieve the rehabilitation goals that we have outlined. The fourth item might be a reference to one of a variety of ways to elicit neuromotor movement or to guide neuromotor movement patterns in physical and occupational therapy. We have elected to discuss certain aspects of this problem later. At this time we wish only to emphasize the importance of a parental counseling and educational program. Parental counseling must be an antecedent to the treatment program. The parents should know enough about cerebral palsy to understand why the child must be evaluated by a variety of trained professional people. They must have a clear idea of the objectives of the program and should take an active part in the therapy themselves.

It is important that the counseling be done principally by the physician but equally important that all concerned with the child's evaluation and treatment know what the counseling program will be. Conflicting advice from two different treatment centers can be understood by the parent, but conflicting advice from the same staff group is exceedingly difficult for the parent to handle.

Parental counseling is not a "one-shot affair" as Reed[77] has described genetic counseling but is a continuous and time-consuming process that has been described effectively by McDonald.[78] Not infrequently, it is unwise to give all of the advice at one time—even though it is clear what the final recommendation will be. Repetition and appropriate timing in counseling are essential to achievement of parental acceptance and cooperation.

TREATMENT

In a broad sense treatment would include everything that one does on behalf of the patient. This includes physical treatment, psychosocial, speech, parental counseling, and special educational technics. We shall refer here only briefly to certain aspects of the physical treatment because this aspect of treatment is often given emphasis by referring to it as a "system of treatment," with implication that there is something unique in one or another system that gives it great superiority over another method or aspect of therapy. We shall not take any protagonistic or antagonistic viewpoint toward any of those mentioned or describe them in any detail. Semans[79] and Gillette[80] have reviewed most of them at length. However, we shall express some points of view in the discussion of the prognosis of cerebral palsy that are applicable to all methods of treatment.

In 1895 Sherrington and Mott[81] described experiments that showed that stimulation of the skin would induce reflex movement. Hagbarth[82] showed that stimulation over most of the skin of the limb facilitates flexor monosynaptic reflexes and inhibits extensor ones. A flexor monosynaptic reflex was inhibited if the skin area over its antagonist was stimulated. The elicitation of reflex motor movement by skin stimulation is well established, and there appear to be clear-cut relationships of skin areas to flexor and extensor responses. As a general rule, the data suggest that skin over an antagonistic pair of muscles is receptive for an ipsilateral extensor, rather than flexor, reflex.

Lack of space precludes discussion of crossed reflexes of cutaneous origin, but we should observe that reflex movement elicited by skin stimulation is also related to the strength of the stimulus through the involvement of fibers of different diameter in the reflex. There also appear to be important relationships to the level of the excitatory state in the internuncial pool. There are many descending cortical and subcortical fiber tracts that terminate in the motoneuron pool, some of which provide excitatory and others inhibitory influences. In addition to the rather discrete pathways, there is a phylogenetically ancient, diffuse reticular system that has its own level of internal excitability and exerts either excitatory or inhibitory influences on the internuncial motoneuron pool that modifies the motor outflow to the muscles. There is also evidence that, through its widespread cortical and subcortical connections, the reticular system influences many higher, integrated neural functions that may be disturbed in children with cerebral palsy. The tremendous literature on the reticular system and its neurophysiologic importance is reviewed interestingly by Magoun.[83]

In 1955 Granit[84] demonstrated some of the neurophysiologic activities and neuro-anatomic connections of muscle spindles and tendon organs, which suggested their role in modifying muscle activity. The central connections made by afferent fibers from muscle spindles and tendon organs together with the small fibers that terminate on the intrafusal fibers of the muscle spindles compose the so-called gamma system. The role of the gamma system in movement and posture has been summarized recently by Boyd and co-workers.[85] Rood[86] has made these neurophysiologic concepts the basis of her recommended technic for eliciting movement of the extremities in children with cerebral palsy.

Rushworth[87] has reviewed neurophysiologic considerations of spasticity and rigidity and the modification of these states by intramuscular procaine injection. Recently Tardieu and Hariga[88, 89] reported extensive experimental animal and clinical studies on the effect of injecting 30% ethyl alcohol intramuscularly at the motor point. They found the spasticity to be ameliorated for a period of a few days to several months.

Interest in the postural reflexes was greatly stimulated by the important monograph on postural reflexes by Magnus.[90] Bobath[91] emphasized the role of postural reflexes in producing the abnormal postures seen in children with cerebral palsy. Dr. and Mrs. Bobath have based their physical therapeutic treatment technic on effort to influence the disturbed postural reflex.

Kabat[92, 93] wrote a series of papers in which he gave special attention to the role of proprioceptive input in the production of voluntary movement. Gellhorn[94] showed experimentally in the monkey that application of resistance to limb movement, which was induced by electric stimulation of the cortex, increased the strength of the movement. As an outgrowth of these studies, Knott and Voss[95] developed patterns and technics of proprioceptive neuromuscular facilitation.

The late Temple Fay[96, 97] suggested that during the development of the human nervous system, the primitive neurologic activity reflects the phylogenetic motor activity of the amphibian and reptilean forms. He described postural positioning that he thought modified the spasticity of muscles, referred to use of these positions as unlocking reflexes, and based his treatment on these suppositions. Delacato[98] has taken Fay's speculations and broadened them with additional assumptions from which he has elaborated a treatment technic. There are many other variations of manual physical therapeutic technics that are reviewed elsewhere.[79, 80]

The advocates of some of the above-mentioned so-called systems of treatment have become so emotionally involved in the treatment theory and technic that they lose proper perspective for the total needs of the child. For example, Paine[99] commented objectively about certain of these considerations and made a plea for objective evaluation of presently used physical therapeutic technics. Paine's remarks elicited the reminder[100] that since 1950 physical therapists have become increasingly interested in the neurophysiological basis of their technics and now look to neurophysiologists for the rationale of their therapeutic approach.

It is likely that there is some merit in most of the so-called systems of treatment. However, the various technics of manual physical therapy compose only one phase of the overall treatment program and should be kept in proper perspective with other aspects of therapy. There is need for careful, unbiased assessment of the effect these technics have on the extent and circadian aspect of maturation of organically impaired brains.

No matter what method of treatment is used, the objective is to develop the patterns of movement necessary for performing independently the Activities of Daily Living. We believe that while the physical therapeutic methods are proceeding, children who are far delayed in sitting, standing, and walking achieve physical and psychologic value from standing in the erect position by means of braces. Weight bearing through appropriate skeletal parts engenders better bone formation and stimulates growth. Assumption of the vertical position also provides more normal circumstances for circulation and prevents the development of postural hypotension. The braces also have the value of guiding the movement of the extremities into a more normal pattern and, by alignment of skeletal components, minimize the production of skeletal deformities.

We consider the technics of training in ambulation, hand activities, and bed and wheelchair activities to be a vital part of the rehabilitation program and will briefly describe the training technics.

Ambulation. In using this method to develop the walking pattern, the following procedures are necessary to control the lower extremities.

The child is fitted with double, long leg braces with a pelvic band or a Knight spinal brace, with ring locks at the hips and knees and with 90-degree ankle stops. He is then placed in parallel bars, with the joints locked, to develop standing balance. (Fig. 25-1.) When he has learned standing balance, the hip locks are opened, which allows flexion and extension at the hip joints. The therapist, standing behind the child, grasps his wrists. At the count of one the therapist moves the child's left arm forward and the child grasps the bar, at the count of two the right extremity is pushed forward with the therapist's right knee, at the count of three the right arm is moved forward, and at the count of four the left extremity is moved.

When the child is able to walk in the parallel bars unassisted, he is fitted with crutches and placed against the wall to learn standing balance. With his crutches under his arms and the hip locks open, he is taught how to shift his weight and raise the crutches. The same procedure is now followed moving away from the wall. The next procedure is to have the child learn to walk unassisted with the four-point crutch gait. When this has been accomplished, one knee lock is opened; later the other one is opened so that the child may develop a normal reciprocal pattern of

Fig. 25-1. Increasing independence—walking for the first time in parallel bars.

walking. As the child progresses he is given new movements to control. When the pelvic band is removed, it is necessary for him to control abduction and adduction and the rotation movements of the hip.

After years of practice some may be able to ambulate with a fairly normal gait without braces and crutches. To ambulate with braces and crutches necessitates the use of the upper extremities. There is no problem with the paraplegic or the hemiplegic cerebral palsied child, who seldom needs crutches. In teaching the athetoid quadriplegic patient to ambulate, it is necessary to limit the movements of the upper extremities to flexion and extension of the shoulder joint. This often can be accomplished by using axillary crutches, adjusted so that the hands can grasp the hand bars with the elbows fully extended, with restraining straps or metal bands attached to the crutches to control the movements of the arm and forearm. The crutches should have suction rubber tips to prevent slipping and should be weighted with lead. With all the movements controlled except flexion and extension at the shoulder and hip joints, the child is started on his program of ambulation.

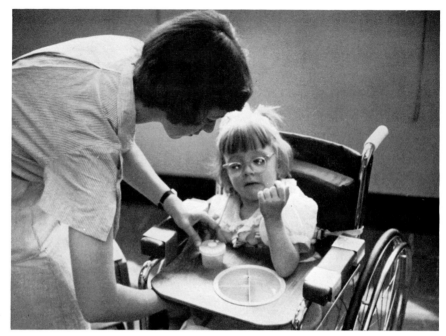

Fig. 25-2. Feeding with utensils devised for the patient helps to promote a sense of independence.

As he attains each objective, new movements are added to his program, such as removing the restraining bars for the arms, flexing the elbow, and removing weights from the crutches. After many years of training, a number of athetoid patients are able to ambulate with a fairly adequate four-point crutch gait with double, long leg braces, open knee locks, and Lofstrand crutches.

Hand activities. The paraplegic cerebral palsied child needs no training in hand activities and the hemiplegic child can usually function adequately with one normal hand. The athetoid and spastic quadriplegic patient, however, has great difficulty in controlling the seven movements of the shoulder, the four movements of the elbow and wrist, and the many movements of the fingers and thumbs. By following the same procedures as outlined for the control of the lower extremities, that is, eliminating all but two movements, it is often possible to teach the child many self-care activities. By tying the arm to the body and placing a cock-up splint on the wrist, the movements of the upper extremity are limited to flexion and extension at the elbow joint. By placing a spoon under the cock-up splint, which extends under the palm of the hand, the patient can be taught to feed himself. He can also learn many of the self-care activities such as washing his face and cleaning his teeth. When these movements are controlled, the arm band should be removed so that the child may learn to control the shoulder movements. At a later period it may be possible to gain control of the wrist, fingers, and thumb (Fig. 25-2).

Bed and wheelchair activities. The most important part of the rehabilitation program for the cerebral palsied child is learning bed and wheelchair activities

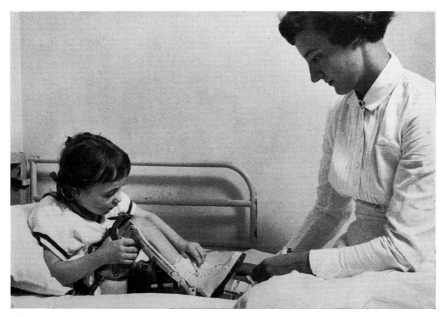

Fig. 25-3. Members of the nursing staff encourage all the child's attempts to achieve independence.

(Fig. 25-3). What is the value of teaching a child to ambulate with braces and crutches if the mother has to dress the child, put on his braces, and lift him to the standing position? If the parent becomes sick, the child cannot move out of bed. Practically every child in our children's service has a tiny-tot or junior size wheelchair prescribed to meet his needs. Bed and wheelchair activities are taught in the ward by the rehabilitation nurse. Teaching a child these activities gives him strength and develops coordination. It is surprising how quickly a child with athetosis or spasticity of the arms learns to roll his wheelchair. In the beginning he may hit one wall and then the other, but he soon learns that forward progress can be attained only by producing a coordinated movement. The coordinated movements learned in rolling his wheelchair may be utilized in many other activities. A great number of cerebral palsied children will never be able to ambulate adequately with braces and crutches; however, if they can be taught to get out of bed into a wheelchair and to roll themselves to the bathroom to perform their necessary toilet activities and to the dining room table to feed themselves, a great deal will have been accomplished for these children and for their parents.

Role of surgery

In the early days of comprehensive approaches to treatment of cerebral palsy, frequent use was made of a large variety of surgical procedures to correct the orthopedic deformities that commonly occur. The results of the surgical approach that dominated the treatment plan 25 years ago were generally unsatisfactory, largely because of failure to select appropriate cases for operation and failure to have an effective postsurgical rehabilitation program that would maintain the gains achieved by the surgical procedure.

As a result of the indifferent success that attended the earlier surgical procedures, the frequency of surgical approaches to treatment declined between 1945 and 1955. During the past decade, orthopedic surgeons have more clearly delineated the indications for surgery and, through organization and provision of appropriate postsurgical care, a variety of surgical procedures have proved to be very vital in the rehabilitation process.

Physical therapeutic treatment and appropriate bracing can minimize and frequently prevent the need for a surgical procedure. On the other hand, there are a number of orthopedic measures that can quickly and easily correct a deformity that either is not responsive to physical therapy or would require many months of treatment to achieve the desired result.

When the orthopedist is confronted with a deformity that is technically amenable to surgical correction, he evaluates not only the deformity but also the neuromotor abnormality (with reference to its effect on the procedure he contemplates) and the intellectual capacity of the child and his ability to profit by deformity correction. In many instances, the orthopedic surgical procedure is the key to the child's rehabilitation, and an effective rehabilitation program is dependent on surgical correction of the deformity.

Our objectives would not be served by discussing any specific orthopedic procedure or the indications for it. These details have been effectively described by Baker[101] and by Silver and Simon[102] with respect to the lower extremity, and by Goldner[103] for the upper extremity.

Drug treatment

The objectives of drug therapy are the amelioration of behavioral and neuromotor problems and the control of seizures. There is a wide range of effective anticonvulsant drugs, and the control of seizures has been mentioned previously. Some of the antiepileptic drugs also have psychotherapeutic effect. Psychopharmacologic agents are frequently of value in regulating restless or impulsive behavior, but one should not rely on drugs alone but should use them only as an adjunct to the rehabilitation program. The amphetamines (Benzedrine, Dexedrine) appear to be the most reliable. The barbiturates are usually ineffective and sometimes aggravate the problem:

Perlstein[104] has discussed the details of drug treatment. During the past 15 years the following new groups of compounds have been synthesized that appear to be useful in selected patients: meprobamate (Equanil and Miltown), chlordiazepoxide hydrochloride (Librium), mephenesin (Tolserol), and mephenesin carbamate (Tolseram). Recently, a new psychotherapeutic agent, benzodiazepine derivative (Valium), has been introduced after extensive animal and human trial. This compound is claimed to be more effective than other compounds in reducing spasticity and rigidity and diminishing athetotic movement.[105, 106] There is great need for compounds with greater effect on neuromotor problems. When using long-term drug therapy, one must remain alert to the side effects that occur with all of the available compounds.

PROGNOSIS

By definition and general agreement the brain lesion in cerebral palsy is a static one. This does not mean that the neurologic deficit is constant and unchanging. For example, Perlstein and Barnett[107] and Paine[99] have pointed out that there is a small percentage of infants who definitely show classic signs of central nervous system impairment in early infancy and gradually improve over a period of months and on subsequent examination appear to be normal. On the other hand, there are some children who appear to develop signs of neurologic deficit as they grow older. This change, however, is usually not marked.

These observations of a changing clinical picture in the presence of a static neurologic lesion provide eloquent testimony that the human infant normally is born with a relatively undeveloped nervous system. This undeveloped functional state is particularly evident with reference to neuromotor function. Some of the vertebrates are able to ambulate within hours of birth and within a few days are very effective at walking, running, and jumping—in contrast to the human infant, who requires 12 to 24 postnatal months to develop these motor capacities. The functions of speech and perception require several years for development. For example, visual depth perception appears not to be fully developed until the child is 6 years old.[105]

Late onset of seizures may also provide further evidence of the dynamic processes that must be ongoing in the human brain. This occurrence could be due to the general scarring effect of glioses at the injury site or could relate to the injury site influence on other maturational processes of the brain. The concept of gradual neuroanatomic and neurophysiologic maturation of the human brain is supported by clinical observation and testing and also by cytologic studies. The magnificent cytologic studies of cortical development by Conel[108] (seven volumes) are not sufficiently well known, nor are their implications adequately appreciated. In Conel's last two volumes (vol. 7 in 1963) he illustrated the very striking differences in cortical cytoarchitectonics that occur between the ages of 2 years (vol. 6) and 4 years (vol. 7). These changes are also reflected in changes in weight in the postnatal period. The human brain increases 360% in weight between the newborn period and 6 years of age but only 8% in the subsequent 12 years.

These cytologic, neurophysiologic, and clinical evidences of the striking changes in brain development, which probably extend over the first 5 or 6 years of life, have very important implications for the cerebral palsied child and the methods used in treatment. Reparation of cellular and tissue damage or disruption is an inherent biologic capacity of all nucleated cells. This capacity is greater in some cell types than in others. The severed axons of peripheral nerves regenerate, but functional restoration is achieved only if the fibers are guided into proper channels. There is evidence that central nervous system axons also have some inherent regenerative capacity. We have learned something of the factors that regulate their growth but have not yet been able to provide the proper environment to obtain maximal benefit from the growth potential. There is little doubt that from the functional viewpoint a static lesion produced at birth does not exert precisely the same effect throughout the 5 or 6 years of structural and functional development of the brain. The lesions may delay development of function or even suppress it, but there may also be developed some reorganized neural patterns that to some extent circumvent total loss of a function. When a given function returns, in part or with qualitative difference, we have an obligation to see to it that imbalance of muscle power, abnormal posture, or inactivity has not resulted in secondary mechanical difficulties (contractures) that prevent the child from taking advantage of the neuromotor capacity that eventually may be available. It is in this area that the physical therapist and occupational therapist play a very important role in the treatment program.

It is also pertinent at this point to ask ourselves what, if any, special neurophysiologic benefits are derived by a physical therapy program in which movements are induced reflexly, by sensory stimulation, versus the same amount of motor neuron activity produced voluntarily by the patient with motivation. Does the local excitatory state around the motor neuron that may be induced by sensory stimulation endure long enough for volitional effort to become more effective on a permanent basis? Does repetitive reflex motor activity induced by sensory stimulation enable the child to perform these movements better volitionally than he could after the same amount of voluntarily induced movement or of passive movement, extending over the same period of time? One might also ask another question in a negative way: If sensory input from muscles provides the central nervous system with continuous information about muscle length and position, would it not be helpful to guide motor patterns into normal channels, with bracing, which might enable the sensory input to be more consistent?

There have not yet been any properly designed controlled experiments that demonstrate the superiority of one physical therapeutic treatment method over another. It is quite possible that all of us are inclined to attribute to our therapeutic technic much of the functional capacity that we see during the 6 years of altered structural and functional maturation of a handicapped brain. We like to think of the totality of these dynamic changes as being the natural history of cerebral palsy and regard it in a sense as effort of the brain to complete its maturation by a positive though limited capacity to circumvent its handicap. On the other hand, the functional deterioration that we frequently see in the adolescent child with cerebral palsy is psychosocial and emotional in origin. This apparent deterioration may occur as the adolescent becomes acutely aware of the social deprivations that face him at this critical phase of life. These changes are important and need greater attention. They are not reflections of organic brain function deterioration and a component of the intrinsic natural history of the disease, but they are vital aspects of the natural history of the person with cerebral palsy.

REFERENCES

1. Little, W. J.: On the influence of abnormal parturition, difficult labours, premature birth and asphyxia neonatorum; on the mental and physical condition of the child, especially in relation to deformities, Trans. Obstet. Soc. London 3:293, 1861-1862.
2. Phelps, W. M.: Cerebral palsy. In Whipple, A. O., editor: Nelson's new loose leaf surgery, New York, 1947, Thomas Nelson & Sons.
3. Cruickshank, W. R., and Raus, J. M.: Cerebral palsy. Its individual and community problems, Syracuse, N. Y., 1955, Syracuse University Press.

4. Strauss, A. A., and Lehtinen, L. E.: Psychopathology and education of the brain injured child, vols. 1 and 2, New York, 1955, Grune & Stratton, Inc.
5. Denhoff, E.: Diagnostic techniques for children with cerebral palsy, Rhode Island Med. J. **32**:483, 1949.
6. Paine, R. S.: Minimal chronic brain syndromes in children, Develop. Med. Child Neurol. **4**:21, 1962.
7. Prechtl, H. F., and Stemmer, J.: The choreiform syndrome in children, Develop. Med. Child. Neurol. **4**:119, 1962.
8. Perlstein, M. A.: Infantile cerebral palsy, classification and clinical correlations, J.A.M.A. **149**:30, 1952.
9. Mac Keith, R. C., Mackenzie, I. C. K., and Polani, P. E.: Definition of cerebral palsy, Cereb. Palsy Bul. **5**:23, 1958-1959.
10. Mac Keith, R. C., Mackenzie, I. C. K., and Polani, P. E.: Memorandum on terminology and classification of cerebral palsy, Cereb. Palsy Bull. **5**:27, 1958-1959.
11. Phelps, W. M.: The rehabilitation of cerebral palsy, Southern M. J. **34**:770, 1941.
12. Levin, M. L., Brightman, I. J., and Burtt, E. J.: The problem of cerebral palsy, New York J. Med. **49**:2793, 1949.
13. Asher, P., and Schonell, F. E.: A survey of 400 cases of cerebral palsy in childhood, Arch. Dis. Child. **25**:360, 1950.
14. Woods, G. E.: A lowered incidence of infantile cerebral palsy, Develop. Med. Child Neurol. **5**:449, 1963.
15. Denhoff, E., and Robinault, I.: Cerebral palsy and related disorders, New York, 1960, McGraw-Hill Book Co., Inc.
16. Polani, P. E.: Effects of abnormal brain development on function, Cereb. Palsy Bull. **1**:27, 1958-1959.
17. Swinyard, C. A., Swenson, J., and Greenspan, L.: An institutional survey of 143 cases of acquired cerebral palsy, Develop. Med. Child Neurol. **5**:615, 1963.
18. Freud, S.: Die infantile Cerebrallaehmung. In Nothnagel, H., editor: Specielle pathologie und therapie, vol. 9 (part 3), Vienna, 1897, Holder.
19. Phelps, W. M.: Etiology and diagnostic classification of cerebral palsy, Proceeding of the Cerebral Palsy Institute, New York, 1950, Association for the Aid of Crippled Children.
20. Minear, W. L.: A classification of cerebral palsy, Pediatrics **18**:841, 1956.
21. Crothers, B., and Paine, R. S.: The natural history of cerebral palsy, Cambridge, Mass., 1959, Harvard University Press.
22a. Balf, C. L., and Ingram, T. T. S.: Problems in the classification of cerebral palsy, Cambridge, Mass., 1959, Harvard University Press.
22b. Balf, C. L., and Ingram, T. T. S.: Problems in the classification of cerebral palsy in childhood, Brit. Med. J. **2**:163, 1955.
23. Courville, C. B.: Cerebral anoxia, Los Angeles, 1953, San Lucas Press.
24. Towbin, A.: The pathology of cerebral palsy, Springfield, Ill., 1960, Charles C Thomas, Publisher.
25. Woods, G. E.: Cerebral palsy in childhood, Bristol, 1957, John Wright & Sons, Ltd.
26. Perlstein, M. A., Gilbs, H., and Gilbs, F. A.: The electroencephalogram in infantile cerebral palsy, Proc. Ass. Res. Nerv. Ment. Dis. **26**:377, 1946.
27. Yannett, H.: The etiology of congenital cerebral palsy, J. Pediat. **24**:38, 1944.
28. Melin, K. A.: EEG and epilepsy in cerebral palsy, Develop. Med. Child Neurol. **4**:180, 1962.
29. Illingworth, R. S.: Convulsions in mentally retarded children with or without cerebral palsy: their frequency and age incidence, J. Ment. Defic. Res. **3**:88, 1959.
30. Livingston, S.: Living with epileptic seizures, Springfield, Ill., 1963, Charles C Thomas, Publisher.
31. McFie, J.: The effects of hemispherectomy on intellectual functioning in cases of infantile hemiplegia, J. Neurol., Neurosurg. Psychiat. **24**:240, 1961.

32. Bender, L.: Psychopathology of children with organic brain disorders, Springfield, Ill., 1956, Charles C Thomas, Publisher.
33. Strauss, A. A., and Lehtinen, L. E.: Psychopathology and education of the brain-injured child, 2 vols., New York, 1947 and 1955, Grune & Stratton, Inc.
34. Laufer, M. W., and Denhoff, E.: Hyperkinetic behavior syndrome in children, J. Pediat. 50:463, 1957.
35. Birch, H. G.: Brain damage in children—the biological and social aspects, Baltimore, 1963, The Williams & Wilkins Co.
36. Hopkins, T. W., Bice, H. V., and Colton, K. C.: Evaluation and education of the cerebral palsied child—New Jersey study, Washington, D. C., 1954, International Council for Exceptional Children.
37. Gerrard, J.: Kernicterus, Brain 75:526, 1952.
38. Rosen, J.: Rh Child: deaf or "aphasic"? Variations in the auditory disorders of the Rh child, J. Speech Hearing Dis. 21:418, 1956.
39. Hannigan, H.: Rh Child: deaf or "aphasic"? Language and behavior problems of the Rh "aphasic" child, J. Speech Hearing Dis. 21:413, 1956.
40. Dublin, W. B.: Neurologic lesions of erythroblastosis fetalis in relation to nuclear deafness, Amer. J. Clin. Path. 21:935, 1951.
41. Haymaker, W., Margoles, C., Pentschew, A., Jacob, H., Lindenberg, R., Saenz Arroyo, L., Stochdorph, O., and Stowens, D.: Pathology of kernicterus and posticteric encephalopathy. In American Academy for Cerebral Palsy, Swinyard, C. A., editor: Kernicterus and its importance in cerebral palsy, Springfield, Ill., 1961, Charles C Thomas, Publisher.
42. Lucey, J. F., Hibbard, E., Behrman, R. E., Esquivel de Gallardo, F. O., and Windle, W. F.: Kernicterus in asphyxiated newborn rhesus monkeys, Exp. Neurol. 9:43, 1964.
43. Gesell, A., Ilg, F. L., and Bulles, G. E.: Vision: its development in infant and child, New York, 1959, Paul B. Hoeber, Inc., Medical Book Department of Harper & Bros.
44. Guibor, G. P.: Some eye defects seen in cerebral palsy with some statistics, Amer. J. Phys. Med. 32:342, 1953.
45. Breakey, A. S.: Ocular findings in cerebral palsy, Arch. Ophthal. 53:852, 1955.
46. Henderson, J. L.: Cerebral palsy in childhood and adolescence, Baltimore, 1961, The Williams & Wilkins Co.
47. Abercrombie, M. L. J.: Perception and eye movements: some speculations on disorders in cerebral palsy, Cereb. Palsy Bull. 2:142, 1960.
48. Phelps, W. M.: The management of the cerebral palsies, J.A.M.A. 117:1621, 1941.
49. Tizard, J. P. M., Paine, R. S., and Crothers, B.: Disturbance of sensation in children with hemiplegia, J.A.M.A. 155:628, 1954.
50. Hohman, L., Baker, L., and Reed, R.: Sensory disturbances in children with infantile hemiplegia, triplegia and quadriplegia, Amer. J. Phys. Med. 37:1, 1958.
51. Mecham, M. J., Berko, M. J., and Berko, F. G.: Speech therapy in cerebral palsy (Amer. Lect. Series), Springfield, Ill., 1960, Charles C Thomas, Publisher.
52. Birch. H. G., and Lefford, A.: Intersensory development in children, Monogr. Soc. Res. Child Develop. 28:1, 1963.
53. Vernon, M. D.: The psychology of perception, London, 1962, Penguin Books, Ltd.
54. Illingworth, R. S.: The development of the infant and young child—normal and abnormal, Edinburgh, 1960, E. & S. Livingston, Ltd.
55. Wepman, J. M.: Speech therapy for cerebral palsy patients, Physiother. Rev. 21:82, 1941.
56. Gratke, J. M.: Speech problems of the cerebral palsied, J. Speech Dis. 12:129, June, 1947.
57. Hedges, T. A.: The relationship between speech understandability and diadochokinetic rates of certain speech musculatures among individuals with cerebral palsy, Speech Monogr. 23:144, 1956.
58. Ingram, T. T. S.: A description and classification of the common disorders of speech in children, Arch. Dis. Child. 34:444, 1959.

59. West, R., editor: Childhood aphasia, Proceedings of The Institute of Childhood Aphasia, San Francisco, 1962, California Society for Crippled Children and Adults.
60. Ingram, T. T. S.: Assessment of the results of speech and other therapies in cerebral palsy, Develop. Med. Child Neurol. 2:149, 1960.
61. Hawk, S. S., and Young, E. H.: Moto-kinesthetic speech training, Palo Alto, Calif., 1955, Stanford University Press.
62. Mountcastle, V. B., editor: Interhemispheric relations and cerebral dominance, Baltimore, 1962, The Johns Hopkins Press.
63. Money, J., editor: Reading disability—progress and research needs in dyslexia, Baltimore, 1962, The Johns Hopkins Press.
64. Palmer, C. E.: Speech and hearing problems. A guide for teachers and parents, Springfield, Ill., 1961, Charles C Thomas, Publisher.
65. Van Riper, C.: Your child's speech problems, New York, 1960, Harper & Brothers.
66. Lewis, M. M.: How children learn to speak, New York, 1959, Basic Books, Inc.
67. Phelps, W. M.: Characteristic psychological variations in cerebral palsy, Nerv. Child 7:10, 1948.
68. Ingram, T. T. S., and Barn, J.: A description and classification of common speech disorders associated with cerebral palsy, Cereb. Palsy Bull. 3:57, 1961.
69. Taylor, E. M.: Psychological appraisal of children with cerebral defects, Cambridge, Mass., 1959, Harvard University Press.
70. Terman, L., and Merrill, M. A.: Measuring intelligence, London, 1937, Harrap.
71. Piaget, J.: The psychology of intelligence, London, 1950, Routledge & Kegan Paul, Ltd. (French edition, 1949.)
72. Piaget, J.: The child's conception of number, London, 1952, Routledge & Kegan Paul, Ltd. (French edition, 1952.)
73. Piaget, J., and Inhelder, B.: The child's conception of space, London, 1956, Routledge & Kegan Paul, Ltd. (French edition, 1948.)
74. Hersoy, L.: Emotional factors in cerebral palsy, Develop. Med. Child Neurol. 5:504, 1963.
75. Floyer, E. B.: A psychological study of the city's cerebral palsy children, Manchester, 1955, British Council for the Welfare of Spastics.
76. Ironside, R.: Disorders of laughter due to brain lesions, Brain 79:589, 1956.
77. Reed, S. C.: Counseling in medical genetics, Philadelphia, 1955, W. B. Saunders Co.
78. McDonald, E. T.: Understand those feelings, a guide for parents of handicapped children, Pittsburgh, 1962, Stanwix House.
79. Semans, S.: Physical therapy for motor disorders resulting from brain damage, Rehab. Lit. 20:99, 1959.
80. Gillette, H. E.: Changing concepts in the management of neuromuscular dysfunction, Southern Med. J. 52:1227, 1959.
81. Sherrington, C. S., and Mott, F. W.: Experiments upon the influence of sensory nerves upon movement and nutrition of the limbs. Preliminary communication, Proc. Roy. Soc. London 57:481, 1895.
82. Hagbarth, K. E.: Excitatory and inhibitory skin areas for flexor and extensor motoneurones, Acta Physiol. Scand. 26(supp. 94):1, 1952.
83. Magoun, H. W.: The waking brain, ed. 2, Springfield, Ill., 1963, Charles C Thomas, Publisher.
84. Granit, R.: Receptors and sensory perception, New Haven, 1955, Yale University Press.
85. Boyd, I. A., Eyzaguirre, C., Matthews, P. B. C., and Rushworth, G.: In Swinyard, C. A., editor: The role of the gamma system in movement and posture, New York, 1964, Association for the Aid of Crippled Children.
86. Rood, M. S.: Neurophysiological mechanisms utilized in the treatment dysfunction, Amer. J. Occup. Ther. 10:220, 1956.
87. Rushworth, G.: Spasticity and rigidity: an experimental study and review, J. Neurol. Neurosurg. Psychiat. 23:99, 1960.

88. Tardieu, C., Tardieu, G., Hariga, J., Gagnard, L., and Velon, G.: Fondement experimental d'une therapeutique des raideurs d'origine cerebrale, Arch. Franc. Pediat. **21**:5, 1964.

89. Tardieu, G., and Hariga, J.: Traitement des raideurs musculaires d'origine cerebrale par infiltration d'alcool dilue, Arch. Franc. Pediat. **21**:25, 1964.

90. Magnus, R.: Körperstellung. Monographie aus den Gesamtgebiet der Physiologie der Pflanzen und der Tiere, vol. 6, Berlin, 1924, Julius Springer Co.

91. Bobath, B.: A study of abnormal postural reflex activity in patients with lesions of the central nervous system, Physiotherapy **40**: 4 parts, Sept., Oct., Nov., Dec., 1954.

92. Kabat, H.: Studies on neuromuscular dysfunction. 1. Neostigmine therapy of neuromuscular dysfunction resulting from trauma, Public Health Rep. **59**:1635, 1944.

93. Kabat, H.: Studies on neuromuscular dysfunction. XIII. New concepts and techniques of neuromuscular re-education for paralysis, Permanente Found. Med. Bull. **8**:121, July, 1950.

94. Gellhorn, E.: Proprioception and the motor cortex, Brain **72**:35, 1949.

95. Knott, M., and Voss, D. E.: Proprioceptive neuromuscular facilitation: patterns and technics, New York, 1956, Paul B. Hoeber, Inc., Medical Book Department of Harper & Bros.

96. Fay, T.: The neurophyiological aspects of therapy in cerebral palsy, Arch. Phys. Med. **29**: 327, 1948.

97. Fay, T.: Neuromuscular reflex therapy for spastic disorders, J. Florida Med. Ass. **44**:1234, 1958.

98. Delacato, C. H.: The diagnosis and treatment of speech and reading problems, Springfield, Ill., 1963, Charles C Thomas, Publisher.

99. Paine, R. S.: On the treatment of cerebral palsy, Pediatrics **29**:605, 1962.

100. Levitt, S.: Physical therapy in the management of cerebral palsy, Develop. Med. Child Neurol. **5**:75, 1963.

101. Baker, L. D.: A rational approach to the surgical needs of the cerebral palsy patient, J. Bone Joint Surg. **38-A**:313, 1956.

102. Silver, C. M., and Simon, S. D.: Operative treatment of cerebral palsy involving the lower extremities, J. Int. Coll. Surgeons **27**:457, 1957.

103. Goldner, J. L.: Upper extremity reconstructive surgery in cerebral palsy or similar conditions, American Academy of Orthopaedic Surgeons Instructional Course Lectures, vol. 18, St. Louis, 1961, The C. V. Mosby Co.

104. Perlstein, M. A.: The current status of drug therapy in cerebral palsy, Crippled Child **27**:8, 1949.

105. Keats, S., Morgese, A., and Nordlund, T.: The role of diazepam in the comprehensive treatment of cerebral palsied children (special supplement: Symposium on diazepam), Western Med. **4**:22, 1963.

106. Carter, C. H.: Diazepam in management of spasticity and related symptoms in brain-damaged patients (special supplement: symposium on diazepam), Western Med. **4**:54, 1963.

107. Perlstein, M. A., and Barnett, H. S.: Nature and recognition of cerebral palsy in infancy, J.A.M.A. **148**:1389, 1952.

108. Conel, J. R.: The postnatal development of the human cerebral cortex, 7 vols., Cambridge, Mass., 1939-1963, Harvard University Press.

CHAPTER 26

PULMONARY PROBLEMS

In recent years as the field of rehabilitation medicine has expanded, it is apparent that its responsibility for the management of respiratory dysfunction embraces two quite distinct areas: the first, the more traditional one, is the management or prevention of respiratory insufficiency in patients with neuromuscular and skeletal disorders involving the thorax; the second, the rapidly growing area, is the application of physical medicine and rehabilitation to patients with chronic obstructive lung disease. Since the pathogenic mechanisms and treatment goals of the two areas differ, the chronic obstructive lung diseases will be considered in the first part of this chapter and respiratory insufficiency in neuromuscular and skeletal disorders will be considered in the second part.

Application of rehabilitation medicine to chronic obstructive lung disease

The modern era of the management of respiratory diseases could not open until René Laennec (1781-1826) described the basic concepts of pulmonary pathology and the auscultatory system (treatise on emphysema). His invention of the stethoscope led to new diagnostic technics that opened new horizons in the detection and description of respiratory ailments. After his pioneering work, it became possible for the first time to differentiate between normal and abnormal breathing sounds and to evaluate the clinical significance of anomalous sounds.

More than a century later, with the development of new surgical methods and the advent of antibiotic drugs for the treatment of tuberculosis, the disease that had plagued mankind since the days of Hippocrates[1] was under relative control. Physical and vocational rehabilitation became an important part of the treatment of patients with tuberculosis.[2] Patients were at last able to return to society, and often to their former jobs, while still on antimicrobial drugs. No longer was it necessary for all tuberculosis patients to receive care in a sanatorium and prolonged bed rest.

With tuberculosis more or less under control among most segments of modern industrialized societies like that of the United States, other respiratory diseases caused by environmental factors prevalent in such societies have become major health problems.[3] The millions of tons of hydrocarbons and other pollutants poured into the environment annually from this nation's more than 100 million internal combustion engines and the generally simultaneous and synergistic effects of the smoke of the cigarettes Americans consumed at the rate of 4280 cigarettes per capita for the year 1967 were not without their predictable results in terms of the prevalence and the chronicity of various respiratory diseases.[4] In a nation that produced nearly 580 million pounds of cigarettes at a moment in time (1967) when, according to a United States government document,[4] "each automobile in the country will discharge in a single year over 1600 pounds of carbon monoxide, 230 pounds of hydrocarbons, and 77 pounds of oxides of nitrogen,"* it is hardly surprising that the prevalence of chronic obstructive pulmonary diseases (COPD) such as chronic bronchitis, emphysema, and allergic bronchitis climbed so dramatically after World War II.[5]

Emphysema deaths among American males, for example, climbed from a little over 1 per 100,000 population in 1950 to close to 15 per 100,000 by 1964. All of the COPD diseases, according to a recent U. S. Public Health Service survey,[5] showed a 190% increase in prevalence between 1965 and 1969. That environmental factors play a significant role in the etiology of the COPD conditions may be seen in the data on lung cancer deaths, which stand at 15 per 100,000 population in rural communities; 18 deaths per 100,000 in cities of under 250,000 people; 22 deaths per 100,000 in cities with populations of between 250,000 and 1 million people.

Accurate statistics on the total number of people with one or another of the COPD conditions are not available. It has been conservatively estimated that between 4 and 12 million people in the United States have from a mild to a severe degree of one of the chronic obstructive pulmonary diseases. Social Security Administration studies[3] of work disability allowances indicate that chronic obstructive pulmonary diseases rank second only to coronary artery disease as the cause of permanent disability in men over 40 years of age covered by social security. Emphysema is now the fastest growing fatal disease in the United States.

During the post–World War I years, the development of more sophisticated tools —such as the spirometer for measuring pulmonary volumes and capacities and the Van Slyke apparatus for determining blood gas composition—made it possible for physicians to arrive at more accurate determinations of a patient's pulmonary performance.[6] After World War II the microelectrodes that allow the measurement of blood gases without recourse to arterial puncture were introduced to clinical medicine.

The new tools and technics of physiologic measurement were made available at roughly the same time that the iron lung and the intermittent positive pressure breathing (IPPB) machine were introduced to assist in mechanical hyperventilation, and the antibiotics and other drugs made the management of COPD systems pos-

---

*From A strategy for a liveable environment, a report to the Secretary of Health, Education, and Welfare by the task force on environmental health and related problems, June, 1967.

sible. Rehabilitation became an integral part of the overall treatment of these patients.[7] The modalities generally used in COPD rehabilitation are (1) relaxation exercises, (2) postural drainage, (3) breathing exercises, (4) and oxygen reconditioning exercises.

Three distinct conditions in which chronic respiratory cripples can benefit from such rehabilitation measures are (1) obstructive airway disease, (2) restrictive respiratory diseases due to skeletal or neuromuscular involvement, and (3) restrictive parenchymal disease such as pulmonary fibrosis, granulomatosis, and pneumoconiosis. The first two conditions are discussed in this chapter. The third, although not discussed, should be considered for rehabilitation measures.

REHABILITATION IN CHRONIC OBSTRUCTIVE PULMONARY DISEASE (COPD)

One of the most difficult problems in clinical medicine is the management of patients with chronic obstructive pulmonary disease conditions such as chronic bronchitis, emphysema, bronchial asthma, and bronchiectasis. Although the pathogenesis of different COPD entities may vary, the rehabilitation modalities are applicable to all obstructive pulmonary conditions, including allergic asthma and cystic fibrosis.

These diseases are characterized by partial or total irreversibility.[8] Their etiology is unclear. Although environmental and internal pollutants such as soot and cigarette smoke are known to be related to the prevalence of COPD conditions, genetic factors are also believed to be involved. Whereas roughly 9 in 10 patients with emphysema turn out to have been cigarette smokers, only 1 cigarette smoker out of every 10 develops emphysema. The role of endocrinologic systems has to be investigated. Exposure to industrial chemicals is known to contribute to the incidence of COPD conditions. A history of repeated low-grade respiratory infections that might have been inadequately treated or not treated at all is fairly common to COPD conditions. It might or might not be true that the 6:1 incidence of COPD in males as against females was that, until recently, cigarette smoking was more common among men than among women. Nor is enough yet known about the etiologic significance in COPD conditions of psychosocial and emotional factors.

Because the underlying causes of the COPD conditions are far from clear, opinions as to their pathogenesis and management are greatly divided. There is no controversy about the physiologic and pathologic effects of the various COPD conditions (Fig. 26-1).

Our concern in this chapter is the rehabilitation of patients with chronic obstructive airway diseases such as chronic bronchitis, chronic endobronchial disease, bronchiectasis, bronchial asthma, and cystic fibrosis. All these diseases have certain attributes in common: (1) they are chronic diseases that once established continue indefinitely; (2) the symptoms may come and go, but the pathologic and anatomic changes, which occur and accumulate, remain and grow worse[9]; (3) they are characterized by slowly progressive symptoms that are not amenable to any known cure and that result in respiratory insufficiency (hypoxia, hypoxemia, and carbon dioxide retention), acidosis, pulmonary hypertensioin, and finally cor pulmonale (right heart failure).

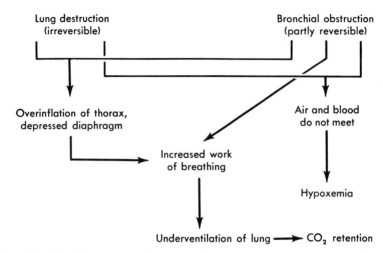

Fig. 26-1. Physiologic and pathologic pathway of obstructive pulmonary disease.

The bronchial infections associated with COPD conditions can be controlled by the newer antibiotics,[10] but the underlying physiologic and pathologic changes that occur are irreversible. While the symptomatic suppression of respiratory infections may relieve some of the patients' discomfort and, indeed, prevent secondary pneumonias from killing as many COPD victims as died of pulmonary infections before the antibiotic era, it also breeds pulmonary cripples who survive as burdens on their families and their communities.

The successes achieved by post-World War II rehabilitation medicine in patients with neuromuscular and skeletal involvement encouraged the physiatrist to attempt rehabilitation of COPD patients. Rehabilitation soon became as important in the management of COPD patients as it had become in patients handicapped by neuromuscular or skeletal impairment.

The clinical course of chronic obstructive pulmonary diseases is characterized by periods of exacerbation and quiescence. During periods of exacerbation the clinical symptoms are (1) severe dyspnea at rest or at the slightest exertion and (2) increased cough and abundant expectoration (the varying colors of the sputum reflecting the bacterial infections that are predominantly due to staphylococci, pneumococci, and *Haemophilus influenzae*). During quiescence the cough may be associated with mucoid sputum (abundant, little, or none). Moderate dyspnea, fatigue of the limbs in the early morning, some degree of stupor, and a tendency to ease irritability also characterize COPD conditions.

On physical examination the patients evidence forward bending of the upper trunk with raised shoulders, giving an impression of physical deformity (kyphosis). One may see distended veins on both sides of the neck. Pitting edema of the ankles is found during far-advanced stages of COPD. Clubbing fingers are not unusual in COPD patients. The facial expression frequently reveals anxiety. There is rapid respiratory movement, a gasping for air, and an inability to speak long sentences without gasping, thus giving the impression of choking.

PULMONARY FUNCTION TESTS

While spirometric pulmonary function tests (PFT) are mandatory in the examination and monitoring of COPD patients, the treating physician must at all times be aware of the limitations of the tests. To begin with, the technics used and the personality of the person who administers the spirometric pulmonary tests have a great influence on the results. The tests must be administered by a well-trained technologist or physician, who must achieve the total cooperation of the patient. Unless the patient is fully confident of the skill of the tester, he might not be fully cooperative, and his spirometric tracings might indicate false values.

The commonly used predicted values (norms) of pulmonary function tests are, in most instances, of little more than historic value. The Cournand-Richards scale, for example, was the product of longitudinal studies conducted in sick people with all kinds of disorders except chronic pulmonary disorders at New York's Bellevue Hospital in 1948.[11] Many of these hospitalized chronic patients had been at rest for long periods of time and were subject to deconditioning. At that time there were considerably less than 50 million motor vehicles registered in the United States. Between 1950 and 1970 the number of cars, trucks, and buses was more than doubled, climbing from 49.3 million to over 105 million. According to the U. S. Public Health Service,[4] by 1970 motor vehicles alone were ejecting 86 million tons of dirt into the environment every year. This load of pollutants includes 1 million tons of particulates, 12 million tons of hydrocarbons, 6 million tons of nitrogen oxides, and 1 million tons of sulfur oxides. All of these environmental pollutants are known to be causative factors and/or to exacerbate COPD disorders, and automobiles are now delivering more than twice as many of them to the ambient atmosphere as they delivered in 1948. The 56 million tons of a similar mix of fossil fuel and chemical pollutants delivered to the environment annually by American industrial plants, power-generating stations, home and office heating units, and garbage burning also add up to considerably more than twice the environmental pollutant load they emitted in 1948.

Spirometry, at best, has a margin of error of ±20%. In addition to this factor, there are great variations of findings even when skilled testers carry out spirometric tests of people in homogeneous, healthy populations. Therefore, the assessment of PFT results has to take into consideration such factors as the following:

1. Females tend to have lower pulmonary function values than males.
2. Obesity tends to decrease and interfere with proper pulmonary function.
3. Competitive athletes and people engaged in heavy manual labor will as a rule have higher pulmonary reserves than are called for in standard tables of predicted values.
4. People engaged in sedentary jobs will more often than not have lower values.
5. Since pulmonary function tests change with body posture, it is important to determine whether the spirometry was carried out standing, sitting, slightly reclined, or supine. (See Fig. 26-2.)

Since the spirometer shows gross but not lower limits of pathophysiology, even if a patient tests out to show 100% of his predicted values he might still have an obstructive pulmonary condition. The only real control for the COPD patient is

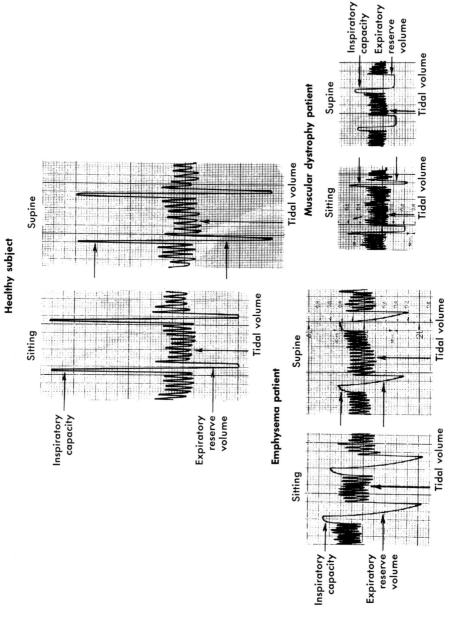

Fig. 26-2. Comparative positional changes of vital capacity and its subdivisions and tidal volume in healthy subjects and patients with emphysema and muscular dystrophy.

the individual himself, and the best frame of reference for his pulmonary values would be his own test results of 2, 3, and even 5 years ago.

One of the main areas of differences in pulmonary function test results can be seen when the test results of patients with obstructive pulmonary disease are compared with those of patients with neuromuscular skeletal disorders (Fig. 26-3). A typical comparative series would show, for example, that, although one of the main characteristics of COPD patients is hyperinflation of the lungs with the total capacity well over 100% of predicted values, in the patients with kyphoscoliosis[13] the total lung capacity—by virtue of the compression caused by their skeletal deformities—is considerably under 100% of predicted values. Similar differences in measurements of the timed vital capacity of both types of patients are also found. The emphysematous patient is usually able to exhale only one third of his vital capacity, whereas the vital capacity of the patient with kyphoscoliosis will test out as being within the normal ranges of predicted values (Fig. 26-3).

The standard pulmonary function tests consist of (1) the determination of lung volumes and capacities, (2) airflows, (3) ventilation, and (4) pulmonary diffusing capacities. They are calculated by body surface area, age, and in some instances (as in vital capacity) only by height and age.[13] As in other laboratory calculations, there is a comparative base line in healthy people. In spirometry the observed values are compared with the predicted or normal base line values and then expressed either in volumes or percentages.

One index number does not make a diagnosis. All pulmonary function test indices have to be evaluated separately at first and subsequently as integrated parts of the whole.

In obstructive airway diseases the pulmonary function tests indicate a great respiratory effort. They also reveal impaired lung volumes and diminished expiratory flow rates. There is slight decrease in vital capacity; drastic decrease occurs only in far-advanced stages. The inspiratory capacity is the vital capacity subdivision most likely to be impaired.

The functional residual capacity and residual volume are moderately to considerably increased. The maximum breathing capacity and timed vital capacity are greatly diminished. The diffusing capacity and carbon monoxide uptake are impaired. In gas exchange, oxygen transport is impaired first, and as the COPD condition advances, the impairment of carbon dioxide transport develops and is characterized by carbon dioxide retention in the arterial blood.

There is poor distribution and uneven alveolar ventilation conducive to hypoxia, hypoxemia, and carbon dioxide retention. The oxygen tension at rest can be low, with consequent oxyhemoglobin desaturation, and a further drop occurs in oxygen tension and saturation at the slightest effort.

Pathologic findings are characterized by structural changes in the lung and bronchi.[9] The changes observed are increased size of the lungs, cysts, blebs, rupture and collapse of the small bronchi, diminished alveoli interface, impairment of the airway ciliary activities, fibrosis of the small bronchi, and destruction of the pulmonary vasculature. Roentgenograms show poorly aerated lungs with hyperinflation and flattened diaphragm and no significant diaphragmatic mobility. (See Fig. 26-4.)

The complex and interrelated function of the lung is that of taking in a sufficient

CARDIOPULMONARY LABORATORY

PULMONARY FUNCTION TEST

Date___7/15/66___

Name___W. C._____   Referred___A. H._____   Height___5'7"___   Weight___119___

Clinical diagnosis___C O P D_____   Age___54___ Sex___M___   B.S.A.___1.61___

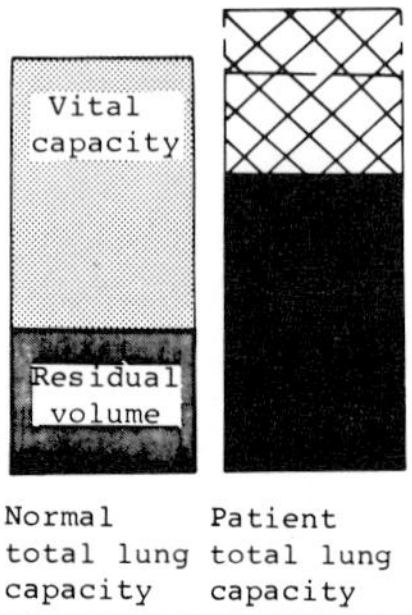

		Predicted value	Before bronchodilator		After bronchodilator	
			Observed value	Percent of predicted value	Observed value	Percent of predicted value
LUNG VOLUME	1 Inspiratory capacity (L.)	2.741	1.178	43%	1.370	50%
	2 Expiratory reserve volume (L.)	0.914	0.767	84%	1.041	114%
	3 Vital capacity-- sum of 1 and 2 (L.)	3.655	1.945	53%	2.411	66%
	4 Standard vital capacity (L.)	3.655	1.754	48%	2.219	61%
	5 Functional residual capacity (L.)	2.193	6.573	300%	5.837	266%
	6 Residual volume (L.)	1.279	5.806	454%	4.796	375%
	7 Total lung capacity (L.)	4.934	7.751	157%	7.207	146%
	8 RV/TLC $\left(\frac{7}{6}\right)$	26%	75%		67%	
	9 FRC/TLC $\left(\frac{7}{5}\right)$	44%	85%		81%	
AIR FLOW	Maximum expiratory air flow (L./min.)		27.6		31.6	
	Maximum mid-expiratory air flow		4.1		9.5	
	Timed vital capacity 1 sec.	83%	31%		24%	
	2 sec.	87%	42%		33%	
	3 sec.	93%	53%		41%	
	Maximum breathing capacity	94 L.	14 L.	15%	18 L.	19%
	Maximum inspiratory air flow (L./min.)					
AIR VELOCITY	Index $\frac{MBC\%}{VC\%}$	1.0	0.28		0.29	
SPIROMETRY						
VENTILATION	Oxygen consumption	0.250	0.113		0.192	
	Respiratory frequency	12-14	8		10	
	Tidal volume	0.500	0.805		0.712	
	Minute ventilation		6.439		7.124	

Pulmonary diffusion capacity for carbon monoxide ___6.6 ml./min./mm.Hg___
(Normal 15 to 30)

Fractional CO uptake___26%___
(Normal 30 to 50)

A

Fig. 26-3. Pulmonary function test with predicted and observed values in, **A**, chronic obstructive pulmonary disease compared to, **B**, neuromuscular skeletal involvement.

CARDIOPULMONARY LABORATORY

PULMONARY FUNCTION TEST

Date___4/13/67_____

Name__J. C._____ Referred___Dr. A. H._____ Height__4'7"__ Weight__72__

Clinical diagnosis___Kyphoscoliosis_____ Age__47__ Sex__F___ B.S.A.__1.14__

		Predicted value	Before bronchodilator		After bronchodilator	
			Observed value	Percent of predicted value	Observed value	Percent of predicted value
LUNG VOLUME	1 Inspiratory capacity (L.)	1.275	0.460	36%	0.490	38%
	2 Expiratory reserve volume (L.)	0.425	0.245	58%	0.220	52%
	3 Vital capacity-- sum of 1 and 2 (L.)	1.700	0.705	41%	0.710	42%
	4 Standard vital capacity (L.)	1.700	0.710	42%	0.740	44%
	5 Functional residual capacity (L.)	1.020	0.870	85%	1.080	106%
	6 Residual volume (L.)	0.595	0.625	105%	0.860	145%
	7 Total lung capacity (L.)	2.295	1.330	58%	1.570	68%
	8 RV/TLC $\left(\frac{7}{6}\right)$	26%	47%		55%	
	9 FRC/TLC $\left(\frac{7}{5}\right)$	44%	65%		69%	
AIR FLOW	Maximum expiratory air flow (L./min.)		25.5		27.5	
	Maximum mid-expiratory air flow		5.7		4.8	
	Timed vital capacity 1 sec.	83%	86%		91%	
	2 sec.	87%	95%		95%	
	3 sec.	93%	100%		100%	
	Maximum breathing capacity	56 L.	16 L.	29%	15 L.	27%
	Maximum inspiratory air flow (L./min.)					
AIR VELOCITY	Index $\frac{MBC\%}{VC\%}$	1.0	0.69		0.62	
SPIROMETRY						
VENTILATION	Oxygen consumption	0.250	0.149		0.180	
	Respiratory frequency	12-14	25		25	
	Tidal volume	0.500	0.170		0.160	
	Minute ventilation		4.360		4.090	

Vital capacity

Residual volume

Normal total lung capacity

Patient total lung capacity

Pulmonary diffusion capacity for carbon monoxide ___6.2 ml./min./mm.Hg___
(Normal 15 to 30)

Fractional CO uptake ___39%_____
(Normal 30 to 50)

B

Fig. 26-3, cont'd. For legend see opposite page.

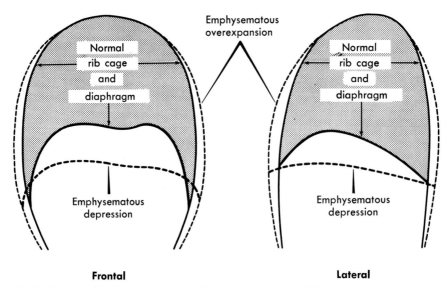

Fig. 26-4. Changes in the rib cage and the diaphragm in healthy and emphysema subjects.

amount of oxygen and eliminating carbon dioxide. This process is called gas exchange. An adequate supply of oxygen is necessary for the metabolic events from which energy is derived. Whenever more energy is required, metabolism is increased; hence the need for oxygen increases.

If faulty mechanics limit the oxygen supply, physical and mental functions will be impaired. Carbon dioxide is an end product of the oxidative process and is almost wholly eliminated through the lung, but it is also a chemical constituent that plays an important part in the regulation of respiration. If carbon dioxide elimination is impaired, this will also contribute to physiologic imbalance, since carbon dioxide is an important factor in regulating the acid-base balance of the body.

Gas exchange involves an intricate and complicated process[13] that includes:

1. *Ventilation,* which is composed of volumes of air sufficiently large to reach the multitude of alveoli in each inspiratory cycle and a *distribution* of this air equally throughout the millions of alveoli in the lung
2. *Diffusion,* by which oxygen and carbon dioxide pass across the alveolar capillary membrane
3. *Pulmonary capillary blood flow,* which must be large enough in volume and be distributed evenly to all the functioning alveoli

It is very important that this exchange should be accomplished by the respiratory and circulatory systems with a minimum expenditure of energy. Mechanical factors in ventilation are of critical significance; in patients with pulmonary insufficiency, an adequate pulmonary gas exchange may be achieved only by a great increase in the work of the respiratory muscles. Whether maximum effort can produce adequate ventilation and oxygenation may prove to be the critical factor for survival.

The ventilatory apparatus is composed of the chest wall and the diaphragm that convert metabolic energy into mechanical work, and the lung that is responsible

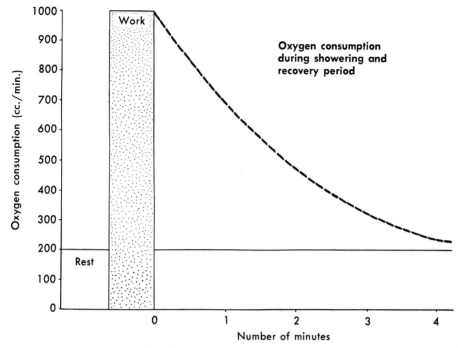

Fig. 26-5. Energy cost of showering.

for the cyclical exchange of air. Breathing is carried out in a ration of 65% for the diaphragm and 35% for the accessory and respiratory muscles.[11-15] These muscle systems include the scalenus muscles, the sternocleidomastoid muscles, the pectoralis muscles, the posterior and superior serratus muscles, parts of the erectors of the spine, and the external and internal intercostal muscles. In quiet breathing the abdominal muscles do not participate in the work of respiration. They participate only in exercise.

In the patient with COPD, the diaphragm is low and flattened, and the chest wall may be expanded in a hyperinflated position. The patient habitually assumes the kyphotic bent-forward posture. This allows him to breathe more easily because it sets the midposition of the diaphragm in the semiexpiratory position. However, the muscular effort necessary to maintain the hunchback position adds an increased load to the COPD patient's already high expenditure of energy; therefore oxygen demand is further increased. The abdominal muscles tend to lose their tonicity and become flabby and atrophied. This state militates against adequate abdominal breathing. Weak abdominal musculature is unable to push up the viscera that would normally assist the diaphragm to ascend against gravity. This is particularly true when the pathologically distended lung does not recoil on exhalation.

In healthy persons the oxygen uptake depends on the metabolic demand for the functioning of vital organs. If faulty mechanics prevent adequate oxygen supply, vital organs exposed to insufficient oxidation may suffer irreversible damage. For example, when pulmonary emphysema restricts the vascular bed, the heart's right

Subject	O$_2$ cons. (cc./min.)	Rise over basal (%)	cc./kg./min.	O$_2$ debt (cc. - min.)
P. B.	590	184	8.10	0.4 - 4
G. G.	813	297	12.29	0.8 - 4
S. M.	723	189	9.42	0.7 - 4
J. K.	604	230	9.99	0.8 - 4
G. X.	713	244	9.95	0.9 - 3.5
G. S.	892	202	9.12	1.0 - 3
J. R.	862	349	15.28	1.5 - 3
E. L.	986	381	14.18	1.3 - 2.75
L. R.	834	278	11.99	1.2 - 4
J. D.	1008	285	13.95	1.1 - 4

Fig. 26-6. Oxygen consumption and debt in showering.

ventricle must increase its work considerably to supply an adequate pulmonary blood flow. The increased strain that this added work load creates may result in cor pulmonale, or right heart failure.

Also, in COPD disorders the lung tissue loses its elasticity and much of its ability to recoil. Consequently, the diaphragm will be unable to overcome the resistance produced by the pathologically distended lung; therefore its normal excursion will be greatly diminished. This means that the accessory and intercostal muscles will be called on to make up this deficiency. The ratio between the diaphragm and the thorax will be reversed; that is, the accessory and intercostal muscles will take over the work with little or no participation of the diaphragm. As a result, breathing becomes more laborious, which causes increased metabolic demands that cannot always be met. This burden, combined with the activities of daily living, produces a lingering oxygen debt.

The term oxygen debt is used by A. V. Hill to describe a physiologic stage during strenuous muscular activities in which the deficit of oxygen intake represents a debt that must be repaid during recovery.[16] In continuous exercise lasting more than a few minutes, the oxygen intake must be adequate to meet the oxygen requirement. When this condition exists, the subject is said to be in a steady state (physiologic equilibrium), which means that he is able to supply sufficient oxygen to keep pace with the processes of breakdown and recovery in his muscles. However, if the exercise is too strenuous and the oxygen intake and recovery cannot keep pace with the muscular activities, the subject is no longer in a steady state and the oxygen debt has been incurred (Figs. 26-5 and 26-6). This is true in healthy people, but in patients with pulmonary emphysema and other COPD disorders, tolerance to muscular activity is considerably decreased, and an oxygen debt may be incurred on the slightest exertion. This oxygen debt can be repaid only after a long rest.

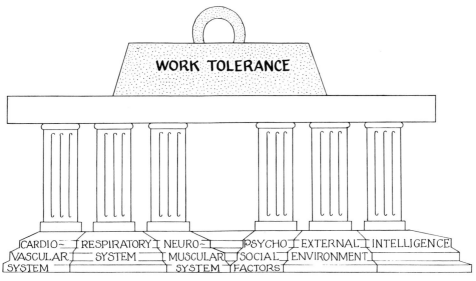

Fig. 26-7. Factors affecting work tolerance.

PRESCRIPTION OF ACTIVITIES

Modulated grading of physical activity is an important principle in the management of chronic diseases.[17] It applies particularly to conditions in which the distribution, diffusion, and transport of oxygen are deficient, with consequent diminution in work capacity. Regulation of activity may be advisable in convalescence when, after prolonged bed rest, deterioration of cardiovascular, respiratory, neuromuscular, and other physiologic function (deconditioning) temporarily renders the patient unfit for physical activities ordinarily undertaken without difficulty.

The stress load must vary with the individual patient's tolerance level. Unfortunately, fitness or tolerance to stress or work cannot be readily estimated by the physician, since it is an intangible summation of many individual physiologic and psychologic factors (Fig. 26-7). On the other hand, the approximate energy demand of any activity can be defined in metabolic terms,[18] that is, the energy expended to accomplish it. In this sense, work consists of two components: intensity and duration.

In any environment, duration can be readily controlled by prescription. The intensity factor, however, must be established for different activities by measuring the rate of oxygen consumption during their execution.[19] Although the usefulness of such data to estimate stress potentials may be admittedly limited, yet in conjunction with other clinical knowledge these data may contribute to a decision regarding the wisdom of allowing a given type or intensity of activity for a patient.

Because of the individual reaction to work, activities cannot be prescribed on a group basis. The prescription should be individualized, taking into consideration six factors: (1) severity (extent) of the lesion or lesions, (2) laboratory findings, (3) chronicity of the lesion or lesions, (4) history of regression, (5) sequelae after treatment or surgery, and (6) type of treatment.

In view of the patient's (1) inability to keep up with his metabolic demand and

(2) his ability or inability to either hold a job or maintain self-care, the first step in the rehabilitation procedure is to determine the status of his pulmonary performance, including arterial blood gas sampling. After this is ascertained, the physician should measure the patient's oxygen consumption under varying conditions of daily activity, such as walking, stair climbing, and, if possible, simulating working conditions of his job.

The treatment of COPD patients is usually oriented around obtaining symptomatic relief with antibiotics, bronchodilators, and oxygen therapy with intermittent positive pressure breathing (IPPB) machines. These modalities relieve some of the symptoms but cannot influence the pulmonary changes, which are, as previously noted, irreversible. Surgical dissection of the diseased part of the lung may be helpful in selected cases, particularly when bronchiectasis is present. Tracheal fenestration with artificial fistula is said to be helpful occasionally to maintain good bronchial hygiene and to decrease the dead space. However, this procedure is of questionable value.

Previously, we described the unusual and psychologic stress-inducing body posture of emphysematous patients, in which the kyphotic, quasi-hunchback position contracts not only the respiratory muscles but also all other muscles of the body. The patient's anxiety state is further exacerbated by the recurrent fear of choking to death. Therefore, one of the first steps in the rehabilitation of the patient is educating him about the nature of his disease. The more the patient and, whenever possible, his family understand about COPD disorders, the more they can help themselves by cooperating with their physicians. Family participation in all phases is paramount for successful rehabilitation.

The second step in the rehabilitation program consists of relaxation exercises as described by Jacobson.[20] Relaxation is essential to any program of rehabilitation.

For maximum relaxation, three basic conditions have to be realized:

1. That the patient is in his most comfortable position: supine with a comfortable pillow supporting his head, a round pillow under his knees, and his arms resting at the sides of his body. In view of positional changes of pulmonary values in the supine position, some patients, particularly the orthopneics, might have more breathing difficulties than relief in a supine position. For these patients, the fetal position will prove more relaxing and should be assumed with a pillow under the head and the feet elevated 5 to 10 inches so that the patient is in a slight Trendelenburg position.

2. That the room should be as quiet as possible, except for some soft music, which can be piped in.

3. That it must be kept in mind that, although these are called relaxation exercises, they are still active exercises, especially during the contraction phases, and the cost in oxygen can be relatively high for a patient who is chronically hypoxic. A low flow of about 2 or 3 L./min. of 100% oxygen through a *double trunk nasal catheter* should be used. This will reassure the patient and, more importantly, it will do no harm to the respiratory drive. If a mask is used, *100% oxygen should not be employed;* a mixture enriched to about *25% to 30% oxygen* will be sufficient to meet the patient's needs in maintaining an adequate oxyhemoglobin saturation.

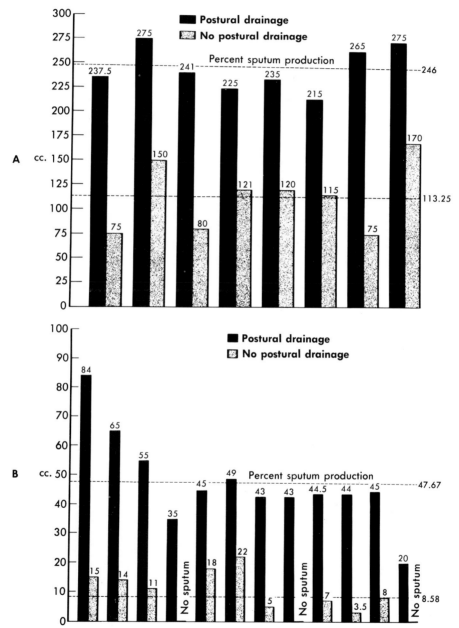

Fig. 26-8. Sputum produced in 24 hours with and without postural drainage in patients with, **A,** bronchiectasis and, **B,** emphysema.

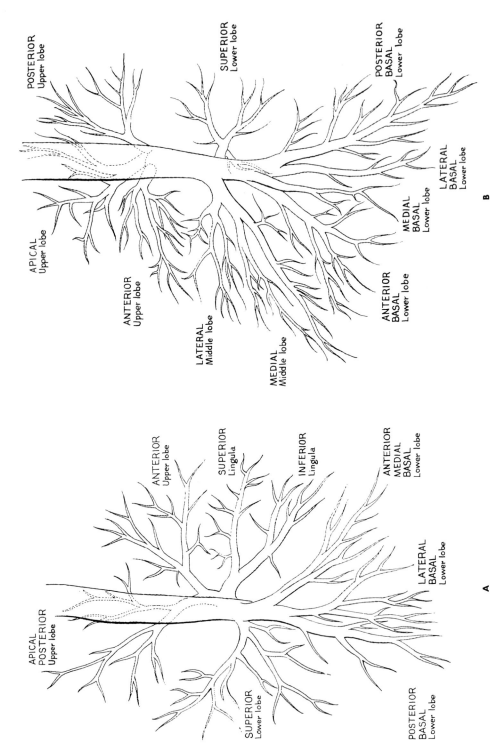

Fig. 26-9. A, Lateral view of the distribution of the bronchopulmonary segments of the left lung. **B,** Lateral view of the distribution of the bronchopulmonary segments of the right lung.

As described earlier, the COPD patient's muscles are tensed in chronic contraction, especially the muscles of the upper trunk. The first step in the relaxation exercises is to explain to the patient and to demonstrate what is asked of the patient in the way of exercise movements. For example, the easiest demonstration is either to lift the shoulder (shrug) or to contract the pectoralis muscles. These are visible and easily palpable muscles and can be contracted so that the patient's understanding will be achieved.

The contraction phase of relaxation exercises should be held for a count of from three to five, followed by a relaxation period for a count of ten. A metronome is a helpful device to provide the count. As a rule, the relaxation phase should be double the contraction phase in duration.

More strenuous relaxation exercises, for example, raising the unbended legs and dropping them, should not be given to patients whose clinical status is precarious.

Once education and these relaxation exercises teach the patient how to be more relaxed, the third rehabilitation step—postural drainage—can be undertaken.[21]

The majority of COPD patients produce sputum most of the time, but only a portion of the secretion is coughed up. This is especially true when the COPD condition is so far advanced that the cough mechanism and ciliary action of the pulmonary system are impaired. Although the accumulated secretions can be removed by suction or by bronchial washing, postural drainage causes the patient the least discomfort. Another advantage of postural drainage is that, unlike the other bronchial hygiene modalities in which the patient's participation is a passive one, in postural drainage the patient takes part in his own treatment. This helps relieve the tedium of the inactivity to which many COPD patients are often condemned.

This helpful technic is also used in postoperative patients. Because most patients with COPD conditions are able to evacuate only the accumulated secretions from the larger bronchi and cannot cleanse the small bronchi where the true obstructions prevent proper alveolar ventilation, postural drainage should be considered to be the bronchial hygiene method of choice (Fig. 26-8).

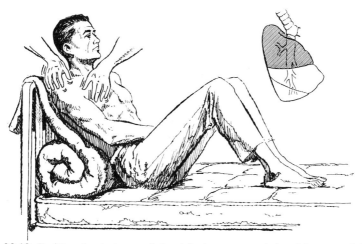

Fig. 26-10. Position 1—drainage of the apical segments of the right upper lobe.

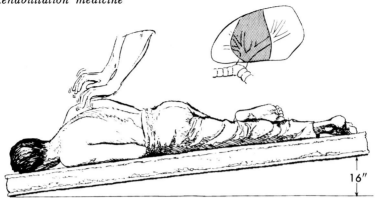

Fig. 26-11. Position 2—drainage of the lateral segments of the right middle lobe.

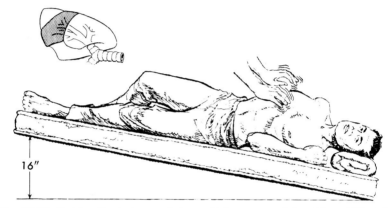

Fig. 26-12. Position 3—drainage of the medial segment of the right middle lobe.

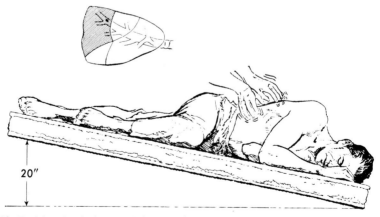

Fig. 26-13. Position 4—drainage of the anterior basal and medial basal segments of the right lower lobe.

Fig. 26-14. Position 5—drainage of the apical segment of the left upper lobe.

To obtain spontaneous drainage of the bronchial tree, the patient should first be helped to break up the secretions by inhaling from a heated aerosol container of a bronchodilator in saline solution. After this preliminary conditioning, the patient should be placed in such a position that gravity will help the cough to drain the site of the lesion through its tributary segmental bronchi (Fig. 26-9). *(It is important that the patient should not simply be instructed to lie in a vertical position, as is often practiced.)*

Postural drainage should entail precise knowledge of the bronchopulmonary segments and the relevant positions necessary for drainage by gravity (Fig. 26-9). It is important that during exercises to stimulate better drainage the physical therapist should tap the patient's upper torso with cupped hands.

The ten basic positions for postural drainage are as follows:

Position 1—drainage of the apical segment of the right upper lobe (Fig. 26-10)
Position 2—drainage of the lateral segment of the right middle lobe (Fig. 26-11)
Position 3—drainage of the medial segment of the right middle lobe (Fig. 26-12)
Position 4—drainage of the anterior basal and medial basal segments of the right lower lobe (Fig. 26-13)
Position 5—drainage of the apical segment of the left upper lobe (Fig. 26-14)
Position 6—drainage of the superior segment of the left lower lobe (Fig. 26-15)
Position 7—drainage of the inferior segment of the left upper lobe (lingula) (Fig. 26-16)
Position 8—drainage of the posterior basal segment of the left lower lobe (Fig. 26-17)
Position 9—basic position for drainage of the trachea (Fig. 26-18)
Position 10—forty-five degree lateral and 90-degree head-down positions for evacuation of the main bronchi (Fig. 26-19)

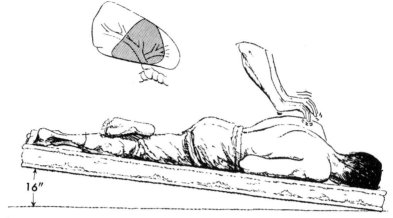

Fig. 26-15. Position 6—drainage of the superior segment of the left lower lobe.

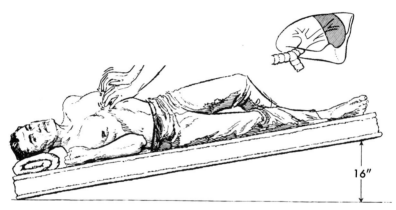

Fig. 26-16. Position 7—drainage of the inferior segment of the left upper lobe (lingula).

RECONDITIONING OXYGEN THERAPY

The major reasons for employing supplementary oxygen to facilitate the rate and extent to which patients with various forms of lung disease can be rehabilitated stem from clinical experience.[22] It has been commonly noted in previous investigations of hypoxic patients that many of the exercises and treatments prescribed in the course of intensive physical medicine and rehabilitation have to be deferred because of easy fatigability and shortness of breath observed in the patients. In other individuals relatively simple exercise and direction were only partly understood because of the effects of hypoxia on both brain function and muscular activity.

The rate of oxygen administration should be carefully regulated. The flow should be adjusted to the individual patient's needs and not according to the tables provided by equipment manufacturers. For patients with alveolar hypoventilation, the oxygen should be administered via an IPPB machine.

Moreover, it has long been noted that patients with chronic lung disease frequently become worse when hypoxia interferes with the function of the respiratory

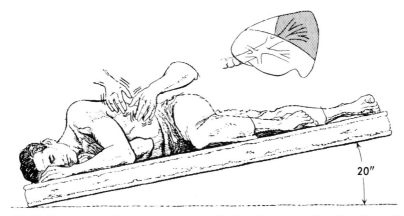

Fig. 26-17. Position 8—drainage of the posterior basal segment of the left lower lobe.

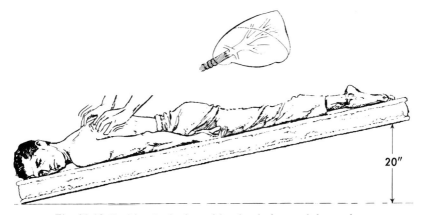

Fig. 26-18. Position 9—basic position for drainage of the trachea.

muscles.[22] A vicious cycle is then started in which chronic lung disease leads to alveolar hypoventilation, which in turn causes hypoxia. The hypoxia in turn interferes with the function of the respiratory muscles, which then aggravates the degree of hypoventilation so that a further degree of hypoxia is introduced. This cycle perpetuates itself so that lung function continuously deteriorates as a result of deteriorating respiratory muscle function.

However, before any oxygen-assisted therapy is initiated, care should be taken to determine whether the patient shows any signs of either carbon dioxide retention or alveolar hypoventilation. Only after these important factors are ascertained should the patient be started on oxygen therapy. It is also necessary to obtain simple and quick methods of evaluating the carbon dioxide tension of blood and/or alveolar gas in patients with pulmonary emphysema who are medicated with oxygen.

It has been known for some time that COPD patients must be subjected to oxygen therapy with great caution to avoid aggravating the degree of carbon dioxide retention and not to kill the respiratory drive. It appears that in these patients the

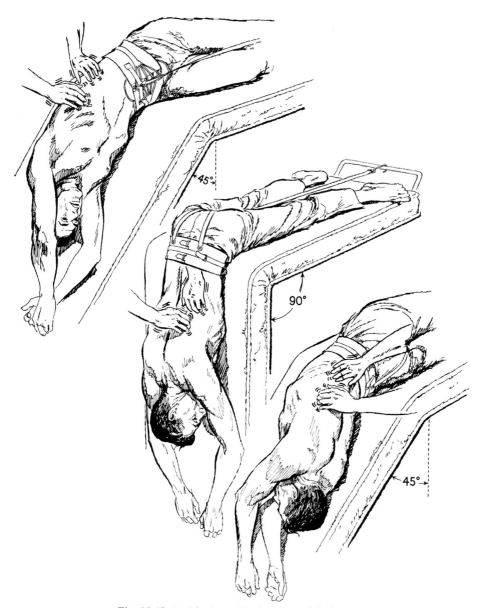

Fig. 26-19. Positioning table for postural drainage.

major stimulus to ventilation is the degree of hypoxia that they possess, rather than the carbon dioxide tension to which they have become accustomed. When hypoxia is eliminated in an oxygen-enriched atmosphere and the last major stimulus to ventilation is reduced, marked underventilation occurs with further carbon dioxide retention and eventual carbon dioxide narcosis. For this reason, daily measurement of the carbon dioxide tension by means of capillary blood, using the microtonometric technic of Astrup[23] or using the end-tidal sample technics and rapid infrared analyzer to obtain an index of alveolar air, should be made. Both of these methods of analyzing blood gas composition eliminate the necessity for arterial punctures.

The carbon monoxide steady state pulmonary diffusing technic, which bypasses the need to make blood gas analyses, can be utilized to sample the end-tidal volume. This technic permits measurements of the pulmonary diffusing capacity to be made every day for periods of as long as 4 to 6 weeks.

THERAPEUTIC BREATHING EXERCISES

Muscle reeducation of patients with musculoskeletal disorders has resulted in spectacular benefits, and our own experience has shown that similarly encouraging results can be obtained with the respiratory cripple. The purposes of respiratory exercises are to teach the patient how to breathe properly and to relax in normal body posture by making use of diaphragmatic and abdominal muscles and to physiologically accommodate to his limited cardiorespiratory reserve. These exercises, diligently carried out, will increase the strength of the abdominal muscles and the diaphragm, thus removing the unnecessary burden from the upper costal and accessory muscle groups that is otherwise so costly in energy expenditure in COPD patients. The economic use of the ventilatory muscles will reduce the oxygen uptake, often so limited, and play an important role in the survival of the patient.

For a proper program of prescribed therapeutic breathing exercises, the patient should first have a full set of tests of pulmonary function and energy studies in daily activities, with viable determinations of oxygen consumption, oxygen debt, and recovery[21] (Fig. 26-20). He should also be given blood gas studies, both at rest and during activity, to evaluate the adequacy of gas exchange.

The technic of diaphragmatic breathing exercises can be summarized as follows: (1) general skeletal muscle retraining during plain muscular activity, attempting to simulate work and vocational requirements, (2) weighted diaphragmatic exercises,[25] utilizing abdominal weights, proceeding from 5 to 15 pounds, and (3) observation of costal motion between thoracic and diaphragmatic breathing.[26]

Exercise 1

Position: Supine, with head lowered to about 15 degrees below level of body (Fig. 26-21).
Instructions: Left hand placed on thoracic cage just below sternum to limit excursion of chest and thus to force abdominal breathing. Right hand placed on abdomen over umbilicus. Patient is then instructed to inhale through nose, pushing and distending abdominal muscles outward, and then to exhale through mouth, pulling in and contracting abdominal muscles. Exhalation rather than inhalation should be emphasized.

Exercise 2

Position: Sitting, slightly reclined in completely relaxed position.
Instructions: Same as for Exercise 1.

Exercise 3

Position: Standing relaxed.
Instructions: Same as for Exercise 1.

Exercise 4

Position: Walking.
Instructions: Same as for Exercise 1, synchronizing steps with metronome and endeavoring to time step first with inhalation and next with exhalation.

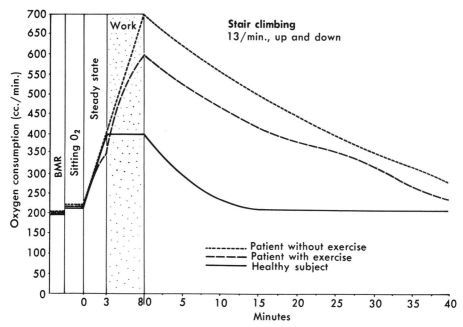

Fig. 26-20. Comparison of oxygen consumption, oxygen debt, and recovery rate from stair climbing in 10 healthy subjects and 10 subjects with emphysema—before and after therapeutic exercises.

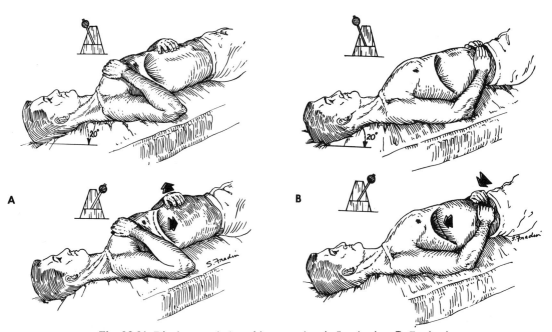

Fig. 26-21. Diaphragmatic breathing exercise. **A,** Inspiration. **B,** Expiration.

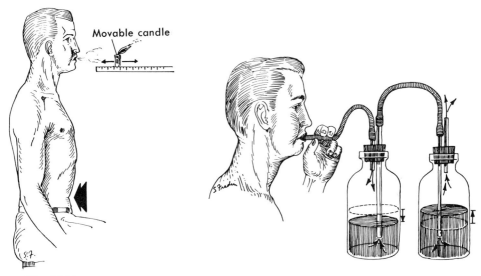

Fig. 26-22. Candle blowing.　　　　　　　　　**Fig. 26-23.** Bottle blowing.

Exercise 5

　　Position: Elevation or walking up stairs.
　　Instructions: Same as for Exercise 1.

As the patient masters the essential breathing pattern, the exhalation period can be gradually extended by means of candle blowing and bottle blowing.

Candle blowing exercises (Fig. 26-22)

　　Use lighted candle set at gradually lengthened distances on yardstick graduated in inches. Candle is placed at level of patient's mouth. Patient, in sitting position, is asked to blow out candle. Exercise is repeated from standing position.

Bottle blowing exercises (Fig. 26-23)

　　Use two half-gallon bottles, half filled with water and connected with three rubber tubes and two glass pipes. Patient, in sitting position, is instructed to blow water from one bottle to the other, with same breathing pattern as taught in candle blowing. Exercise is repeated in standing position.

These exercises should be limited initially to a total of 5 minutes, being gradually increased to 30 minutes. No changes in basic position (lying, sitting, standing, or walking) should be made until the patient is able to do the prescribed exercises without difficulty. Finally, he should be encouraged to increase the frequency of the exercises himself until he has completely changed his breathing pattern.

All of the different types of exercises prescribed for COPD patients should be performed with or without supplementary oxygen, according to the treating physician's judgment.

In planning a program of rehabilitation exercises, it must be borne in mind that one of the problems confronting COPD patients is their tendency to develop anxiety and panic states, most notably during periods of severe dyspnea. Clinical experiences have demonstrated that when a patient can be taught to overcome his fears and relax, his discomfort is greatly alleviated and his rehabilitation will be more successful.

The value of rehabilitation maneuvers has long been recognized in the management of patients with COPD conditions. Unfortunately, these measures cause much skepticism and misgivings because they have not been supported by sufficient scientific measurements.[27] Spirometric measurements not being selective enough to detect any significant changes, the results have to be judged by the subjective impressions of the patients rather than by the measureable increased tolerance of daily activity.

If the patient is taught appropriate breathing patterns, he will make the best use of his diminished cardiopulmonary reserve and relief will be physical as well as psychologic.

SOCIOVOCATIONAL REHABILITATION

The basic philosophy of rehabilitation is to do everything that can be done to return the COPD patient to a useful, self-sufficient role in society.[2] This means that once the physician has helped the patient function at his permanently reduced physiologic capacity, the various other members of the rehabilitation team—from the psychologists and the vocational counselor to the social workers, the occupational therapist, and the job counselors—must undertake the complex tasks of reinstating the COPD patient to as close to his premorbid status as is physically possible. Therefore, the *earlier* in the course of the COPD disorder the vocational rehabilitation program is instituted, the better will be the potential results.

Ideally, the time to start rehabilitation of the COPD patient is while he is still able to be treated on an outpatient basis and is still able to keep his own job. However, since the majority of COPD patients do not seek medical help for their disorders until they are no longer able to hold their jobs or even manage for themselves, the problems and prognosis of vocational rehabilitation are grossly magnified by this lack of understanding.

The success of rehabilitation depends on certain salient factors that can be influenced and often controlled by either the physicians or the members of the rehabilitation team. For example, are the social climate and the family situation such as can motivate the patient to want to help himself? Here the social worker and the psychologist should collaborate in making the needed evaluations.

Does the rehabilitated COPD patient have enough cardiorespiratory reserve to return to his former job on a full-time basis? Or can he only assume part-time work in his previous skill? Here the physician's judgments are paramount.

Or had the patient already been reduced to part-time employment status by the time his clinical and rehabilitation therapy started? Here the physiatrist must alert the vocational counselors to the altered status of the patient's work tolerance.

Or did rehabilitation enable the originally helpless patient to perform sedentary work now? If, as is often the case, the physically rehabilitated patient requires training in new vocational skills commensurate with both his educational level and his reduced pulmonary capacity, care must be taken to train him for jobs that will be available to him in our changing employment markets. Here the guidance of skilled occupational placement counselors must be sought to make the rehabilitation complete.

In some instances the physician might conclude that the patient perform only in sheltered workshops or at home, where IPPB machines and other equipment needed for the immediate relief of symptoms are always available. Here, too, the occupational therapist and placement counselors can help find such sources and settings for productive work.

As the population of COPD victims continues to grow in our nation, the improved clinical modalities are increasing the individual rates of survival in pulmonary disorders. The physically salvaged COPD patients, however, are often unable to meet the cardiorespiratory demands of their former jobs. This need not mean that they should be doomed to become public charges, leading aimless and economically nonproductive lives. Rehabilitation can assist the human beings of this rapidly growing segment of our total population to lead meaningful, self-sufficient, and economically productive lives.

POSTOPERATIVE MANAGEMENT

With the development of improved methods and the availability of an effective group of antibiotic drugs, thoracic surgery has made remarkable advances during the last decade. However, the most skillfully performed thoracic surgery is bound to leave serious pathologic anatomic and physiologic residua.

In opening the chest wall the surgeon is forced to cut through the large muscles of the operative area, traumatizing muscles that are vitally important in shoulder girdle mobility and trunk posture. During the surgery the scapula on the operated side is elevated to a position that tends to overstretch the scapular (both anterior and posterior) muscles. Furthermore, the fibers of the trapezius, rhomboid major, latissimus dorsi, and serratus anterior muscles are sectioned and, although ultimately repaired, still inevitably show varying degrees of posttraumatic atrophy and dysfunction conducive to progressive decreases in range of motion and structural changes. The severed muscles are usually sutured, resulting in their being shortened and weakened.

Studies in the physiology of muscular activity have demonstrated that muscle functions best when contraction is initiated with the muscle in its normal resting length (Sterling's law). Traumatic shortening, therefore, is conducive to inefficient function. Thus with the muscles of the involved side anatomically and physiologically impaired, the contralateral muscles become relatively stronger, pulling and twisting the spinal column into possible deformity.

The operative trauma leaves a trail of fibrosis, matted soft tissues, and atrophy, plus considerable discomfort and pain. These sequelae often encourage dysfunction, malposition, and ankylosis of varying degrees.[28] (See Fig. 26-24.) The functional complexity of the back muscles is such that trauma inflicted on any one of the muscles seriously upsets the synergistic action of the whole group, and consequently results in an imbalance of the upper trunk, and in varying limitations of the range of motion of the arm on the affected side. To anticipate and to counteract, as far as possible, the permanent postoperative sequelae, the physiatrist must consider the whole functional complexity and interrelationship of all back muscles on both sides and apply the therapeutic measures accordingly.

In some instances, however, additional postoperative complications arise. The discussion of several follows.

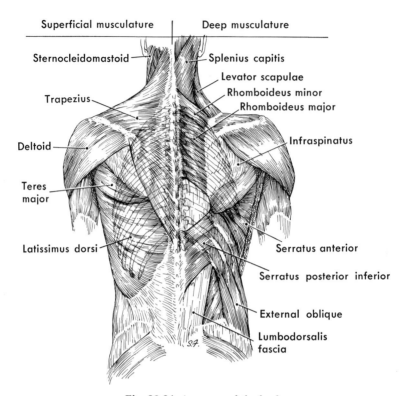

Superficial musculature | Deep musculature

Sternocleidomastoid — Splenius capitis
Levator scapulae
Rhomboideus minor
Trapezius Rhomboideus major

Deltoid Infraspinatus

Teres
major

Latissimus dorsi Serratus anterior
Serratus posterior inferior

External oblique
Lumbodorsalis
fascia

Fig. 26-24. Anatomy of the back.

In closed thoracotomy (a procedure frequently used in thoracic surgery, in explorations, or postsurgery) a tube is inserted into the pleural cavity through the middle fibers of the serratus anterior muscle on the edge of the latissimus dorsi muscle. The irritation caused by the tube and the resultant pain produced by the movement of the arms frequently force the patient to lock his arm to the side in an effort to immobilize the shoulder girdle for prolonged periods of time, leading to varying degrees of fibrotic ankylosis. Because of the same factors (the irritation of the tube and the pain), the latissimus dorsi muscle, instead of relaxing when the arm is fully flexed, remains in a contracted stage, imbalancing the synergistic action of the serratus, the trapezius, and the pectoralis muscles (Fig. 26-25).

It is essential that when applying appropriate exercises, it be emphasized that the inserted tube remain in situ. Forward flexion of the arm with gradual forward stretching is actually the only safe exercise under the circumstances. A bed pulley can be installed, and the patient should be encouraged to use this helpful device. The pain cannot be entirely avoided. Pain can, however, be substantially reduced with heat and infrared diathermy.

Possible complications in thoracic surgery include damage to the nerve supply as a result of unintential trauma to the subclavicular part of the brachial plexus. The trauma may be caused by undetected pressure while the patient is on the oper-

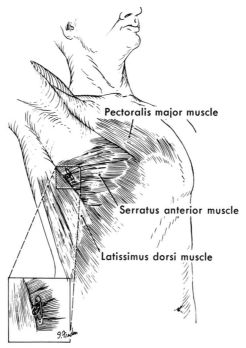

Pectoralis major muscle

Serratus anterior muscle

Latissimus dorsi muscle

Fig. 26-25. Closed thoracotomy.

ating table, or it may be due to accidental stretching of a branch of the brachial plexus during the procedure. Since the patient lies in a lateral position on the un-affected side during surgery, the pressure neuritis occurs on that side; this may result in a palsy of one of the branches of the brachial plexus—axillary, radial, median, or ulnar nerve. Physical therapy measures, plus heat, interrupted galvanic currents, and exercise, speed recovery to normal unless the trauma and the inci-dental damage have been severe and irreversible. In such cases, recovery will be quite slow or even impossible.

Similarly, injury of a nerve due to accidental stretching during surgery varies in intensity with the severity of the trauma. It is desirable to detect such nerve injury early so that effective therapy can be instituted to prevent secondary degeneration and crippling changes in the affected limb.

Diagnosis of the nerve involvement is clinically relatively simple. Electrodiagnostic measures (galvanofaradic measures and electromyography) help in evaluating the extent of the trauma and the rate of recovery of nerve function.

A frequent and feared complication of thoracic surgery is bronchopleural fistula. Failure to recognize this complication in its early stage can seriously compromise the outcome of the operation itself.

There are two ways of dealing with bronchopleural fistula:

1. A tube is inserted into the pleural cavity and is attached to an under-water suction bottle to drain the fistula and thus hasten its closure. The rehabilitation exer-

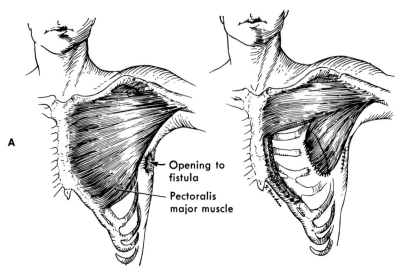

A

Opening to
fistula

Pectoralis
major muscle

Fig. 26-26. Pectoralis myoplasty. **A,** First step. **B,** Second step.

cise used is the same as that outlined for a thoracotomy, and the same special care
must be taken not to dislodge the draining tube.

2. If the above-mentioned maneuver proves to be unsuccessful and the fistula
remains patent, a pectoral myoplasty has to be performed. (On very rare occasions
the latissimus dorsi is used.) The technic of this operation is quite complex and
often quite devastating to the muscular relationship of the upper chest. The whole
muscle girdle is separated from its attachment to the sternum and the anterior chest
wall, and then the mass of muscle is displaced to the approximate midaxillary line.
Next an attempt is made to obliterate the troublesome fistula with the detached
muscles. Postoperatively the arm of the affected side is strapped to the chest for
7 to 10 days, until satisfactory obliteration of the fistula is attained. (See Fig. 26-26.)

Although this procedure effectively solves the surgical problem, it also creates a
chain of severe and complex sequelae. The operative procedures (frequently multiple) leave weak, atonic, atrophied, and imbalanced muscles, plus a great amount
of crippling fibrotic ankylosis that may well be severe and in some instances irreversible.

The rehabilitative measures should start immediately after surgery. Immobilization of the arm is unnecessary, since the pectoralis muscles once severed from their
origin do not readily regain the function as prime adductors and/or internal rotators
of the arm and therefore will not interfere with the healing process. Passive, assistive,
or active exercises to avoid ankylosis of the shoulder after a closed thoracotomy should
be skillfully and individually applied. They cannot possibly cause any harm and will,
in fact, effectively minimize postoperative discomfort and pain. Exercises can also
help prevent any permanent crippling sequelae.

Although extensive thoracoplasty is no longer routine, there is still considerable
surgery performed in which bone structure is altered. Proper supervised exercises
for postthoracoplasty patients remain the treatment of choice for the physiatrists.

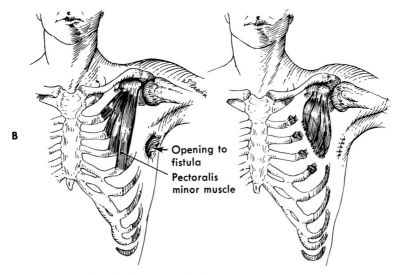

B

Opening to
fistula

Pectoralis
minor muscle

Fig. 26-26, cont'd. For legend see opposite page.

THERAPEUTIC EXERCISES

Abdominal breathing exercises, as prescribed for chronic obstructive pulmonary diseases, and general preventive and corrective exercises should be taught preoperatively and postoperatively and continued faithfully after the completion of surgery.

The necessity of oxygen therapy (tent, IPPB, etc.) is no barrier to an early reconditioning program. Experience has taught that the following seven exercises are effective in achieving the goal of rehabilitation. The exercises should start, if possible, with the breathing exercises prescribed on pp. 523-525 to induce mild coughing to free the air passage. These should be started, if possible, as soon as the patient is no longer under anesthesia.

The seven basic preoperative and postoperative exercises

Exercise 1 (Fig. 26-27)

Position: Supine.
Instructions: Patient is instructed to level pelvis, avoiding any rotation or lifting of one hip, and then to level shoulders, which are held parallel to level hips.
Objective: To emphasize correct bed posture.

Exercise 2 (Fig. 26-28)

Position: Supine, with pelvis and shoulders level and head normal.
Instructions: Patient places left hand on chest and right hand on abdomen and is instructed to breathe abdominally (diaphragmatically), right hand moving with each inspiration and expiration. Taught not to move chest or left hand.
Objective: To avoid thoracic breathing as far as possible.

Exercise 3 (Fig. 26-29)

Position: Supine.
Instructions: Patient bends trunk laterally toward nonoperative side and is instructed to breathe diaphragmatically.
Objective: To hasten mobilization of affected side of diaphragm, and thus help avoid para-

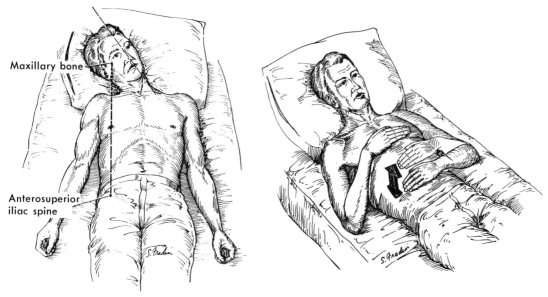

Fig. 26-27. Exercise 1.   **Fig. 26-28.** Exercise 2.

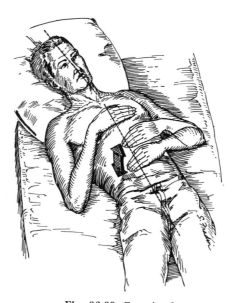

Fig. 26-29. Exercise 3.

doxical breathing after thoracoplasty, and to speed pulmonary reexpansion after excisional surgery. Incidentally, this exercise also induces cough reflex, of considerable importance in evacuating the pathologic contents of the bronchi.

Exercise 4—Active head extension (Fig. 26-30)

Position: Supine, with pelvis, shoulders, and head level.

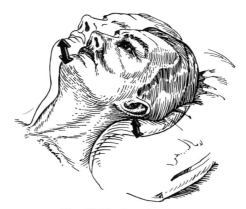

Fig. 26-30. Exercise 4.

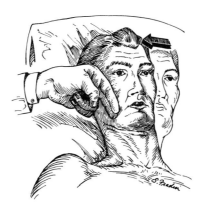

Fig. 26-31. Exercise 5.

Instructions: With chin depressed, patient presses head and neck backward for count of two and relaxes for count of three. To be repeated three times.

Objective: To strengthen primary and synergistic extensor of head and cervical spine (upper fibers of trapezius, splenius longus colli, etc.).

Exercise 5—Active shift of head to operative side (Fig. 26-31)

Position: Basic supine (bed mirror tilted over patient).

Instructions: Head moved laterally (like a Bali dancer) only to one side. Head, held at right angle to body, slide toward operative side for count of two and returned to normal position for count of three. To be repeated three times.

Exercise 6—Active adduction and depression of scapula (unilateral) (Fig. 26-32)

Position: Supine.

Instructions: Therapist places hand under scapula. Patient now instructed to press backward medially and downward on supporting hand. To be repeated three times and held for count of three, with rest periods of 3 seconds between exercises.

Objective: To teach the patient specific coordinate exercises that will be required postoperatively to restore the function of the adductors and the depressors of the affected scapula (rhomboids, middle and lower fibers of trapezius, latissimus dorsi, eat.). This exercise, inci-

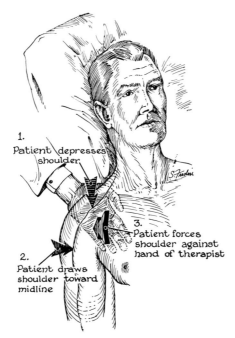

Fig. 26-32. Exercise 6.

1.
Patient depresses
shoulder

2.
Patient draws
shoulder toward
midline

3.
Patient forces
shoulder against
hand of therapist

Fig. 26-33. Exercise 7.

dentally, stretches the pectoralis muscles, thus combating tendency to forward rotation of the shoulder.

Exercise 7—Active-assistive depression of affected shoulder girdle (Fig. 26-33)

Position: Supine (bed mirror tilted over patient).

Instructions: Forearm flexed on upper arms and fingers on shoulder. Arm held at side of trunk. Therapist "fixes" shoulder girdle and flexed elbow joint with his fingers. Patient depresses flexed arm downward, stretching shoulder capsule, ligaments, and muscles, and holds this position (shoulder not to be raised above normal level) for count of three. To be repeated three times.

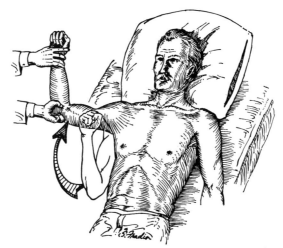

Fig. 26-34. Exercise 8.

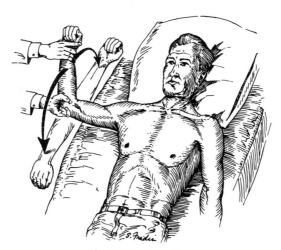

Fig. 26-35. Exercise 9.

Objective: To stretch and strengthen muscles of shoulder girdle (latissimus dorsi, lower fibers of trapezius, pectoralis, etc.) and thus try to prevent postoperative "high shoulder."

Postoperative phase exercise program

Start with the seven basic exercises given preoperatively, following the same sequence and timing. Then continue with the following eleven exercises:

Exercise 8—Active-assistive abduction of the arm at the shoulder (unilateral) (Fig. 26-34)

Position: Supine, with arm of operated side flexed at elbow and held close to chest.

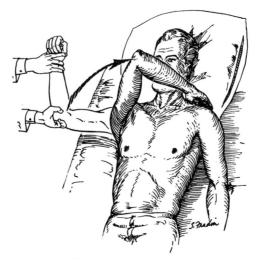

Fig. 26-36. Exercise 10.

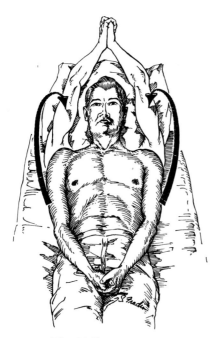

Fig. 26-37. Exercise 11.

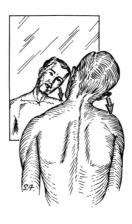

Fig. 26-38. Exercise 12.

Instructions: Therapist supports arm at flexed elbow and hand. Patient is encouraged to abduct arm without strains, with the therapist assisting active effort. To be repeated three times with 3-second intervals of rest.

Objective: To assure maximum possible abduction of the shoulder joint by preventing "freezing" and muscular atrophy.

Exercise 9—Active-assistive rotation of the shoulder joint (Fig. 26-35)

Position: Supine, with arm in same starting position as for Exercise 8. Again the patient is encouraged to carry out this exercise to the point of tolerance, with the therapist supplying the degree of assistance indicated. To be repeated three times, with period of rest to count of three.

Objective: Same as with abduction—to assure maximum mobilization of joint and to hasten the rehabilitation of the muscles of the shoulders.

Exercise 10—Active-assistive "horizontal" abduction and adduction of arm (unilateral) (Fig. 26-36)

Position: Supine, with arm flexed at elbow and abducted to 90 degrees if possible; if not, abducted as close to 90 degrees as possible.

Instructions: Therapist supports flexed arm, as in Exercises 8 and 9. Flexed arm is carried medially, striving to touch opposite shoulder with fingers. Frequency and time intervals same as in previous exercise.

Objective: To stretch postoperative adhesion and to exercise adductors of scapula and extensors of arm during the abduction phase of the procedure.

Exercise 11—Active-assistive elevation (flexion) of both arms (Fig. 26-37)

Position: Supine.

Instructions: Patient clasps hands with arms extended and resting on his trunk, and is then instructed to elevate arms slowly overhead to the point of tolerance, which may be quite limited by pain during the early postoperative stages. To be repeated three times, with customary rest periods. During this and all other exercises it is essential to be on guard for substitution movements such as elevation of the scapula and forward rotation of the shoulder; desirable to adhere to physiologically normal joint movements.

Objectives: To mobilize shoulder joints and exercise muscles of the shoulder girdle and the arm.

●　　●　　●

The first phase of the postoperative exercise routine is diligently carried out in the lying position for a period of 1 week. At the start of the second week after surgery, if the condition of the patient permits, he is seated in front of a full-length mirror, and the rehabilitation routine is continued with the following exercises:

Exercise 12—Active unilateral flexion of head (Fig. 26-38)

Position: Sitting, with pelvis, shoulders, and head level.

Instructions: Patient bends head to affected side, trying to touch ear to shoulder and avoiding rotation of head, holds for count of two, returns to normal position, and rests for count of three. To be repeated three times.

Objective: To strengthen residual flexors of affected side and to stretch those of the symmetric (unoperated) side.

Exercise 13—Active circumduction of both shoulders (Fig. 26-39)

Position: Sitting, with arms flexed at elbows and held at sides of body and finger tips touching shoulders.

Instructions: Patient carries flexed arms forward and upward to right angles with trunk; abducts arms horizontally, winging out to sides of body; and returns to starting position. To be repeated three times with essential rest periods. If the patient finds this procedure too taxing, limit procedure to affected shoulder only.

Objective: To keep shoulder joints freely mobile and to stretch matted soft tissues.

Exercise 14—Abduction and adduction of flexed arms (Fig. 26-40)

Position: Sitting, hands clasped back of neck, flexed arms winged at sides, shoulders high.

Instructions: Patient adducts flexed arms, touching elbows at front of body, and then carries arms horizontally back to original position. To be repeated three times with usual rest intervals.

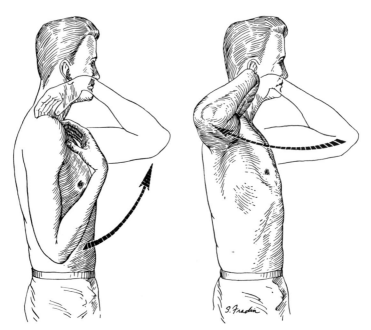

Fig. 26-39. Exercise 13.

Objective: To stretch postoperative matted soft tissues and to exercise the adductors of the scapula.

Exercise 15—Adduction and depression of the scapula (Fig. 26-41)

Position: Sitting, hands clasped over umbilicus, arms held at sides, and shoulders depressed.
Instructions: Patient is instructed to approximate scapulas by contracting adductor muscles. To be repeated three times with suitable rest intervals.
Objective: To strengthen the adductors of the scapula and to stretch the pectoral muscles.

Exercise 16—Flexion and extension of the trunk (Fig. 26-42)

Position: Sitting erect, arms handing loosely at sides, knees together.
Instructions: Patient slowly lowers trunk until head touches knees and hands reach the floor. Repeat three times with appropriate rest intervals.
Objective: To mobilize spine and further stretch matted tissues.

Exercise 17—Active-assistive elevation (flexion) of both arms (Fig. 26-43)
Position: Basic sitting.
Instructions: Same as for Exercise 11.
Objective: Same as for Exercise 11.

Exercise 18

Position: Sitting, with forearms at right angle to upper arms and supinated.
Instructions: Patient carries flexed arm posteriorly while therapist applies resistance against elbow. Arm returned to starting position. To be repeated three times with rest periods. Because of the relatively strenuous nature of this exercise, it should not be utilized until the patient has regained maximum strength and vitality.
Objective: To develop the pectoralis minor to actual hypertrophy in effort to fill in postoperative cavitation of the upper part of the anterior aspect of the chest wall.

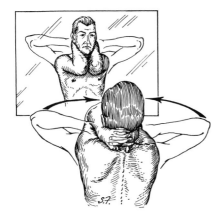

Fig. 26-40. Exercise 14.

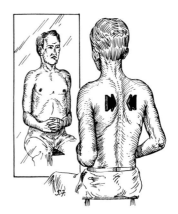

Fig. 26-41. Exercise 15.

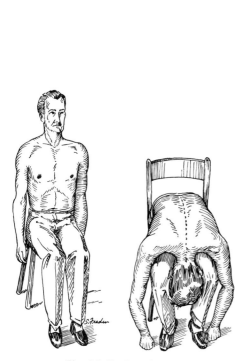

Fig. 26-42. Exercise 16.

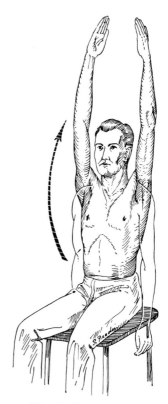

Fig. 26-43. Exercise 17.

The corrective therapy program should be divided into two distinct phases[28]: preoperative and postoperative. The preoperative phase is to educate and strengthen the synergist muscles not directly affected by the operative procedure so that they may take over, as much as possible, the functions of the primary muscles that are to be temporarily disabled by surgical procedures. Teaching in this preoperative phase is greatly simplified by use of mirrors. The patient is carefully trained in the technics of diaphragmatic breathing, which are essential during the postoperative phase. This is combined with emphasis on proper posture—lying, sitting, and standing. Finally, the patient is taught the basic exercises that he will require preoperatively and postoperatively.[28] Within 24 hours after surgery the second phase of corrective or rehabilitation therapy is started and is continued daily for weeks or even months until maximum possible rehabilitation is achieved.

Corrective exercises are customarily divided into four basic types: (1) *passive,* in which the entire movement is carried out by the therapist; (2) *assistive* (active-assistive), in which the patient does as much as he is capable of doing and the therapist completes the indicated range of motion; (3) *active,* in which all the motion is carried out by the patient himself; and (4) *active-resistive,* in which "load" is added to the exercises to increase the power of the muscles.

• • •

Rehabilitation medicine is a newcomer among medical disciplines. It has already been recognized that rehabilitation medicine has made enormous strides in the processes of reintegrating into active economic and social life many people who would hitherto have been condemned to live the life of a helpless cripple. Rehabilitation of COPD and other respiratory disease victims, however, is still not fully recognized as an integral part of the medical management of these conditions.

For some unaccountable reason, the rehabilitation of patients with chronic obstructive pulmonary disease is now lagging behind the rehabilitation of patients with neuromuscular and skeletal disabilities. The reasons for this lag may be that (1) the rehabilitation of the patient with obstructive pulmonary disease is not as spectacular as the rehabilitation of a paralyzed patient; (2) the obstructive pulmonary disease patient has, admittedly, a prognosis of a shorter life-span; (3) because of the COPD characteristic of slow deterioration and of the declining socioeconomic standing of the patient, his self-supporting productivity for the future is questionable, and therefore the financial investment in rehabilitation treatment might not appear to be fully justifiable; and (4) because improvements in these patients cannot be measured by routine laboratory procedures, some physicians believe that they are purely psychologic—due to tender loving care—and at best are only of short duration.

Another objection to ordering rehabilitation programs rests in the high cost of such treatment. The typical rehabilitation program requires much more than the presence of the patient's own physician and the use of expensive laboratory and therapeutic equipment. (See Chart 19.) Skilled professional and paraprofessional people such as psychologists, vocational and physical therapists, and social workers are all intimately involved in the various aspects of rehabilitation therapy. These complex software and hardware requirements generally mean that rehabilitation programs for most patients work best when conducted under one roof. When the

CHART 19

Modality	First pulmonary function tests (pretherapy) →	First arterial blood gas analysis →	Pretreatment instruction (patient and family)
Personnel	Physician, technician	Physician, technician	Physician, psychologist, physical therapist
Time	One hour	30 minutes	Variable
Equipment	Spirometer, helium meter, capnograph, co diffusion equipment	pH meter, Cournand needle, Van Slyke apparatus, or CO_2 and O_2 electrodes	Available literature

Modality	Vocational counselling ←	Psychiatric counselling ←	Breathing exercises, postural drainage
Personnel	Physician, counselor, job placement counselor	Psychiatrist, psychologist	Physician, physical therapist
Time	Variable	Variable	One hour
Equipment			$IPPB$ machine, tilting table, ultrasonic nebulizer

Modality	Second pulmonary function test →	Second arterial blood gas analysis →	Prevocational rehabilitation
Personnel	Physician, technician	Technician	Occupational therapist, psychologist, vocational counselor
Time	One hour	30 minutes	One week
Equipment	Spirometer, helium meter, capnograph, co diffusion equipment	pH meter, Cournand needle, Van Slyke apparatus, or CO_2 and O_2 electrodes	

Summary of the rehabilitation management of COPD patients. Shows needs of hospitals in terms of staff and equipment.

rehabilitation program is carried out in a hospital setting, the patient's own physician should continue to administer his day-to-day treatment.

Respiratory insufficiency in neuromuscular and skeletal disorders

Periodically during the course of a variety of neuromuscular or musculoskeletal diseases, respiratory insufficiency supervenes. It may be coincident with the acute onset of the illness, as in infectious poliomyelitis or cervical cord transection; it may occur intermittently through many years, as in the chronic aftermath of quadriplegia; or it may be the end result of an inexorably progressive situation, such as muscular dystrophy. In these situations the physiatrist will often play two therapeutically different roles: one, in the rescue of patients already under his care from acute respiratory failure; the other, in the application of physical medicine and rehabilitation to a patient disabled by chronic respiratory insufficiency, to restore ambulation, self-care, and work capacity.

Either role requires a working familiarity with the pathogenesis of respiratory insufficiency in the aforementioned situations. Fortunately, these conditions, although numerous, may be assembled together according to the similarity of their pathogenesis, since they lend themselves to the following physiologic classification into three broad categories along with an indication of the approximate frequency of occurrence. Although residual poliomyelitic paralysis and quadriplegia represent large groups of patients within a rehabilitation setting and therefore will be dealt with most frequently, outside such a setting, kyphoscoliosis and obesity will present the most frequent problems.[29]

I. Loss of motor power
   Poliomyelitis*   Myotonias
   Muscular dystrophy†   Chronic brain injury†
   Guillain-Barré syndrome†   Hemiplegia
II. Mechanical overload
   Kyphoscoliosis*   Pectus excavatum
   Obesity*   Constrictive pleuritis
   Ankylosing spondylitis   Thoracoplasty*
III. Mechanical overload and motor-sensory loss
   Quadriplegia*

PATHOGENESIS OF THE RESPIRATORY FAILURE
Physiologic common denominator for all three groups

For any one of the neuromusculoskeletal groups under discussion, it appears that all roads lead to Rome: despite great diversity in pathogenesis and development, each disease entity can eventually end in alveolar hypoventilation. Thus in the absence of intrinsic lung disease a crippled thoracic cage can create a situation characterized by the failure to balance the rate of alveolar ventilation with the metabolic

*Frequent occurrence, seen by every physiatrist or rehabilitation service; †infrequent occurrence but not rare; the rest are uncommon.

rate of the individual, so that alveolar CO_2 tensions become higher and alveolar O_2 tensions become lower than in the normal state.[30] This balance can be stated simply and arithmetically: the alveolar CO_2 tension = K × metabolic rate ÷ alveolar ventilation. Moreover, within this relationship, it has been shown that metabolic rate and alveolar ventilation are varied in such a fashion in the normal person as to maintain the alveolar CO_2 tension at 40 mm. Hg. There are several conclusions regarding these patients that can be drawn from that relationship: Whenever the metabolic rate rises, a similar rise in alveolar ventilation must take place or the alveolar CO_2 tension will rise. Conversely, decreasing the alveolar ventilation, with a constant metabolic rate, will also cause the alveolar CO_2 tension to rise. In the normal situation the alveolar CO_2 tension (and the arterial blood CO_2 tension with which it is in equilibrium) is maintained within a very narrow range of 40 ± 2 mm. Hg through peripheral and central nervous chemoreceptors. These receptors in turn respond to the acid-base changes that alterations in the blood CO_2 produce and they regulate the thoracic muscles and the level of ventilation by way of the central nervous system.

A failure in this system is therefore signaled by the onset of alveolar hypoventilation, and the presence of alveolar hypoventilation is traditionally defined in diagnostic terms as a rise in alveolar CO_2 tension. It must be emphasized, however, that as the alveolar CO_2 tension rises, there is a reciprocal fall in alveolar O_2 tension. In fact, abnormalities of both alveolar gases in disease entities involving only the thoracic cage always go hand in hand, and when the alveolar gas levels are imposed on the blood perfusing the alveoli, the arterial blood emerging from such an underventilated lung is both hypoxic and hypercapnic. Thus these entities comprise an important contrast when compared with intrinsic lung diseases, such as diffuse interstitial disease or chronic pulmonary emphysema, which work in different fashions to alter the blood gases. For example, diffuse interstitial disease is noted for its tendency to impair pulmonary gas diffusion between alveolus and capillary, and chronic bronchitis and pulmonary emphysema begin their course by the failure to bring together in proper proportions air in alveoli with blood in pulmonary capillaries. As a result, in these two forms of intrinsic lung disease, both because CO_2 is easily diffusable and because some areas of the diseased lung are still hyperventilated, during the early stages, CO_2 elimination remains normal, whereas O_2 exchange is impaired. Thus oxyhemoglobin saturations in the arterial blood fall below the normal values of 96%, either at rest or during exercise, relatively early in the course of the lung disease.

By way of contrast, a crippled thoracic cage creates a situation in which alveolar ventilation decreases in a relatively homogeneous fashion throughout the entire lung and, although some areas of the lung, because of thoracic cage distortion, may be more poorly ventilated than others, it seems likely that there are no areas of the lung that can be hyperventilated in the same compensatory fashion as may occur in the obstructive lung diseases.[31] Consequently, in contradistinction to pulmonary emphysema, the alveolar O_2 and CO_2 tensions in thoracic cage disease become abnormal at precisely the same time. Since in most instances these abnormalities are only slowly progressive, a situation is set up for a relatively distinctive clinical syndrome: CO_2 tensions in arterial blood often reach extremely high values before the degree of arterial blood hypoxemia is sufficient to be incapacitating. As a rule, in entities affecting the thoracic cage, by the time clinical symptoms become pressing, the arterial CO_2 tension is markedly elevated and symptoms referable to the central nervous system effects of CO_2 are far more noticeable than in chronic bronchitis and pulmonary emphysema. For these reasons and because there are so many other causes of hypoxia, the arterial CO_2 tension is the main diagnostic criterion for the presence of alveolar hypoventilation.

Once alveolar hypoventilation is established, hypoxemia and hypercapnia may be accompanied by respiratory acidosis. The degree of acidosis will depend on the rapidity of rise of the arterial CO_2 tension versus the ability of the kidney to retain base bicarbonate as a compensatory device.[32] Thus acute and suddenly precipitated alveolar hypoventilation in such patients will result in marked respiratory acidosis, with all of its deleterious effects on the maintenance of blood pressure, cardiac tone, and cardiac rhythm. The general effects of these abnormal blood gases on other aspects of organ function have been summarized and are similar to those seen in other forms of respiratory failure. Thus muscular weakness, diminished mental acuity, altered liver function, increased erythropoiesis, and peripheral vasodilation are attributable to the hypoxemia.[33] In the same manner the diminished mental alertness, the elevated cerebrospinal fluid pressures, the choked optic disk, and the narcosis and coma that result from high CO_2 tensions and respiratory acidosis are often even more striking in these patients with thoracic cage disease than in those with obstructive lung disease.[34]

It is important to note that the symptom of dyspnea is far less intense in entities involving the thoracic cage than it is in patients with intrinsic lung disease. Thus patients with obstructive pulmonary disease frequently show the systemic effects of hypoxemia and hypercapnia combined with marked dyspnea on exertion, harassing coughs, and abundant sputum. Respiratory failure in thoracic cage disease is often a more subtle affair; even though these patients may have dyspnea on exertion, the medical literature abounds with descriptions of heavily narcotized, obese patients who are otherwise quite comfortable[35, 36]; in fact, the description of the fat boy in Charles Dickens' *The Pickwick Papers* has lent the name of the pickwickian syndrome to the obese patient with CO_2 narcosis.[37] And sadly amusing, although possibly apocryphal, stories abound in which hunchback and poliomyelitic patients report themselves sufficiently free of the discomforts of dyspnea that they are perplexed by their inadvertent drowsing during critical junctures of poker games.

The development of pulmonary vascular hypertension with subsequent strain of the right ventricle, its enlargement, and potential right-sided congestive heart failure (consistent with the diagnosis of cor pulmonale) is a common entity in patients with respiratory insufficiency due to neuromuscular disorders. There appear to be two major factors in the development of the pulmonary hypertension: anatomic restriction of the pulmonary vascular bed and functional vasoconstriction due to hypoxia and hypercapnia (Fig. 26-44). The development of the anatomic restriction seems clear; kyphoscoliosis, thoracoplasty, and constrictive pleuritis are typical examples of entities that are characterized by lungs diminished in volume, distorted in shape, and partially or completely atelectatic in some areas. Such forms of relative atelectasis may be expected to alter the volume of the pulmonary vascular tree and the resistance to pulmonary blood flow. In fact, recent experimental evidence in the normal lung has been able to demonstrate a direct relationship between the state of lung inflation and the volume of its vascular bed. However the functional factors, hypoxemia and respiratory acidosis, seem to be of considerably greater importance in the genesis of pulmonary vascular hypertension. The role of hypoxemia seems particularly pertinent because it has long been known to cause vasoconstriction in the pulmonary circulation of man or animals when acutely exposed to a hypoxic situation, or of man chronically at high altitudes. Moreover, the relief of hypoxemia in patients with chronic obstructive lung disease is usually attended by a prompt and significant fall in pulmonary arterial pressure. In addition, in the last few years the role of systemic acidosis or respiratory acidosis has been delineated and suggests that

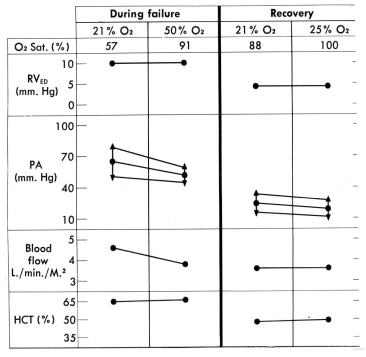

Fig. 26-44. Effect of oxygenation on the course of congestive heart failure in a patient with kyphoscoliosis. Patient underwent cardiac catheterization, and the initial data, shown on the first column, revealed an arterial oxygen saturation depressed to 57% while breathing air, a right ventricular end-diastolic pressure (RV_{ED}) elevated to 10 mm. Hg, indicating the presence of congestive heart failure, and a marked elevation of pulmonary arterial pressure. When 50% oxygen was given and the arterial oxygen saturation thereupon rose to 91%, there was a prompt fall in pulmonary arterial pressure, suggesting the reversibility of the pulmonary vasoconstriction due to hypoxia. The measurements during recovery followed 2 weeks of hyperventilation in a mechanical respirator. As shown, the right ventricular end-diastolic pressure is now normal and the pulmonary arterial pressure is virtually normal. Even so, correcting the relatively mild degree of arterial oxyhemoglobin unsaturation (88%) by breathing 25% oxygen caused a further fall in pulmonary arterial pressure. Such measurements emphasize that a great deal of the pulmonary hypertension, cor pulmonale, and congestive heart failure is due to hypoxia and is potentially reversible.

pulmonary vasoconstriction occurs in proportion to the degree of acidosis. Finally, there is increasing evidence that the combination of hypoxemia and hypercapnia synergizes their vasoconstrictive effects and produces elevations of pulmonary arterial pressure that are greater than the sum of the responses produced by either the hypoxemic stimulus or the hypercapnic stimulus alone. The most noteworthy therapeutic implication of these advances has been the clear demonstration that the bulk of the pulmonary hypertension observed in these patients can be reversed if hypoxemia and hypercapnia are ameliorated, and a persuasive example of this demonstration appears in Fig. 26-44. Furthermore, if hypoxemia and hypercapnia are not treated, the usual therapeutic regimen used in congestive heart failure, featuring digitalis, salt restriction, and diuretic therapy, will be far less effective than when combined with procedures to increase alveolar ventilation. In fact, many patients have been observed in which all of the symptoms of right-sided congestive heart failure, including distended neck veins, hepatomegaly, polycythemia, and peripheral edema, have been reversed, without the use of a cardiotonic-diuretic regimen, by bringing the arterial oxyhemoglobin saturation and CO_2 tension to normal.

Specific development of respiratory insufficiency

Loss of motor power. The loss of motor power category includes a wide variety of lesions responsible for loss of motor power, ranging along the entire neuromyal pathway from cerebrum to muscle itself. It comprises entities that run an acute course, such as poliomyelitis[38, 39] and Guillain-Barré syndrome; entities that are chronic and static, such as hemiplegia and brain injury; and those that are slowly progressive, such as muscular dystrophy.

In general, paralysis or weakness of both the diaphragm and intercostal muscles is a factor in any of the stages, but it is the chronic or slowly progressive forms with insidious presentations that require the closest scrutiny, and these will receive the greatest emphasis.

The approach to acute respiratory insufficiency, as in acute poliomyelitis or the Guillain-Barré syndrome, can be summarized briefly: It includes ascertaining the level of alveolar ventilation by means of measurement of arterial blood gas, instituting a controlled form of assisted ventilation through an indwelling tracheal tube and mechanical respirator (positive pressure device, piston, or volume respirator), and, by means of frequent arterial or alveolar CO_2 analysis, maintaining the proper level of ventilation and a patent airway free of secretions until the acute episode of paralysis has subsided.

But once the patient with poliomyelitis, brainstem trauma, or stroke has recovered from this acute episode, he is susceptible to a far more chronically developing syndrome of alveolar hypoventilation during the rest of his life. Warning signs of exertional dyspnea are particularly absent in this group because they are confined to wheelchairs. Moreover, the precise reason for worsening thoracic cage failure years after the initial spinal cord insult is uncertain; increasing degrees of scoliosis due to muscle imbalance, osteoporosis of spine and rib cage, and fibrosis and calcification of thoracic cage joints and ligaments may all enter into this equation, as well as the coup de grace administered by an abrupt respiratory infection or other forms of acute stress. In fact, how these latter ingredients are mixed will determine whether any patient's career will feature either slowly progressive alveolar hypoventilation or acute respiratory failure with hypoxemia, heart failure, and hypercapnic narcosis during a period of stress, or admixtures of both syndromes.

Hemiplegia is another recently identified cause of abnormal respiratory performance. Only slight decreases in arterial O_2 tension or saturation have been observed, but these have been with the patient at rest, and even more marked changes might occur if such patients could exercise. Although little data exist to date, the

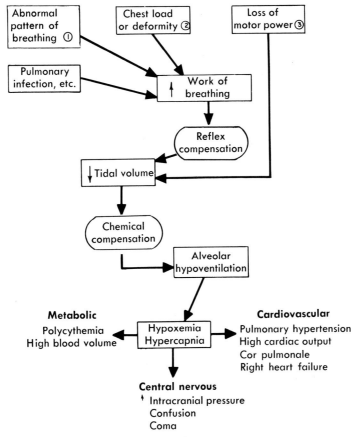

Fig. 26-45. Pathogenesis of alveolar hypoventilation in patients with neuromuscular disease or skeletal deformities of the thorax. Any one of three pathways may be taken, depending on the disease entity: abnormal pattern of breathing, *1,* as in quadriplegia; an augmented resistance to movement of the thorax, *2,* as in obesity or kyphoscoliosis; or loss of motor power, *3,* as in the aftermath of poliomyelitis. In pathways 1 and 2, an increased work of breathing results that, as shown, is temporarily compensated by increased sensory reflex activity; this compensation fails and the subject economizes on the work of breathing by decreasing the tidal volume. The slight CO_2 accumulations and degrees of hypoxemia serve as chemical compensation temporarily to prevent further decreases in tidal volume; but eventually these are lost and severe clinical degrees of alveolar hypoventilation result. Pulmonary infection and other stresses are capable of aggravating the problem of the work of breathing. The final result, hypoxemia and hypercapnia, has far-reaching metabolic, central nervous, and cardiovascular effects.

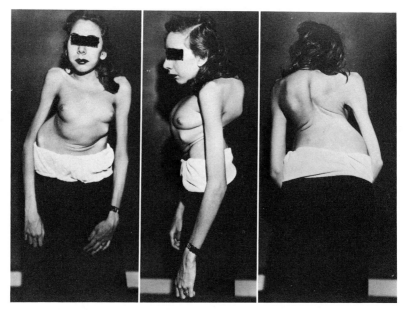

Fig. 26-46. Extent and type of deformity required to produce respiratory insufficiency in kyphoscoliosis. Both scoliosis and kyphosis are present, so that considerable dwarfing occurs, as indicated by the relatively long length of the arms. Despite the patient's youth, this deformity was severe enough to produce alveolar hypoventilation. Any patient with a typical "hunchback" deformity is a potential candidate for respiratory insufficiency.

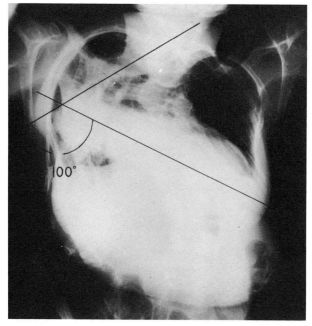

Fig. 26-47. Radiologic assessment of typical kyphoscoliosis. Intersecting angle indicates the usual orthopedic assessment of severity; the greater the angle, the worse the deformity. Note displacement of the heart, the hyperexpansion of the left thorax, and the contraction of the right thorax.

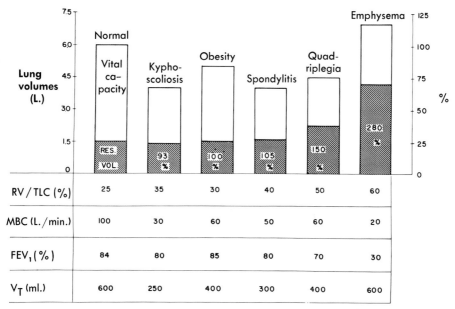

Fig. 26-48. Lung volumes and maximal ventilation in the neuromuscular and skeletal diseases compared with that of a normal subject and an emphysematous patient. Each compartment represents total lung capacity and its subdivisions, the vital capacity (*unshaded* compartment) and the residual volume (*shaded* compartment). The ratio of residual volume to total lung capacity (RV/TLC), the maximum breathing capacity (MBC), the forced expiratory volume in 1 second (FEV_1), and the tidal volume (V_T) are also shown.

arterial O_2 tensions seem attributable to underventilation of the lung affected by paralyzed respiratory muscles on the hemiplegic side.[40] To date the remaining normal side seems capable of being ventilated sufficiently to maintain normal arterial CO_2 tensions.

Mechanical overload. The pathogenesis of alveolar hypoventilation in the mechanical overload group, embracing the kyphoscoliotic,[41-43] the obese,[44, 45] and the spondylitic patient,[46, 47] has been studied the most thoroughly in the past. Although some controversy remains, the events leading up to respiratory failure have been well mapped out. As Fig. 26-45 indicates, the entire process appears to be initiated by the chest deformity or, as in the case of obesity, by the added massive fat depots overlying the rib cage or underlying the diaphragm. Ample observations in the past have clearly demonstrated that there is a direct relationship between the degree of respiratory failure and the severity of the deformity in kyphoscoliosis (Figs. 26-46 and 26-47). A similar relationship appears to apply in patients with ankylosing spondylitis or pleural fibrosis.[48] On the other hand, there does not appear to be a direct relationship between the degree of obesity and the occurrence of alveolar hypoventilation. For this reason, additional factors have been indicated, such as a primary insensitivity of the medullary respiratory centers.

Even though some relationship may exist between the degree of deformity and the severity of respiratory failure, the mechanism whereby alveolar hypoventilation develops is complex and cannot be defined simply in terms of classical pulmonary function. For instance, Fig. 26-48 illustrates the lung volumes, their pattern and distribution, as well as the maximum breathing capacity in kyphoscoliosis, obesity, and ankylosing spondylitis. For the sake of comparison the lung volumes of a normal subject of the same size and sex is also illustrated, along with the subdivisions of the total lung capacity, such as the inspiratory capacity (IC), the expiratory reserve volume (ERV), both of which comprise the vital capacity, and the residual volume (RV). In addition, the ratio of residual volume to total lung capacity of about 25% and the maximum breathing capacity are indicated. For comparison the typical lung volumes of patients with chronic pulmonary emphysema, including the greatly enlarged residual volume and total lung capacity, are shown. They reflect the hyperinflated pulmonary tissue of emphysema and are in sharp contrast to the lung volumes in the musculoskeletal conditions in question. The clearest testament to this fact is the normal residual volume values for kyphoscoliosis, obesity, and ankylosing spondylitis. Thus the increase in the ratio of the residual volume to total lung capacity, ranging from 30% to 40% in the latter group as illustrated, is due to an unchanged residual volume in the presence of marked encroachment on the vital capacity. Fig. 26-48 illustrates that the ratio of residual volume to total lung capacity is, in itself, not a meaningful diagnostic value for the presence of pulmonary hyperinflation in musculoskeletal disorders. The values shown for the maximum breathing capacity, the maximal rate at which the lung can be voluntarily ventilated, indicate moderate reductions in this component of lung function, although in most instances the reductions are not so severe as those seen in chronic obstructive lung disease. However, the reductions in maximum breathing capacity observed in the neuromuscular disorders have an origin quite different from that in chronic obstructive pulmonary disease. In obstructive diseases, airway resistance is the paramount reason for the decrease in maximum breathing capacity and, although it can be measured in many different ways, the simplest indicator, the forced expiratory volume in one second (FEV_1), has been earmarked by long and traditional use and, as indicated in Fig. 26-48, is markedly impaired in chronic obstructive lung disease. By

Table 26-1. Differential diagnosis of respiratory insufficiency due to neuromuscular disorders and that due to bronchial obstruction or interstitial lung disease

	Vital* capacity	Residual* volume	Maximum* breathing capacity	Forced expiratory volume ($_1$)
Neuromuscular disorders	Decrease	Normal	Decrease	Normal
Chronic obstructive lung disease	Normal, slight decrease	Increase	Decrease	Decrease
Diffuse interstitial lung disease	Decrease	Decrease	Normal	Normal

*Normal, increased, or decreased from the value predicted for age, sex, and size of the individual. Vital capacity = maximal volume expirable from lung; residual volume = volume remaining in lung after maximal expiration; maximum breathing capacity = maximal ventilatory rate from voluntary effort; forced expiratory volume ($_1$) = percentage of vital capacity expired in 1 second by maximal effort.

way of contrast, in the neuromuscular disorders the FEV_1 is comparatively normal and indicates that the impaired maximum breathing capacity is attributable to neuromuscular and skeletal dysfunction of the thorax. Thus a clear-cut difference exists between the respiratory insufficiency observed in neuromuscular dysfunction and that due to chronic obstructive lung disease or diffuse interstitial lung disease. The differential diagnostic points for these three types of respiratory insufficiency are incorporated in Table 26-1, which is a simplified summary of only the pertinent pulmonary function tests.

It is important to point out that the reductions in lung volume and maximal ventilation that we have indicated do not in themselves explain the respiratory insufficiency. Within the limits of these admittedly decreased values, the patient would be expected to be able to maintain normal ventilation during rest and moderate

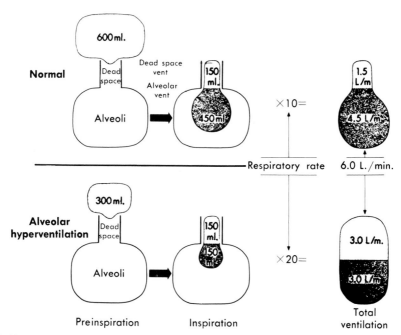

Fig. 26-49. Pattern of breathing that leads to alveolar hypoventilation in the neuromuscular and skeletal entities associated with increased resistances to thoracic expansion. In the normal situation a 600-ml. breath is shown in the cloud waiting to enter the lung, which is diagrammatically depicted as a tracheobronchial "dead space" and a parenchymal portion, the alveoli; about 450 ml. is inspired into the alveoli with 150 ml. remaining behind to fill the tracheobronchial dead space. In the subject with alveolar hypoventilation, who decides to breathe with a breath of only 300 ml., it will be distributed during inspiration in such a way that only 150 ml. of tidal volume penetrates into the alveoli, since the tracheobronchial dead space requires 150 ml. to fill it. As shown in the bags labeled "total ventilation," even though the respiratory rates are increased in alveolar hypoventilation so that total minute ventilation is equal to that of the normal subject, there is a marked difference in alveolar ventilation. In the particular example illustrated, the normal subject gets 4.5 L./min. of alveolar ventilation, whereas the subject with alveolar hypoventilation gets only 3.0 L./min.

activity. A more pertinent explanation for the development of respiratory failure in neuromuscular diseases is suggested by the extremely small tidal volumes, the averages of which are shown in Fig. 26-48. These markedly reduced tidal volumes occur despite the fact that the vital capacity (that is, the total inspirable breath) may be more than five times the size of these tidal volumes. Moreover, as if in some attempt to compensate for the small tidal volume, most of these subjects exhibit a spontaneous increase in the frequency of breathing. As a result, the total ventilation breathed per minute in these subjects may be normal or even greater than normal; however, because of the shallow breaths, an inordinantly large portion of each breath washes back and forth in the tracheobronchial tree without ever reaching the parenchymal gas-exchanging surfaces of the lung. The economic disadvantage suffered by such patients with neuromuscular disease is illustrated in Fig. 26-49.

The manner in which this peculiar and deleterious pattern of breathing comes about is of considerable interest because the principle is applicable in some degree to all of the neuromuscular and skeletal diseases. The main prerequisite appears to be a chest deformity, added chest load, or restriction that results in abnormally large resistances to the expansion of the thoracic cage or depression of the diaphragm. When the respiratory muscles are called on to overcome these resistances during breathing, measurable increments occur in the amount of work done for a given amount of alveolar ventilation. Fig. 26-50 indicates the amount of work required in

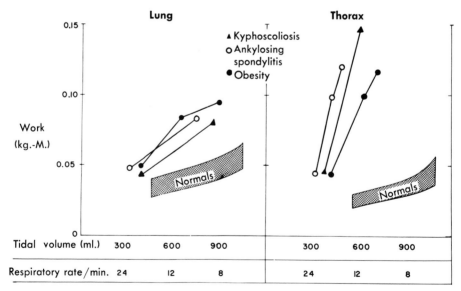

Fig. 26-50. Work done to move the lungs and the thoracic cage in kyphoscoliosis, ankylosing spondylitis, and obesity. Measurements were made using the different patterns of breathing so that the tidal volume and frequency could be altered without changing the total minute ventilation. As shown, for comparable tidal volumes near the usual range (about 600 ml.), the patient group does four to five times as much work to move the thoracic cage as does the normal subject group. Moreover, there is a smaller, but still significant, increment in the work done to expand the lung in the patient group.

three typical subjects with kyphoscoliosis, obesity, or ankylosing spondylitis.[49, 50] In a study of this nature it is possible to measure separately the work required to mobilize the thorax and that required to mobilize the lung[51]; passively breathing these patients in a body respirator, while simultaneously measuring the cyclically induced negative pressure in the respirator, along with the tidal volume inspired as a result of the negative pressure, yields a value, when integrated, for the total work required to move the thorax and lungs together. In a similar manner, if the intrapleural pressure (in practice, esophageal pressure) developed during a breath is integrated with the volume of the breath that is inspired, values may be obtained for the work done to expand the lung alone. As is shown in Fig. 26-50, measurements were made using different tidal volumes, and it is apparent that the larger the tidal volume, the greater the work done per unit of ventilation. The differences between the normal subject and the patient group are dramatic and striking; for comparable tidal volumes of about 600 ml., the patients require about five times as much work to move the thorax as does the normal person. Although not as striking, it is also apparent that the work to expand the lung is increased in patients with obesity, kyphoscoliosis, and ankylosing spondylitis.

As expected, the patients refuse to pay such an inflated price for the energy of breathing and prefer instead to economize. The most apparent way in which an economy can be effected is illustrated by the slopes of the values for work in Fig. 26-50; moving down these slopes toward the abscissa makes it apparent that the work per liter of ventilation can be markedly reduced if the individual breathes with a smaller tidal volume. The mechanism whereby these marked savings in work can be achieved, if the tidal volume is reduced, is a relatively simple one and depends on the established fact that the work done to overcome the elastic resistances to breathing (any elastic resistance, but here particularly those of a rigid thorax) increases by the square of the increase in the tidal volume.[51] Thus, should the tidal volume double, the work against the elastic resistances would quadruple (2^2); should the tidal volume triple in size, the work done against the elastic resistances would increase ninefold (3^2). Stated another way, two breaths, each of 300 ml., still cost together only one-half the work of one breath of 600 ml. Although the principle is true in normal persons, it takes on added importance in the patients with neuromuscular and skeletal disorders, in whom the elastic resistances to breathing at all tidal volumes is greatly increased and in whom large savings in energy cost and work of breathing can be effected when the tidal volumes are reduced.

There are still several unexplored areas regarding the developmental chain that are only suggested in Fig. 26-45. The major one concerns the pathways whereby an increased work and energy cost of breathing becomes translated into decreases in tidal volume in the process of conserving energy. From the teleologic standpoint the mechanism works to diminish the sensation of dyspnea; from the mechanistic standpoint the mechanism appears to involve the performance of the sensory reflex arc and the chemical stimulation regulating respiration.[52-54] Thus, when the mobility of the thorax is first impaired, it is still possible to compensate for the increased tension that the respiratory muscles must develop in the process of producing a normal tidal volume. The compensation apparently involves sensory muscle spindles in the intercostals or diaphragm; these are believed to be responsible for enhanced motor neuron activity when the length of the muscle does not correspond to the usual standards of tension, that is, when the mobility of the thorax is so impaired that shortening of the respiratory muscles to their usual length has not occurred despite the development of the usual muscle tension. This form of compensation eventually fails, perhaps because of some form of adaptation of the muscle spindle; or perhaps, with the worsening degree of thoracic deformity, the usual number of motor impulses will no longer excite the muscle to increasing the tension development that is now required. Thus, in the face of increasing resistances to thoracic mobility, the tidal volume falls sufficiently to induce a modest degree of alveolar hypoventilation and thereby to raise the arterial blood CO_2 tension slightly. This increase in CO_2 tension in turn provides a second compensation that (probably through the hydrogen ions that accompany CO_2 retention) resets activity in the respiratory centers of the brainstem to a higher level. Eventually even this second compensation fails[55] and more profound degrees of alveolar hypoventilation supervene. The reasons for the failure of chemical compensation have been ascribed to a variety of factors, including (1) increasing resistances of the thoracic cage as deformity worsens, (2) a form of pharmacologic insensitivity of the respiratory center to small changes in blood CO_2 tension because of chronic exposure to high concentrations of CO_2, or (3) disappearance of the true respiratory stimulus, that is, acidosis, as the kidney compensates for the carbonic acid formed during the carbon dioxide retention by a process of retaining bicarbonate.

The special case of obesity. Since obesity represents a chronic and often intractable disease, it is of potential and particular interest to the physiatrist. For this reason it is not enough to relegate the respiratory insufficiency of the obese patient to the mechanical overload group in a simplistic fashion. On the contrary, a sufficient number of questions have been raised regarding this entity, so that it deserves special consideration. The major problem is the lack of correlation between the degree of alveolar hypoventilation and the degree of obesity. Unfortunately for the theory, no such relationship occurs: a 500-pound circus fat lady may be normal, whereas a 250-pound individual of similar height may demonstrate profound alveolar hypoventilation. Nor is there any definite relationship between the presence of alveolar hypoventilation and actual measurements of the work of breathing, although in almost case the work of breathing is generally increased. This riddle has led to the belief that other contributing factors must be present in the obese individual, in addition to the augmented work load, before alveolar hypoventilation can supervene. One such major factor implicated has been the sensitivity of the neural respiratory centers in the brainstem or the peripheral chemoreceptors. This hypothesis holds that respiratory centers of low-normal sensitivity or centers that have been damaged by disease fail to produce compensatory increments in the neural output of the obese thorax and diaphragm when they have become burdened by an increased work of breathing. The onset of alveolar hypoventilation occurs when a normal person with low-normal sensitivity of his respiratory center, by chance, also develops obesity.

The special case of quadriplegia. The sequence of events that leads from the acute respiratory failure immediately after the onset of traumatic quadriplegia though a period of recovery, followed by the potential development of chronic respiratory insufficiency, now seems reasonably clear. After an acute onset of the quadriplegia, since most spinal cord damage is below the fourth cervical segment, some or all diaphragmatic activity remains intact, whereas there is total paralysis of the intercostal muscles because of their thoracic innervation. There appear to be two possible reasons for the acute respiratory failure that so often necessitates tracheostomy and support by mechanical respirators: one is the possible presence of hemorrhagic edema in areas of the cord adjacent to the trauma so that motor drive via the phrenic is temporarily diminished; the other is a mechanical dysfunction arising from the initial flaccid paralysis of the intercostal muscle during the period of spinal shock, so that the development of the normal, negative intrathoracic pressures during the process of inspiration is accompanied by paradoxical retraction of the rib cage.[56] Within a few weeks, however, recovery from the respiratory failure usually occurs—perhaps at the same time that hemorrhagic edema recedes or that the initial spinal shock is succeeded by a period of spastic paralysis of the intercostal muscles, which produces fixation of the rib cage.

Despite the usual recovery from this early form of respiratory failure, some of these patients are heir to two forms of respiratory insufficiency: (1) the very gradual onset of a moderate degree of alveolar hypoventilation, characterized by blood CO_2 tensions as high as 55 mm. Hg, or (2) intermittent episodes of severe alveolar hypoventilation precipitated by stress, trauma, or infections on a background of apparently normal levels of alveolar ventilation. Fig. 26-48 shows that the lung volumes of the quadriplegic subject are not markedly reduced[57]: the vital capacity, as shown, is usually 60% of the predicted value; the residual volume shown at 150% of the predicted does not represent hyperinflation of the lung, but rather the fact that the paralyzed muscles of expiration do not permit expiration of that portion of the vital capacity known as the expiratory reserve volume. Thus the residual volume shown includes not only the residual volume itself but the expiratory reserve volume. Despite the loss of virtually every muscle of expiration, the quadriplegic subject is able to do surprisingly well in terms of maximal performance. For instance, the maximum breathing capacity is 60% or more of the predicted value in the usual case and, even without the muscles of expiration, a forced expiratory volume in 1 second achieves more than 70% expulsion of the vital capacity. This astonishing performance is a reminder that much of expiration even in the normal individual is carried out by the recoil properties of the lung rather than the muscles of expiration. The small tidal volume (Fig. 26-48) for the quadriplegic subject again suggests that alveolar hypoventilation in this group comes about through a pattern of breathing similar to that of the kyphoscoliotic or obese patient. Thus small tidal volumes and rapid frequencies of breathing produce normal minute ventilation but waste so much ventilation on the tracheobronchial dead space that little is left over to ventilate the alveoli.

As in kyphoscoliosis and obesity, it appears that the excessive work of breathing accounts for the tendency of these patients to underventilate their lungs. However, in contrast to kyphoscoliosis and obesity, there is no distortion of rib cage mechanics or added fat load to account for the increased work. Instead, studies of respiratory mechanics in quadriplegia have shown that a special pattern of breathing, in which diaphragmatic movement predominates, results in a marked increase in the work of breathing.[58] Typical values for respiratory mechanics in a quadriplegic patient

Table 26-2. Work of breathing in a quadriplegic patient compared with that of a normal subject

	Tidal volume (ml.)	Volume moved by diaphragm (ml.)	Compliance (ml./cm. H_2O)		Work per breath (gm.-cm.)		
			Lung	Diaphragm-abdomen	Lung	Diaphragm-abdomen	Total
Quadriplegia	400	400	100	100	800	800	1600
Normal	500	200	200	100	625	200	825

and a similar-sized normal subject are shown in Table 26-2. The first two columns indicate the tidal volume and that portion of the tidal volume inspired by movement of the diaphragm. Although the exact portion of the resting tidal volume moved by the diaphragm in a normal subject is uncertain, a range of 200 to 300 ml. seems probable; on the other hand, for the typical C-5 quadriplegic subject, the entire tidal volume must be moved by the diaphragm. Even though the compliance of the diaphragm-abdomen system is the same for quadriplegic and normal subjects, the quadriplegic subject does more work (as indicated in the last two columns of Table 26-2) simply because the diaphragm moves a greater distance for each breath.[59] That the work per breath for the diaphragm-abdomen system is so much greater in the quadriplegic subject than in the normal subject (800 versus 200 gm.-cm.) is a testament to the relationship previously described, which says that the work done against elastic resistances increases by the square of the increase in the tidal volume. In this example, although the tidal volume itself has not increased, the volume moved by the diaphragm and hence the work of diaphragmatic excursion have increased.

In addition, it is apparent that lung compliance in quadriplegic subjects is moderately reduced also. Such a reduction adds to the work required to stretch the lung and has been attributed to the inability of these subjects to take the deep, sighing breaths that reexpand alveoli undergoing continuous atelectatic collapse even in the normal lung. Finally, as indicated, as a result of the diaphragmatic movement and decreased lung compliance, the work of breathing in the quadriplegic subject may be twice normal, even when he is in the most advantageous postural position, that is, supine. To add to his burden, recent studies have also indicated that the work of breathing increases in the sitting position because of crowding of the viscera in the abdomen and a greater resistance encountered by the descending diaphragm.[58] For instance, these data suggest that the compliance of the diaphragm-abdomen system decreases by 50% in going from the supine to the upright sitting position (Fig. 26-51).

The second major problem in the genesis of respiratory insufficiency in the quadriplegic patient is the impaired cough reflex. Although this impairment may not in itself lead to chronic alveolar hypoventilation, it may be expected to produce retention of secretions sufficient to lead to acute respiratory infections and either aggravate alveolar hypoventilation or induce it in the susceptible patient. The problem faced by the quadriplegic patient in executing a cough is illustrated in Fig. 26-52, which shows the pleural pressure developed as a result of his inspiration and the

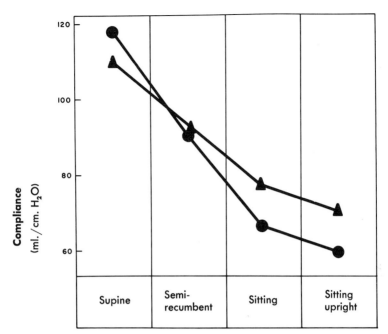

Fig. 26-51. Effect of body position on the ease of diaphragmatic descent in 2 quadriplegic patients. Diaphragmatic descent is assessed by the compliance of the diaphragm-abdomen system, so that the more compliant the system, the easier the descent. The semirecumbent, the sitting, and the sitting upright positions all compromise diaphragmatic descent, as these positions make for crowding of the abdominal viscera and their resistance to displacement by the diaphragm. In one of the patients it requires twice as much work to move the diaphragm (one half of the compliance) in the upright position as in the supine position.

tidal volume that he inspires to begin the cough.[59] At the arrow he closes his glottis and relaxes his muscles of inspiration so that the intrapleural and intra-alveolar pressure become slightly positive as a result of the recoil properties of the lung acting on the air enclosed by a closed glottis. When the glottis opens, the inspired air is forced out and the pleural pressure comes back to its usual value, slightly less than 0. In contrast, the normal subject, using the same size breath with which to initiate his cough, is able to produce markedly positive pressures within his pleural and alveolar spaces because of his ability to contract the anterior abdominal wall and other muscles of expiration. As a result, the tidal volume that was inspired is expelled with a far greater velocity, along with tracheobronchial secretions entrained in the air jet. Of interest is the effect of the sitting position on the quadriplegic patient. Here it is apparent that he is able to develop larger positive pressures at the height of the cough when his glottis is closed, because of the upward recoil of viscera crowded into a smaller abdominal space. This recoil acts with a pistonlike effect on the diaphragm to push air out more rapidly. So, for the quadriplegic patient, less respiratory work is done in the supine position but better cough occurs in the sitting position.

A recent study has detected a third important component in the genesis of respiratory insufficiency in the quadriplegic subject: the lack of sensory appreciation of lung and thoracic loads. Thus, in one study,[60] repeated respiratory complications tended to occur on an episodic basis to those quadriplegic patients who did not sigh and were unable to detect an elastic load on respiration. Such observations indicate that in normal individuals important information is constantly being conveyed by sensory ascending pathways regarding the resistances to breathing, the compliances of the rib cage and lung, and, perhaps, the condition of the tracheobronchial tree. These patients could not tell when the effort required to inspire a given breath had been deliberately and artificially increased. As a result it could be anticipated that the advent of atelectasis was unlikely to elicit the automatic sighing mechanism, which has been found regularly in normal persons and which acts to correct alveolar atelectasis. The discussion of the role of this sensory deficit in the management of these patients follows.

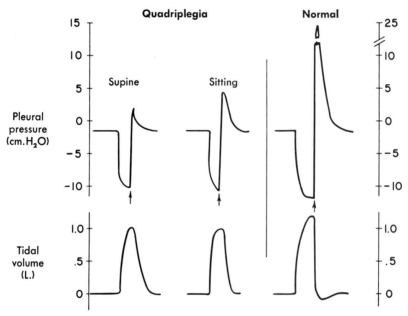

Fig. 26-52. Mechanism of cough in a normal subject and a quadriplegic patient. In the supine position, when the quadriplegic patient expands his thorax, a negative pleural pressure of about −10 cm. H_2O produces the inhalation of a 1 L. tidal volume; at the arrow, the muscles of inspiration are relaxed with the glottis closed so that a moderate positive pressure can be attained amounting to about 3 cm. H_2O; opening the glottis permits the lung to recoil because of its own elastic properties and the tidal volume is expelled in a passive fashion and with a slow velocity. In the sitting position, the quadriplegic subject develops a greater positive pressure and a higher velocity of expiration because of the abdominal compression. These measurements are in marked contrast to the normal subject who is capable of developing enormous positive pressures inside the thorax, as shown at the arrow, when he closes his glottis and uses both the recoil properties of the lung and muscles of expiration to expire; as a result, a high velocity of expiration, as shown, is capable of efficiently entraining particles and secretions in the jet.

PRINCIPLES OF THERAPY AND REHABILITATION

Acutely ill patient with alveolar hypoventilation. The sudden onset of respiratory failure, with hypoxemia and hypercapnia, is a frequent accompaniment of acute respiratory infection, excessive fatigue, or other forms of traumatic or operative stress. The alleviation of stress and the antibiotic therapy of respiratory infections that have been adequately cultured and investigated need no emphasis here except to point out the importance of proceeding rapidly. During the waiting period for such measures to take effect, however, the hypoxemia and hypercapnia present problems that require immediate and continuous attention. It is clear that the most dangerous element of these two abnormalities is the hypoxemia, since it is responsible for not only disorientation, but also pulmonary hypertension, congestive heart failure, cardiac arrhythmias, and sudden death. It is also clear that hypoxemia could be relatively simply corrected by the breathing of an enriched oxygen atmosphere by nasal catheter, face mask, or oxygen tent. A problem arises, however, that illustrates a well-known lesson in the treatment of respiratory disease: that breathing an enriched oxygen mixture is dangerous for patients with alveolar hypoventilation in whom tolerance to chronic hypercapnia has dulled the ventilatory response to small increases in blood CO_2 (so that the existing hypoxemia is the only stimulus to ventilation). Under these circumstances the relief of hypoxemia by breathing an enriched oxygen mixture may be associated with further underventilation and increasing degrees of hypercapnia. Before the principles of the chemical regulation of ventilation were well understood, such oxygenation procedures elicited relief from shortness of breath and cyanosis but inexorably produced a steadily rising CO_2 level and falling blood pH with resulting somnolence, coma, and CO_2 narcosis.

Therefore, the most effective technic of oxygenating involves the use of the mechanical respirator, which is capable of both raising alveolar oxygen tensions and at the same time decreasing alveolar CO_2 tensions. In many cases, periods of mechanical hyperventilation of this sort on an hourly basis for the few days of the time of crisis will suffice to maintain the overall level of alveolar ventilation and blood gas composition at reasonable values. Even in these cases, however, a severe degree of hypoxemia may exist during the intervals between the treatment periods of mechanical hyperventilation. These intervals may be effectively handled by administration of measured amounts of enriched oxygen so that oxygen tensions and saturations of arterial blood are raised above the critical levels but not brought completely to normal. Such an approach would have the advantge of avoiding the catastrophic cardiovascular effects of severe hypoxemia, while at the same time maintaining a degree of hypoxia that would be sufficient to stimulate ventilation. Thus a patient whose oxygen saturation is 75% would benefit from the use of a Venturi-type face mask delivering 24%, 28%, or 32% oxygen, depending on the situation, any of which mixtures would be sufficient to raise the arterial oxygen saturation to slightly less than 90% (6% or 7% below the normal value of 96%).

Often, if hypoxemia and hypercapnia have endured long enough, these two chemical derangements will have caused sufficiently severe pulmonary vasoconstriction, pulmonary hypertension, and increases in blood volume so as to result in cor pulmonale and right heart failure. Digitalis, salt restriction, and diuretic therapy are as indicated in this form of congestive heart failure as in other more conventional forms due to heart disease. However, one critical difference should be emphasized: the conventional diuretic produces a renal loss of potassium and a consequent hypo-kalemic alkalosis. This alkalosis removes the needed hydrogen ions that drive the respiratory center and can result in an aggravation of the alveolar hypoventilation For this reason, adequate potassium replacement is essential, and diuretics such as carbonic anhydrase inhibitors (acetazolamide), which result in a moderate metabolic acidosis, are to be preferred in this form of cor pulmonale.

Patient with chronic alveolar hypoventilation. The form of chronic alveolar hypoventilation that features moderately severe degrees of hypoxemia and hyper-capnia and that has been building up in a progressive fashion for months is seen most often years later in the chronic aftermath of poliomyelitis, in patients with muscular dystrophy, in the kyphoscoliotic and obese subjects, and occasionally in the quadriplegic. In each of these types of presentation, one can legitimately ad-vocate a period of hospitalization, preferably in a rehabilitation setting, during which intensive efforts are made to increase the level of alveolar ventilation and stabilize it at a more nearly normal level for a period of 7 to 10 days. The goal of this form of therapy is to produce a normal alveolar and arterial blood CO_2 tension, at 40 mm. Hg, if possible, to allow the arterial blood pH to return to normal by the process of excreting buffer bicarbonate through the kidney. The goal of this form of therapy is to maintain a normal milieu of stimuli for the respiratory centers in the brainstem so that they reestablish normal sensitivity to CO_2 and hydrogen ion stimuli and lose what has been considered to be a pharmacologic tolerance to these stimuli brought on by slow accumulation of CO_2 over a period of months. Such therapy predicts that once sensitivity has been reestablished, patients will reset their level of alveolar ventilation at a higher value and maintain their period of normal oxygenation and normal CO_2 levels for a considerable period of time.

Aside from resetting the respiratory center to a higher degree of sensitivity, the hospitalization may conceivably produce other beneficial effects: (1) a loss of fluid from the chest wall through diuresis with a consequent decrease in the work of breathing, (2) mechanical stretch of the elastic resistances of the thorax or lung through positive pressure breathing or other forms of mechanical ventilation, or (3) the relief of hypoxia that may be providing a self-perpetuated cycle of respira-tory failure by producing respiratory muscle weakness, which in turn diminishes lung function, which in turn aggravates hypoxia, which further weakens respiratory muscles, and so forth; thus round and round the cycle goes, growing worse and worse. unless it can be broken at some point. The easiest target point from the standpoint of therapy appears to be the relief of hypoxia for a period of time long enough to permit recovery of function by the respiratory muscles. Whatever the mechanism involved, the observation itself seems valid: even though no obvious changes oc-curred in the thoracic deformity or in the function of muscles paralyzed by polio-myelitis or muscular dystrophy or in spinal cord function in the quadriplegic, after a period of oxygenation and normalization of CO_2 levels, patients remain well for surprisingly long periods of time at improved rates of alveolar ventilation, even when artificial forms of oxygenation and ventilation are not applied.

The therapeutic approach that utilizes respiratory stimulation is still a contro-versial area. In fact, this form of therapy has been considered ineffective and equated to "beating a tired horse." This viewpoint states that the alveolar hypoventilation originates from paralyzed respiratory muscles or the inability of these muscles to overcome greatly augmented resistances to breathing. For this reason potent respiratory stimulants, such as ethamivan, which have so many important adverse side effects on the central nervous system, have generally been discarded. On the other hand, the viewpoint has recently evolved that milder forms of respiratory stimulation may be effective in certain cases. Such individuals would compose a group, found even in normal population, with respiratory centers in the brainstem that are relatively insensitive when a standard test is applied with a breathing mixture of 5% CO_2. The range of responses is so great in the normal population (150% to 400% increase in ventilation) that the hypothesis has arisen that some patients represent a combination of low-normal sensitivity of the respiratory center in addition to paralysis of respiratory muscles and increased work of breathing. It is this group of patients that might benefit by moderate degrees of respiratory stimulation. Three forms of respiratory stimulation seem reasonably free from side effects: (1) production of a mild metabolic acidosis by acetazolamide or any other carbonic anhydrase inhibitor, (2) pharmacologic stimulation by aminophylline products by the oral or rectal routes, or (3) a combination of the two.

Patient with resistant chronic alveolar hypoventilation. Despite the measures previously outlined to restore alveolar ventilation to normal, not infrequently chronic alveolar hypoventilation becomes an established fact and requires a disciplined program either to stabilize the blood gases or to return them toward normal, insofar as it is possible. The reasons for this detailed program have become evident in recent observations in the course of these patients. Hypercapnia and hypoxemia in a chronic situation almost never are spontaneously stable, and almost inevitably, once high levels of CO_2 have blunted the sensitivity of the respiratory center to the CO_2 stimulus, further increases in CO_2, such as occur in sleep, do not call forth increases in minute ventilation that are required to prevent such a rise. In almost every case the program to stabilize alveolar hypoventilation involves the use of some form of mechanical assistance to ventilation. The type of mechanical ventilator will depend on the condition of the musculoskeletal system of the thorax (Fig. 26-53).

For those patients with paralysis of the muscles of respiration, such as in the aftermath of poliomyelitis, muscular dystrophy, amyotrophic lateral sclerosis, or even the chronic quadriplegic, relatively simple, low-efficiency devices can be used. Their use is possible because the compliance of the lung and thorax in such patients is usually not greatly diminished. One such simple unit is the rocking bed. This bed consists essentially of a board resting on a fulcrum near the middle of its length so that a seesaw motion is developed by an electric motor that is capable of producing excursions through a 40-degree arc at both the head and feet during a single cycle (Fig. 26-53, *C*). As a result the abdominal viscera move both craniad and caudad during a single respiratory cycle according to the effects of gravity and in the process push the diaphragm craniad and caudad, in a similar fashion, so as to ventilate the lung. This device is easily capable of eliciting tidal volumes of 500 ml. and respiratory rates of more than 22 per minute as long as the compliance of the thorax and lung are near normal. Another device useful in this type of patient is the chest cuirass, which consists of a plastic dome overlying the rib cage and sealed posteriorly so that

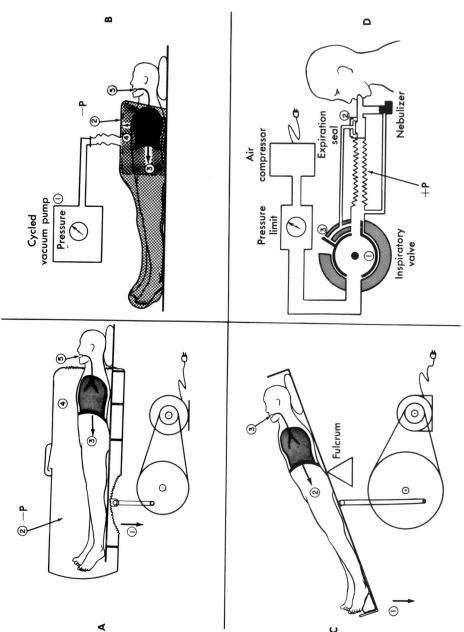

Fig. 26-53. For legend see opposite page.

Fig. 26-53. Principles of mechanical assistors to ventilation.

A, Body respirator. The "iron lung," developed by C. K. Drinker from an already well-known principle, requires the incorporation of the patient from the neck down in the sealed tank, while leaving the head on the outside to have access to atmospheric air. The respirator is motor-driven through a belt drive to an eccentric cam that at *1* moves a large flexible leather membrane in and out of the respirator to produce alternately negative and positive pressures in the respirator, or, as is usual, by means of the escape valve shown at the top, *2,* only a negative pressure cycle. As a result the diaphragm is pulled down and the rib cage expanded, *3* and *4,* and the expansion of the lung induces a negative pressure within the alveoli that in turn pulls air, *5,* at atmospheric pressure through the trachea and into the lung. This respirator is capable of inducing negative pressures up to 30 to 35 cm. H_2O and of a variable respiratory rate ranging from 14 to 24/min. It is capable of ventilating most patients with neuromuscular or skeletal disease with marked increases in the resistances to breathing.

B, Cuirass respirator. This type is a variation of the body respirator, but with more freedom for the movement of arms and legs, which are left free. A rigid, tunnel-shaped steel latticework is placed over the chest and abdomen, with tapering to avoid pressure in the armpits. A flexible plastic "raincoat" then encases the entire body with the exception of the head and neck. A cycled vacuum pump produces rhythmic negative pressures that are confined to the chest and abdominal region and produce negative pressures, *2,* to pull down the diaphragm and expand the rib cage, *3* and *4,* and induce the same type of negative alveolar pressure that pulls in atmospheric air, *5,* as seen with the body respirator. However, it has considerably less capability than the rigid body respirator because the area of negative pressure applied leaves a portion of the chest out of the subatmospheric area.

C, Rocking bed. This respirator consists of a flat bed balanced on a fulcrum and rocked up and down by an electric motor attached by a belt to an eccentrically connected cam. A cycle consists of both a foot-down and head-down movement so that the total excursion at either end may approach 35 to 45 degrees from the parallel. When the foot moves down at *1,* the weight of the abdominal viscera moves by gravity to pull the diaphragm down, *2,* to expand the lung with a subsequent subatmospheric pressure and thereby to pull in atmospheric air as shown at *3.* By a series of gear reductions the cycling rate may vary between 14 and 24/min., and the degree of excursion can be varied also, along with the point of the fulcrum, to increase the relative excursion at either the foot or head end.

D, Typical intermittent positive pressure device. During inspiration the rotary valve shown at *1* is opened by slight inspiratory negative pressure; air from the compressor is delivered to the patient without leak because the expiration seal that consists of a flat rubber balloon prevents expiration, since the balloon is inflated by air pressure from the inspiratory valve. A pressure-limiting device stops flow when a positive pressure ranging from 10 to 40 cm. of water has been reached; the spring-loaded inspiratory valve closes; and, as shown at *3,* the pressure within the expiration seal is released to atmosphere and expiration proceeds through the expiratory portion of the valve, *2.* This type of IPPB device depends on the patient's initiating a slight negative or suction pressure during inspiration to open the inspiratory valve. Other types provide the option of automatic cycling so that the patient cannot slow his respiratory rate during a treatment or maintenance. Also shown is the scheme of a nebulizer that provides aerosol medication.

a cycled vacuum pump connected to the air enclosed within can produce negative pressures for the purposes of inspiration. In general, these two devices can be utilized for the entire night's sleep, with additional supplemental use as needed during the weekends and holidays to maintain reasonable stability of the blood gases. A trial-and-error method is required to ascertain the length of periods of assisted respiration that are required, and this trial-and-error technic is expedited by weekly or biweekly analysis of arterial or capillary O_2 and CO_2 tensions. In many cases this regimen acts only to prevent the retention of CO_2 that occurs during the sleep period when the respiratory center in the region of the reticular substance becomes depressed. In the more optimal situations the assisted respiration during the night actually lowers the CO_2 tension and raises the O_2 tension toward more normal values. Thus a patient arises in the morning reasonably refreshed from a night of normal blood gases and is often able to do a day's work with far less fatigue, sleepiness, and memory lapse. Even though the blood gases are gradually becoming abnormal during the course of the day, a night of assisted ventilation often can be expected to restore them.

In those patients in whom the compliance of the lung and thorax is markedly decreased and in whom the work of breathing constitutes a burden, ventilatory assistance with the simple forms of respirators is usually not possible. For this reason, patients with kyphoscoliosis, ankylosing spondylitis, and pleural fibrosis require respirators that can impose much greater cyclic negative pressures, of the order of 20 to 30 cm. of water, to move a relatively immobile thorax. The Drinker-type body respirator, or iron lung (Fig. 26-53, *A*), fulfills this need and patients can also be taught to enter this respirator for a night's sleep in much the same way as they use the rocking bed. Although there is much initial discomfort and apprehension, these may largely be overcome even in a home setting. More recently, a cycled negative respiratory appliance has been developed, entirely analogous to the iron lung, that utilizes a rigid framework overlying the chest and abdomen, encased in a soft plastic fabric that permits considerably greater mobility of the arms and legs and can produce negative pressures of almost the same level as the iron lung (Fig. 26-53, *B*). In general, the regimen for patients with marked reduction of thoracic and lung compliance is the same as that described for the iron lung.

Prophylaxis against respiratory insufficiency in patients with neuromuscular disease. Several prophylactic measures have long been employed in the prevention of respiratory insufficiency. There are long-term measures such as prevention of the scoliotic deformity in the idiopathic form of scoliosis, as well as in various forms of neuromuscular disease such as poliomyelitis or muscular dystrophy. The bracing and surgical procedures required early in the course of these situations is beyond the scope of this text, but their importance cannot be overemphasized.

Another important prophylactic measure is the prevention and early treatment of respiratory infections. The early use of antibiotics in acute or subacute bronchitis or in bronchial pneumonia can avert a disaster in patients with borderline levels of respiratory insufficiency. The prophylactic clearing of bronchial secretion in more chronic forms of bronchitis when neuromuscular impairment is present can also be a great advantage in preventing the retention of secretions and secondary infection. This therapy usually takes the form of steam aerosol inhalation, mucolytic aerosols, such as acetylcysteine, and postural drainage, on a continuous basis usually by a chest physical therapy group.

It has recently become evident that periodic lung inflation may be one of the newest and most important prophylactic agents available for respiratory insufficiency due to neuromuscular disease. This advance represents an outgrowth of observations made in normal man, breathing for long periods of time under controlled ventilation, such as occurs under general anesthesia. Under these circumstances there occurs progressive atelectasis, as manifested by steadily decreasing lung compliance, which is correctable by periodic artificial or manual hyperinflations of the lung to at least half the vital capacity. Such observations are apparently applicable to normal individuals without anesthesia, in whom it is now believed that the periodic sighing that occurs approximately every 5 to 10 minutes as a universal observation is a reflex mechanism that prevents spotty, progressive, generalized atelectasis of small units of the lung.

The patient with neuromuscular or skeletal disorders who has a marked impairment of his inspiratory capacity is in the same position as the anesthetized normal subject. Neither can produce the deep inspiration characteristic of a sigh to prevent the natural tendency of the lung to undergo alveolar atelectasis. As a corollary, recent clinical observations have suggested that the moderately reduced lung compliance so often seen in patients with kyphoscoliosis, obesity, or quadriplegia may be rectified at least in part by repeated hyperinflations utilizing the mechanical assistance of an intermittent positive pressure breathing device (Fig. 26-53, *D*). Accordingly, a regimen employing such a mechanical device, utilizing the pressures ranging from between 25 and 35 cm. of water, on a periodic basis throughout the day, may be instrumental in maintaining a near-normal compliance of the lung and thereby diminishing the work of breathing. Even moderate decreases in the work of breathing may spell the difference between the energy to obtain normal alveolar ventilation and the fatigue that brings about underventilation of the lung.

REFERENCES

Chronic obstructive lung disease

1. The genuine works of Hippocrates, Baltimore, 1939, Francis Adams.
2. Haas, A., Rusk, H. A., and Zivan, M.: The results of combined medical and rehabilitation programs in tuberculosis, Arch. Phys. Med. 35:77, 1954.
3. Social Security studies of work distribution allowances, 1958.
4. A report to the secretary of Health, Education, and Welfare, by the task force on environmental health and related problems, June, 1967.
5. U. S. Pub. Health Surv. 1959-1964.
6. Consolazio, C. F., Johnson, R. E., and Marek, E.: Metabolic methods: clinical procedures in the study of metabolic functions, St. Louis, 1951, The C. V. Mosby Co.
7. Haas, A., and Cardon, H.: Rehabilitation in chronic obstructive pulmonary disease, Med. Clin. N. Amer. 53:593, 1969.
8. Mitchell, R. S.: Theories of the pathogenesis of emphysema. Symposium on emphysema and "chronic bronchitis" syndrome, part 2, Amer. Rev. Resp. Dis. 80:2, 1959.
9. Lyne, R.: The pathology of emphysema, London, 1967, Lloyd-Luke Medical Books, Ltd.
10. Hallett, W. Y., Beall, G. N., and Kirby, W. M.: Chemoprophylaxis in chronic obstructive pulmonary emphysema: a twelve-week study with erythromycin, Amer. Rev. Resp. Dis. 80:716, 1959.

11. Baldwin, E. deF., Cournand, A., and Richards, D. W., Jr.: Pulmonary insufficiency; physiological classification, clinical methods of analysis, standard values in normal subjects, Medicine 27:243, 1948.
12. Bergofsky, E. H., Turino, G. M., and Fishman, A. P.: Cardiorespiratory failure in kyphoscoliosis, Medicine 38:263, 1959.
13. Comroe, J. H., Jr.: The lung; clinical physiology and pulmonary function tests, Chicago, 1955, Year Book Publishers, Inc.
14. Karpovich, P.: Physiology of muscular activities, ed. 5, Philadelphia, 1959, W. B. Saunders Co.
15. Campbell, E. G. M.: The respiratory muscles and the mechanics of breathing, London, 1958, Lloyd-Luke Medical Books, Ltd.
16. Edwards, H. T., Margaria, R., and Dill, D. B.: The possible mechanism of contracting and paying the oxygen debt of the role of lactic acid in muscular contractions, Amer. J. Physiol. 106:689, 1933.
17. Deitrick, J. E., Whedon, G. D., and Shorr, E.: Effects of immobilization upon various metabolic and physiologic functions of normal men, Amer. J. Med. 4:3, 1948.
18. Darling, R. C.: Some metabolic observations during and after exercise in patients convalescing from acute illness, Arch. Phys. Med. 30:71, 1949.
19. Boldman, E. E., and Haas, A.: Energy cost during various physical activities in convalescing tuberculosis patients, Amer. Rev. Tuberc. 71:722, 1955.
20. Jacobson, E.: Progressive relaxation. A physiological and clinical investigation of muscular states and their significance in psychology and medical practice, Chicago, 1938, University of Chicago Press.
21. Haas, A., and Luczak, A.: The application of physical medicine and rehabilitation to emphysema patients, Rehabilitation monograph 22, New York, 1963, Publications Unit, Institute of Physical Medicine and Rehabilitation, New York University Medical Center.
22. Haas, A., Rusk, H. A.: Rehabilitation of patients with obstructive pulmonary disease. "The role of enriched oxygen," Postgrad. Med. J. 6:612, 1966.
23. Anderson, O. S.: The acid-base status of the blood, ed. 2, Baltimore, 1964, The Williams & Wilkins Co.
24. Haas, A., and Luczak, A.: Importance of rehabilitation in the treatment of chronic pulmonary emphysema, Arch. Phys. Med. 42:733, 1961.
25. Gayrard, P., and Bergofsky, E. H.: The effects of abdominal weights on diaphragmatic position and excursion in man, Clin. Sci. 35:589, 1968.
26. Haas, A., and Dani, A.: Rehabilitation of patients with chronic obstructive pulmonary disease, GP 31:92, 1965.
27. Haas, A.: When to hospitalize the emphysema patient, Hosp. Practice 1:No. 3, p. 59, 1966.
28. Haas, A., Rusk, H. A., and Goodman, W.: Rehabilitation in thoracic surgery, J. Thorac. Surg. 24:304, 1952.

Respiratory insufficiency in neuromuscular and skeletal disorders

29. Bergofsky, E. H.: Cor pulmonale in the syndrome of alveolar hypoventilation, Progr. Cardiovasc. Dis. 9:414, 1967.
30. Fishman, A. P., Turino, G. M., and Bergofsky, E. H.: The syndrome of alveolar hypoventilation (editorial), Amer. J. Med. 23:333, 1957.
31. Ratto, O., Briscoe, W. A., Morton, J. W., and Comroe, J. H., Jr.: Anoxemia secondary to polycythemia and polycythemia secondary to anoxemia, Amer. J. Med. 19:958, 1955.
32. Robin, E. D.: Abnormalities of acid-base regulation in chronic pulmonary disease with special reference to hypercapnia and extracellular alkalosis, New Eng. J. Med. 268:917, 1963.
33. Bates, D. V., and Christie, R. V.: Respiratory function in disease. In chapter "Primary alveolar hypoventilation syndrome," Philadelphia, 1964, W. B. Saunders Co.
34. Sieker, H. O., and Hickam, J. B.: Carbon dioxide intoxication: clinical syndrome, its etiol-

ogy and management with particular reference to use of mechanical respirators, Medicine **35**:389, 1956.

35. Sieker, H. O., Estes, E. H., Jr., Kelser, G. A., and McIntosh, H. O.: A cardiopulmonary syndrome associated with extreme obesity, J. Clin. Invest. **34**:916, 1955.

36. Auchincloss, H., Jr., and Gilbert, R.: The cardiorespiratory syndrome related to obesity. Clinical manifestations and pathologic physiology, Progr. Cardiovasc. Dis. **1**:423, 1959.

37. Burwell, C. S., Robin, E., Whaley, R. O., and Bichelman, A. G.: Extreme obesity associated with alveolar hypoventilation—a Pickwickian syndrome, Amer. J. Med. **21**:811, 1956.

38. Lukas, D. S., and Plum, F.: Pulmonary function in patients convalescing from acute poliomyelitis with respiratory paralysis, Amer. J. Med. **12**:388, 1952.

39. Cherniack, R. M., Ewart, W. B., and Hildes, J. A.: Polycythemia secondary to respiratory disturbances in poliomyelitis, Ann. Intern. Med. **46**:720, 1957.

40. Haas, A., Ben-Yishay, Y., and Diller, L.: Hypoxemia in hemiplegia, Fifth International Congress of Physical Medicine, Montreal, 1968.

41. Bergofsky, E. H., Turino, G. M., and Fishman, A. P.: Cardiorespiratory failure in kyphoscoliosis, Medicine **38**:263, 1959.

42. Hanley, T., Platts, M. M., Clifton, M., and Morris, T. L.: Heart failure of the hunchback, Quart. J. Med. **27**:155, 1958.

43. Chapman, E. M., Dill, D. B., and Graybiel, A.: The decrease in functional capacity of the lungs and heart resulting from deformities of the chest, Medicine **18**:167, 1939.

44. Bedell, G. N., Wilson, W. R., and Seebohm, P. M.: Pulmonary function in obese persons, J. Clin. Invest. **37**:1049, 1958.

45. Hackney, J., Crane, M., Collier, C., Rokaw, S., and Griggs, D.: Syndrome of extreme obesity and hypoventilation. Studies of etiology, Ann. Intern. Med. **51**:541, 1959.

46. Renzetti, A. D., Nicholas, W., Dutton, R. E. J., and Jikoff, E.: Some effects of ankylosing spondylitis on pulmonary gas exchange, New Eng. J. Med. **262**:215, 1960.

47. Travis, D. M., Cook, C. D., Julian, D. G., Crump, C. H., Helliesen, P. K., Lobin, E. D., Bayles, T. B., and Burwell, C. J.: The lungs in rheumatoid spondylitis. Gas exchange and lung mechanics in a form of restrictive pulmonary disease, Amer. J. Med. **29**:623, 1960.

48. Feltman, J. A., Newman, W., Schwartz, A., Stone, D. J., and Lovelock, F. J.: Cardiac failure secondary to ineffective bellows action of the chest cage, J. Clin. Invest. **31**:762, 1952.

49. Sharp, J. T., Henry, S. P., Sweeney, S. K., Meadows, W. R., and Pietras, R. J.: The total work of breathing in normal and obese men, J. Clin. Invest. **43**:728, 1964.

50. Fishman, A. P., Turino, G. M., and Bergofsky, E. H.: Disorders of the respiration and circulation in subjects with deformities of the thorax, Mod. Conc. Cardiovasc. Dis. **27**:449, 1958.

51. Rahn, H., Otis, A. B., Chadwick, L. E., and Fenn, W. O.: The pressure-volume diagram of the thorax and lung, Amer. J. Physiol. **146**:161, 1946.

52. Von Euler, C., and Fritts, H. W., Jr.: Quantitative aspects of respiratory reflexes from the lungs and chest walls of cats, Acta Physiol. Scand. **57**:284, 1963.

53. Campbell, E. J. M., and Howell, J. B. L.: Proprioceptive control of breathing. In Ciba Foundation Symposium: Pulmonary structure and function, Boston, 1962, Little, Brown & Co.

54. Cherniack, R. M., and Snidal, D. P.: The effect of obstruction to breathing on the ventilatory response to CO_2, J. Clin. Invest. **35**:1286, 1956.

55. Alexander, J. K., Wood, J. A., and West, J. R.: Chronic hypercapnia: a specific respiratory depressant in chronic pulmonary disease and other conditions, J. Clin. Invest. **33**:915, 1954.

56. Boshes, B.: Trauma to the spinal cord. In Baker, A. B., editor: Clinical neurology, New York, 1962, Paul B. Hoeber, Inc., Medical Book Department of Harper & Bros.

57. Stone, D. J., and Keltz, H.: The effect of respiratory muscle dysfunction on pulmonary function: studies in patients with spinal cord injuries, Amer. Rev. Resp. Dis. **88**:621, 1963.

58. Bergofsky, E. H.: Mechanism for respiratory insufficiency after cervical cord injury, Ann. Intern. Med. **61:**435, 1964.
59. Bergofsky, E. H.: Quantitation of the function of respiratory muscles in normal individuals and quadriplegic patients, Arch. Phys. Med. **45:**575, 1964.
60. McKinley, A. C., Auchincloss, J. H., Gilbert, R., and Nicholas, J. J.: Pulmonary function, ventilatory control, and respiratory complications in quadriplegic subjects, Amer. Rev. Resp. Dis. **100:**526, 1969.

CHAPTER 27

REHABILITATION OF PATIENT WITH CARDIOVASCULAR DISEASE

T he philosophic concept of rehabilitation has gained wide recognition over the past 10 years. Among the reasons for the rapid advance in this comparatively new discipline are the numerous studies that have begun to define scientifically the meaning and practice of rehabilitation. The scientific knowledge thus attained in concepts of rehabilitation, coupled with the increased physiologic and biochemical knowledge of congestive heart failure and a better understanding of the meaning of cardiac functional classification in terms of ergometry and ergonomics, has at long last permitted a more scientific as well as clinical basis for rehabilitation.

The projected population in the United States in 1980 is estimated to be about 225 million. There appears to be no doubt that by 1980, 35% of the individuals in this population will be over 45 years of age. Accordingly, nearly 68,000,000 individuals will be in an age group wherein cardiac disease will be the most prevalent.[1] At the present time, the cost of heart disease to the nation—in terms of merely dollars and cents—exceeds 4 billion dollars annually. This does not include, of course, the intangible losses in happiness, comfort, and equanimity of the individual and his family. It is therefore imperative to recognize the present as well as the future needs for rehabilitation of the patient with cardiac disease.

The technics and practice of rehabilitation procedures have proved to be of inestimable value for the hemiplegic patient and the patient with coronary heart disease. Results in various industries prove conclusively that many patients with coronary heart disease not only resume their original occupation but also continue their duties with fewer absentee days and with at least as much efficiency as shown by the healthy, noncardiac group. In addition, it has been demonstrated in a group of in-

569

dustrial cardiac patients in functional classes I and II that over 70% returned to their former occupations and maintained an equal work status with their noncardiac colleagues.[2]

Rehabilitation of the patient with heart disease involves the same team approach, cooperation, and principles established in other areas of rehabilitation. Specifically, for the patient with cardiac disease, the physician—either alone or in cooperation with the other disciplines—must endeavor to obtain the maximal cardiac function with minimal risk to the individual patient, enabling him to perform his role with the most efficiency and acceptability in the environment in which he lives and works.

COST OF HEART DISEASE

Heart disease costs in terms of dollars and cents are enormous. This problem will tend to increase, not decrease, unless neglect is avoided in at least three separate areas of research and medical care: (1) detection and prevention of heart disease, (2) care of the acute cardiac problems, and (3) rehabilitation of the existing cardiac problems.

There is no absolute method of estimating the direct or indirect costs of cardiac disease. However, certain estimates appear to be valid. There are 51.9 million working days lost by individuals with cardiovascular conditions. This represents 12% of the total working days lost by the usual working population. It is estimated that cardiovascular problems in this country annually cause the death of 875,000 individuals, of whom 250,000 are members of the working population. The total gross income loss from this group is 1 billion dollars, and the annual federal income tax revenue loss would be 137 million dollars. In addition to this annual loss of federal taxes, there is a total of 125 million dollars lost annually to industry due to cardiac deaths. The above figures give merely a portion of the real costs of cardiovascular diseases in this country. Table 27-1 shows the truer economic loss. Although this information is somewhat outdated, these figures are the best currently available.

Table 27-1. Total cost of heart disease*

Disability resulting from heart conditions	
Industry's production loss from absenteeism	$1,660,800,000
Industry's loss in benefit payments for absenteeism	$1,200,000,000
	$2,860,800,000
Deaths resulting from heart conditions	
Labor's loss in income	$1,050,500,000
Federal government's loss in income tax revenue	$ 137,300,000
Industry's loss in turnover costs	$ 124,700,000
	$1,312,500,000
Total loss	
Loss from disability	$2,860,800,000
Loss from death	$1,312,500,000
Total cost	$4,173,300,000

*From Felton, J. S., and Cole, R.: The high cost of heart disease, Circulation **27:**957, 1963 By permission of the American Heart Association, Inc.

CARDIAC REHABILITATION

The first necessity in cardiac rehabilitation is to recognize the limitations with which one is beset clinically in the diagnosis and prognosis of cardiac disorders. An understanding of the pathologic classification and the therapeutic classification of all cardiac disease, in accordance with the standards set by the New York Heart Association,[4] is of prime importance. Without a detailed knowledge and basic understanding of these, there should be no attempt at cardiac rehabilitation.

The basic pathologic classification is as follows: (1) arteriosclerotic heart disease, (2) hypertensive heart disease, (3) rheumatic heart disease, (4) congenital heart disease, and (5) miscellaneous heart disease (viral, etc.).

Knowing precisely the degree of therapy to which the patient is being subjected is an important aspect of cardiovascular rehabilitation. This is a practical consideration because a patient may do very well in his everyday activity on no medication, but with his increased activity in a job situation digitalis may be necessary, thus altering the therapeutic category in response to myocardial demand. The therapeutic classification is tabulated below:

> *Class A*—Patients with cardiac disease whose physical activity need not be restricted in any way
> *Class B*—Patients with cardiac disease whose ordinary physical activity need not be restricted, but who should be advised against severe or competitive efforts
> *Class C*—Patients with cardiac disease whose ordinary physical activity should be moderately restricted, and whose more strenuous efforts should be discontinued
> *Class D*—Patient with cardiac disease whose ordinary physical activity should be markedly restricted
> *Class E*—Patients with cardiac disease who should be at complete rest, confined to bed or chair*

These classifications are important because the limitations and hazards of employability may differ in individuals with coronary artery disease or rheumatic fever even though in each the heart muscle is clinically adequate. For example, one would never permit an individual with coronary artery disease who has recovered from a myocardial infarction to engage in public transportation as a bus driver, a train engineer, or a commercial pilot,[5, 6] whereas those occupations may not be hazardous—for either the patient or the community—for a patient with mild mitral disease.

In addition to these obvious factors, the whole concept of preventive rehabilitation must also be considered.

The next phase in the preliminary evaluation of the cardiac patient for rehabilitation is to categorize the patient into proper physiologic or functional classification. This will require the utmost of clinical acumen and careful evaluation. The guidelines for this classification appear quite simple at the outset, but in reality much

---

*From Diseases of the heart and blood vessels—Nomenclature and criteria for diagnosis, by the Criteria Committee of the New York Heart Association, ed. 6, Boston, 1964, Little, Brown & Co., pp. xxiii and 463.

Table 27-2. The interrelationship of cardiac functional classification, physiologic symptoms, and ergometric values*

Cardiac functional classification	Physiologic symptoms	Maximum cal./min.	
		Sustained	Intermittent
I	Patients with cardiac disease but without resulting limitations of physical activity. Ordinary physical activity does not cause undue fatigue, palpitation, dyspnea, or anginal pain	5.0	6.6
II	Patients with cardiac disease resulting in slight limitation of physical activity. They are comfortable at rest. Ordinary physical activity results in fatigue, palpitation, dyspnea, or anginal pain	2.5	4.0
III	Patients with cardiac disease resulting in marked limitation of physical activity. They are comfortable at rest. Less than ordinary physical activity causes fatigue, palpitation, dyspnea, or anginal pain	2.0	2.7
IV	Patients with cardiac disease resulting in inability to carry on any physical activity without discomfort. Symptoms of cardiac insufficiency or of the anginal syndrome may be present even at rest. If any physical activity is undertaken, discomfort is increased	1.5	

*Based on data from A. M. Jones (personal communication to H. K. Hellerstein and and A. B. Ford).

judgment is required. For example, in Table 27-2 the physiologic symptoms for class II are listed as follows: "Patients with cardiac disease resulting in slight limitation of physical activity. They are comfortable at rest. Ordinary physical activity results in fatigue, palpitation, dyspnea, or anginal pain." It is soon recognized that a truck driver may be distressed if he has to drive a truck but that he may not be disturbed if he drives an elevator, whereas the elevator operator may be distressed even at this occupation.

The physiologic classification has recently been translated into more meaningful terms by Jones,[7] who has attempted to give an ergometric definition to the four categories in the functional classification (Table 27-2). These values are excellent guideposts with which to match ergonomic values of various occupations.

Accordingly, the first major step in evaluating a cardiac patient for rehabilitation is to classify him—pathologically, physiologically, and therapeutically. Thus the patient should be described in the following terms: (a) rheumatic heart disease, (b) mitral stenosis, (c) atrial fibrillation, and (d) ii: B.

The cardiologist would then (1) know that this patient must be placed in an environment where there is a minimal chance for reinfection, (2) maintain the patient on digitalis and other adjuvant therapy, and (3) be constantly vigilant concerning his cardiac condition.

Progress in the various fields of cardiac rehabilitation has been advanced by including the studies of ergonomics (for an example see Table 27-3) so as to match

Table 27-3. Energy costs of various activities and occupations*

Activity or occupation	Energy (cal./min.)	Activity or occupation	Energy (cal./min.)
Washing and dressing	1.6	Printer	2.0
Washing face and combing hair	2.5	Shoe repairing	3.0
Sitting	1.2-1.6	Shovelling	10.0
Standing	2.0	Postman	10.0
Sewing, 30 stitches/min.	1.1	Planing hardwood	9.1
Polishing floor	4.5	Grass cutting	4.3
Peeling potatoes	2.4	Gardening	4.4-5.6
Bedmaking	5.4	Hoeing	4.4
Beating and brushing carpets	7.8	Driving a car	2.8
Climbing stairs	6.0-10.0	Golfing	5.0
Walking 3.0 m.p.h.	5.6	Tennis	7.1
Typing	1.5	Cycling	5.0-10.0
Inspector	1.2	Dancing	5.2

*From Passmore, R., and Durnin, J. V.: Human energy expenditure, Physiol. Rev. 35:801, 1955.

each cardiac disorder with the job requirements and the cardiac patient's ability to sustain the effort of the job.

WORK EVALUATION CLINIC

The American Heart Association, in association with its local chapters, has been instrumental in setting up work evaluation units to help assess the functional capacity of the heart and its relationship to work capacity. This type of clinic basis its appraisal of the patient on the principles of rehabilitation espoused by Rusk.[2] These are as follows:

1. Evaluate the patient's medical, psychologic, social, and vocational status.
2. Individualize management, including diet, drugs, physical activity, emotional stress, environmental stress, etc.
3. Discuss with the patient the nature of his disease and the prognosis and treatment, as well as the assessment of his capacities and how he can best arrange his life to fit his needs.
4. Make every effort to eliminate the patient's fear of heart disease.
5. Encourage the patient to live the best life possible within the limits (if any) imposed by his disease.*

Personnel of a work evaluation unit and their responsibilities

Nurse. The nurse is responsible for the patient's being available for the medical examination, the cardiac appraisal, and the various tests to be accomplished. In addition, she should make the patient feel at ease and encourage him. The nurse should correlate the working of the entire clinic.

*From White, P. D., Rusk, H. A., Lee, P. R., and Williams, B.: Rehabilitation of the cardiovascular patient, New York, 1958, McGraw-Hill Book Co., Inc.

Social service worker. A social service worker should gain the patient's confidence so that she will be able to learn much of his personal problems, particularly those that are apparent to the patient. The social service worker should learn of the family interrelationships, the economic needs and aspirations, and the social needs and desires. Another important area, which is often neglected, is the important task of learning the patient's attitudes toward heart disease. These facts, in association with the other aspects of the examinations, are very important in the final appraisal and recommendations to the patient.

Vocational counselor. The responsibilities of the vocational counselor are likewise very important. The vocational counselor must define in great detail the patient's responsibilities while he is at work, going to work, and returning from work. This work description must not only appraise the physical aspects of the work but also give a clue to the social and psychologic impact of the job. In addition, the vocational counselor must be prepared to reevaluate the patient's potential for job retraining according to the possible overt or—perhaps—covert skills possessed. And, finally, the vocational counselor should have wide and varied contacts so that job placement will be possible when he finally receives the patient's job prescription.

Psychologist. The responsibilities of the psychologist in the work evaluation unit include assessing the patient's ability and appraising the patient's psychologic makeup with reference to the immediate circumstances resulting from his cardiac condition in relationship to his previous attitudes and behavior. The psychologic appraisal should complement the appraisals given by the social service worker and the vocational counselor. The importance of this will be evident when taken in conjunction with the medical and physiologic evaluation.

Cardiologist. The cardiologist in charge of a work evaluation unit must be a man who is clinically oriented to the problems of cardiovascular disease and is aware both of the limitations and of the possibilities in the laboratory facilities that are available for the appraisal of a patient with heart disease. Above all, he must be practical in his appraisal of the suggestions of the complete staff and must weigh the entire clinical assessment of the patient by each discipline prior to making his final recommendations.

CONCLUSION—EVALUATION OF THE PATIENT

The evaluation of the patient with cardiac disease—for rehabilitation—requires the formation of an opinion based on the combined appraisals of the individual specialists mention. This overall appraisal is fused into a comprehensive practical and scientific recommendation, which is then tested by repeated follow-up evaluations by the group to see whether the recommendation is satisfactory or should be altered to suit the patient or his work environment.

The form adopted by the New York Heart Association is helpful in guiding vocational counselors and work agencies in placing the cardiac patient in a job environment (Chart 20).

ASPECTS OF PREVENTIVE REHABILITATION

It has been thought that cardiac rehabilitation begins after the patient has his cardiac illness. However, it should be stressed that preventive cardiology is, in the

CHART 20

MEDICAL EVALUATION FORM

To _____
(Name of agency)

    Address _____

From _____
(Name of physician, hospital, or other referring agency)

    Address _____

    Telephone No._____ Extension_____

Re_____ Hospital No._____
(Name of applicant)   O.P.D. No._____

Address_____

Birth date_____ Sex_____ Marital status_____

Occupation_____

Patient has been under care   From _____ To _____

Date of last examination_____

If hospitalized, number of hospitalizations _____

Most recent hospitalization   From_____ To_____

MEDICAL DIAGNOSIS

_____

_____

_____

_____ New York heart classification: _____
(See attached Formula for Job Prescription)

    Associated condition (noncardiac):
      Describe resulting disability (if any):

    Necessary appliances that must be worn or used:

PROGNOSIS

When can applicant go to work? _____

Hours of sustained activity (present time): Hours per day_____ Days per week_____

Hours of sustained activity (in 3 months): Hours per day_____ Days per week_____

Any contraindication for day, night, or overtime work (specify):

Does applicant need medical follow-up?   Yes_____ No_____

    How often?_____ Any special time: A.M. _____ P.M. _____

    Where?_____

Continued.

CHART 20, cont'd

ACTIVITIES: Can applicant do the following activities without undue distress?

1. Walking: No. of blocks _____

2. Standing: Unlimited_____ Limited_____ (No. of hours_____ Frequency_____)

3. Stair climbing: No. of flights_____ Occasional _____ Frequent_____

4. Ladder climbing: None_____ Occasional_____ Frequent _____

5. Stooping and bending: None_____ Occasional_____ Frequent_____

6. Lifting and carrying: No. of lb. _____ Occasional_____ Frequent _____

7. Pushing and pulling: No. of lb. _____ Occasional_____ Frequent _____

8. Working speed—use of arms constantly
<blockquote>
at high speed   Yes _____   No _____
in raised position   Yes _____   No _____
</blockquote>

9. Driving commercial vehicles—light (1½ ton) truck   Yes_____ No _____
    Driving commercial vehicles—heavy truck   Yes_____ No _____
    Driving transport vehicles   Yes_____ No _____

10. Other activities that should be limited:

WORKING CONDITIONS: Is applicant limited as to following working conditions (check):

	YES	NO
1. Outside	_____	_____
2. High temperature	_____	_____
3. Low temperature	_____	_____
4. Sudden temperature changes	_____	_____
5. Humid, wet	_____	_____
6. Dusty	_____	_____
7. Fumes	_____	_____
8. High places	_____	_____
9. Noisy atmosphere	_____	_____

10. Other limitations (such as allergens or skin irritants). Explain:

Signature_____M.D.

Date_____

CHART 20, cont'd

FORMULA FOR JOB PRESCRIPTION

ACCORDING TO CARDIAC CLASSIFICATION

	CLASS I	CLASS II	CLASS III	CLASS IV
A	Walk—unlimited Stairs—unlimited Lifting—unlimited Standing—unlimited			
B	Walk—unlimited Stairs—4 flights Lifting—40-60 lb. Standing—unlimited	Walk—1 mile Stairs—3 flights Lifting—25-40 lb. Standing—unlimited		
C		Walk—½ mile Stairs—2 flights Lifting—15-25 lb. Standing—unlimited	Walk—5-10 blocks Stairs—1 flight Lifting—10-15 lb. Standing—unlimited	
D			Walk—less than 5 blocks Stairs—less than 1 flight Lifting—5-10 lb. Standing—limited, 50% of time	Walk—less than 1 block Stairs—less than 1 flight Lifting—5 lb. Standing—limited, 75% of time (sheltered workshop)
E	Unemployable			

Continued.

CHART 20, cont'd

SAMPLE JOB PRESCRIPTION

(In certain common functional capacity and therapeutic classifications of patients with diseases of the heart)

I-B 1. Walking — Unlimited (1 mile)
2. Stairs — 4 flights
3. Weight lifting — 40-60 lb.
4. Standing — Unlimited
Cardiac types — Rheumatic heart disease with grade 1 enlargement
Myocardial infarct, asymptomatic and over 6 months
Essential hypertension up to 200/100

II-B 1. Walking — 1 mile
2. Stairs — 3 flights
3. Weight lifting — 25-40 lb.
4. Standing — Unlimited
Cardiac types — Rheumatic heart disease with grade 2 enlargement
Myocardial infarct, 4 to 6 months
Hypertension with left ventricular hypertrophy 2 plus

II-C 1. Walking — ½ mile
2. Stairs — 2 flights
3. Weight lifting — 15-25 lb.
4. Standing — Unlimited
Cardiac types — Rheumatic heart disease with grade 3 enlargement
Myocardial infarct at 3 months, or coronary sclerosis with mild angina
Hypertensive heart disease without failure

III-C 1. Walking — 5-10 blocks
2. Stairs — 1 flight
3. Weight lifting — 10-15 lb.
4. Standing — Unlimited
Cardiac types — Rheumatic heart disease with failure
Moderate angina pectoris
Hypertensive heart disease with failure or fairly severe hypertension with diastolic over 120 and left ventricular hypertrophy 3 plus

III-D 1. Walking — Less than 5 blocks
2. Stairs — Less than 1 flight
3. Weight lifting — 5-10 lb.
4. Standing — Limited, 50% of time
Cardiac types — Moderately advanced cardiac failure
Severe angina pectoris
Severe hypertensive vascular disease

IV-D 1. Walking — Less than 1 block
2. Stairs — Less than 1 flight
3. Weight lifting — 5 lb.
4. Standing — Limited, 75% of time
Cardiac type — Fairly advanced heart disease with cardiac insufficiency at rest (sheltered workshop case)

essence, the first phase of cardiac rehabilitation. This not only resembles early detection of overt disease but often is more important and difficult than the detection of overt disease. After the disease has been detected, the institution of preventive measures will certainly delay and may even obviate the disease processes. What technics are there available that will afford the individual such information?

HYPERTENSIVE CARDIOVASCULAR DISEASE

The esoteric quality of this disease is gradually being reduced. In the natural history of hypertension as it is known today the following sequelae are observed:

1. Cerebral arterial disease
2. Cardiomegaly with congestive heart failure
3. Coexisting coronary artery disease
4. Renal failure

All of these conditions will require rehabilitation when apparent. Accordingly, preventive rehabilitation must be directed not only at the primary cause—that is, the cause of the hypertension—but also at the early detection of the known sequelae of hypertensive cardiovascular disease, as listed above.

Following is a partial listing of the etiology of hypertension.

1. Central
2. Endocrine
   a. Hyperthyroidism
   b. Hyperaldosteronism
   c. Pheochromocytoma
   d. Carcinoid disease
   e. Cushing's syndrome
   f. Eclampsia (?)
3. Renal
   a. Arteriosclerotic renal artery disease
   b. Nephritis
   c. Pyelonephritis
   d. Glomerulonephritis
4. Hemodynamic
   a. Atherosclerosis
   b. Coarctation of aorta
   c. Aortic insufficiency
5. Peripheral
   a. Vasoconstriction

Early detection of hypertension is now being accomplished through regular physical examinations by private physicians or private clinics or by personnel in industry and public health stations, to mention a few of the major areas. However, the early diagnosis is now being considered in terms of etiology, for many cases of hypertension are now permanently cured by surgery—through surgical removal of diseased renal arteries, certain renal tumors, and adrenal tumors. Thus it is apparent that when an exact diagnosis is made, the treatment may, in the majority of instances, also be exact. However, there are many instances of diagnoses that are not precise or exact. These fall into the categories of (1) neurogenic disorders, (2) peripheral arterial constriction, and (3) arteriosclerotic conditions, either alone or combined with (1) or (2). In these latter instances there exists sufficient therapeutic knowledge to lower the blood pressure to within normal limits in nearly all instances. The principles in this approach lie in the adoption of the following basic therapies, used either singly or in combination:

1. Salt removal
   a. Salt restriction
   b. Salt removal with chlorothiazide derivatives

    c. Rule out possible primary or secondary hyperaldosteronism
  2. Drugs
    a. Those that increase renal blood flow (hydralazine hydrochloride)
    b. Those that decrease sympathetic tone (guanethidine sulfate)
    c. Cerebrovascular drugs that affect blood pressure
      1. Reserpine (Serpasil) derivatives
      2. Monamine oxidase inhibitors

ARTERIOSCLEROTIC HEART DISEASE

Although Gertler, Garn, and White[9] had reported a possibility of classifying the candidate for coronary artery disease, this was not feasible until the multiple factors known to be associated with the disease had been assessed in their proper perspective. There is still some discussion as to the relative importance of certain parameters in the selection of coronary artery disease patients, but the best-known factors are as follows:

1. Serum cholesterol
2. Serum uric acid
3. Body build (particularly the degree of mesomorphy)
4. Family history
5. Hypertension
6. Diabetes
7. Triglycerides
8. Height
9. Phospholipids (total)

The first seven are positive factors, while 8 and 9 are considered negative factors. The method employed for prediction is beyond the scope of this chapter, but the details have been published elsewhere.[10] Cigarette smoking is also considered to be a risk factor.

PRINCIPLES OF PREVENTION OF CORONARY HEART DISEASE

Diet. Two aspects of diet are of importance: (1) limiting the caloric intake to maintain an isocaloric diet and (2) limiting the fat intake, from 45% of total calories of saturated fat intake to 30% of total calories of fat, and substituting unsaturated fat for the saturated fat.

Weight control. The available evidence today indicates that overweight per se is not useful as a predictive index for coronary heart disease.[9, 11, 12] The question of relative overweight is important only in the fact that this may be associated with hypertension, or be reflected in the serum by producing hypercholesterolemia or hyperuricemia. The important element in weight control is maintenance of a body weight suitable for one's somatotype. Thus the first step in weight control is to recognize the individual variation in weight and not to assume or match one's weight from age-height-weight charts. If variation in weight will lower an abnormally high blood pressure, or will lower serum cholesterol or serum uric acid, then one should maintain one's weight to keep these three factors at desirable levels. If, however, this is not achieved by merely a hypocaloric diet, then one may have to resort to a special diet, which would be an unsaturated fat diet with or without sodium restriction. In essence, weight control is important only to normalize the body's physiologic state and should be strictly an individual prescription employing general principles.

Exercise. There is an abundant body of evidence that is suggestive that exercise or manual labor is an effective antidote against coronary heart disease. Studies by Taylor and Keys[13] on railroad workers showed that switchmen have a lower incidence

Table 27-4. Energy expenditure and physical activity*

Physical activity	Energy expended (cal./min.)	Rate per 100,000	
		All causes (except arteriosclerosis)	Coronary heart disease
Sedentary (judges)	3.0	75	102
Light (engineers)	4.0	82	94
Medium (surveyors)	5.0	102	81
Heavy (laborers)	6.0+	140	47

*From Buell, P., and Breslow, L.: Mortality from coronary heart disease in men who work long hours, J. Chronic Dis. 11:615, 1960.

of coronary heart disease than do conductors. The evidence from studies in Israel similarly shows that the field workers in the kibbutz or collective farm have fewer cases of coronary artery disease than the more sedentary workers.[14] Additional evidence, included in Table 27-4, also tends to support the same thesis. The differences in death rates from all causes (except arteriosclerosis) and from coronary heart disease in various occupations, between ages 45 and 64, are indicated.

A further word may be said in favor of calculated exercise in coronary heart disease. Exercise has been shown recently to be effective as a protection against coronary heart disease in three ways: (1) it increases collateral circulation in the myocardium,[15] (2) it increases the plasmin serum levels and thus aids in the dissolution of blood clots,[16] and (3) it decreases the incidence of atherosclerosis, shown in rabbits fed high-cholesterol diets.[17] The exercise should not be sustained but, rather, intermittent with an output of not more than 4.0 calories per minute.

CONGENITAL HEART DISEASE

There are 30,000 to 40,000 children annually born in the United States who have congenital heart lesions. These babies may have operable or inoperable lesions. For those who have relatively simple abnormalities, such as patent ductus arteriosus or coarctation lesions, operative procedures, if undertaken early, reestablish the child to virtually normal in most cases. In other cases, such as intercommunications between the atria or ventricles, the cure may be as good if undertaken early and if the surgery is successful. As one proceeds to the more complicated congenital cardiac abnormalities, such as Eisenmenger's syndrome, tetralogy of Fallot, transposition of the great vessels, and others, surgery is not as curative, abnormalities in growth develop, hematologic changes occur, and cardiopulmonary problems manifest themselves. Accordingly, it is observed that prevention of cardiac abnormalities would be much simpler than to follow through the possibilities of surgical cure, medical management, and economic disaster in many instances.

The basic principles involved in the prevention of congenital cardiac abnormalities are similar to those observed in the prevention of other congenital lesions. It should be stressed, however, that with all known precautions, there will still be some babies born with defects. The following precautions should be take by all young women prior to pregnancy and during the early months of pregnancy:

1. Avoid excessive use of tobacco and alcohol.

2. Avoid the use of drugs, particularly the drug whose clinical experience is of short duration.

3. Avoid exposure to x-ray, cosmic rays, and other gamma rays of any source. Included here should be alpha and beta rays as well.

4. Avoid exposure to virus disease, particularly in the early months of pregnancy.

5. It would be desirable for every female child to be immunized against German measles (rubella).

RHEUMATIC HEART DISEASE

The incidence of rheumatic heart disease is, happily, decreasing. The immediate care of sore throats in children, teen-agers, and young adults is imperative. The proper use of antibiotics in such cases has been one of the most rewarding facets in preventive medicine. The use of antibiotics and sulfonamides for the prevention of sore throats due to *Streptococcus hemolyticus,* Lancefield type A, has achieved almost miraculous results in prevention of rheumatic fever.

The central core in the assessment for rehabilitation of the patient with cardiac disease is to determine how well the entire heart muscle and accompanying cardiac blood supply will adjust to the physical and emotional demands made on them. How well will the blood be propelled to the other organs so that they can function properly? Accordingly, the true basis of cardiac rehabilitation is to determine how biochemical energy is made available to the heart muscle and converted into mechanical energy.

SPECIAL REHABILITATION FEATURES OF CARDIAC DISORDERS

The rehabilitation of the cardiac patient begins, as stated previously, with preventive technics. If these technics have not been successful and the acute episodes occur, then rehabilitation of the acute cardiac disorders must be instituted. A physician once said that cardiac rehabilitation begins with the moment the patient's family summons advice! Cardiac rehabilitation is in essence divisible into two parts: (1) treatment of acute myocardial infarction and (2) treatment of the patient with various degrees of congestive heart failure.

Acute myocardial infarction

During immediate crisis. Both the patient and his family must be reassured and comforted. For example, the oxygen tent should be explained as a helpful, precautionary measure and should not be interpreted by the family or the patient as a harbinger of doom. The entire course of illness should be projected to the family, whereas the patient should be reassured in general terms, with specifics and details of his illness left to be discussed as the question and the occasion arise.

During course of illness and acute complications. The well-known complications of an acute myocardial infarction should be anticipated. Accordingly, the following should be looked for and guarded against by proper remedial means: (1) shock, (2) cardiac arrhythmias, (3) pulmonary edema, (4) embolic complications, (5) allergic reactions to drugs, (6) poor care of the bowels or bladder, and (7) cyanosis.

Much has been written concerning the energy costs of a patient in bed and caring for his Activities of Daily Living. Thus it has been estimated that the following numbers of calories per minute are expended during the performance of the tasks listed: shaving, 2.5; feeding, 3.0; using bedpan, 4.7; sitting in bed, 1.5; sitting in chair, 1.2; lying in bed, 1.4. These activities may be performed by patients in classifications I to III with ease. Patients in classification IV cannot achieve all of these. Good judgment should be exercised at all times concerning when a patient should (1) feed himself, (2) use a bedside commode, and (3) sit in a chair. It is certainly preferable, all other things being equal, to permit the patient to sit in a chair, for the energy cost of sitting in a chair is less than that of lying in bed or sitting in bed. In addition, there is some suggestion that pulmonary emboli are less frequent. The use of a commode is preferable to use of a bedpan, which is uncomfortable. In addition, the Valsalva maneuver that the patient uses on a bedpan predisposes to pulmonary emboli—many of which have been fatal.[18]

During the first 10 days following a coronary thrombosis, there should be as minimal activity as possible, but the beneficent and salutary effects of the procedures mentioned above are not to be understated. However, if the infarct has been an unusually large septal infarct, caution may be exercised—in view of the possibility of a septal perforation, which occurs usually about 10 to 14 days after the initial damage. It is thus incumbent on the cardiologist to judge the extent of activity he will permit the patient to attempt during his early convalescent period.

It is generally believed that the myocardial necrotic tissue following an infarction is replaced by fibrous tissue in approximately 20 to 25 days. One then asks this question: If no further signs of myocardial damage exist, as indicated by a stabilized electrocardiogram, normal white blood count, serum transaminase, and physical signs such as absent pulmonary signs of congestive heart failure, should one not be more permissive and allow the patient to sit longer, shave himself, walk to the bathroom for every conventional emergency, and gradually increase his physical activity? There certainly is no contraindication to this, provided it is done conservatively and that symptoms such as angina pectoris, dyspnea, and palpitation are not present. In essence, after the 3-week period, activity should be moderatly increased as long as no evidence of impaired coronary blood flow or myocardial compromise is evident.

Congestive heart failure

One of the most important phases in the rehabilitation of patients with congestive heart failure is a proper diagnosis, as to both the functional and the therapeutic classification. With the proper diagnosis, specific medical treatment for the particular type and degree of heart disease may be instituted. In addition to the general supportive and therapeutic measures given to the the patient with congestive heart failure—such as digitalis, a low-sodium diet, diuretics, and vitamins—specific measures may include (1) for coronary disease, anticoagulants and an isocaloric diet or a diet which consists of 30% fat, of which 70% is unsaturated fat; (2) for hypertensive cardiovascular disease, the use of one or more of the many antihypertensive drugs; and (3) in rheumatic heart disease, the use of salicylates and steroids, which during the early phase of failure may be mandatory. In addition, the use of antibiotics for prevention of further rheumatic episodes is in order.

A technic now being evaluated is draining of lymph by cannulating the thoracic lymph duct. This procedure may be very helpful in refractory cases of cardiac edema.[19]

In the final analysis, functional cardiovascular tests attempt to measure the reason or reasons why biochemical energy fails to be converted into mechanical (or pumping) energy. This question may be divided into three parts.

1. Failure of energy production. In this category there are several distinct possibilities.
   a. Failure of adequate coronary blood flow commensurate with myocardial demand
   b. Failure of energy to be converted into utilizable form, for example, thiamine deficiency
   c. Failure of oxygenation, either at the lungs or at the cellular level, for example, carbon monoxide or cyanide poisoning inhibition
2. Failure of energy transference. In this category one might consider the possibility of "uncoupling" of oxidative phosphorylation, either by (1) excess adenosinetriphosphatase (ATPase) such as one might see in muscular dystrophy or by (2) excessive oxidative rates wherein phosphorylation cannot keep pace with the abnormal situation, for example, hyperthyroidism.
3. Failure of energy utilization. This category presupposes the existence of a break in the continuity of high-energy phosphate ATP (adenosinetriphosphate) with the actomyosin of cardiac muscle; the available evidence today does not have any definitive examples of this type of failure.

Accordingly it can be seen that the attempts at measuring the integrity of the cardiovascular system fail to measure all the possibilities. More than one type of test is necessary to fulfill these requirements. Tests that purportedly reveal an inadequate coronary circulation do not reveal the accompanying electrocardiographic changes due to an inadequate blood supply, a state of insufficient phosphorylation accompanying oxidation, or some other phenomenon. Similarly, a test of total oxygen consumption bears virtually no relationship to myocardial efficiency or utilization of oxygen for the purposes of energy production.

The great difficulties that are encountered, both in the practical measurement and in the extrapolation of the results for practical purposes, are indicated in the following paragraphs.

The definition of congestive heart failure[5] must be considered prior to introducing and extending the physiologic concept of ergometrics and ergonomics: "The syndrome, when fully developed in man, is characterized by dyspnea, generalized edema, which is more severe in the dependent portions of the body, generalized and systemic venous hypertension, diffuse hepatomegaly and possible peritoneal or pleural transudates or both. Bilateral basal pulmonary râles, accentuated pulmonic second sound, pulsus alternans, gallop rhythm, orthopnea, weakness and many other manifestations may be present. Although there may be variations in detail, all well-known to clinicians, the general pattern is consistent."* In addition to this clinical description, arteriovenous oxygen difference is of paramount importance.

Eventually, all cardiac patients—whether their disease is hypertensive, arteriosclerotic, rheumatic, or congenital in type—are more likely to develop the syndrome

---

*From Gertler, M. M.: Bioenergetics in cardiac rehabilitation, Conn. Med. 25:697, 1961.

of congestive heart failure than any other complication.[28] The exact number of individuals who are hospitalized or who are ambulatory with congestive heart failure is unknown, for the statistics on this problem are unsatisfactory and have not been considered in great detail. For diverse reasons, it is difficult to know at the present time the exact incidence of congestive heart failure.

Rehabilitation of the patient with congestive heart failure involves the same team approach, cooperation, and principles established in other areas of rehabilitation. Specifically, for the patient with congestive heart failure the physician, either alone or in cooperation with the other disciplines, must endeavor to obtain the maximal cardiac function with minimal risk to the individual patient so that he can perform his role with the most efficiency and acceptability in the environment in which he lives and works.

To understand the limits of work performance of the individual with congestive heart failure various indirect means for assessment of cardiac work have been attempted. These tests are designed to test the efficiency of the cardiovascular system. Originally, many of them were devised as an assessment of physical fitness. However, it was soon recognized that physical fitness is dependent on all muscles—including the heart. Since the heart is the central pump, of necessity good physical fitness depends on the well-balanced integrity of the entire cardiovascular system. Accordingly, for the purposes of cardiovascular rehabilitation there is no single test known today that measures adequately the integrity and response of the cardiovascular system to physical or mental stress. Furthermore, the projection of the results of the tests to natural work situations may be hazardous.

Cardiac work

Energy measurements of cardiac work have been calculated directly and indirectly. The direct measurement exceeds the indirect measurements in accuracy by far, but cannot be employed routinely or continually. These measurements consist of bioenergetic studies of mitochondrial systems prepared from animals in experimental heart failure[20] and coronary sinus catheterization.[21] The indirect measurements are not without criticism but are practical and useful for the evaluation of the patient for cardiac rehabilitation. The methods employed most frequently for the evaluation of energy costs of myocardial work are based on indirect calculation from (1) oxygen consumption of the total individual during a standard sustained exercise, (2) dye-dilution studies, and (3) ballistocardiograms.[26] The results obtained by Bruce,[22] Kottke,[23] Jones,[7] Benton and Rusk,[18] and Hellerstein and Ford[24] have contributed much to our knowledge concerning the functional capacity of the heart during and following illness. This information, coupled with the classic ergometric observations of Passmore and Durnin,[8] has placed cardiac rehabilitation on a scientific and practical basis.

The measurement of energy output for the total body (ergometric studies) is not synonymous with ergometric studies of the heart itself. The work of the heart may be calculated in several ways, but the basic principle involved is to calculate the energy required by cardiac muscle for the creation of sufficient pressure within the myocardium to produce adequate kinetic energy, within the large vessels, to permit continuous body blood supply and flow. Thus the calculation for cardiac work becomes:

$$W = V_1P + \frac{MV_2^2}{2g}$$

W	=	Work in kilogram-meters
V_1	=	Mean aortic pressure in meters of blood
M	=	Weight of blood
V_2^2	=	Square of mean blood velocity
g	=	Acceleration due to gravitational forces

It is obvious that the work of the heart may be measured by combining the mean aortic pressure and the minute volume in liters, that is, cardiac output, since the kinetic factor is virtually constant with each beat.[25]

Accordingly, the continuous measurement of cardiac output during work performance would be of inestimable value in determining cardiac work ability with reference to generalized energy expenditure. Obviously this is virtually impossible, and therefore the indirect method of assessing work performance relative to cardiac efficiency is utilized. Thus oxygen consumption, changing heart rate, and changing blood pressure have been measured before, during, and after work performance, for example, exercise,[26] various working conditions,[8] or disease processes.

The use of oxygen consumption as an index to cardiac work reserve has been employed by several investigators. It matters little whether the work load measurement has been by means of treadmill,[22] two-step test variation,[24] or actual on-the-job work performance, the measurement of total oxygen consumption has little bearing on the energy production within the heart muscle itself, since at best the human heart utilizes 4 to 5% of the total oxygen consumed during any activity. Any measurement thus introduces error in a ratio of at least 20 to 1 prior to revealing any change in total oxygen consumption. The total resting oxygen consumption as measured by the Haldane principle takes into account the oxygen consumed by all organs, including the skeletal muscle, and the efficiency of the pulmonary system. It must naturally follow that the heart therefore is included and not specifically related to the total oxygen measurement. If oxygen consumption during rest is 0.3 L./min. and is increased to 2 L./min., it is obvious that the heart will, on theoretical grounds, increase its oxygen consumption by 0.085 L.[25] It is known that the failing myocardium may increase its oxygen extraction to a greater extent than the normal heart. Thus it is reasonable to suggest that the failing heart may, without resort to increased oxygen consumption but by means of more efficient oxygen extraction (i.e., with arteriovenous difference increased), perform greater work loads without the requirement of increased available oxygen. It is therefore difficult, on basic physiologic grounds, to relate oxygen consumption—which is calculated to determine total body work energy expenditure—to cardiac work performance because it is not only at best only a guess, but also not entirely within physiologic concepts.

This concept of energy requirements may be further extended by reviewing the energy source of the myocardium. It has been well established that the heart is an aerobic organ and derives most of its energy from such sources. The most important source of cardiac energy is from carbohydrates or derivatives thereof. It is also well-known that anaerobic energy will yield only a fraction of the energy that may be derived from the glucose molecule as aerobic energy. The free energy yield from 1 molecule of glucose, if completely combusted as follows ($C_6H_{12}O_6 + 6O_2 \rightarrow 6CO_2 + 6H_2O$), will be 686,000 calories.

Of this total combustion, 57,000 calories are available during glycolysis, which is essentially the anaerobic phase, whereas the remaining 629,000 calories are available via the Krebs or citric acid cycle, which is essentially the aerobic phase. The glycolytic phase ends at the formation of lactic acid and/or pyruvate, whereas the aerobic phase ends at carbon dioxide and water. The anaerobic phase is accompanied by the yield of 2 adenosine triphosphate (ATP) molecules, whereas the aerobic phase yields 36 ATP bonds. The efficiency of each phase for the conversion or trapping of energy as ATP is about 65%.[27]

From the above discussion it becomes apparent that the contribution of glycolysis or the anaerboic phase of oxidation is a small portion of the actual energy for aeorbic cells. The heart utilizes faty acids and proteins, as well as carbohydrates, for energy production. Since the same basic rates of anaerobic versus aerobic oxidation hold for these compounds, the heart must of necessity by oxygen-dependent.

Another important aspect of biochemical processes must be considered, for it is essential to the understanding of cardiac rehabilitation. The major source of energy for cardiac muscle is from the release of the terminal phosphate bond of ATP. The most efficient mechanism for obtaining ATP bonds is via the process of oxidative phosphorylation. By this process the energy available from substrates such as α-ketoglutarate succinate is passed along as electron equivalents to molecular oxygen via a chain of respiration enzymes such as diphosphopyridine nucleotide (oxidized) (DPN), flavin adenine dinucleotide (FAD), and the cytochrome system. The harnessed energy is trapped in various areas along the chain as ATP by the union of inorganic phosphate and adenosine diphosphate (ADP). This fundamental process that takes place in cardiac mitochondria is largely responsible for the energy that is available within the myocardium and is transferred to the actomyosin complex during cardiac contraction. Congestive heart failure may be looked on as a biochemical process, as Bing[21] and others have clearly shown by emphasizing a few of the basic factors involved in congestive heart failure. These observations have been studied and extended by others in the experimental animal.[28]

In an experimental study relating bioenergetics to the failing heart, it was learned that mitochondrial preparations derived from failing guinea pig hearts were as efficient in synthesizing ATP from inorganic phosphate and ADP as were mitochondrial preparations derived from normal guinea pig hearts. The P/O* ratios for the substrate α-ketoglutarate in normal and failing hearts were 2.90 and 2.91, respectively, while the corresponding $Q_{O_2}^N$ † values were 1170 and 1032, respectively.

Under the influence of a combination of cofactors, coenzyme A (CoA), DPN, and TPN, the $Q_{O_2}^N$ values increased 26.5% and 47.5%, respectively, whereas the P/O

---

*P/O $= \dfrac{\text{Micromoles of orthophosphate consumed}}{\text{Microatoms of oxygen utilized}}.$

†$Q_{O_2}^N$ = Microliters of oxygen consumed per hour per milligram of mitochondrial nitrogen.

ratios did the reverse. The latter decreased 33% in the failure group as compared to 20% in the normal group. However, the energy calculations reveal that identical amounts of energy are utilized by each.[29] This observation emphasizes an important concept of cardiac rehabilitation. The heart as an organ extracts more oxygen during failure, in keeping with Starling's observation,[30] because it is still required to supply kinetic energy to the blood for body needs, under conditions of decreased efficiency of oxidative phosphorylation, that is, in providing ATP from oxygen at the electron level.[29] In addition, the heart is able to function at peak myocardial efficiency only to a point, where it begins to operate at decreased efficiency if the demand on it is continued and increased. Thus during performance of increased activity the already failing heart is working at decreased efficiency, that is, it is unable to extract sufficient oxygen to convert ADP to ATP. If the demand is increased, as measured by general oxygen uptake, symptoms of anoxia, such as angina pectoris, or symptoms of failure to provide kinetic energy, such as dyspnea, supervene. Accordingly, the general relationship of cardiac oxygen utilization to total oxygen utilization may be completely unrelated by present standards.

As seen from the foregoing discussion, it becomes increasingly difficult to assume that tests of physical fitness, exercise tolerance tests, or treadmill studies offer a true index of cardiac performance in relationship to body requirements. For cardiac rehabilitation the heart must always be placed in a condition in which it meets its demands easily and efficiently. Consequently, these basic principles of cardiac rehabilitation should be applied: know the exact diagnosis and evaluate the treatment from the viewpoint of the whole patient. Five other basic principles of cardiac rehabilitation have been stated on p. 573.

The energy production in the heart is dependent to a great extent on two factors: (1) the rate of coronary blood flow and (2) the degree of oxygen extracted from arterial blood.

Under normal circumstances, the healthy myocardium will depend more on oxygen extraction than on coronary blood flow because of the large amount of energy to be derived from efficient coupling of oxidative phosphorylation. Thus during rest the extraction of oxygen from the coronary arteries is 70% to 75%. An important contribution to this basic problem has been made by Messer and co-workers,[31] who employed subjects with various types of heart disease under stressful exercise while measuring coronary blood flow by means of the catheterization technic. The results are summarized in Tables 27-5 and 27-6, based on their work, and in the accompanying discussion of their normal subjects and of the patients having coronary insufficiency, congestive heart failure, or mitral stenosis.

Normal subjects

Myocardial oxygen extraction is virtually unchanged during exercise in this group of individuals.

Coronary insufficiency

In the resting state, patients with coronary insufficiency had significantly lower myocardial oxygen extraction rates (66 ± 8), but during standard exercise the myocardial oxygen extraction increased in over 80% of the cases.

Table 27-5. Myocardial oxygen extraction (MOE) and coronary venous oxygen (CVO$_2$) values during resting (R) conditions and following exercise (E)*

		MOE†	CVO$_2$†
Controls	R	70 ± 6	29 ± 6
	E	→	→
Coronary insufficiency	R	66 ± 6	33 ± 7
	E	73	26
Congestive heart failure	R	73 ± 5	26 ± 5
	E	See below	See below
With coronary insufficiency	R	73 ± 5	26 ± 5
	E	↑	↓
Without coronary insufficiency	R	→	→
	E		
Mitral stenosis	R	75 ± 4	24 ± 4
	E	→	→

*Modified from Messer, J. V., and associates: Patterns of human myocardial oxygen extraction during rest and exercise, J. Clin. Invest. 41:725, 1962.
†No change →. Increase ↑. Decrease ↓.

Table 27-6. Total body oxygen consumption (TBOC), mean blood pressure (MBP), mean heart rate (MHR), coronary blood flow (CBF), and myocardial oxygen consumption (MOC) in four clinical groups at rest (R) and after exercise (E)*

Clinical group		TBOC (ml./min./ M.²)	MBP (mm. Hg)	MHR (beats/ min.)	CBF (ml./100 gm./min.)	MOC (ml./100 gm./min.)
Control	R	129	114	80	66	7.8
	E	279	133	102	90	10.8
Coronary insuf- ficiency	R	137	113	78	70	8.8
	E	306	136	100	100	13.8
Congestive failure	R	153	115	85	91	11.0
	E	334	136	113	121	15.0
Mitral stenosis	R	150	98	88	68	9.0
	E	341	116	125	98	13.2

*Modified from Messer, J. V., and associates: Patterns of human myocardial oxygen extraction during rest and exercise, J. Clin. Invest. 41:725-742, 1962.

Congestive heart failure

The resting MOE of 73 ± 5 is significantly increased in this group of patients. The rate of MOE was different according to the cardiac indices. Thus patients whose cardiac index was less than 3.0 had an increase of MOE to 77, whereas for those patients whose cardiac index was more than 3.0 it was reduced to 70.

It is interesting to note that the response of the MOE during exercise was dependent on the degree of coexisting coronary insufficiency. If coronary insufficiency coexisted, the MOE increased.

Mitral stenosis

Patients whose mitral valve diameter was less than 1.5 cm. square had MOE resting values of 75 ± 4, which, as in the congestive heart failure patients, differ from the control values of 70; but values did not differ from those in the congestive heart failure group where the MOE value is 73. In this group of patients, MOE response to exercise appears to be unrelated to the presence or absence of coronary insufficiency.

These results are summarized in Table 27-6.

These data merely emphasize the principle that one must, with great caution, extrapolate the results obtained from total oxygen consumption studies when relating these to cardiac efficiency.

PULMONARY FUNCTION STUDIES

Chart 21 (compiled by Haas and Gertler) indicates the results in 30 patients with myocardial disease consisting of the following illnesses: (1) myocardial infarction, 24 cases; (2) rheumatic heart disease, 4 cases; and (3) hypertensive cardiovascular disease, 2 cases.

These investigations reveal a trend toward a decrease in pulmonary reserve as best evidenced by the vital capacity measurements and by the ratio of expiratory reserve volume and functioning residual capacity. These trends should be taken into consideration when giving the cardiac prescription to a patient undergoing cardiac rehabilitation.

These data again point out the complexity in evaluating the cardiac patient and the folly in attempting to gauge cardiac reserve by simple tests.

SPECIAL ASPECTS OF CARDIAC REHABILITATION

There are many instances wherein cardiovascular disease either coexists with the disease involved in rehabilitation or is actually manifested because of the added energy costs involved in rehabilitation. Special examples will be discussed.

Hemiplegia

There may be a common etiologic denomination in hemiplegia and cardiac disease, particularly if thrombosis and hypertension are involved. In addition, hemiplegia may be secondary to emboli arising from myocardial infarction or rheumatic heart disease. It is thus apparent that when the additional energy cost is placed on a potentially pathologic heart, symptoms and signs such as angina pectoris, dyspnea, and signs of congestive heart failure may occur. Thus each patient with hemiplegia should be assessed from a cardiac viewpoint so that he may be prepared for the sometimes arduous tasks of rehabilitation.

Emphysema and allied pulmonary conditions

There is considerable cardiac involvement in emphysema. Aside from the traditional and classic cor pulmonale, left heart failure will develop if the emphysema

CHART 21

PULMONARY FUNCTION STUDIES*
(Average Values)

	CARDIAC PATIENTS	NORMAL	
AGE	52		
HT.	5'8"		Height
WT.	165		Weight
B.S.A.	1.88		Body surface area
R.R.	14	10-14	Respiration rate
T.V.	0.604	0.500	Tidal volume
M.V.	8.94	6.0	Minute volume
O₂	0.231	0.250	Oxygen consumption
1st V.C.	4.038	4.800	First stage vital capacity
2nd V.C.	4.415	4.800	Second stage vital capacity
% Pred.	88%		Per cent predicted
I.C.	2.998	3.600	Inspiratory capacity
E.R.V.	1.717	1.200	Expiratory reserve volume
F.R.C.	3.297	2.400	Functional residual capacity
R.V.	1.894	1.200	Reserve volume
T.L.C.	6.307	6.000	Total lung capacity
R.V./T.L.C.	29%	20%	
F.R.C./T.L.C.	52%	50%	
M.B.C.	162.47		Maximum breathing capacity
% Pred.	111%		Per cent predicted
T.V.C. 1	74%	83%	Timed vital capacity—first
T.V.C. 2	87%	94%	Timed vital capacity—second
T.V.C. 3	92%	97%	Timed vital capacity—third
A.V.I.	0.98		Air velocity index
Total cases	30		

*Compiled by Albert Haas, M.D., and Menard Gertler, M.D., New York, N. Y.

is uncontrolled. The attendant electrolyte changes associated with respiratory acidosis may produce cardiac changes secondary to involvement of cardiac contractility. Thus it is imperative, when treating emphysema, to ascertain the cardiac status.

There are certain attendant cardiac changes that occur secondary to the chest cage configuration in such conditions as pectus excavatum and marked scoliosis. These too must be recognized during rehabilitation procedures.

Paraplegia and quadriplegia

The increased work output in these conditions, particularly in the older individual, may cause the discovery of occult coronary disease or produce mild congestive heart failure in an already borderline-functioning heart. Thus cardiac evaluation is also imperative in these cases as well.

Telemetry

The technic of telemetry should provide useful and helpful information about the cardiac at work. This technic is becoming more useful and practical because of the miniaturization of the equipment necessary for study of the cardiac at work. It is possible at present to monitor electrocardiographic changes, pulse rates, and respiration rates. However, it is still difficult to measure blood pressure without some encumbrance to the patient. The use of certain physiologic assessment of heart rate changes and blood pressure changes may give some indication of the cardiac stress or adjustment to work. This technic has a definite future in the study of cardiac patients at work.

RESULTS OF CARDIAC REHABILITATION

The final assessment of the cardiac patient from a pathologic, physiologic, and therapeutic viewpoint will provide an educated estimate of his work capacity. This information coupled with the ergometric and ergonomic information will provide, if matched as closely as possible, the best method of cardiac rehabilitation. The results have been most gratifying, as is summarized below.

The experience of individuals engaged in placement of cardiac patients at work is similar to those reported by Franco,[32] Greer and associates,[6] and Hellerstein and Ford.[33] These investigators have observed the following:

1. About 70% to 75% of cardiac patients return to their former employment within a period of 3 to 6 months.
2. In approximately half the cases the emotional impact of the disease is as important or even more important than the disease itself. The emotional aspect of the disease is increased when the patient is unable to return to his former employment and is decreased when he is able to do so.
3. Of cardiac patients who returned to their former jobs, 54% had unblemished attendance records for a year or longer. Their absence frequency is only 25% of the general absentee rate.
4. Among individuals with cardiac classification iiA to iiiB, 30% were able to keep their regular jobs. The remaining 70% were on limited duty in their regular jobs or were transferred to positions requiring less energy cost.

There has been an attempt on an international basis to standardize the evaluation of the cardiac patient in order to offer him a prescription for exercise and return to gainful employment. It should be stressed at this juncture that an exercise prescription is only one phase of the rehabilitation program. The other phases have been discussed earlier in this chapter under preventive rehabilitation.

The testing for cardiac function and performance is usually for patients who have had acute episodes of ischemic heart disease and who have made a satisfactory clinical recovery. Testing procedures for rheumatic heart disease, hypertensive heart disease, congenital heart disease, and other types are also attempted, but the results are not as meaningful. The testing is in reality a measurement of what the patient's heart will permit him to do. The consensus regarding the testing procedure is a twofold test. The first part is a gradual increase in work load with rest intervals until the maximum work load is reached. The second phase is to increase the work load until electrocardiographic signs of cardiac ischemia occur as is evidenced by (1) changes in the repolarization process, that is, S-T segment depressions and reversal in polarity of the T wave, and (2) premature contractions of atrial or ventricular origin. Symptoms of angina pectoris also serve as an order to stop testing. Probably the best example of this type of testing is that of Kellermann and associates.[35] Kellermann's criteria for excluding patients are as follows:

1. Acute coronary insufficiency or intractable angina pectoris
2. Acute cardiac failure
3. Diastolic blood pressure above 115 mm. Hg
4. History of severe arrhythmia

Kellermann starts his spiroergometric tests for the diseased groups at 25 w./min. (150 kg. M./min.) and maintains the patient at this level of activity for 5 minutes' duration with a 5-minute rest period prior to increasing the work load by 25 w./min. This cycle is maintained until the above-mentioned signs and symptoms manifest themselves. Healthy subjects may also be tested by this technic. Kellermann starts healthy subjects at 50 w./min., then proceeds to 100 w./min. after a 5-minute rest period. He then increases the work load by 25 w./min. until the signs and symptoms signaling cessation manifest themselves or, as is usually the case, the patients stop because of dyspnea and tachycardia. It has been found by various investigators that the average poorly conditioned male 40 to 49 years of age, and the postmyocardial infarction patient can tolerate circa 110 w./min. or 660 kg. M./min., whereas the patient with angina pectoris is able to tolerate only circa 70 w./min. The experience of many investigators is that patients with recovered myocardial infarctions are able to return to their former employment in approximately 85% to 90% of the cases. The patients with angina pectoris (severe) return to work in about 50% to 60% of the cases.

The parameter of exercise as a prescription in the treatment of ischemic heart disease takes on various forms in various countries. The variations in technic are dependent on local customs and conditions. The Yugoslavian clinic at Opatija under the direction of Professor Plavšić[36] takes advantage of the natural terrain and seashore. The patients are classified according to their evaluation criteria and then placed into the category of exercises suitable for the individual. Brunner and associates[37] and Kellermann[35] in Israel employ a different type of exercise program that is uniquely suited for Israel. The patient who participates in the exercise phase of the rehabilitation program is permitted to work at various work levels in a garden and agriculture milieu. The duration of working time and the amount of work accomplished per day is gauged by the patient's progress and freedom from symptoms.[35] The rehabilitation center in India under the direction of K. K. Datey[38] employs various Western and special yoga exercises to recondition their cardiac patients. In addition to these exercises, there is an extensive program of graded occupational therapy such as basket weaving, chair making, furniture polishing, and broom making, which are both energy cost situations and useful employment experiences.

It is to be noted that the calisthenic exercises and the additional exercise programs, whether it be gardening, mountain climbing, or in the occupational therapy area, do not exceed 3 to 5 calories/min. This rate and degree of energy expenditure is equal to most and even exceeds some energy costs of certain occupations.

A CRITIQUE OF CARDIAC TESTING PROCEDURES

The main purposes of assessing physical work capacity are (1) to determine the degree and extent of reconditioning the patient can tolerate and (2) to determine whether the patient should return to his previous employment.

It should be emphasized that even after cardiac evaluation and specifically work classification, no guarantee may be given concerning the ability of the patient to work in his previous surroundings or the surroundings that have been suggested to him, based on the test results.

The various standardized cardiac testing procedures have been evaluated by Blackburn.[39] He has pointed out that although the tests are convenient, they do not represent a "steady-state" situation but are rather a single-state or nonsteady state procedure. Therefore, the extrapolation of the results to the steady state and continuous situations must be attempted with great care. It would appear from the data compiled by Blackburn that in the United States the treadmill at a 5-degree pitch and at 3 mph for 3 minutes is best suited for cardiac evaluation. The comparison was made with the standardized 3-minute, 9-inch, double two-step test; a 3-minute two-step test, 9 inches high, performed at a rate of 40 crossings within a 3-minute period by all subjects regardless of age, weight, or sex; the 3-minute single-step test, 12 inches high, at the rate of 20 ascents-descents per minute. The 3-minute, motor-driven treadmill walk at 3 mph at a 5% grade, and a 3-minute, bicycle ergometer test, seated at a pedal rate of 50 rpm against a resistance giving a load of 600 kpm/min. The conclusions stated by Blackburn are worth noting. In spite of these defects, submaximal exercise testing procedures should be attempted and employed not as the sole criterion for evaluation but as one of the parameters. Blackburn states: "Comparison between populations tested with these particular different types of commonly used, short, single-stage submaximal exercise procedures cannot be made with confidence either for distribution of the absolute work heart rate or probably for the frequency of post exercise ECG abnormalities, because the energy cost and circulatory responses to these tests differ significantly. It is possible, of course, to design tests with differing modes of imposing the stress for which the results are generally transposable. Better means of standardizing the physiological load is needed for valid comparisons."*

---

*From Blackburn, H., and others: Exercise tests. Comparison of the energy cost and heart rate response to five commonly used single-state, non-steady-state, submaximal work procedures, Medicine Sport 4:28, 1970.

REFERENCES

1. U.S. Department of Commerce, Bureau of the Census: Population estimates, Series P-25, No. 388, Mar. 14, 1968.
2. White, P. D., Rusk, H. A., Lee, P. R., and Williams, B.: Rehabilitation of the cardiovascular patient, New York, 1958, McGraw-Hill Book Co., Inc.
3. Felton, J. S., and Cole, R.: The high cost of heart disease, Circulation 27:957, 1963.
4. Diseases of the heart and blood vessels—Nomenclature and criteria for diagnosis, by the Criteria Committee of the New York Heart Association, ed. 6, Boston, 1964, Little, Brown & Co.
5. Gertler, M. M.: Bioenergetics in cardiac rehabilitation, Conn. Med. 25:697, 1961.
6. Greer, W. E. R.: In White, P. D., Rusk, H. A., Williams, B., and Lee, P. R.: Cardiovascular rehabilitation, New York, 1957, The Blakiston Division, McGraw-Hill Book Co., Inc.
7. Jones, A. M.: Personal communication to H. K. Hellerstein and A. B. Ford, J.A.M.A. 164:225, 1957.
8. Passmore, R., and Durnin, J. V.: Human energy expenditure, Physiol. Rev. 35:801, 1955.
9. Gertler, M. M., Garn, S. M., and White, P. D.: Young candidates for coronary heart disease, J.A.M.A. 147:621, 1951.
10. Gertler, M. M., Woodbury, M. A., Gottsch, L. G., White, P. D., and Rusk, H. A.: The candidate for coronary heart disease, J.A.M.A. 170:149, 1959.
11. Gertler, M. M., and White, P. D.: Coronary heart disease in young adults. A multidisciplinary study, Cambridge, Mass., 1954, Harvard University Press and The Commonwealth Fund.
12. Dawber, T. R., and Kannel, W. B.: Application of epidemiology of coronary heart disease in medical practice, Rev. Mod. Med. 30:123, 1962.
13. Taylor, H., and Keys, A.: Personal communication, 1962.
14. Brunner, D.: The influence of physical activity on the incidence and prognosis of ischemic heart disease. In Raab, W.: Prevention of ischemic heart disease; principles and practice, Springfield, Ill., 1966, Charles C Thomas, Publisher.
15. Eckstein, R. W.: Effect of exercise on growth of coronary arterial anastomoses subsequent to coronary arterial narrowing in dogs. Paper read before the Twenty-ninth Scientific Sessions, American Heart Association, Cincinnati, Oct. 28, 1956.
16. Ruegsegger, P., Nydick, I., Hutter, R. C., Freiman, A. H., Bang, N. U., Clifton, E. E., and LaDue, J. S.: Fibrinolytic (plasmin) therapy of experimental coronary thrombi with alteration of the evolution of myocardial infarction, Circulation 19:7, 1959.
17. Kobernick, S. D., Niwayma, G., and Zuchlewski, A. C.: The effect of physical activity on cholesterol atherosclerosis in rabbits, Proc. Soc. Exp. Biol. Med. 96:623, 1957.
18. Benton, J. G., and Rusk, H. A.: The patient with cardiovascular disease and rehabilitation: the third phase of medical care, Circulation 8:¹17, 1953.
19. Dumont, A. E., Clauss, R. H., Reed, G. E., and Tice, D. A.: Lymph drainage in patients with congestive heart failure: comparison with findings in hepatic cirrhosis, New Eng. J. Med. 269:949, 1963.
20. Gertler, M. M.: Some biochemical aspects of experimentally produced congestive heart failure, D.Sc. Med. thesis, New York University, Postgraduate Medical School, April, 1958.
21. Bing, R. J., and Daley, R.: Behavior of the myocardium in health and disease as studied by coronary sinus catheterization, Amer. J. Med. 10:711, 1951.
22. Bruce, R. A.: Evaluation of functional capacity and exercise tolerance of cardiac patients, Mod. Conc. Cardiovasc. Dis. 25:321, 1956.
23. Kottke, F. J.: The relationship of muscular work to cardiac work, Pardue Farm-Cardiac Seminar, paper no. 8, Sept., 1958.
24. Hellerstein, H. K., and Ford, A. B.: Energy cost of the Master two-step test, J.A.M.A. 164:1868, 1957.
25. Fulton, J. F.: Textbook of physiology, Philadelphia, 1955, W. B. Saunders Co.
26. Asmussen, E., and Nielsen, M.: Cardiac output during muscular work and its regulation, Physiol. Rev. 35:778, 1955.

27. White, A., Handler, P., Smith, E., and Stetten, DeW.: Principles of biochemistry, ed. 2, New York, 1959, McGraw-Hill Book Co., Inc.
28. Plaut, G. W. E., and Gertler, M. M.: Oxidative phosphorylation studies in normal and experimentally produced congestive heart failure in guinea pigs: a comparison, Ann. N. Y. Acad. Sci. 72:515, 1959.
29. Gertler, M. M.: Differences in efficiency of energy transfer in mitochondrial systems derived from normal and failing hearts, Proc. Soc. Exp. Biol. Med. 106:109, 1961.
30. Patterson, S. W., Piper, H., and Starling, E. H.: The regulation of the heart beat, J. Physiol. 48:465, 1914.
31. Messer, J. V., Wagman, R. J., Levine, H. J., Neill, W. A., Krasnow, N., and Gorlin, R.: Patterns of human myocardial oxygen extraction during rest and exercise, J. Clin. Invest. 41:725, 1962.
32. Franco, S. C.: The cardiac can work, Industri. Med. 23:315, 1954.
33. Hellerstein, H. K., and Ford, A. B.: Rehabilitation of the cardiac patient, J.A.M.A. 164: 225, 1957.
34. Lee, P. R., and others: An evaluation of rehabilitation of patients with hemiparesis or hemiplegia due to cerebral vascular disease, Rehabilitation monograph 15, New York, 1958, Institute of Rehabilitation Medicine, New York University Medical Center.
35. Kellermann, J. J., Modan, B., Feldman, S., and Kariv, I.: Evaluation of physical work capacity in coronary patients after myocardial infarction who returned to work with and without a medically directed reconditioning program, Medicine Sport 4:148, 1970.
36. Plavšić, C.: In Proceedings of The First International Biennial Conference on Cardiac Rehabilitation, Dubrovnik, Yugoslavia, March 26-28, 1969, Dubrovnik, Yugoslavia.
37. Mitrani, Y., Karplus, H., and Brunner, D.: Coronary atherosclerosis in cases of traumatic death. The influence of physical occupational activity on the development of coronary narrowing, Medicine Sport 4:241, 1970.
38. Datey, K. K., Deshmukh, S. N., Dalvi, C. P., and Vinekar, S. L.: "Shavasan"—a yogic exercise in the management of hypertension, Angiology 20:325, 1969.
39. Blackburn, H., Winckler, G., Vilandré, J., Hodgson, J., and Taylor, H. L.: Exercise tests. Comparison of the energy cost and heart rate response to five commonly used single-stage, non-steady-state, submaximal work procedures, Medicine Sport 4:28, 1970.

CHAPTER 28

REHABILITATION OF PATIENT WITH PERIPHERAL VASCULAR DISEASE

The term peripheral vascular diseases, as referred to in this discussion, will encompass diseases of the arteries, veins, and lymph vessels in the extremities. The disease processes described will include not only pathologic conditions within the confines of these vessels but also many conditions due to reflex disturbances in these vessels secondary to sympathetic, parasympathetic, and spinal cord influences. It should be emphasized that this section is not so complete as it might be but will include those diseases in which rehabilitation technics are helpful.

It is quite apparent that the entire complexion of care of obliterative arterial diseases has been radically altered by the success of direct definitive surgical treatment.

The study of peripheral vascular diseases requires essentially the same technics that one generally employs in internal medicine: history, physical examination, and special evaluation technics. There are, however, certain basic aspects of this group of diseases that require special emphasis in the examination and that may be elicited only by a sound history. It is therefore most imperative that a thorough knowledge of the classification of these diseases be known.

EXAMINATION

The physical examination in peripheral vascular disease must be thorough and general prior to concentration on the affected area. Chart 22 depicts, in abbreviated form, the extent of the examination to be carried out. The importance of obtaining a history of trauma, diabetes, and previous venous thrombotic or cardiac diseases is self-evident. Certainly the knowledge that a malignancy with metastases exists in a patient with recurrent venous thrombosis that does not respond to anticoagulants is

Text continued on p. 602.

CHART 22

PERIPHERAL VASCULAR CLINIC

Patient_____ Address_____ Sex_____

Age_____ Telephone No._____ Race_____

Referred by_____ Occupation_____ S.M.W.D._____

Date_____ Place of birth_____ Hospitalization_____

PAST HISTORY

Medical_____

Surgery_____

Obstetric_____

Vascular (thrombosis, ulcers, cold, etc.)_____

PRESENT HISTORY

Chief complaint_____

  Onset, progress, treatment received_____

Symptom and history—review

  1. Claudication_____

  2. Pain_____

  3. Swelling_____

  4. Paresthesias_____

  5. Reactions to environmental temperature_____

  6. Use of elastic support_____

  7. Smoking, alcohol consumption, and allergies_____

  8. Medications_____

CHART 22, cont'd

FAMILY HISTORY

_____

EXAMINATION: Weight_____ Blood pressure_____ Urine_____ Blood_____

1. E N T _____

2. Heart and lungs_____

3. Eye grounds_____

4. Temperature of extremities _____

5. Tenderness_____

6. Discolorations, scars, ulcers_____

_____

7. Trophic changes_____

8. Fungus infection_____

9. Edema_____

10. Veins_____

_____

_____

11. Arterial palpation_____

12. Orthopedic status_____

_____

_____

_____

13. Neurologic_____

_____

_____

_____

_____

Continued.

CHART 22, cont'd

	RIGHT	LEFT
1. Homan's sign _____		
2. Elevation		
Ischemia of hands _____	_____	_____
Ischemia of feet _____	_____	_____
3. Dependency _____	_____	_____
Venous filling time _____	_____	_____
Rubor of hands _____	_____	_____
Rubor of feet _____	_____	_____
Pallor of hands _____	_____	_____
Pallor of feet _____	_____	_____

4. Measurements (in inches):

	RIGHT	LEFT
Foot _____	_____	_____
Ankle (inches from L B P) _____	_____	_____
Calf (inches from L B P) _____	_____	_____
Knee (midpatella) _____	_____	_____
Thigh (inches above U B P) _____	_____	_____
Hand _____	_____	_____
Wrist _____	_____	_____
Forearm (inches below olecranon)	_____	_____
Upper arm (inches above olecranon)	_____	_____

5. Oscillometric readings:

	RIGHT	LEFT
Dorsalis pedis _____	_____	_____
Tibialis posterior _____	_____	_____
Popliteal _____	_____	_____
Femoral _____	_____	_____
Hand _____	_____	_____
Wrist _____	_____	_____
Forearm _____	_____	_____
Upper arm _____	_____	_____

6. Thermometric readings (room temperature):

	RIGHT	LEFT
First toe _____	_____	_____
Second toe _____	_____	_____
Third toe _____	_____	_____
Fourth toe _____	_____	_____
Fifth toe _____	_____	_____
Sole _____	_____	_____
Dorsum _____	_____	_____
Calf _____	_____	_____
Thigh _____	_____	_____
Hand _____	_____	_____
Wrist _____	_____	_____
Forearm _____	_____	_____
Upper arm _____	_____	_____

CHART 22, cont'd

OTHER FINDINGS

_____

_____

_____

_____

_____

_____

_____

DIAGNOSIS, COMMENT, RECOMMENDATIONS

_____

_____

_____

_____

_____

_____

_____

_____

_____

_____

_____

_____

_____

_____

_____

a potent factor governing further therapeutic endeavor. The physical findings of a mitral valvular lesion associated with auricular fibrillation in a patient with evidence of acute arterial occlusion may be of paramount importance in establishing the etiologic basis of the presenting symptoms.

Much information may be gained for the specific diagnosis of peripheral vascular diseases and for the clinical evaluation of the condition by means of tests designed to measure some physiologic function or physiologic capacity. In addition, specific measurements will enable one to evaluate objectively the response to therapy Accordingly, the use of Mays' test for intermittent claudication, oscillometric readings, temperature recordings, and various tests for vasomotor stability should be employed when indicated. Angiography and venography are frequently used; infrared photography may be employed in unusual instances.

The treatment of peripheral vascular disease extends far beyond the confines of rehabilitation medicine, but care for these disorders may be aided with the use of certain technics.

TREATMENT

The scope of treatment of peripheral vascular disorders within the discipline of rehabilitation medicine may be outlined as follows:

1. Thermotherapy
   a. Hyperthermia—direct and indirect heating
   b. Hypothermia
   c. Alternating temperatures—contrast baths
2. Electrotherapy
   a. Muscle stimulation
   b. Iontophoresis
3. Mechanotherapy
   a. Intermittent venous occlusion
   b. Pressure-suction boot
   c. Oscillating bed
   d. Vasopneumatic compression
   e. Syncardial massage
   f. Massage
4. Therapeutic exercises
   a. Buerger's exercises
   b. Therapeutic walking
5. Prophylactic measures
   a. General
   b. Footgear
6. Comprehensive rehabilitation measures

Thermotherapy

Hyperthermia. The inducement of repeated episodes of maximal vasodilatation for relief of angiospasm and for promotion of collateralization is an established objective of treatment in chronic arterial disease and, to a somewhat lesser degree, in venous thrombosis. The effectiveness of the procedures employed, however, depends on the degree of decompensation of the arterial blood flow. In overwhelming arterial embolization and in far-advanced thrombotic disease, therefore, these procedures may be of no avail.

Direct heat. Direct heating of limbs is the most efficient means of effecting vasodilatation. The local rise of temperature results in an increase of cellular activity with an incident increase of concentration of acid metabolites and histamine-like substances. These chemical agents are potent vasodilators. In addition, the rise of the temperature of the blood stimulates medullary centers to bring about reflex vasodilation; increase of the temperature and blood flow in an extremity by heating of some remote part of the body is based on this principle of reflex response.

Direct heat may be applied by use of radiant heat lamps, conduction heating with electric pads, hot-water bottles and fomentation, thermostatically controlled heat cradles, and the conversion heat of diathermy and microwave.

The employment of local heat always involves assumption of a calculated risk. The effective heating of the tissues exposed is at all times greater than the metabolic effect so that, with even moderate heating, there is great danger of burns. Maintenance of safe surface temperature with application of local heat depends on the heat-absorbing capacity of an unimpeded blood flow. Occlusive arterial disease excludes this function of the vascular system and the quantity of heat is permitted to accumulate to the critical point of tissue destruction. In addition to the inadequate dissipation of heat, the concomitant rise of local metabolic processes may be deleterious by compounding the oxygen requirements of already anoxic tissue, as well as by increasing the local accumulation of metabolic end products.

Indirect heat. Application of indirect heating evokes a generalized reflex vasodilation, the extent of which is dependent on the degree of vasoconstriction and arterial blood flow decompensation. An application of the procedure may be exemplified by use of a heating pad to the abdomen for 20 or 30 minutes when desirous of induction of vasodilatation in the lower extremities. The method is safe, usually efficacious, and easily carried out. Reflex heating technics are useful in relieving the secondary vasospasm that is so frequent an accompaniment of venous thrombotic disease.

Hypothermia. It is only recently that general body hypothermia or "artificial hibernation" has gained well-deserved recognition in its use in cardiovascular and neurologic surgery. Its use in less dramatic medical conditions has not as yet been fully exploited.

In acute arterial occlusion the application of hypothermic measures can be readily supported on the theoretic basis of equating tissue metabolic requirements with the diminished blood supply. Practically, however, this therapeutic approach has not been carfeully defined and is deserving of further study. It is at least universally accepted that the application of heat is unphysiologic and most ill advised.

Actual refrigeration of a limb has proved to be extremely useful in securing control of a situation wherein irreversible and extensive ischemic changes dictate ablation, but where some other medical consideration mitigates against proceeding immediately. Packing of the limb distal to the point of contemplated amputation is usually followed by rapid improvement of the patient, subsidence of toxic symptoms, drop of temperature, and abatement of pain. Distal icing is not associated with disturbance of wound-edge healing and secondary wound infection.

Hypothermic studies in the treatment of local tumors have established that cold is effective in diminishing edema and effusion and in relieving pain. These responses appear to be operative in the use of ice bags applied to the calf and thigh by some physicians in the treatment of patients with thrombophlebitis and venous thrombosis. Inflammation, induration, edema, and pain appear to be rapidly controlled, so that cold appears to be much more efficacious than hot fomentations. An occasional patient treated in this manner may find the cold intolerable so that its application must be terminated. Such an individual, however, is the one who manifests unusual degrees of secondary vasospasm that responds best to paravertebral sympathetic nervous system blocks.

Alternating temperatures. The use of contrast water baths, in which the patient alternately submerges his feet in warm and cold water, is now employed to little advantage in the treatment of vascular disorders. With arterial problems, for example, the constrictor of the immersion into cold far outweighs the supposed beneficial effects of alternating the caliber of the vessels.

In the patient allergic to cold, however, it is sometimes beneficial to effect gradual desensitization by daily treatments with immersion of the affected extremity into water that is progressively cooled.

Electrotherapy

Muscle stimulation. Stimulation of muscle with electric current finds no place in the treatment of vascular disorders. Some years ago, however, an electrostimulator designed to prevent venous thrombosis following surgery had been introduced. In this approach the calves and thighs of patients on the operating table were subjected to rhythmic contractions in an attempt to improve venous blood flow and prevent the static influences favoring venous thrombosis. The apparatus has not found widespread application.

Iontophoresis. Ion transfer with histamine and methacholine (Mecholyl), when carried out with due precautions, is a safe, effective means of obtaining local vasodilatation in both arterial and venous disease. Ulcerations, indolent to other forms of treatment, may exhibit remarkable response to direct iontophoresis with methacholine. This type of treatment appears to be particularly helpful in the care of patients with vasospastic disorders, such as Raynaud's disease, with ulcerations of the distal ends of the digits.

Mechanotherapy

In general, mechanical modalities have proved of little value in the treatment of peripheral vascular diseases. Briefly they may be reviewed as follows.

Intermittent venous occlusion. In this procedure occluding cuffs are applied to the extremity under a specific pressure for a fixed interval; the pressure is then released for a short relaxation period. The physiologic basis of the treatment is considered to be an application of the observation that venous occlusion is followed by a period of reactive hyperemia. Several instruments are available in which alternating periods of venous congestion and relaxation are intermittently produced and for which sometimes excessive claims of beneficial results in treatment of ischemic arterial disease have been made. In general, however, observed results are not impressive and have not withstood careful scrutiny of blood flow changes with plethysmographic methods. The recommended course of therapy is lengthy, and the good clinical results reported do not take cognizance of the natural forces building up collateral channels.

Pressure-suction boot. The pavex boot, another example of passive mechanical exercise, employs the application of intermittent suction and pressure to the involved extremity, which is enclosed in a treatment chamber. Although early reports noted rise in temperature and beneficial effects on the ulcerations and ischemic neuritis of patients with occlusive disorders, this procedure has no real proved value and for practical purposes has been relegated to therapeutic antiquity.

Oscillating bed. For practical purposes the oscillating bed may be described as a device that passively administers Buerger's exercises to the recumbent patient. Again, the procedure is far from proved in therapeutic value but appears to be of some benefit in relieving ischemic pain. The continuous oscillating movement of the bed does at least seem to have some sedative effect and also helps control the severe edema frequently seen in patients with arterial disease who maintain the limb in a constant position of dependency in an attempt to obtain relief from discomfort. Reports concerning rise of surface temperature of the limb after use of the rocking bed have been at great variance.

Vasopneumatic compression. The apparatus for this procedure consists of a series of fourteen rubber cuffs that are progressively inflated in either centrifugal or centripetal fashion in an attempt to effect a pressure wave traveling toward the heart or peripherally toward the distal end of the extremity. Cuff pressure and compression rate are controlled. The apparatus has been employed in centrifugal fashion in patients with arterial insufficiency and in a centripetal direction in the treatment of patients with edema resulting from chronic venous insufficiency. A review of results reported lends some support to the employment of the apparatus in treatment designed to reduce edema secondary to chronic lymphatic and venous obstruction.

Syncardial massage. The instrument employed in this procedure represents another mechanical device for the treatment of vascular disorders. In application of the procedure the ventricular complex of the electrocardiogram is detected from electrodes applied to both arms. From the involved extremity, to which an inflatable cuff has been applied, an arterial pressure curve is picked up. In an attempt to augment peripheral blood flow, a cuff, complete with an electronic delay, is assembled on the patient. The time at which the cuff inflates is related to the ventricular complex, as well as the descending limb of the arterial pulse curve. Although several investigators who used such objective means as pressure study in vessels distal to the site of cuff application and plethysmography have reported little physiologic effect, other investigators in the United States and Europe have rendered enthusiastic empirical reports.

Massage. Massage is of little or no value in arterial disease, although light stroking or sedative massage brings about a reflex vasodilatation and increased cutaneous blood flow. Heavier massage maneuvers should not be employed, since they may represent a form of repeated small trauma of sufficient additive magnitude to precipitate destructive skin changes. It may be noted that employment of vibratory devices falls into the same category, even though thermometric and radiocative sodium clearance increases have been demonstrated in the local area of application.

Because of the danger of dislodgment of loosely adherent thrombi, with secondary embolic complications, massage is strictly interdicted in patients with acute venous thrombosis. However, the late complications of this disease (indolent edema, induration, eczema, and ulceration) may be benefited by the judicious application of massage as part of an overall plan to combat the effects of venous insufficiency. In 1939 Disgaard of Denmark introduced an ambulatory treatment for patients with indurated legs, which combines a program of elevation, massage, bandaging, and exercise. Results reported in the United States and in Britain lend support to the efficacy of the program and indicate the desirability of more extensive application.

Therapeutic exercises

Buerger's exercises. In these postural exercises the limbs are elevated until blanched and then placed in a dependent position until beginning rubor. An interval of rest follows. This alternating filling and emptying of the vessels theoretically increases the arterial blood flow. The procedure is widely employed but is of highly questionable value, especially in the dubious light cast by carefully controlled studies on clearance of radioactive sodium from the calves of patients performing this exercise.

Therapeutic walking. Therapeutic exercises in the form of controlled walking represent a most physiologic approach to treatment of arterial occlusive disease, provided that trophic lesions are not present. The latter qualification itself may eventually have to be altered in the light of some recent favorable report on the use of exercise as a means of promoting collateral circulation in patients with limited gangrenous lesions. Walking places a physiologic demand on the muscles of the extremities that can be accommodated only by the appearance of collateral blood channels. The patient with intermittent claudication should be encouraged to walk at a slower but regular cadence, up to the point of pain. It does not appear feasible to force walking beyond this point, since the vasoconstriction incident to increased pain is a detrimental factor. Rather, the patient should be instructed to accept the rest period required for subsidence of discomfort and then to resume ambulation. Collateral studies of radioactive sodium clearance from the calf musculature, referred to in regard to Buerger's exercises, were also carried out following periods of active use of the limb. The preponderance of evidence points to greater clearance and, therefore, greater blood flow following ambulation and active exercise.

Prophylactic measures

General. The prophylactic measures related to the care of patients with arterial diseases should be considered carefully, since approximately 30% of patients reporting to the department with potentially serious conditions have self-inflicted lesions caused by improper shoeing, misapplication of medications, injudicious podiatric care, and generally poor hygienic measures. Enumeration of the precautions to be observed by such patients may be found in one form or another in the various texts on vascular disease, but because of their importance, they bear repetition. Prophylactic measures advised at the Institute of Rehabilitation Medicine are given in Chart 23. Because of the high incidence of trophic lesions initiated by the use of improper footgear, it is pertinent to discuss this measure in somewhat greater detail.

Footgear. Footgear, by definition, includes all coverings of the foot, both hose and shoes.

Hose. The arteriosclerotic patient should use lisle hose, fitted two or three sizes larger than the shoe and manufactured without constricting elastic tops. Lisle is porous and absorbent. Hose manufactured with squared-off toes are most desirable, to avoid constriction of the terminal portion of the digits and to avoid the tendency toward ingrown nail growth. Stretch hose and nylon do not meet these criteria and should not be worn.

Shoes. There is no compromise for comfort, so that shoes must fit properly im-

CHART 23

PERIPHERAL VASCULAR SERVICE

INSTITUTE OF REHABILITATION MEDICINE
New York University Medical Center

CARE OF THE FEET AND GENERAL INSTRUCTIONS

1. Wash feet each night with face soap and warm water; dry gently by patting with clean soft cloth

2. Apply 70 percent rubbing alcohol and allow feet to dry thoroughly; then apply liberal amount of petrolatum, toilet lanolin, or coconut oil and gently massage skin of feet

3. Always keep feet warm; wear woolen socks or wool-lined shoes in winter and cotton socks in warm weather; wear clean pair of socks each day

4. Wear loose-fitting bedsocks at night; apply hot-water bottle or electric pad to abdomen; never put either of these directly to feet or legs

5. Wear shoes of soft leather without box toes; be particularly careful not too tight

6. Cut toenails straight across and only when in very good light and only after feet have been cleansed thoroughly

7. Do not cut corns or calluses; see podiatrist

8. Do not wear circular garters

9. Do not sit with legs crossed

10. Overlapping toes and excessive perspiration of toes should be corrected by inserting lamb's wool between them

11. Never use tincture of iodine, Lysol, cresol, carbolic acid, or other strong antiseptic drugs on feet

12. Call doctor's attention to appearance of troublesome corns, ingrowing toenails, bunions, or calluses; also sores, rashes, or blisters on feet or legs

13. Eat plenty of green vegetables and fruits in a well-balanced diet, unless ordered to follow a special diet

14. Do not use tobacco in any form

mediately; esthetics are secondary. Shoe size is determined by a fit that permits enough room beyond the toes (approximately 1 inch beyond the great toe) on both sides to allow for spread on weight bearing and room over the dorsum to avoid injury by pressure and abrasion. It seems most practical to fit shoes in the early evening to allow for the even slight edema that most vascular patients exhibit. The sole of the shoe should be made of leather rather than rubber or a composition material, since the latter materials are poor heat conductors and increase perspiring.

New shoes should not be worn for more than from 1 to 3 hours a day for the first week. The preferable style is an oxford manufactured from either kangaroo or Vici kid leathers, both of which are supple and porous, permitting ventilation and avoidance of hyperhidrosis. Patients with hallux valgus should be fitted with bunion-last shoes that are designed with a specific pocket to accommodate the deformity of the toes.

The repair of shoes of these patients is also quite important, since reconstruction may alter the fit. Resoling, for example, ordinarily results in narrowing of the shoe because the leather is drawn in when the new sole is stitched on. Ideal conditions exist when the shoe is returned to the factory and repaired on the original last.

Comprehensive rehabilitation measures

The principles of rehabilitation of the patient with vascular disease vary only in minutiae from the general concept of care of patients with disabilities.

Arteriosclerotic patients with compensated arterial blood supplies usually are able to pursue a normal occupation limited in major aspect to avoidance of exposure to conditions precipitating intermittent claudication or trophic disturbance. For example, a patient employed as a waiter in a busy restaurant required vocational reexploration and counseling because of continuous onset of intermittent claudication. A young police officer with thromboangiitis obliterans required transfer to an indoor assignment when it was noted that he suffered onset of extreme vasoconstrictor phenomena when exposed to inclement weather. This type of patient, as well as one with chronic venous insufficiency and vasospastic diseases, represents primarily vocational problems so that normal gainful employment may be maintained or restored within the restrictions of the physical defect. The amputee, however, has a special problem that has been discussed previously. The psychologic and social adjustment of the latter patient to mutilating surgery and the fitting and adaptation to the use of a prosthesis are at least as important in rehabilitation care as are vocational considerations.

REFERENCES

1. Allen, E. V., Barker, N. W., and Hines, E. A.: Peripheral vascular diseases, Philadelphia, 1955, W. B. Saunders Co.
2. Lewis, T.: Vascular disorders of the limbs, New York, 1936, The Macmillan Co.
3. Wright, I. S.: Vascular diseases in clinical practice, Chicago, 1952, Year Book Publishers, Inc.
4. Kramer, D. W.: Peripheral vascular diseases, Philadelphia, 1948, F. A. Davis Co.
5. Samuels, S.: Diagnosis and treatment of vascular disorders, Baltimore, 1956, The Williams & Wilkins Co.
6. Wakim, K. G., and others: The influence of syncardial massage on the peripheral circulation, Arch. Phys. Med. 37:538, 1956.

7. Freeman, N. E.: Influence of temperature on the development of gangrene in peripheral vascular diseases, Arch. Surg. **40:**326, 1940.
8. Wisham, L. H., Abramson, A. A., and Ebel, A.: Value of exercise in peripheral arterial disease, J.A.M.A. **153:**10, 1953.
9. Fay, T., and Smith, L. W.: Temperature factors in cancer and embryonal cell growth, J.A.M.A. **113:**653, 1939.

CHAPTER 29

REHABILITATION OF
PATIENT WITH STROKE

T he general principles underlying the initial evaluation and prescription of a rehabilitation program for the stroke patient are
somewhat different from those procedures outlined previously for other handicapped
persons. The hemiplegia following a cerebral vascular accident (CVA) should not
be considered as a specific clinical entity but rather as a functional deficit involving
one side of the body. The localization and extent of the anatomic defect (cerebral
infarction) produced by the stroke will result in variations in the hemiplegic pattern.
Therefore, there is no standard program for the rehabilitation of hemiplegic patients.
Instead, the prescription of treatment must be based on the specific functional deficit
in the patient.

One must differentiate between the hemiplegia occurring in a child with congenital cerebral palsy and that in an adult following a cerebral vascular accident.
In the child there has been no previous learning process, whereas this experience
has already been acquired by the adult patient with a stroke.

GENERAL PRINCIPLES

The rehabilitation program for the hemiplegic patient is geared to the teaching
and retraining of the individual so that he may reacquire and "relearn" established
skills on an intellectual and functional level consistent with the limitations imposed
by the cerebral dysfunction. The hemiplegic patient performs during this relearning
process in a pattern that is directly related to the extent of the disability.[1] In addition, he learns to substitute for the deficit by acquiring alternate methods of function. Lastly, the cerebral changes may undergo spontaneous reversal. As the pathologic process regresses, the patient is capable of regaining a functional level consistent
with the degree of remission.

At times it is difficult to distinguish between the hemiplegic manifestations and
insidious secondary musculoskeletal and neuromuscular complications that obscure

and often retard recovery of function.[2] Prompt recognition and treatment of these problems may contribute toward a more complete restoration of function.

In evaluating the rehabilitation potential of the hemiplegic patient, the *central* as well as the *extremity* manifestations must be considered.

1. Central manifestations
   a. Psychologic disturbances
   b. Emotional instability
   c. Speech and writing problems
   d. Visual defects
   e. Hearing loss
   f. Incontinence
   g. Pain syndrome (thalamic)
2. Extremity manifestations
   a. Motor function
   b. Spasticity
   c. Rigidity
   d. Ataxia
   e. Clonus
   f. Astereognosis
   g. Other sensory changes
   h. Contractures
   i. Pain and other peripheral limb syndromes

The various psychologic and emotional problems and their evaluation have been described elsewhere in this text. They are essentially the same as for other brain-damaged individuals. Profound, often unrecognized disturbances in spatial perception, particularly in patients with a left hemiplegia, may prevent any effective rehabilitation despite relatively good motor function.[3, 4] Suitable placement of such patients should be arranged after an adequate period of observation rather than subjecting them and staff personnel to a protracted, unrewarding rehabilitation program.[5]

The same is true for the stroke patient with a severe organic brain syndrome with memory lapses and inability to respond effectively to the training program.[6] This is particularly true in patients with parietal lobe involvement. On the other hand, the depression and lack of motivation frequently encountered during the early convalescent period should not be confused with an organic brain syndrome. The elderly patient with cerebral arteriosclerosis, abandoned by the family, will often respond favorably when exposed to the stimulating social contact in a rehabilitation setting.

The evaluation of the speech and writing problems of the hemiplegic patient has been discussed in Chapters 5 and 14. In contrast to the relatively poor prognosis in the presence of perceptual defects, a patient with a severe aphasia may be successfully rehabilitated to the point of complete physical independence, despite the inability to communicate or understand (global aphasia). The agitated aphasic patient who is depressed because of inability to communicate is frequently considered to be psychotic, only to improve when his problem is accurately evaluated and managed.

Hemianopsia and blindness in one eye secondary to a retinal artery thrombosis are visual defects not included under perceptual difficulties. Other causes of visual impairment include cataract, double vision secondary to extraocular palsies, and central blindness. From the practical point of view, the management of these problems in terms of rehabilitation consists of teaching the patient substitute compensatory movements of the head to offset the specific visual defect. Improvement in specific areas of impaired function must be attributed to the resolution of the central lesion rather than to the specificity of the "treatment."

A severe hearing loss constitutes a serious handicap for the patient in a rehabilitation setting. It is often overlooked in the absence of an adequate medical history, and the lack of cooperation by the patient is attributed to poor motivation, organic brain syndrome, or even aphasia. The prescription of a suitable hearing aid may produce remarkable personality changes in the elderly stroke patient.

Urinary and fecal incontinence in the hemiplegic patient cannot be explained on the basis of a sphincter paralysis, such as encountered in spinal cord lesions. Persistent incontinence in the absence of urinary tract infection or obstruction is frequently a manifestation of an organic brain syndrome and indicates a poor prognosis for rehabilitation.

The so-called "thalamic syndrome" occurring in the hemiplegic patient is associated with personality disorders and emotional problems. The diagnosis should be made by exclusion after all other possible "peripheral" causes have been ruled out. A characteristic feature of the thalamic syndrome is the excruciating pain with a burning quality involving the hemiplegic side of the body. The pain is refractory to all analgesics.[7]

The motor function of the hemiplegic extremities can be readily evaluated in terms of degree of paralysis and joint mobility. However, it should not constitute the sole basis for determining the rehabilitation potential of the stroke patient. Relatively good muscle power, when demonstrated while the patient is in bed, does not necessarily indicate good functional performance. Conversely, patients with a flaccid lower extremity can be trained to walk with the aid of only a short leg brace.

Spasticity in the hemiplegic upper extremity is usually indicative of a poor prognosis for return of effective function when it occurs after the initial flaccid stage. In the lower extremity, spasticity may assist in stabilizing the knee joint but accentuate the hip and ankle dysfunction. The control of this problem is discussed elsewhere in this chapter.

Rigidity occurs in those patients with preexisting basal ganglion disease, such as parkinsonism, or those in whom the cerebral infarction or hemorrhage has been extensive, resulting in a more diffuse encephalopathy. The prognosis is less favorable when rigidity is present, since it prevents effective voluntary function. Careful supervision of such a patient is necessary to avoid mishaps.

Ataxia constitutes the most serious extremity defect and may prevent effective function despite relatively good motor power. It is particularly disabling when associated with perceptual difficulties.

Astereognosis and other defects in sensation are difficult to evaluate, particularly in the aphasic patient, and should not be overlooked. This is likely in the patient who has had several so-called "small strokes" involving both sides. In such instances the motor deficit may be relatively mild; yet the bilateral unrecognized sensory impairment can preclude any effective ambulation. In the upper extremity the astereog-

CHART 24

PATIENT PROFILE*

P. *Physical* condition including diseases of the viscera (cardiovascular, pulmonary, gastrointestinal, urologic, and endocrine) and cerebral disorders which are not enumerated in the lettered categories below.
   1. No gross abnormalities considering the age of the individual.
   2. Minor abnormalities not requiring frequent medical or nursing supervision.
   3. Moderately severe abnormalities requiring frequent medical or nursing supervision yet still permitting ambulation.
   4. Severe abnormalities requiring constant medical or nursing supervision confining individual to bed or wheelchair.
U. *Upper* extremities including shoulder girdle, cervical and upper dorsal spine.
   1. No gross abnormalities considering the age of the individual.
   2. Minor abnormalities with fairly good range of motion and function.
   3. Moderately severe abnormalities but permitting the performance of daily needs to a limited extent.
   4. Severe abnormalities requiring constant nursing care.
L. *Lower* extremities including the pelvis, lower dorsal and lumbosacral spine.
   1. No gross abnormalities considering the age of the individual.
   2. Minor abnormalities with fairly good range of motion and function.
   3. Moderately severe abnormalities permitting limited ambulation.
   4. Severe abnormalities confining the individual to bed or wheelchair.
S. *Sensory* components relating to speech, vision, and hearing.
   1. No gross abnormalities considering the age of the individual.
   2. Minor deviations insufficient to cause any appreciable functional impairment.
   3. Moderate deviations sufficient to cause appreciable functional impairment.
   4. Severe deviations causing complete loss of hearing, vision, or speech.
E. *Excretory* function, i.e., bowel and bladder control.
   1. Complete control.
   2. Occasional stress incontinence or nocturia.
   3. Periodic bowel and bladder incontinence or retention alternating with control.
   4. Total incontinence, either bowel or bladder.
S. *Mental* and *emotional status.*
   1. No deviations considering the age of the individual.
   2. Minor deviations in mood, temperament and personality not impairing environmental adjustment.
   3. Moderately severe variations requiring some supervision.
   4. Severe variations requiring complete supervision.

PROFILE

P	U	L	S	E	S

*From Moskowitz, E., and McCann, C.: Classification of disability in the chronically ill and aging, J. Chron. Dis. 5:342, 1957.

nosis may impair effective function and produce serious emotional problems in the frustrated patient who cannot understand why he is unable to use a limb that is mobile and grossly unimpaired.

Severe clonus in the ankle is usually associated with spasticity and may produce mass reflex action when triggered by external stimuli and "throw" the patient. It can usually be controlled by bracing, phenol block, or surgery.

Other peripheral limb syndromes can really be considered sequelae and will be discussed among the complications. They should be recognized and treated during the early convalescent period. When the disability resulting from these complications is first recognized in the rehabilitation center, it is often irreversible and may prevent effective restoration of function.

The summation of the various central and extremity manifestations provides a comprehensive evaluation of the hemiplegia. However, one must consider the entire physical and mental profile of the patient with the hemiplegia, rather than just the disability itself. Chart 24 illustrates a method of functional evaluation of the total patient.[8]

The profile will naturally change during the rehabilitation program and serve as an effective yardstick for measuring the progress or regression in the functional capabilities of the patient. The various components of the profile are interdependent so that any change in one area may have a profound effect on the other. For instance, dyspnea in the hemiplegic patient with congestive heart failure may preclude ambulation; conversely, ambulation may enable the patient to resume normal bowel and bladder function. In essence, the rehabilitation of the hemiplegic patient is a "functional readjustment" consistent with the limitations expressed in the profile, in contrast to a "muscle reeducation," as commonly prescribed for patients with lower motor neuron disorders.

MANAGEMENT

The early institution of the rehabilitation program should coincide with and be an effective part of the total management of the hemiplegic patient and should not be delayed until the medical condition is stabilized.[9] An exception must be made in those cases with subarachnoid bleeding and relatively mild paralytic involvement. The presence of nuchal rigidity in an anxious, restless patient with a flushed face, associated perhaps with alternating periods of confusion and clarity, is indicative of continued cortical irritation and bleeding. A persistent diastolic hypertension is frequently evident in such an apoplectiform episode. It is advisable to maintain such a patient at absolute bed rest until all the signs of bleeding have subsided.

Early rehabilitative measures, on the other hand, should be instituted in those patients with long-standing hypertensive cardiovascular disease without evidence of bleeding. Not infrequently, vigorous treatment with antihypertensive drugs produces a physiologic hypotension in these patients and prevents any effective mobilization. The severe lassitude induced by the drug therapy deters the patient from standing or even sitting in a chair. The hypertensive patient can fully participate in the rehabilitation program under appropriate medical supervision.

Similar problems may arise when there is evidence of myocardial insufficiency or recent infarction, particularly when noted on routine electrocardiograms. Practical experience has demonstrated that such ECG changes should be evaluated with considerable leeway, particularly in the elderly patient. One must consider the alternatives—to maintain the hemiplegic patient at absolute bed rest or to proceed with the rehabilitation program under close clinical observation. An episode of acute myocardial insufficiency rarely occurs during any form of physical or occupational therapy for the simple reason that these patients are observed closely and their physical activity graded to their tolerance. As a matter of fact, when a hemiplegic patient on a rehabilitation service does have another stroke or myocardial infarction, it usually occurs during the night or weekend when there is no scheduled treatment.[2]

The hemiplegic patient with chronic obstructive pulmonary disease is similarly treated. Extensive pulmonary function studies do not afford any practical answer to the question as to how much physical activity the patient can tolerate. By gradually increasing the physical activities in the rehabilitation program, the physician can observe the patient and plan the future program accordingly.

In controlling the diabetes of the patient after a stroke, one must consider the variability of the dietary intake in prescribing hypoglycemic drugs. This is particularly significant in patients with pseudobulbar palsy who cannot chew or swallow certain types of food. Other patients are unable to cut their food or feed themselves when their dominant hand is affected. This poses a serious problem during evenings and weekends when staff personnel in the hospital are not always available. Training in self-feeding should be undertaken at the bedside as soon as the patient is alert and responsive to prevent hypoglycemic episodes.

Dehydration is a frequent result of these feeding difficulties in the stroke patient. The problem is further complicated in restless elderly individuals by the prescription of sedatives and tranquilizers. This causes a reversal of their sleeping habits, making it impossible to carry on any effective rehabilitation activities. If the patient is unable to eat or drink and the routine medication is continued, further dehydration ensues with elevated blood urea nitrogen and higher drug concentrations in the blood. The patient becomes stuporous and the diagnosis of another cerebral vascular accident is made. Hydration of such a patient and the elimination of all sedation will produce a prompt and dramatic remission.

Malnutrition secondary to pseudobulbar palsy requires painstaking care, particularly in the patient who has had more than one stroke with protracted recovery. It is time-consuming and is beset with the hazards of aspiration. In selected cases the patient can best be managed with a gastrostomy rather than nasogastric tube feeding. If there is any return of the gag reflex, the patient can be started on small quantities of semisolid food as an oral feeding supplement; if successful, the gastrostomy can be gradually clamped and finally closed.

The dental care of the patient with a severe central facial or pseudobulbar palsy may pose serious problems, particularly if the individual has dentures. The imbalance and paralysis of the oropharyngeal muscles produces a malocclusion of the bite interfering with the chewing of food and the speech mechanism. The dentures are forgotten or discarded and subsequently cannot be tolerated when inserted after a lapse of weeks due to oral tissue changes.

During the early convalescent period, the emphasis is directed toward maintaining full range of motion in the involved limbs. Passive exercises are instituted to

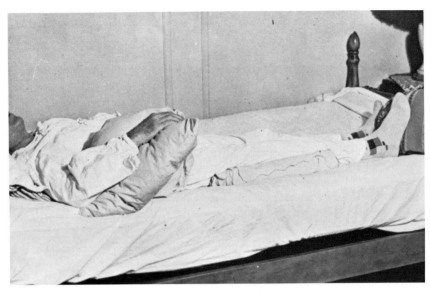

Fig. 29-1. Positioning of the hemiplegic patient during the early convalescent period. (From Moskowitz, E.: Rehabilitation in the home, Postgrad. Med. 23:71, 1958.)

prevent postural contractures.[10, 11] Since the flaccid upper extremity usually lies at the side of the patient, it is necessary to stress those joint motions that tend to overcome this malposition, as well as the effect of gravity. In the shoulder, abduction and external rotation are emphasized and the patient is instructed to duplicate this maneuver by raising the flaccid arm overhead with the uninvolved extremity. Flexion and extension in the elbow, as well as rotation in the forearm, are performed in the routine maneuver. Dorsiflexion of the wrist and extension of the thumb and fingers are emphasized. Proper positioning of the patient while in bed is illustrated in Fig. 29-1. The hand should be elevated to prevent edema. In the early phase a folded pillow is adequate to maintain the position of the hand and is preferred to a splint, which requires frequent reapplication and may only increase the edema in the fingers as a result of compression.

The paretic lower extremity tends to lie externally rotated, with flexion and adduction at the hip joint, especially exaggerated when the patient lies on his side. Similarly, the knee is flexed and the ankle is in an equinus (drop foot) position. This postural attitude is aggravated by the pressure of tight bed covers, as seen in Fig. 29-2. Passive exercises are prescribed to counteract these potential deformities; they consist of (1) abduction, extension, and internal rotation of the hip, (2) extension of the knee to prevent shortening of the hamstring muscles, and especially (3) dorsiflexion of the ankle to prevent shortening of the calf muscles. A footboard, padded with folded blankets, is used to support the drop foot and to relieve the pressure of the bed covers. In a hospital bed the foot support must be so positioned that the patient can be comfortably gatched for meals without being pulled away from the board. Pillows placed under the knee will produce a flexion contracture. The posterior surface of both heels should be inspected daily for early pressure areas.

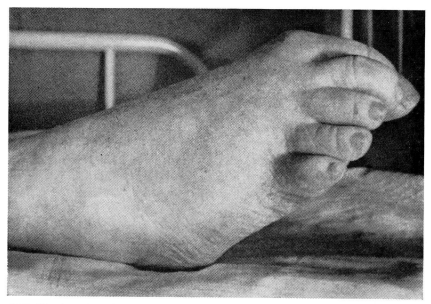

Fig. 29-2. Fixed equinus deformity resulting from malposition while in bed. (From Moskowitz, E.: Rehabilitation in the home, Postgrad. Med. **23:**71, 1958.)

Active exercises should be carried out on the unaffected side to prevent the disuse atrophy secondary to immobilization, as well as to prevent postural contractures. The circulation in both lower extremities should be closely observed. An overhead trapeze bar is invaluable in assisting the patient to perform a few simple exercises for the upper extremity as well as to make frequent changes of position in bed, thus helping to prevent decubiti.

The conscious hemiplegic patient without medical complications should be out of bed within 24 to 48 hours after a cerebral thrombosis. However, placing or lifting him into a wheelchair is in itself not rehabilitation unless the patient actively participates by assisting in the process. In effecting this transfer from the bed to the wheelchair, one must teach the patient to utilize the uninvolved side in standing and in supporting himself. The nursing staff and ancillary personnel rendering the day-by-day care to the patients require training and orientation in these simple procedures.

Similarly, the stroke patient should be taught and encouraged to feed and toilet himself with the uninvolved hand. Self-care activities must become part of routine bedside nursing care and be readily incorporated into the total medical management of the patient. Although the physical therapist initiates the training of the patient in the Activities of Daily Living, it is essential that there be effective communication with nursing and other hospital personnel so that they will encourage and permit the patient to perform those activities that he has relearned.

Immobilization in the wheelchair may produce sitting contractures in all four extremities. Frequent examination of the patient in the supine position should be carried out to detect early signs of malposition. This is particularly true in the hip and knee joints, in which there may be an underlying osteoarthritis.

As the patient improves, he gradually assumes his place in the organized re-habilitation activities as a member of a group. Participation in the group setting tends to motivate and stimulate the individual, in addition to giving him the first exposure to "the outside." He is encouraged to wear regular clothing and is taught to dress himself as part of getting ready for the "class." This promotes more regu-larity in patient care. Many stroke patients can be successfully rehabilitated in the community hospital when there is close cooperation among the various paramedical personnel. The more severely involved hemiplegic patient requires the comprehensive services of the rehabilitation center.

Further treatment in this specialized facility is based on the functional potential of the patient consistent with the extent of the hemiplegia, taking into consideration the various extremity and central manifestations previously discussed. For example, the patient with severe ataxia and incoordination requires an entirely different pre-scription for gait-training than does the one with a simple flaccid paralysis. The functional pattern may change and improve with the resolution of the cerebral edema. Ultimately, the patient is left with a residual disability that is irreversible and is taught substitute patterns of function to compensate for the peripheral defect.

Various neuromuscular facilitation technics have been described for the treat-ment of the stroke patient. Since the localization and extent of the cerebral infarc-tion is so varied and often ill-defined, it is difficult to match comparable patients for controlled studies of the value of any specific system of neuromuscular "thera-peutic exercises." Cerebral edema may play a major role in producing transient ischemia, yet cause the same degree of hemiplegia as an infarct. The prognosis is obviously better in the former patient, yet on initial examination they both have the same "peripheral" defect. Until well-documented, controlled objective studies are available, intensive and protracted treatment of the stroke patient with these technics does not seem justified at this time.

Assistive devices and braces are prescribed to help compensate for the functional deficit. For instance, the use of a short leg brace with a 90-degree posterior stop at the ankle is quite effective in controlling a drop foot. A sling is indicated in the patient with a flaccid, relaxed shoulder joint to prevent the upper extremity from dangling at the side and throwing the patient off balance. It also prevents excessive traction on the shoulder joint and stretching of the relaxed capsule. The sling should be discarded when the limb becomes spastic, as there is sufficient tension in the shoulder muscles to prevent any subluxation and the spastic biceps muscle prevents the lower arm from dangling. Protracted immobilization of the arm in the sling, particularly when the patient is not ambulatory, may only increase the tendency to an adduction–internal rotation contracture at the shoulder.

A simple platform splint, with the wrist in a neutral or a slightly flexed position, is sufficient for the hand. Any attempt at positioning the spastic hand in the func-tional attitude of dorsiflexion only elicits a stretch reflex in the spastic flexors, causing the fingers to contract into the classic hemiplegic attitude with the thumb adducted beneath them. A C bar may be added to the splint to abduct the thumb and release some of the tension in the flexor groups. Various plastic materials can be utilized, particularly if an occupational therapy department is available. Otherwise, plaster of paris molded to the hand is quite adequate. Edema of the fingers is avoided by careful application of the cuffs or elastic bandage used to support the hand splint.

The activities in the occupational therapy department are geared to the functional deficit in the upper extremity. In many instances, it is necessary to change the handedness of the patient when the dominant limb is affected. The prescription by the physician for treatment must be specific in this regard and is based on the prognosis. The patient is taught to utilize the hemiplegic upper extremity in an assistive capacity consistent with the return of function. The training of the disabled homemaker is discussed in Chapter 10.

Speech therapy is an integral part of the rehabilitation program (Chapter 14). Some aphasic patients can be taught to communicate in writing, provided there is no agraphia present. In such instances the occupational therapist is concerned with increasing the overall functional dexterity of the hand in preparation for the more complex writing activities, which are supervised by the speech therapist. There must be frequent exchange of information among the nursing and other paramedical personnel, physicians, and the speech therapist so that everyone is informed as to the extent of the speech and communication deficit. It is not uncommon for the patient to lose the ability to communicate in an acquired language yet retain this function in a native tongue that may not have been used for years. A careful language history may establish effective communication with a person who seems to be totally aphasic and who is quite frustrated in his efforts to be understood. At times it is difficult to distinguish between a rather severe organic brain syndrome and pure global aphasia. A hemiplegic individual in such cases should not be transferred to a psychiatric institution unless there has been an adequate attempt made to evaluate the reason for his emotional instability, which may be based entirely on his inability to communicate.

COMPLICATIONS IN THE HEMIPLEGIC PATIENT

Reflex sympathetic dystrophy. Many hemiplegic patients develop pain in the upper extremity. Pitting edema, cyanosis, and limitation of extension in the wrist and finger joints, associated with pain, are not uncommon and can be attributed to malpositioning and lack of therapy. These findings should not be confused with the shoulder-hand syndrome (reflex dystrophy), which occurs in about 5% of the stroke patients. Although both conditions may be associated with a painful shoulder, the dystrophic hand has the more characteristic appearance.[12] There is slight swelling, without any pitting edema. The skin temperature is increased and the hand has a pink glossy texture. There may be considerable atrophy present in the more-advanced cases. Instead of the usual flexion contractures, these patients are prone to develop extension deformities in the metacarpophalangeal joints (Fig. 29-3). X-ray examination of the hand in the advanced case reveals rather marked spotty demineralization, particularly in the carpal bones, in contrast to the homogenous osteoporosis seen in the uncomplicated case (Fig. 29-4). The pain may be so intense, particularly when associated with causalgia, that it prevents the administration of an effective exercise program. The patient becomes quite depressed and resentful. It is difficult to determine whether the emotional instability is an etiologic factor in the development of the dystrophy or is simply a sequela.

The pain and limitation of motion may obscure latent function in the hand.

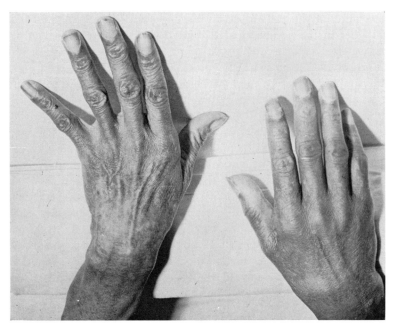

Fig. 29-3. Posthemiplegic reflex sympathetic dystrophy, right hand. Left hand is shown for comparison.

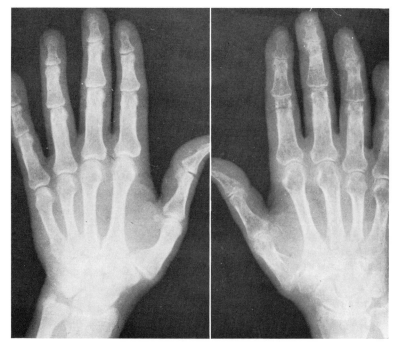

Fig. 29-4. Spotty demineralization of the right hand in posthemiplegic reflex sympathetic dystrophy. Left hand is shown for comparison. (From Moskowitz, E., and others: Posthemiplegic reflex sympathetic dystrophy, J.A.M.A. **167:**836, 1958.)

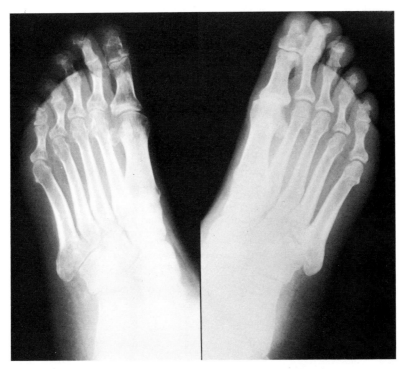

Fig. 29-5. Spotty demineralization of the left foot in posthemiplegic dystrophy. Normal right foot is shown for comparison.

In such instances the loss of function is due to the immobilization rather than to the hemiplegia. Early recognition of the dystrophy is therefore essential so that prompt treatment may be instituted to prevent progression of the disability to the atrophic stage with fixed, irreversible deformities.

Therapy consists of vigorous mobilization of the fingers, emphasizing flexion rather than extension, particularly in the metacarpophalangeal joints. Paraffin baths may be utilized prior to the passive and active-assistive exercises, provided the paraffin glove is molded with the fingers in maximum flexion. Stellate ganglion blocks should be employed if there is no progress after a few days of therapy or when the pain is so intense that it prevents any range-of-motion activities. It should be noted that the sympathetic blocks have no effect on the shoulder disability. Manipulation of the fingers should be carried out immediately following the blocks. The use of a platform splint, with the fingers in extension, is obviously contraindicated. The use of corticosteroids with a varying degree of success has been reported. The painful shoulder is discussed later in this chapter.

Dystrophic changes are seldom encountered in the lower extremity. This may be due to early weight-bearing activities that stimulate active motion in the limb. Progressive loss of motion in the foot and pain during ambulation may be indicative of an early reflex dystrophy. The changes in the foot are similar to those described in the hand. X-ray examination of the foot (Fig. 29-5) reveals the same spotty demineralization. Paravertebral blocks will afford relief of the symptoms and permit early resumption of ambulation.

Peripheral nerve lesions. Lower motor neuron lesions in the hemiplegic upper extremity may occur in patients who were unconscious after the onset of the cerebral vascular accident. This superimposed neurologic deficit has been attributed to a traction or pressure neuropathy incurred during the unconscious period.[13] The paralysis may be segmental, as in an Erb's palsy (Fig. 29-6), or follow a peripheral nerve pattern, in contrast to the limb type that is seen in hemiplegia. Sensory changes, when present, follow a similar pattern. Return of function in the uncomplicated hemiplegic upper extremity usually occurs initially in the shoulder and extends distally. Any deviation from this pattern may be the result of an extrinsic factor, such as a traction neuropathy. Electrodiagnostic testing is necessary to confirm the lower motor neuron involvement.

Prompt recognition and treatment of this usually reversible complication can

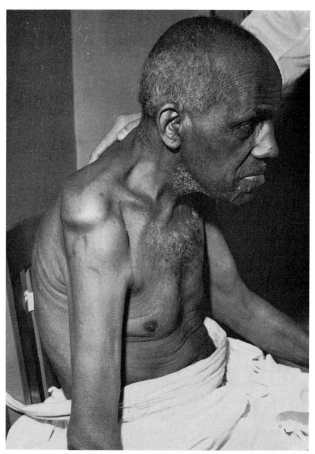

Fig. 29-6. Upper brachial plexus palsy (Erb type) in a hemiplegic upper extremity. (From Moskowitz, E., and Porter, J. I.: Peripheral nerve lesions in the upper extremity in hemiplegic patients, New Eng. J. Med. 269:776, 1963.)

prevent the development of superimposed contractures with loss of function that cannot be explained by the hemiparetic pattern. Passive exercises to maintain the range of motion in the involved segments must be carried out, as in other lower motor neuron disorders, and are followed by active exercises when there is return of function. In the upper brachial plexus type of lesion (Erb) the use of a sling may increase the tendency to shoulder contracture. Sling suspension in a wheelchair (Fig. 29-7) will overcome this disadvantage and provide a means of active-assistive exercises.

Prolonged pressure over the sciatic notch in a debilitated patient may cause a sciatic neuropathy producing a compound neurologic deficit with spasticity at the knee and flaccidity at the ankle level. Likewise, pressure over the head of the fibula will cause the classic peroneal nerve palsy with a dangle type foot. Flaccidity in the calf muscles and an absent ankle jerk is definite evidence of sciatic rather than peroneal nerve involvement. Lastly, bilateral neuropathies should not be overlooked in the diabetic patient.

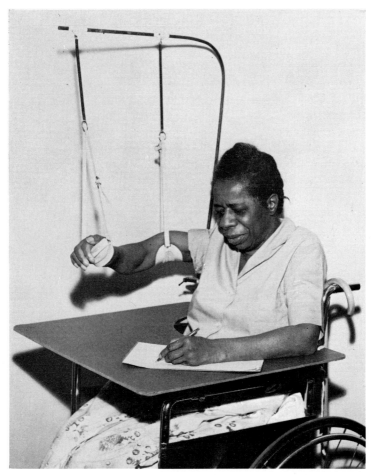

Fig. 29-7. Overhead sling suspension to support the flail shoulder of a hemiplegic patient with an associated traction neuropathy.

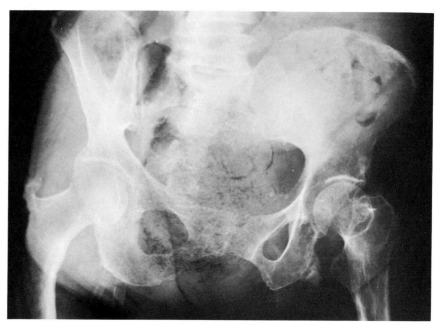

Fig. 29-8. Heterotopic calcification adjacent to the lesser trochanter of the left hip, associated with extensive osteoporosis.

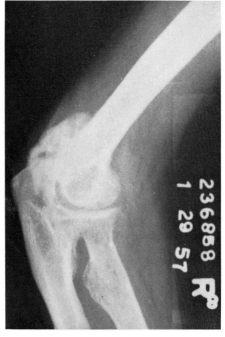

Fig. 29-9. Heterotopic "ossification" in the elbow of a hemiplegic patient. (From Moskowitz, E., and others: Posthemiplegic reflex sympathetic dystrophy, J.A.M.A. **167**:836, 1958.)

Heterotopic calcification. Extra-articular calcification has been described in other neurologic disorders and it may occur in hemiplegic patients.[14] It is seldom encountered in the patient who has been quite active and conscious during the early period following the stroke. Repeated insignificant trauma may play a role in the development of this progressive refractory complication. It may be associated with parietal lobe dysfunction causing the patient to neglect the limb, predisposing it to repeated injury.

The extra-articular calcification occurs about the lesser trochanter of the femur (Fig. 29-8), producing a severe adduction-flexion contracture of the hip. Similarly, it can produce a complete extra-articular bony ankylosis of the shoulder joint.

The extensor surface of the elbow just above the olecranon process may also be the site of involvement.[15] The elbow is held in an atypical attitude of relative extension rather than flexion, as commonly seen in the hemiplegic patient. Initially, the elbow is quite painful, warm, and slightly swollen, without any joint effusion. A small amorphous calcific deposit at the insertion of the triceps tendon, seen in the lateral x-ray film, may be the initial finding. In other patients this complication may not be recognized until the acute phase has subsided. The calcification may progress to complete ossification (Fig. 29-9).

Treatment of this complication is very unsatisfactory, particularly when it is not recognized. During the acute painful phase, it is advisable to immobilize the elbow in flexion and avoid all motion until the swelling and pain have subsided. Removal of the ossification during the late stage has been recommended in traumatic conditions but has not been described in hemiplegic patients.

Painful shoulder. Degenerative and traumatic disorders involving the extra-articular structures about the shoulder joint are quite common among individuals past 40 years of age. One can therefore anticipate at least a similar incidence among hemiplegic patients. The painful shoulder is probably the most frequent and most disturbing complication encountered in the stroke patient. It can be quite disabling, since it interferes with active function, including the performance of simple dressing activities and may even secondarily immobilize the hand.

The cause of the painful shoulder in the hemiplegic patient has been the subject of considerable discussion. Basically, several factors may be involved: (1) preexisting degenerative changes in the rotator cuff; (2) repeated small trauma to the periarticular structures, incurred as a result of traction on the shoulder when the patient lies unconscious on the hemiplegic arm or when he tries to move in bed with the arm flaccid at his side; (3) excessive traction on the shoulder capsule, with marked inferior displacement of the humeral head in a flaccid limb; (4) vigorous stretching of the "tight" spastic shoulder during physical therapy, producing further trauma to the cuff and long head of the biceps tendon; (5) unrecognized trauma such as an impacted fracture of the surgical neck of the humerus, incurred as a result of falling during the onset of the stroke; and (6) contractures that occur in the untreated patient or develop as a result of peripheral nerve lesions and heterotopic calcification.

The treatment of the painful shoulder should be predicated on an accurate diagnosis rather than consist of routine procedures and exercises. Although the use of a sling may be of considerable value in supporting the flaccid shoulder, it may only aggravate the adduction–internal rotation contracture in a painful spastic shoulder. Local infiltration of the tender long head of the biceps tendon with procaine and corticosteroids may afford considerable relief when the diagnosis is peritendinitis. In some instances overhead sling suspension may produce sufficient relaxation to permit effective mobilization. Oral corticosteroids have been used extensively on an empirical basis, and the conflicting results can be explained only by the variability of the underlying shoulder disorder. This is equally true in the nonhemiplegic patient.

Physical therapy procedures are prescribed to relieve the pain and increase the range of motion. The application of hot packs over the deltoid region has become more or less a routine procedure. Frequently it is more desirable to apply the packs over the spastic or contracted pectoral muscles while the shoulder is abducted and in some external rotation rather than adducted and internally rotated at the side. The physical therapist can then initiate the shoulder exercises from a more advantageous position. Diathermy should be used with considerable caution, particularly in patients having a sensory deficit. Ultrasound therapy has been employed with varying results but may be indicated in treating a persistent localized tendinitis.

The application of a modality to the painful hemiplegic shoulder should always be followed by range-of-motion exercises. The patient must be in a supine position so that the shoulder can be adequately stabilized and the effects of gravity eliminated. Vigorous stretching should not be performed, and adequate time and patience must be devoted to this phase of the treatment to permit the pectoral and other adductor muscles to relax. Stretching will only elicit a stretch reflex and increase the spasticity in these muscles.

Pulley exercises should not be prescribed unless there is at least 60 to 70 degrees of passive abduction present; otherwise they will only increase the scapulothoracic substitution. The patient must be properly positioned so that the overhead pulley is slightly behind him to effectively abduct and externally rotate the shoulder during this exercise.

Vascular complications. Thrombophlebitis in the hemiplegic lower extremity occurs not infrequently during the early convalescent period and may be overlooked. This could explain the persistent edema in the lower leg of some patients after they are started on ambulation activities. A simple elastic compression bandage or stocking applied before the patient gets out of bed is quite adequate. Ample provision must be made for the swelling about the ankle when fitting the patient for shoes or a short leg brace. The T strap on the brace must be broader than usual and well padded to avoid any constriction about the ankle. The lower leg should be elevated when the patient is sitting in a wheelchair.

An incipient occlusion of the femoral artery may simulate a thrombophlebitis. Evaluation may be difficult, particularly when the patient is aphasic. Vascular changes may have been present prior to the onset of the hemiplegia. Pain in the lower limb is persistent, and there may be evidence of an ischemic neuropathy particularly involving the deep branch of the peroneal nerve, which is most sensitive to arterial insufficiency. The foot drop, if present initially, will become more pronounced, the hyperactive tendon reflexes may be lost, and a segmental sensory deficit involving especially the first web space may signal the onset of a major arterial occlusion. Prompt surgical intervention is indicated at this stage rather than waiting for the appearance of the cyanosis of the foot and demarcation of the ischemic skin.

A discussion of anticoagulants is beyond the scope of this chapter. However, many hemiplegic patients are on anticoagulant therapy and should be carefully observed for complications, particularly after trauma. A relatively minor bump of the head against a sill or shelf may be followed, in hours or even days, by a so-called "relapse," which may be caused by a latent subdural hematoma. A relatively insignificant trauma incurred while walking in the parallel bars may produce an extensive hematoma of the thigh. The patients on anticoagulant therapy should be identified so that all personnel treating them are alerted to the dangers.

Seizures. Epileptic seizures may occur as a late complication of a cerebral vascular accident.[16] The incidence may be as high as 10% to 15% among those patients who survive beyond the first year. Since the focal signs are localized in the hemiplegic extremities and in the speech mechanism, they often lead to the diagnosis of another stroke. However, these patients recover quite rapidly within the first 24 hours and resume their previous level of activity. Other patients, without convulsions and perhaps with only transient aphasia or psychomotor manifestations, may be diagnosed as having "small strokes." Small doses of diphenylhydantoin with or without phenobarbital, depending on the severity of the focal manifestations, effectively control this posthemiplegic complication.[17]

Decubiti. The older hemiplegic patient with peripheral vascular disease must be carefully observed during the early phase of the illness to prevent pressure sores, particularly over the posterior aspect of both heels. A "sheet burn" over the heel of the uninvolved leg may develop into a large decubitus and prevent ambulation despite relatively good return of function in the hemiplegic limb. However, ambulation should not be deferred unless the ulcer is on a weight-bearing surface.

Trauma. The hemiplegic patient is accident-prone, especially when ataxia is the predominant peripheral defect. Fracture of the femoral neck is a frequent mishap, particularly in those ataxic patients who walk with a spastic adducted gait and invariably fall on the hemiplegic side. This injury may be associated with an unrecognized impacted fracture of the surgical neck of the humerus.

Careful consideration should be given to the management of the hip fracture. Replacement of the femoral head with a prosthesis, followed by immediate ambulation with full weight bearing, has become the accepted procedure in fractures of the femoral neck in the elderly hemiplegic patient. More recently, consideration is being given to simply pinning the fracture after closed reduction, followed by immediate progressive weight bearing as tolerated by the patient.[18] In choosing between these two procedures one must consider the alternatives—the appreciably greater operative morbidity after the insertion of a prosthesis (infection, dislocation, and periarticular calcification) and the high incidence of nonunion and avascular necrosis in the pinned cases.

The management of the patient with a comminuted intertrochanteric fracture is a more critical problem, since weight bearing must be deferred. Many of these hemiplegic patients are transferred to nursing homes and confined in a wheelchair to await healing of the fracture. Joint mobility should be maintained during this period to prevent flexion contractures of both hips and knees.

The marked adductor spasticity in many such patients accentuates the difficulty in both types of management. In such cases it is prudent to recommend that an adductor tenotomy be performed at the time of surgery. It not only facilitates the technical procedure but also removes a disabling feature of the gait. A dislocation of the prosthesis may occur as a complication, but it is less likely when the spastic and contracted adductor muscles have been tenotomized.

Early mobilization of the impacted fracture of the surgical neck of the humerus, within the limitation of pain, is indicated. Protracted splinting is to be avoided.

BOWEL AND BLADDER TRAINING

Bowel and bladder "training" is an integral part of the rehabilitation program. The peripheral innervation of the bowel and bladder is unimpaired in the hemiplegic patient. The incontinence may be a manifestation of the organic brain syndrome. Other older patients are simply confused as they would be after any other illness and become incontinent, by what is called "habitation," when insufficient personnel is available to respond to the frequent calls of the anxious patient fearful of wetting the bed. This is particularly true during the evening and night hours. The use of the indwelling catheter during the so-called "acute phase" in an incontinent patient who is confused or unconscious is justified. However, bladder training should be initiated as soon as the patient is responsive and cooperative. Communication with the aphasic patient at this primitive level is not too difficult. The use of a repeated gesture or jargon to indicate the need for the bedpan or urinal is usually effective.

Adequate fluid intake must be maintained. The catheter is clamped and the bladder is emptied at hourly intervals. The interval is gradually increased during a period of 3 to 5 days, to 1½ hours and then to 2 hours. At this time it should be noted whether the patient can void around the catheter and if he experiences a sense of fullness or discomfort. The catheter is removed before breakfast and the time schedule for opening the catheter clamp is continued, substituting the bedpan or urinal at the appropriate time. Bladder training is a nursing procedure and is successful when there is a continuity in the schedule under the supervision of the same individual who is familiar with the patient. Continuous catheter drainage should be maintained during the night hours when personnel is not available for the periodic emptying of the bladder. The clamping of a catheter and institution of bladder training is contraindicated in the presence of a urinary tract infection or evidence of renal failure.

Failure to establish urinary and bowel continence in the patient who has other symptoms of an organic brain syndrome is usually indicative of a poor prognosis for the entire rehabilitation program. However, this may be a transient phase, particularly in the older patient who is toxic and dehydrated after a urinary tract infection. Rather striking reversal of these manifestations can occur after the infection is controlled and the patient adequately hydrated. A similar dramatic reversal can also occur in an overmedicated, restless, and confused patient after the sedatives or tranquilizers have been discontinued.

SPASTICITY

Spasticity occurring in the later stages of the hemiplegia presents many problems that interfere with the successful completion of the rehabilitation program. Various neuromuscular facilitation technics have been described and are based, in part, on the utilization of pathologic reflexes to produce relaxation in the spastic muscle groups. Considerable attention has been devoted to these technics in the treatment of the spastic child who has had no previous learning experience. The application of the same treatment principles to the adult hemiplegic patient is debatable at the present time and requires further confirmation.

Many tranquilizers and so-called "muscle relaxant" drugs have been used extensively on an empiric basis, since they are not selective in their effect but act through the central nervous system. The spasticity in emotionally unstable hemiplegic patients may be influenced by their mood and may change from day to day. Such individuals may get some relief from small doses of tranquilizers, provided they are not used for protracted periods. These drugs may produce secondary disturbances such as rigidity in a patient who already has some parkinsonian characteristics.

Motor nerve[19] and motor point[20] blocking of spastic muscles with dilute solutions of phenol has been proved to be effective in releasing the spasticity. The results are uniformly better in the lower than in the upper extremity. Repeated injections may be necessary, since the spasticity does recur in many patients. It may be advisable to apply a plaster cast to the forearm and wrist immediately after the block to stretch out the spastic muscles. This is not essential in the lower extremity, since immediate weight bearing accomplishes the same purpose.

Palliative surgical treatment for the relief of spasticity in the hemiplegic patient is indicated when other methods have failed. It consists essentially in the interruption of either the afferent or the efferent pathways of the spinal reflex arc, to produce flaccidity in the involved muscle groups. Surgical intervention for the relief of spasticity is indicated for the following purposes: (1) to improve or permit continued function, (2) to prevent or correct deformity, and (3) to enable the patient to wear a brace or shoe.

In recommending these procedures one must be reasonably certain that the spasticity, rather than some other associated manifestation such as ataxia, is primarily responsible for the deficit. Peripheral nerve blocking is a valuable diagnostic procedure and should be employed to produce a temporary interruption in the innervation of the offending spastic muscles. The patient should be tested during the effective period of the nerve block by evaluation of the anatomic and functional gain resulting from the relief of the spasticity. The patient can also get some idea of the result to be expected from the surgical procedure.

Tenotomy is a relatively simple procedure that can in many instances be performed under local anesthesia. As previously discussed, it is quite effective in relieving severe adductor spasticity in the hip. Preliminary evaluation should be performed by ambulating the patient immediately after blocking of the obturator nerve. Section of the obturator nerve itself is also performed. It is generally more effective but involves a more complex operative procedure.

Severe spasticity in the hamstring muscles with flexion contracture of the knee is very incapacitating and prevents effective ambulation. The use of a long leg brace is not always desirable under such circumstances. Blocking of the sciatic nerve at a high level will establish whether the knee deformity is solely a result of the hamstring spasticity or is complicated by irreversible changes in the posterior joint capsule. Subcutaneous tenotomies would be effective only in the former situation.

Selective tenotomies may also be performed in the patient with a severe spastic foot and a predominant varus deformity that cannot be controlled by a short leg brace with a pronator strap. Such a patient can be quite uncomfortable from the callus that usually develops on the lateral border of the foot. The gait becomes unstable, since the entire lower leg tends to roll outward and the patient walks on the outer border of the shoe despite the short leg brace. Blocking of the sciatic nerve in the posterior thigh, followed by ambulation, can provide some idea as to the feasibility of performing a posterior tibial tenotomy. At times it may be difficult to rule out the gastrosoleus as the offending muscle, particularly when there is an associated equinus, since both muscles are innervated by the sciatic nerve. If the equinus is fixed, it must be corrected before considering any other procedure. Subcutaneous tenotomy of the heel cord may be indicated in those patients who cannot be subjected to the more extensive procedures. In such cases a double stop at the ankle joint of the brace may be required. In other instances, section of the motor branches of the tibial nerve to the calf muscles[21] (Stöffel procedure) may relieve the spastic equinus and troublesome ankle clonus, provided there is no fixed deformity in the ankle.

In selected patients who are relatively young and in good medical condition, the varus type of spastic foot without an equinus can be corrected by an arthrodesis of the subtalar and midtarsal joints. This is highly desirable, since it eliminates the need for the brace and gives the individual a stable foot. The arthrodesis can be combined with a neurectomy for the spastic equinus with a varus deformity.

Selective tenotomies in the spastic hand have been enthusiastically recommended by some authors but have not, as yet, been widely accepted. Considerable investigation is still necessary to evaluate the results of surgery for the spastic hand. Arthrodesis of the wrist should not be considered unless the wrist is in extreme hyperflexion with dislocation that cannot be controlled by a splint. If arthrodesis is performed, the wrist should be fused in a neutral position to avoid severe flexion in the spastic fingers.

AFTERCARE

The aftercare of the stroke patient who has achieved maximum benefit from a rehabilitation program with only partial recovery from the hemiplegia can create a serious dilemma. Once the prognosis for recovery has been delineated, it is essential that the patient be oriented during the active phase of the rehabilitation program in such a way that he learns to substitute for the functional deficit. As he adjusts to this compromise, he also begins to accept the residual disability. Otherwise, abrupt termination of treatment without such preparation may produce a rather profound antagonistic reaction and feeling of rejection. The highly intellectual patient is particularly prone to such a depression.

Continuation of physical therapy for such patients can be tapered to a reduced schedule over an indefinite period of time on a so-called "maintenance" basis. It is intended primarily to recognize and prevent contractures and the other complications previously described. A maintenance program can then be carried out with the assistance of visiting nurse or public health nursing services available in the community. Many such agencies have a physical therapist on their staff. The purpose of the treatment must be clearly defined and not predicated on any anticipated improvement. An overhead pulley installed in a doorway or closet should be standard equipment in the home, and other simple devices can be improvised to maintain the range of motion in the upper and lower extremities.

Communication with the aphasic patient in the home may be extremely difficult. The family of such a patient should be adequately oriented by the speech therapist during the active rehabilitation program so that the level of the receptive and expressive deficit is clearly understood. Not infrequently, hospital personnel develop little tricks of communication with the patient, and they should transmit this information to the members of the family prior to discharge from the rehabilitation facility.

The aged hemiplegic patient responds more readily in a familiar setting and has considerable difficulty in adjusting to a strange environment, particularly in a large institution. Whenever practical, it is important that he be transferred back to his own home from the hospital with provision for continued treatment. The patient becomes confused and insecure when transferred to a nursing home and may regress rather than improve after completion of an organized rehabilitation program.

Any intercurrent illness or a minor injury followed by a few days of bed rest may cause a regression in the functional gains that were achieved during the active rehabilitation program. In such instances it is necessary to place the patient on a "reactivation" program to bring him back to his previous functional level. It then becomes essential again to keep him at this level through what is termed a "maintenance" program. The elderly hemiplegic patient with a multiplicity of medical conditions must be maintained at maximum functional capacity, with provision for preventive and restorative activities as the need arises. This cycle of observation

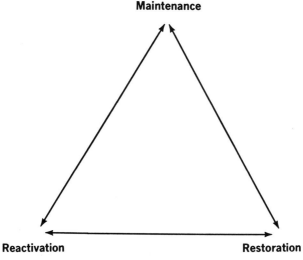

Fig. 29-10. Graphic representation of "dynamic custodial care" of hemiplegic patients. (From Moskowitz, E., and others: A controlled study of the rehabilitation potential of nursing home residents, New York J. Med **60:** 1439, 1960.)

and treatment has been termed "dynamic custodial care" and is illustrated in the triangle presented in Fig. 29-10. Restoration refers to the more complex program within the rehabilitation center.[22] Careful reevaluation at periodic intervals can be facilitated by establishing a Hemiplegia Registry for the follow-up of such patients.

The reassimilation of the hemiplegic patient into the family and community requires careful evaluation by the physician with the help of the social worker and psychologist. Psychiatric consultation may be indicated in the case of the elderly patient with an organic brain syndrome associated with incontinence. One must consider the entire family as a unit, rather than just the patient who is being discharged. Nursing facilities for such patients must be carefully evaluated to make certain that the gains achieved by the rehabilitation program are not lost after active supervision has terminated.

VOCATIONAL REHABILITATION

In general the stroke patient with a technical or professional background has the more favorable prognosis for vocational rehabilitation. However, it must be recognized that such an individual, who is independent in all the Activities of Daily Living and has no speech defect, is not necessarily qualified to resume previous responsibilities. Latent psychologic deficiencies may become manifest only when independent judgment is required.

Serious consideration must be given to the granting of permission to drive a motor vehicle, and evaluation in each case must be on an individual basis. Disturbances in visual perception and the presence of homonymous hemianopsia must be carefully ruled out before clearing such a patient for a driver's permit.

In some families it is possible to effect a reversal of role in the vocational activities within the household. The hemiplegic husband may be taught to be a homemaker just as in the case of the handicapped housewife. The wife is then encouraged to seek appropriate employment within the community. The physician must be the prime mover in getting the husband to accept such a turnabout.

REFERENCES

1. Twitchell, T. E.: The restoration of motor function following hemiplegia in man, Brain **74:**443, 1951.
2. Moskowitz, E.: Complications in the rehabilitation of hemiplegic patients, Med. Clin. N. Amer. **53:**541, 1969.
3. Bruehl, J. H., Peszczynski, M., and Volk, D.: Disturbances of perception of verticality in patients with hemiplegia, Arch. Phys. Med. **38:**776, 1957.
4. Birch, H. G., Proctor, F., Bortner, M., and Lowenthal, M.: Perception in hemiplegia, Arch. Phys. Med. **41:**19, 1960.
5. Lorenze, E. J., and Cancro, R.: Dysfunction in visual perception with hemiplegia: its relation to activities of daily living, Arch. Phys. Med. **43:**514, 1962.
6. Peszczynski, M.: Prognosis for rehabilitation of the older adult and the aged hemiplegic patient, Amer. J. Cardiol. **7:**365, 1961.
7. Haymaker, W.: Bing's local diagnosis in neurological diseases, ed. 15, St. Louis, 1969, The C. V. Mosby Co.
8. Moskowitz, E., and McCann, C. B.: Classification of disability in the chronically ill and aging, J. Chronic Dis. **5:**342, 1957.
9. Buchanan, J. J.: Rapid mobilization of cerebrovascular accident patients, Arch. Phys. Med **37:**150, 1956.

10. Strike back at stroke, Public Health Service Pub. No. 596, Washington, D. C., 1961.
11. Rusk, H. A., Deaver, G. G., Covalt, D. A., Marks, M., Benton, J., and Turnblom, M.: Hemiplegia and rehabilitation, seminar, vol. 14, no. 1, West Point, Pa., 1952, Sharpe & Dohme, Inc.
12. Moskowitz, E., Bishop, H. F., Pe, H., and Shibutani, K.: Posthemiplegic reflex sympathetic dystrophy, J.A.M.A. **167:**836, 1958.
13. Moskowitz, E., and Porter, J. I.: Peripheral nerve lesions in the upper extremity in hemiplegic patients, New Eng. J. Med. **269:**776, 1963.
14. Voss, H.: Ueber die parostalen und paraarticulären Knochenneubildungen bei organischen Nervenkrankheiten, Fortschr. Roentgenstr. **55:**423, 1937.
15. Moskowitz, E., and Steinman, R. I.: Heterotopic calcification in the hemiplegic upper extremity, New York J. Med. **64:**432, 1964.
16. Louis, S., and McDowell, F.: Epileptic seizures in nonembolic cerebral infarction, Arch. Neurol. **17:**414, 1967.
17. Fine, W.: Post-hemiplegic epilepsy in the elderly, Brit. Med. J. **1:**199, 1967.
18. Moskowitz, E.: Rehabilitation in extremity fractures, Springfield, Ill., 1968, Charles C Thomas, Publisher.
19. Khalali, A. A., and Betts, H. B.: Peripheral nerve block with phenol in the management of spasticity, J.A.M.A. **200:**1155, 1967.
20. Halpern, D., and Meelhuysen, F. E.: Phenol motor point block in the management of muscular hypertonia, Arch. Phys. Med. **47:**659, 1966.
21. Stoeffel, A.: The treatment of spastic contracture, Amer. J. Orthop. Surg. **10:**611, 1912.
22. Moskowitz, E., Goldmann, J. J., Randall, E. H., Fox, R. I., and Brumfeld, W. A., Jr.: A controlled study of the rehabilitation potential of nursing home residents, New York J. Med. **60:**1439, 1960.

CHAPTER 30

REHABILITATION OF PATIENT WITH CANCER-RELATED DISABILITY

The challenge in rehabilitation of the cancer patient lies in the extent and the variety of the disease-related and treatment-related disabilities presented. In the specific treatment for cancer three principal approaches are followed, either separately or in combination: surgery, radiation therapy, and chemotherapy. In the hope of securing a cure, the approach is usually radical unless evaluation has determined that only palliative treatment can be offered. Radical procedures, however necessary, exact a toll from the patient and may leave extensive physical deficit to which he and his family must make adjustment.

One in every 3 of the over 600,000 people in the United States who will develop cancer each year eventually will be considered to be cured. The rest will develop recurrent or metastatic cancer and will require additional treatment. However, time of survival is lengthening with the increases in methods of control of disease. Both "cured" and "controlled" patients have frequent need for treatment of disability to improve the quality of survival for each individual.[16]

The disabilities presented by the cancer patient represent a general category of all disabilities and do not differ essentially from those arising from other diagnostic backgrounds. In addition, there are disabilities particularly related to cancer and its treatment, such as the neuromyopathies secondary to the remote effects of cancer on the nervous system, the secondary effects of treatment agents used, and focal effects such as pathologic fractures. The older cancer patients, in particular, may also be victims of additional forms of chronic disability, the result of other disease prevalent in this age group and adding to the need for care.[4]

The first step in the provision of rehabilitation care is the prompt recognition by the treating clinician of the presence of existing disability and, ideally, the recognition of potential disability, such as may follow scheduled treatment. Initial examination and evaluation of the patient and the provision of the first orders for care can be made at the bedside without waiting for the patient to reach a status of readiness for being moved to a rehabilitation service area.

The disability encountered may be either acute or chronic. Acute disability is found in the immediate postoperative period when there may be enforced immobilization, pain, fear, and confusion. Chronic disability is found when there is a long-lasting handicap. Prompt attention for the patient in the form of preventive as well as definitive rehabilitation therapy can reduce the degree of disability and also the time needed for recovery. This emphasizes the advantage of early and dynamic assistance and readjustment. For surgical cases, preoperative as well as postoperative training and counsel are of benefit. Dependent on each patient's needs and findings, there should be an individually prescribed care program. Provision should be made for physical and occupational therapy, training in the Activities of Daily Living, psychosocial evaluation and counseling, as well as the supply of such orthotic or prosthetic devices as may be required.[3]

A cooperative service has been conducted between the Institute of Rehabilitation Medicine and Memorial Hospital for Cancer and Allied Diseases in New York City to develop and demonstrate an active program in the rehabilitation care of

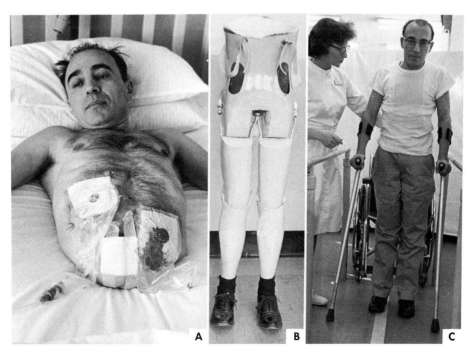

A   **B**   **C**

Fig. 30-1. **A,** Translumbar amputation between L-4 and L-5 (hemicorporectomy). **B,** Full prosthesis for translumbar amputation. **C,** Ambulation with Lofstrand crutches after translumbar amputation.

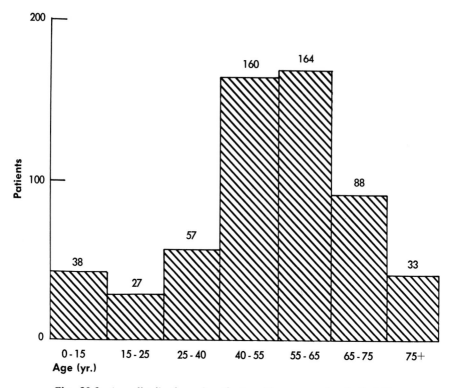

Fig. 30-2. Age distribution of patients with cancer-related disability.

cancer patients. The first patient treated in this program was a 49-year-old male who underwent a hemicorporectomy (a translumbar amputation at the L-4 to L-5 level) in September, 1964. The surgery was performed at Memorial Hospital for treatment of inoperable epidermoid carcinoma of the urinary bladder, which had spread locally to involve the pelvis and was causing intractable pain. Postoperatively the patient was transferred to the Institute of Rehabilitation Medicine (Fig. 30-1, *A*). Rehabilitation was started with bed exercises for strengthening the upper extremities. The patient was prepared for eventual sitting in a prosthetic jacket, wheelchair propulsion, independence in Activities of Daily Living, and ultimately he was provided with a full prosthesis in which he was trained to stand and ambulate (Fig. 30-1, *B* and *C*). Finally, he received driver training and obtained his driving license after passing his road test. He is active, out and about, 6½ years after his surgery. This cooperative program has encouraged increasing referral of patients for rehabilitation, so that the visits can be made as early in the disability as possible and a constructive program developed for ongoing care.

A goal for rehabilitation should be selected for each patient at onset of care to establish a realistic program. Suggested are the categories of "restorative," if the patient can be expected to become able to return to the premorbid status without essential residual handicap; "supportive," if ongoing disease or handicap must persist, but the patient can expect to achieve elimination of as much disability as possible by proper training and treatment; and "palliative," if there is increasing disability to be expected from progress of disease, but where appropriate provision of treatment will eliminate some of the complications that might otherwise ensue. These latter would include bedsores, contractures, problems in personal hygiene, and emotional deterioration secondary to inactivity and depression. Regardless of the classification of the type of disability, early availability of care is important. Rehabilitation should not wait to follow the completion of definitive treatment but should start with onset or recognition of disability.

When a status of disability has become permanent, the social service worker can provide detailed inquiry into the family structure and develop intimate knowledge of the living and possible working conditions to be faced by the patient after discharge from the hospital. The psychologist and the vocational counselor can provide valuable assistance and should be called for guidance and instruction whenever indicated.

The ultimate goal of job placement for the disabled patient with a cancer diagnosis is often a difficult objective to attain. Positive understanding is needed by the employer to modify problems of acceptance of the employee's cancer diagnosis. Planning helps circumvent limited ability for the patient in the use of public transportation and aids in the facing of competition in the labor market and the possible need for specialized job placement within the limitations of residual disability.[8, 12d, 25] The majority of patients referred for care have been in the employable age group, with an average age of approximately 49 (Fig. 30-2). The disabilities depend on both the anatomic or organ system affected by the cancer and on given treatment. Rehabilitation treatment is varied according to individual case findings and need and can best be categorized by disease region and cancer treatment procedure.

BONE AND SOFT TISSUE MALIGNANCY, AMPUTATION AND DISARTICULATION, PATHOLOGIC FRACTURES

Tumors of the bone and soft tissues affect individuals of all ages but are found most commonly in the younger group, between 10 and 30 years of age. Definitive treatment demands that amputation be performed for almost all cases of malignant tumors of bone, unless treatment is for a pathologic fracture.

Primary or metastatic tumors involving bone may destroy the bony architecture, and pathologic fractures may threaten or occur. These require appropriate orthopedic and radiation therapy. Depending then on the individual disability and tolerance and the surgery performed, treatment begins with active range-of-motion exercises, and later, if the patient is able to participate, training and assistance in ambulation with appropriate supports to mobilize him. Otherwise, a goal of wheelchair mobilization should be sought.

Soft tissue tumors differ from bone tumors in that they arise in the tissues around bone and not from bone itself. If the diagnosis has been made early, there is the opportunity to do a surgical excision of the tumor and surrounding tissues, usually removing large muscle groups, but still leaving limb support and enough muscle control to be useful to the patient. The patient may, of course, need a brace to stabilize a lower extremity or a static or functional splint for upper extremity support and use.

Before surgery the patient should be told of the implications of amputation, including phantom sensations and of what to expect from surgery and rehabilitation, taking into account the level of amputation, the patient's general physical status, his intelligence, athletic ability, and premorbid way of life. It can be helpful to have a rehabilitated amputee with similar problems demonstrate to the patient what can be expected after amputation. Preoperative consideration and discussion of the provision of an artificial limb for the patient increases his understanding, confidence, and motivation. For the prospective lower extremity amputee, preoperative training in crutch use is of great value. During the preoperative period the patient has no discomfort from a surgical wound, has not been weakened by a period of immobilization, is not affected by medication for pain, and has not developed a fear of falling due to lack of limb support. The security gained at this time from training in crutch ambulation or the use of a walker is of great assistance during postoperative recovery and speeds the progress of rehabilitation.

Postoperative exercise instruction for the lower extremity amputee should include progressive resistive exercises to the opposite lower extremity and both upper extremities and the shoulder depressors. The upper extremity amputee needs one-handed Activities of Daily Living training, especially if he has lost a dominant upper extremity. Occupational therapy should be included for the improvement of dexterity and handwriting when appropriate.

It must be noted that amputations for cancer infrequently leave a limb stump. Upper extremity amputations are either shoulder level disarticulations or interscapulothoracic amputations (in which there is removal of the collar bone and shoulder blade along with the entire upper extremity). For the patient with lower extremity cancer the surgeon's choice is usually either a hip disarticulation or a hemipelvectomy (in which the ipsilateral pelvic bony ring is removed with the limb).

If there is an amputation stump present, stump conditioning should be performed in accordance with standard rules of shaping and bandaging, strengthening, and toughening. Repeated evaluation of stump condition must be made. Postoperative stump exercises, range-of-motion exercises, and activities for prevention of contracture should be ordered.

The weeks immediately after surgery are the most important in the care of the amputee. This is the period when he will or will not develop the proper attitude for the successful use of the prosthesis and his return to society. His strengths and weaknesses physically, emotionally, vocationally, intellectually, and socially will determine whether a prosthesis should be prescribed and if so what type.[17] Consideration of a suitable prosthesis or none at all comes up for every amputee. There may be conditions apart from the stump or level of amputation that would affect the use of a prosthesis or eliminate its prescription. These include severe cardiorespiratory disease, contralateral hemiplegia, prior radical mastectomy, old preexisting physical disability, senility, and others. If the patient is an accepted candidate, a prosthesis can be fitted immediately after operation or as soon as the wound is healed if there has been an active program from the start and no contraindications exist. The use of a temporary articulated prosthesis will help to avoid or to reverse many of the problems of the the new amputee.[17]

The approach of fitting a prosthesis either immediately or early in the postoperative period is of particular value for the cancer patient.[15, 17, 19] In both instances there is immediate postoperative application of a light, thin wound dressing covered by a snug cast of elastic plaster of paris, which conforms exactly to the contour of the stump. This material, in contrast to elastic bandaging, ceases to create any elastic constrictive pressure as soon as it has hardened, and creates sufficient support to prevent edema and swelling of the stump. Also, it appreciably cuts down on postoperative pain and discomfort by its splinting effect. In case of immediate fitting, a pylon prosthesis is then added by the surgeon and the prosthetist while the patient is still on the operating table. The device is in parts, uses coupling devices, and can be adjusted both immediately and subsequently for alignment and length. In case of early fitting, rather than immediate, the prosthesis is applied 10 to 14 days after surgery, when the operative dressing is first changed and the skin sutures are removed. If no complications are present at that time and the condition of the stump is satisfactory, an elastic plaster mold is immediately taken of the stump for attachment to the pylon, and a tight plaster dressing is reapplied to maintain stump shape and prevent edema.

Use of these forms of total contact prosthesis as soon as possible after amputation creates an increased proprioceptive sense and utilizes the phantom sensation present. The properly snug application of the elastic plaster of paris is essential to either method. Subsequent treatment continues with general strengthening exercises to the other extremities, stump conditioning exercises, and training in ambulation with the new prosthesis.

It must be emphasized that success in early prosthesis application depends on the proper and secure attachment at the time of amputation surgery of the muscles and fascia to or over the bone end, so as to afford good stump control.[15, 19]

For lower extremity amputees, prostheses of functional value may be provided for patients with any amputation from a maximum of hemicorporectomy to a residual stump at any level.

Amputations for cancer of the lower extremities in growing children are more frequently done at the knee or intercondylar level in an effort to retain the distal epiphysis and thus preserve growth potential. This is in contrast with amputations for adults.

In the management of children who are amputees, the restored function and cosmesis afforded by a prosthesis are of practical and psychologic value to both the child and his parents. Data from recent studies have indicated that a significant percentage of child amputees with malignancies survive 1 to 5 years postoperatively and wear their prostheses successfully for a year or longer, the majority of them full time. Delay in noncontraindicated prosthetic restoration is therefore unjustifiable, since the expenditure of time and money involved is vindicated by the rehabilitation results.[12b] Successful independent ambulation may be accomplished by a 3-year-old child even after a hemipelvectomy if a proper prosthesis and training are provided.

Although the cost of an artificial limb is high, it is only equivalent to the cost of about 10 additional days of hospitalization, and this emphasizes the importance of early fitting. Changes that become necessary in the socket shape and size are also less costly than keeping the patient in the hospital for a prolonged period waiting for the stump to shrink.

As with lower extremity amputation, upper extremity loss also needs to be treated in a positive approach by the provision of an appropriate prosthesis and training in its use. As noted previously, amputations for cancer rarely leave a stump on which to fit a prosthesis. For the interscapulothoracic amputation of the upper extremity, the earliest and most useful prosthesis to be recommended is a shoulder-cap cosmetic prosthesis to support the clothing on the amputated side and to provide the patient with symmetry in appearance. Little can be said for any form of functional prosthesis after this type of operative procedure, as it is too difficult to anchor the prosthesis on the cone of the chest wall because of rotation and displacement when either traction or pressure is exerted. However, a purely cosmetic full-arm-and-hand prosthesis of light weight and good appearance may be provided for those patients desiring it. A shoulder disarticulation does not so radically disturb stability and in such cases a prosthesis can be functionally useful. Depending on residual stump lengths of the upper extremity, progressively more useful prostheses can be made that are proportional to the amount of stump available.

Immediately after operation, the upper extremity amputee should begin occupational therapy for training in dexterity, technics of one-handed activities in daily living, and strengthening exercises for any residual stump to maintain its maximal activity potential. This training should continue after fitting of the prosthesis, with emphasis then on such use of the new replacement appliance as it will allow.

Following amputation most patients have the sensation that the limb is still present. This is called phantom sensation. If the sensation of the limb is painful or disagreeable, it is referred to as phantom pain. Phantom pain may be constant or intermittent and may be of varying severity. It is sometimes relieved by simultaneous efforts in bilateral exercises of the phantom and the normal limb.

Neuromas forming in the amputation stump may be the source of pain for the patient. The cut end of every nerve becomes a neuroma, which is usually painless if adequately protected. Neuromas subjected to pressure may give great pain and require surgical revision. (Capping at surgery of the cut nerve ends with a substance such as Millipore or Silastic is of some benefit.)

Clothing for amputees should be as similar as possible to clothing worn by the nonamputee. It should however be looser than usual where it covers the prosthesis to allow for the frequently bulky nature of the sockets and joints.

Psychologic adjustment to a prosthesis depends on the realization that the prosthesis is a tool for performing certain activities. The artificial arm cannot be used above the head or behind the back because of the harnessing, nor can it be used where there is no vision. The lower limb prosthesis is effective for one gait cadence only. Almost all units present diffculties when the patient tries to go up or down stairs or inclines. It is important to mention these facts to the amputee so that he is not deluded into believing that he will be "almost normal" after fitting and training with the artificial limb.[17]

The amputee must understand that prostheses have an unstable attachment to the skeletal system, that they feel heavier than the normal limb, that there is a false joint between the prosthesis and the body leading to socket instability and bell clapper type of motion that may lead to skin irritation, and that perspiration in the socket can result in maceration of the skin. More important for the upper extremity than the lower extremity amputee is the loss of sensory feedback, and the upper extremity patient must use visual control at all times.[17]

With respect to vocational consideration, amputation may make little difference in the working life of a person whose job was primarily sedentary, but for manual workers a complete change in vocation and living may be necessary. Such patients should be referred at the earliest possible moment to a vocational rehabilitation counselor for evaluation and direction.

Pathologic fractures occur when the bone has been replaced and weakened by a neoplastic lesion, most frequently a metastatic focus. All bones can be affected, but the fractures occur in the long bones, spine, pelvis, and ribs. Long bone fractures are treated by intramedullary nailing or joint replacement, prosthetic insertions, such as at the hip, with the use of Austin Moore, Thompson, Marcove, and similar units.

Mobilization of the patient is important and active range of motion of the involved extremities should be started as soon as permitted by the operating surgeon. Passive or more than gentle assistance for range of motion is never to be given in the presence of metastatic disease of bone with or without treated or impending pathologic fractures. The patient may need a sling to support an upper extremity or a brace for a lower extremity. Ambulation should be instructed and assisted with degrees of weight bearing allowed by the surgeon for the type of procedure used.

Radiation therapy may be administered to sclerose a neoplastic focus in bone and may precede or follow surgery. Cryosurgery, which is the removal of tumor tissue by the freezing effect of liquid nitrogen, is also used, and postoperative supportive physical therapy and active exercises are of value, but their use depends on the allowance of the surgeon in accordance with the patient and procedure.

CANCER OF THE BREAST
Mastectomy

The patient with cancer of the breast can be considered a candidate for rehabilitation attention in both the preventive and restorative areas. Orientation to mastectomy should begin if possible before the surgical procedure. The patient can benefit from proper preoperative counsel and instruction in the hope of reducing the psychologic impact of the postoperative discovery of her radical mastectomy. A program in this sense can be started before surgery, but in usual practice it begins postoperatively because the establishment of diagnosis most frequently accompanies the operative procedure.

The surgical procedure of radical mastectomy involves removal, along with the breast, of the pectoralis major and pectoralis minor muscles. In almost all cases the long thoracic and thoracodorsal nerves are preserved, but occasional sacrifice of the long thoracic nerve occurs. Postoperative physical disability is related to the loss of the pectoral muscles, and the patient has decreased efficiency in horizontal adduction at the shoulder, in adduction above 90 degrees, and in chopping motion. For example, golfing and tennis will be affected. The muscle that is most effective in substitution is the deltoid in its anterior and middle portions. Some weakness will result in shoulder depression also, and the patient may have a problem in crutch ambulation, should this need arise. Winging of the scapula will occur when the long thoracic nerve has been divided.

Modified radical mastectomy is a term usually given to procedure in which there is preservation of the pectoral muscle. The breast is removed in its entirety, and a complete dissection of the axilla is performed to remove the axillary pad and lymph nodes. Little if any physical function deficit follows this procedure. The patient will respond promptly to postoperative exercises and range of motion.

Extended radical mastectomy is performed when the cancer of the breast has invaded the chest wall or requires intrathoracic extension of the procedure for internal mammary node dissection or for superior mediastinal surgery. Resections of segments of the chest wall in these procedures result in local instability and create increased problems in range of motion and strength in the shoulder on the operated side. Response to a physical therapy program is slower than after the standard radical mastectomy. The patients tolerate full activity programs, however, with attention only to individual needs and discomfort.

Rehabilitation has three main goals: restoration of external appearance, maintenance of range of motion and function in the operated arm and shoulder, and aid in psychologic adjustment. The patient's severe reactions of disappointment, depression, anxiety, and fear require support, reassurance, and encouragement from the surgeon, the rehabilitation team, and the family. The patient must be instructed in the performance of early routine exercises in the preservation and restoration of function, and she should be promptly given an appropriate prosthesis to provide for symmetry in appearance and some balance.

The early institution of a rehabilitation program after surgery assures best results in restoring physical function and preventing frozen shoulder. Postoperative dressings and positioning should keep the arm in abduction and slight elevation. Early exercises of the forearm and hand assist in the control of edema and the improvement of general circulation.

Exercises should start as promptly postoperatively as permitted by the operating surgeon. The initial program will be influenced by certain variations in the operative procedure, including skin grafts, chest wall resections, thoracotomy, and wound closures under relative tension. Exercises should include range-of-motion activities and isometric and isotonic strengthening activities, actively performed by the patient and, when of value, assisted by use of a rope with pulley or substitute. Effort by the patient and understanding are needed to avoid or eliminate substitute movements at the waist, wrist, and elbow when shoulder range of motion is being sought.

Encouragement is given to perform all regular Activities of Daily Living, especially at home and at work. Exercises started in the hospital should be continued at home.

If the therapeutic approach has been by radiation therapy, any related complications of paresthesia or lymphedema in the arm on the operated side develop late. Skin reactions are now seldom seen, and axillary web formation in the scar is also infrequently seen, with present surgical technic keeping the incision line out of the axilla itself. Should such a web develop, the patient may need a plastic repair for release.

Approximately 50% of mastectomy patients at some point experience some degree of postoperative arm edema on the operated side. There is marked variation in the areas of the arm that become involved, in the time of initial occurrence of the edema, and in its persistence. Slight swelling that may occur in the early postoperative period most frequently subsides. Swelling that occurs weeks or months after surgery is more likely to be persistent or progressive. Lymphedema disability is proportional to the extent of the edema and the disfigurement it creates. It may be a later sequel due to fibrosis in subcutaneous tissues after supervoltage radiation therapy and occurring a year or more following the institution of this therapy.

Treatment for lymphedema has been varied. Proper positioning of the extremity should be taught. The use of salt-free diets, diuretics, massage, mechanical aids for intermittent compression, elastic sleeves, and surgical procedures all have value. Massage, however, should not be used where there is recurrent carcinoma in the tissues to be handled. The intermittent compression must be frequent and daily to be in any way effective and needs the concomitant use of a properly fitted elastic sleeve, often with a separate elastic gauntlet to support the hand. Intermittent compression may be effected by the use of a mechanical apparatus or by a simple pneumatic sleeve or long boot with attached blood pressure gauge operated by the patient. Antibiotic courses may be of value if subclinical infection is suspected.

Goldsmith[5] has had encouraging success with the control of postmastectomy lymphedema by the transposition of the partially detached omentum beneath the superficial tissues of the chest wall to the axilla and upper arm.

In the prevention of lymphedema, infection must be meticulously avoided and antibiotics used more liberally, especially if there is any necrosis of the wound margin, fluid beneath the wound flaps, or the suggestion of infection in the arm on the operated side. These patients should never be given any form of injection, vaccinations, or infusions in the arm on the operated side, and they should be warned about the danger of infection in the fingers and hand, burns, and sunburns, for which they should seek immediate care.

Satisfactory prostheses can be purchased in most department stores and at corsetieres for prices ranging between $3 and $15. More elaborate custom-made units start at $35 and on occasion can cost in the hundreds of dollars. Immediately postoperatively the patient can be given a nylon lace "sleep bra," stretch type unit. Next in the early postoperative period the use of light fluff or air-filled plastic units are recommended for combined comfort and cosmesis.

There is a prosthesis for any type of female figure. Some are simple rubber forms for small-breasted women and others are weighted with various materials to simulate and balance the heavy breast. Costly molded rubber forms can be obtained. All patients can have an appropriate prosthesis for swim wear.

Clothing alterations can usually be minimal. Capes, stoles, short jackets, shoulder decorations, and proper choice of garments at time of purchase will provide a full and satisfactory wardrobe. Sleeves should be chosen in relatively loose and comfortable styling, and the presence of lymphedema will demand more careful choosing and planning for allowance of letting out of the sleeve. Bathing suits with halter-neck fronts are recommended, and slight modifications will make many other styles entirely satisfactory.

In an effort to answer the many questions that arise postoperatively in the patient's mind regarding her new situation and fears, a small 16-rpm record has been devised by Dr. Guy Robbins of Memorial Hospital to be given to a patient a few days after surgery. This recording is of an informal panel discussion between three mastectomy patients, a nurse, and a physician. The patient may play and replay it to herself to obtain answers to her own questions and problems. This avoids embarrassment for the patient and saves much time for the doctor.

Advice to husbands in promotion of understanding of the operative procedure and its affects will reduce much of the shock and distress otherwise possible. Such understanding helps in improved emotional and physical recovery for the patient. Printed brochures suggesting approaches in this area have been set up by Memorial Hospital and by the American Cancer Society.

An occasional patient may show evidence of an underlying psychologic instability and inability to face reality; these will require psychiatric advice and therapy. The necessity to advise a young patient to avoid pregnancy and occasionally of the recommendation for castration must be presented to the patient with sympathetic understanding and careful explanation. Group therapy provides additional motivation and understanding for these particular patients.

Palliative programs, in accordance with the needs or disabilities presented, can be offered to the patient who has developed advanced cancer of the breast with metastases.

Increasing lymphedema may respond to proper positioning and support of the extremity, as well as to reduced dietary salt intake and the use of diuretics. Amputation occasionally has to be performed to relieve a patient of an intolerable and painful upper extremity. Training in independence in the one-handed performance of Activities of Daily Living will greatly free the patient from dependence on others, whether she is suffering from disability from lymphedema or has had an amputation performed.

Spine lesions due to metastatic deposits in the vertebras, at whatever level, can be made less painful with the use of cervical collars or appropriate back braces to limit range of motion. On occasion, a bivalved, laminated, lightweight plastic jacket can be molded to incorporate the patient from chin and occiput to pelvis to permit mobilization out of bed, which would otherwise be impossible.

Pathologic fractures, treated by intramedullary nailing, need bracing and ambulation training or upper extremity active and assistive exercises. Hip and shoulder joint replacement procedures done for the treatment of local bone metastases with or without pathologic fractures create the requirement that the patients be provided with programs of activity and appropriate support as promptly as possible following surgery.

Neurologic deficits, such as paralysis and pain due to brachial plexus involvement by tumor, should receive sling support and assistive exercise prescription. Patients with paraplegia or paralysis from spinal cord compression by metastatic lesions, whether treated definitively by radiation therapy or decompression laminectomy, will respond to training in transfer activities, wheelchair independence, and whatever ambulation remains possible to the patient with the aid of parallel bars, walker, crutches, or canes. Extensive bracing is not recommended for these patients.

Occasional patients, who have undergone hypophysectomy or those with intracranial metastatic lesions, will develop hemiparesis or hemiplegia and will respond to appropriate training with improved function and independence.

Those patients who have developed advanced pulmonary complications will benefit from training in adequate voluntary breathing and coughing control as well as postural drainage.

Wheelchairs, walkers, crutches, canes, and other indicated devices should be provided for the patient as appropriate for reduction of her handicap. The patient who is confined to bed should be trained for independence in turning and self-care in the bed.

When the patient can be discharged home, the medical social worker can assist in outlining plans for improvement in home setup and facilities, particularly the bathroom and the kitchen, to increase utilization and independence.

CANCER OF THE BOWEL

Resection of the bowel for cancer frequently entails a permanent or temporary colostomy. Ileostomy is performed less frequently, but an ileal conduit bladder commonly results from radical bladder and genitopelvic surgery. These stomas leave the patients with loss of sphincter control in elimination. Each type of stoma requires different technics of care and rehabilitation assistance.

Colostomy

The presence of a colostomy may develop great adjustment problems for the patient.[29] There may be periods of leakage, noisy expulsion of gas, and odor, with resultant anxiety and embarrassment and often marked social withdrawal. Bladder control may be lost either temporarily or, in occasional patients, permanently after surgery. Impotence occurs about 50% of the time in males. Perineal area discomfort occasionally may result from the scar tissue formed.

The time consumed to care for a colostomy, and to a lesser degree the other stomas, may demand a change in the patient's and his family's routine and time-table. Varying degrees of dietary change may be required for a few patients. Many patients report a reduced ability to work; they often change jobs, and some stop work completely.

Rehabilitation is of great importance to these patients and should start whenever possible during the preoperative stage with appropriate reassurance and advance knowledge to help lessen the shock of finding the artificial stoma. Attempts also should be made to prevent the early postoperative distressing experience for the patient of massive evacuation and soiling. The latter can be minimized in the operating room by inclusion in the immediate postoperative dressing of a disposable plastic collecting bag unit.

Continuity of postoperative colostomy care should be provided daily by the same nurse, or enterostomal therapist, trained especially in colostomy handling. The patient requires consistent emotional support. His response to training depends on how he has been introduced to his diagnosis. Great attention must be paid by the people who care for the patient to provide him with sympathy and guidance and to avoid any apparent distaste, hostility, or impatience.[20]

The use of temporary ostomy bags provides early containment of fecal drainage during training in control. They are easier to handle than rubber bags and they decrease odor. The patient's initial reaction to his colostomy is related to his need for cleanliness. Early bowel control is therefore a primary goal.

It is important to select the method of care and appropriate equipment best suited to each patient. Management is generally better by irrigation than by the less frequent nonirrigation technic. In management by irrigation a regular "enema," using a catheter with a special colostomy cover and lead-off apparatus, or the simpler bulb syringe technic[22] is usually administered. Commercially made enemas and evacuant suppositories are not consistently satisfactory.[12f]

Because of its great impact, the first irrigation should be done in private and under the best circumstances by an experienced individual with ample time and sympathetic understanding. A nurse, specially trained and regularly assigned, or an enterostomal therapist will be of greatest value. The North American Association of Enterostomal Therapists maintains a listing of qualified enterostomal therapists who are available for instruction of patients. Increasing numbers of medical centers are training such therapists.

A toilet or commode should be used unless the patient must remain lying in bed, when the leadoff can be to a bedside receptacle. As the patient's strength increases, he is taught to take over increasing amounts of daily care until he assumes complete care and has given himself several irrigations before leaving the hospital.

The nonirrigation technic is not widely used but may have merit when toilet facilities are not satisfactory, when no regulation has been accomplished after trial of the irrigation method, or when debility, handicap, or old age would make any active procedure impossible.

Irrigation equipment appropriate for the method chosen for the individual patient should be provided promptly postoperatively. The patient is then instructed with this equipment, which is his personal property, and he takes it home with him for continued use.

Time for colostomy irrigation is chosen best with consideration of the patient's living habits and his preoperative bowel evacuation routine. The best colostomy response to irrigation is about 30 minutes after a meal, and an hour of uninterrupted time should be available to the patient. Once a time is chosen for a patient for irrigation, that should be his regular hour and effort should be made to adhere to the same time schedule both in the hospital and later at home if satisfactory control is to be maintained.

Usually 7 to 10 days is required to achieve the start of control. Daily irrigation is needed consistently by some patients. Others may be trained to an alternate day routine.

The amount of water used depends on the method. One to two quarts is sufficient for catheter equipment. Less is needed with the bulb syringe technic. Tepid plain water is usually adequate, but hard water is helped by addition of bland soap or bicarbonate of soda, 1 teaspoonful per quart.

The catheter should be inserted only about 6 inches. Deep insertion is not necessary, but gentle in and out movement of the catheter during irrigation can assist the flow into the bowel and also gently stimulate return. Insertion of the catheter should never be done with force. Return of water may be aided by slight change of position and also by gentle massage or pressure on the abdomen.

Early independence in self-irrigation should be taught in combination with proper stoma cleansing and bag application. The stoma and surrounding skin should be gently wiped with paper to remove feces and mucus, and the surrounding skin should be washed with warm water and soap and gently dried. Irritated skin about the stoma may be treated with ointment such as Desitin or aluminum powder in mineral oil. When there is no irritation and protection is desired, tincture of benzoin may be applied to the skin both to protect against moisture and also to enhance the attachment of the adhesive facing about the bag opening. The bag opening should be trimmed to about 1⁄8 inch larger than the colostomy itself to permit adequate adherence and also to prevent bare skin being irritated by the fecal drainage.

There are many available types of ostomy bags. They may be sealed units or openable at the unattached end for emptying purposes. Most are disposable. Many have adhesive applied about the opening. For patients sensitive to adhesive the karaya gum or karaya ring can be helpful. Surgical adhesive may also be used. Bags that are not adhesive can be worn with a belt, but these tend to shift position and, therefore, move away from the stoma and fail their purpose. Selection of equipment for the individual patient should include consideration of his physical limitations and his other functional needs in daily living.

Ease of control of the colostomy is related to its level. Sigmoid and descending colon colostomy control usually is gained more promptly and with more dependability than when the stoma is in the transverse colon. Ascending colon colostomy can occasionally be controlled with a daily irrigation, but like ileostomies there is liquid loss usually, and an open-ended disposable plastic bag cover is appreciated by the patient.

The patient may benefit from the visits of a volunteer who has successfully managed a colostomy of his own. Such a visit should only be made with the operating surgeon's permission. The help of a fellow ostomate is greatest preoperatively when the patient may fear for his survival in the future, and again when he begins to learn to handle his own care and needs an example to follow for reassurance of his own activity potential. Much help may also be obtained from the "ostomy" club in the patient's community, and the patient can be encouraged to visit this unit if it is available to him.

Staff education materials include the American Cancer Society's film *Caring for the Patient With a Colostomy,* which is available on loan through all unit offices.

Prior to discharge from the hospital, discussion should cover home plans with the family so that when the patient reaches home adequate preparations will have been made and needed equipment secured. A visiting nurse may be called in to give a helping hand during the first few days at home to eliminate some concern and confusion.

Regular follow-up facilities should be provided for all patients, as problems may arise and functional changes occur, such as stricture at some level of the stoma. The latter may require local surgical revision to restore adequate functioning or it may respond to dilation. The patient can be instructed by the surgeon or physiatrist in the technic of dilation of the stoma with a finger covered with a finger cot and lubricated. Dilation may be performed daily if necessary and should be checked regularly by the physician.

There may be small amounts of bleeding noted from the colostomy. These result from minor irritation and are not in need of specific attention, but if bleeding is persistent or greater than spotting, medical investigation must be sought.

If the patient has to skip irrigation, he should wear a bag to protect himself from soiling. If he travels, he will need to adjust the timing of his irrigation initially, more in relation to the time interval since his last irrigation than with respect to the time of day at his new location. Air travel may require a single dose of paregoric or dyphenoxylate (Lomotil) to reduce the tendency toward increased peristalsis and passage of gas or fecal material stimulated by change in atmospheric pressure, even in "pressurized" aircraft. The traveler should also be instructed that water in a foreign country that is unsafe to drink is also unsafe to use for irrigation unless it has first been boiled.

Control problems include constipation and diarrhea. Constipation responds usually to increase in fluid intake and the use of small amounts of well-diluted milk of magnesia sipped slowly on an empty stomach about 8 hours before irrigation time. Stool softeners are not recommended for colostomy control. Diarrhea may require reduced fluid intake and change of diet or discontinuation of medication for other problems, but which cause the diarrhea as a side effect (diuretics, antibiotics, antihypertensives, and tranquilizers).[21]

The patient will need understanding and assistance from his family. However, assistance should be provided only when actually needed. Sexual relations can be maintained within the individual's potential. A small cover for the ostomy may be all that is needed. Mutual support and affection between the ostomate and his family are critical to the success of his tolerance and control of his colostomy.

Ileostomy

Ileostomy care differs from that for colostomy. The ileostomy drains semiliquid material in frequent spurts throughout the day and night. The drainage contains digestive juices that are irritating to surrounding skin. An adequate collecting device must be worn at all times both to collect drainage and to protect the skin surfaces surrounding the stoma. The collecting unit may be plastic or rubber. It is part of or can be detached from a faceplate. The faceplate attachment to the skin about the stoma must be watertight. Bonding to the skin is effected by a waterproof adhesive. The fit must be exact and the diameter of the opening, like that of the colostomy, should be no more than 1/4 inch greater than that of the stoma. The device should be comfortable, light, and easy to empty, clean and change.

The faceplate is either plastic or rubber and is cut by or for the patient to fit for the individual stoma. The technic of proper application is first to wash about the stoma with mild soap and water, rinse, wipe with alcohol, pat dry, brush a light coat of liquid adhesive both on the skin and faceplate, and then after a short drying period (2 to 3 minutes) place the faceplate in position. A double-faced adhesive disk can be obtained for use instead of the liquid adhesive. Plain tincture of benzoin is used on the skin prior to the disk placement, and the disk is attached first to the ileostomy faceplate and then to the skin.

Irritation of the skin should be treated by careful cleansing with soap and water and dusting with karaya gum powder or the use of a karaya gum ring between the appliance and the skin. Such rings dissolve and will require frequent changes of the bag.

The ileostomy pouch will require emptying every 4 to 6 hours or according to patient need. Removal should only be done with the use of adhesive solvent to prevent skin irritation. No restrictions in activity or living are required by the ileostomy patient. He should eat a diet relatively low in residue and should chew all food thoroughly to avoid tendency of blockage by lumpy residue. Individually poorly tolerated or gas-forming foods should be avoided by the patient. Odor can be controlled by placing a few drops of chlorophyll solution (Airwick) in the pouch and by the use of vinegar or liquid bleach when cleansing the pouch. Airing of the pouch and intermittent use by interchange helps also.

Thorough cleansing and drying of the skin surrounding the stoma helps in the prevention and removal of keratotic plaques that form about both ileostomy and ileal bladder openings. The same principles of care apply to ileal bladder stomas as to ileostomy. Leakage is prevented by adequate seal, and emptying depends on frequency of filling of the bag. No one type of appliance is suitable for all patients and selection for the individual patient may need to be made by trial and error.[28]

CANCER OF THE LARYNX
Laryngectomy

Cancer of the larynx treated by total laryngectomy results in loss of speech, entailing a major deficit in communication. Speech training becomes the most pressing need in rehabilitation either through the development of esophageal speech or with the use of an artificial larynx. Esophageal speech is a preferred method, for it is much more satisfactory than a prosthetic appliance insofar as articulation, intelligibility, and phonation are concerned, and it can be successfully taught in the majority of cases. Poor communication, by comparison, follows with any form of buccal speech or whispering.

Prior to surgery the patient should be advised that the expected loss of speech will be temporary and that there will also be temporary loss of taste, which will return as he learns to talk. In the immediate postoperative period and until he has learned to talk, the patient should communciate by writing and signs and thus avoid development of bad communication habits, facial grimacing, and whispering. The importance of correct instruction cannot be overemphasized. Only a trained speech therapist or another laryngectomee, a successful speaker, should undertake this training. If such a trained instructor is not available in the patient's home area, the patient should travel to the nearest center. Preoperative speech training and the ability to "belch" are not related to success in becoming an esophageal speaker. It is generally felt that active speech instruction should start at about the time of discharge from the hospital. Instructions in learning esophageal speech production can be obtained at many larger universities and colleges, most medical centers, some private hospitals, private speech clinics, through private instructors, and by arrangement with Lost Chord Clubs. Information is readily obtainable from the International Association of Laryngectomees and the American Cancer Society for names and places to contact. The American Cancer Society publishes an instruction manual titled *Your New Voice,* and Lauder is the author of a book titled *Self-help for the Laryngectomee.*[26] Available material for instruction of the laryngectomee also includes *Helping Words for the Laryngectomee* from the International Association of Laryngectomees and the American Cancer Society.

The patient should be encouraged to resume as many of his usual activities as possible. Some things, however, will be changed. He cannot strain or lift heavy loads, being unable to hold his breath. Laughter and singing ability are lost. Swimming cannot be safely tried because of the tracheostomy, but nearly all other regular activities can be enjoyed.

Following laryngectomy the patient benefits from breathing properly humidified air and should use a humidifier or vaporizer or place open pans of water and plants in living and working quarters. It is valuable at the time of discharge to provide each patient with a postlaryngectomy kit containing the main essentials required for good care: a tracheotomy tube, gauze, petrolatum, a bib, a shower cover, and saline or Zephiran in saline.[12e]

Social service and vocational training play large rehabilitation roles. Training for changes in occupation may be necessary, especially if previous employment required verbal communication or exposure to fumes, dust, or underwater work. The patient and the family require instruction in the care of the stoma, the use of the bib, and general hygienic care.

Medical identification and information cards are distributed through the International Association of Laryngectomees, 219 East 42nd Street, New York, N. Y. 10017, to be carried by the laryngectomee, and that instruct others as to his stoma and special needs, including methods for possible artificial respiration administration. Detailed first aid manuals covering care for the laryngectomee are published by and obtainable from the same source.

Radical neck dissection frequently is part of the surgical procedure for treatment of cancer involving neck structures. Rehabilitation procedures helpful after such surgery are discussed in the next section.

Radical neck dissection

Radical neck dissection results in both cosmetic defect and disability in the shoulder on the operated side, secondary to section of the accessory nerve. The patient develops varying degrees of trapezius muscle paralysis with a dropped painful shoulder and a rotated scapula, depending on the amount of primary innervation carried to the trapezius by the accessory nerve.

Treatment requires support of the arm and shoulder initially with a sling to prevent overstretching of the trapezius and an exercise program to include strengthening and conscious utilization of the rhomboids and levator scapulae for training in movements to facilitate abduction of the arm at the shoulder. A certain number of patients with direct innervation of the upper trapezius by cervical roots will retain function lost by the others. In "preventive" surgical rehabilitation of patients, attempts have been made to use nerve grafts to replace the removed segment of the accessory nerve. The greater auricular nerve can be used for this purpose. Reports of results are varied, but it is considered worthy of trial if the operative opportunity presents easily for the surgeon concerned. During the healing after this procedure, muscle stimulation electrically may be beneficial. Use of supportive fascial slings and muscle transplant of the levator scapulae have also been considered.

These patients should also be taught to support the shoulder and upper arm when seated, on a chair arm or pillow, to prevent stretching of the trapezius. A booklet for instruction in exercises, similar to that given to breast patients, is now distributed to patients by the Head and Neck Service at Memorial Hospital.

CANCERS OF THE FACE AND MOUTH

The treatment of cancers of the face and mouth may be followed by severe cosmetic problems as well as functional defects in speech, mastication, swallowing, and salivary control. Between one third and one half of the patients treated for oral cavity cancer will remain cured following surgical excision. Successful radical oro-facial surgical procedures have increased both the number of patients and their extent of facial disfigurement.

Because of the physical area of disfigurement, it is extremely difficult for the patient to mask or hide his defect, and this can create severe psychologic problems, greater in many ways than the problems from the physical deficit itself. Rehabilitative care should start ideally in the preoperative period and should include counseling of the patient and the family, in close communication and cooperation with the surgeon. The patient should be informed of the type of surgery planned and of the expected result to assist him in long-term planning. Smoking and alcohol should be prohibited during treatment.

Enucleation of an eye creates a reduction in total visual field and change in the accuracy of depth perception, as well as need for cosmetic prosthetic replacement. The patient should be given training in occupational therapy in eye-hand coordination for improved depth perception accuracy.

The patient with oral cancer will need adequate presurgical dental care with elimination of sepsis by appropriate restoration or extraction. Proper oral hygiene and care are necessary before, during, and after radiation therapy to the mouth and face to reduce the incidence of jaw infection and to increase comfort.[23]

A bland mouthwash of sodium bicarbonate or 50% milk of magnesia is soothing and effective. Use of a toothbrush should be avoided for about two weeks following surgery. Dry mouth due to decreased salivation after radiation therapy is relieved by discontinuation of smoking and alcoholic drinks and by using gelatin and glycerin fruit-flavored pastilles or a mixture of glycerin and proprietary mouthwash. Increase in environmental humidity also helps, and increased fluid intake should be advised when possible. Effective relief for several hours follows the application of small local quantities of liquid silicone. Dry, bulky foods should be avoided. When dryness of the nasal cavity is a problem, daily irrigation with corn syrup provides relief.[23]

Dental appliances constructed for oral cancer patients should provide maximum coverage, have multiple clasps, permit modification, and be compatible with underlying tissues. Facial prostheses can be constructed as soon as edema has subsided and healing is adequate, usually after 6 weeks from the date of surgery. Provision of dentures may have to wait until 3 to 12 months have elapsed, when tissues that have been irradiated have recovered. Appropriate construction of obturators and other appliances will minimize speech problems associated with palatal defects.

Resection of the tongue will result in difficulties in speech and swallowing. The problem is increased by the amount of tongue tissue resected. Speech training and instruction in mastication and swallowing, using small amounts of clear water for practice, can increase proficiency and avoid aspiration problems.

Swallowing difficulty and aspiration pneumonia result when one or both superior laryngeal nerves have had to be sacrificed. Correction of this problem may be effected by splitting (myotomy) of the cricopharyngeus muscle. The patient can then be taught to exhale while swallowing, clearing the laryngeal vestibule of material that otherwise would be aspirated.

Late surgical reconstruction of orofacial defects depends on factors of extent of functional loss, degree of deformity, and patient reaction to his disability. Reconstruction procedures include flap closures for fistulas, mobilization or reconstruction of the tongue, bone graft mandibular replacement, muscle transfers for facial nerve palsies, implants, and autografts for correction of contour defects.

Surgical procedures of several stages are often necessary for these patients, and there are tedious waiting periods. Rehabilitation can be greatly assisted by the supplying from the surgical and dental team of either a temporary or a permanent maxillofacial prosthesis, functional if possible, to lessen disfigurement and restore all possible useful activity. Reconstructive plastic surgery requires patience and understanding on the part of the patient himself and the team rendering him care and support.[12a]

Social service should work extensively with the patient and the family. Vocational counseling should be obtained to encourage the patient, to arrange for continuation of his regular job, or to retrain him if necessary in a new field of work. Every effort should be directed to prevent the patient from adopting an attitude of social withdrawal.

Recurrent tumor can cause fear and anxiety as well as pain. If there is gradual upper airway obstruction, a tracheostomy may be indicated to relieve choking sensation. If swallowing is blocked, a cervical esophagostomy for feeding is advised in preference to a gastrostomy.

Cachexia develops in the debilitated and malnourished cancer patient. Cachexia is characterized by weakness, depletion, and anorexia. There are electrolyte and water abnormalities, and vital functions fade progressively. The patient becomes detached and apathetic and may be restless and anxious, representing mental effects of metabolic disruption. [23]

Protein loss is a major factor in cachexia, and attempt at replacement is difficult. Frequent small meals are better tolerated than intense efforts or large meals. Parenteral feeding can be helpful, but at best it is not sufficient. Supplemental or tube feeding will be needed to supply a high-calorie, high-protein liquid intake.

CANCER OF THE LUNG

The rehabilitative measures concerned with cancer of the lung entail consideration of the effects created by thoracotomy, associated pulmonary resection, pneumonectomy, and other procedures. There may be postoperative restriction of pulmonary function, furthered by bandaging as well as pain; decreased respiratory reserve, depending on the amount of lung tissue removed; and impairment of the mechanics of breathing due to chest wall resection.

The rehabilitation program should include preoperative training in breathing control and proper coughing technic, taught to the patient as early as possible to promote understanding of directions and to create familiarity of the patient with the chest team. These procedures prior to surgery can assist in clearing the tracheobronchial tree of undesirable secretions. Breathing exercises include segmental control, especially of the basal segments, for aid in ventilation as well as in expansion. Proper bed position and posture is stressed. Effective coughing is taught by correct control of respiration rather than by force or volume of expelled air. Activity of the accessory musculature of the chest, neck, and back is reduced to minimize discomfort. Manual splinting of the chest, assisted by the patient, helps to reduce both discomfort and apprehension.

Technics of chest percussion and shaking are not added for the early postoperative patient because of the presence of the wound. If postural drainage is ordered, it should be continued until the cough is nonproductive and the patient is ambulatory.

Nonvigorous exercises of the arms, trunk, and lower extremities are taught, starting on the first postoperative day, for the purpose of restoring or maintaining full range of motion, strength, and circulation in the extremities, especially on the operated side, and to assist in stimulating coughing.[6, 24]

CENTRAL AND PERIPHERAL NERVOUS SYSTEM TUMOR INVOLVEMENT

Malignancies, whether primary or metastatic, involving the brain, spinal cord, and peripheral nerves create motor and sensory deficits and affect coordination. Treatment for the patient should begin as soon as the disability is diagnosed or the surgery is completed. The general treatment is the same as the standard recognized care program for such disabilities unrelated to cancer.

Focal weakness may result from the excision or transection of a major nerve. Treatment consists of range-of-motion exercises for the affected extremity and, if an arm, provision of a sling for support to limit subluxation and pain at the shoulder. A functional splint or ADL splint may answer an individual need, especially when the loss has been that of function in a dominant upper extremity. Special exercise and practice in the unilateral use of the normal arm and hand are especially important. In lower extremity unilateral paralysis, bracing with a long or short leg brace, depending on the need, may render the patient ambulatory. Muscle reeducation should be carried out when there is any evidence of reinnervation, either functionally or by electrodiagnostic testing.

When pain is the patient's complaint, a search should be made for its cause. Eradication of infection may eliminate pain. Pain from nerve compression or invasion by tumor may respond to radiation therapy or surgical decompression. The same is true for bone tumor, in which decompression may be combined with cryosurgery for removal of the tumor from its site by the freezing action of liquid nitrogen.

Local nerve block with alcohol or 5% to 7% phenol is useful for focal pain. Narcotic drugs, when required, should be started with small doses with attempts made to potentiate effects with promethazine hydrochloride (Phenergan) or chlorpromazine hydrochloride (Thorazine).

In addition to the use of medication and radiation therapy, control of intractable persistent pain may require a neurosurgical approach. Procedures include nerve block, rhizotomy, cordotomy, medullary tractotomy, thalamotomy, and lobotomy. Hypophysectomy or adrenalectomy may be useful, especially in the presence of breast carcinoma with multiple metastases. Unfortunately, after cordotomies, which are usually bilateral when pain is persistent, the pain may recur even though there is residual cutaneous anesthesia. The lobotomy procedures are done more for anxiety relief than for relief of pain per se.

Causalgia and paresthesia are occasionally helped by the application of hot packs or cold packs and by gentle massage. Sympathectomy may also help. It has been recommended in some centers that for the persistent case of causalgia, without known etiology or therapeutic response, the use of hypnosis administered by an accredited hypnotist may be helpful.[11]

Problems related to sensory losses are not responsive to direct therapy. The patient can be assisted, however, by training and substitution of other sensory modalities, or other patterns of activity, to minimze the effect of the loss. Impairment of motor coordination also may be improved by special training.

LEUKEMIA, THE LYMPHOMAS, AND HODGKIN'S DISEASE

The lymphomas, either during treatment or after therapeutic control, may leave the patient with residual weakness and diminished tolerance for stress and fatigue. With these as well as with metastatic carcinomas of other origin, pathologic fractures may threaten or occur. These require appropriate orthopedic and radiation therapy. Depending then on the individual disability and tolerance, treatment begins with active range-of-motion exercises and later, if the patient is able to participate, training and assistance in ambulation with appropriate supports to mobilize him. Otherwise, a goal of wheelchair mobilization should be sought.

Neuromyopathies, weakness, debility, and the inanition secondary to anemia all respond to supportive care with an active program of bed exercises and training in independence, progressing to out-of-bed activities.

REMOTE EFFECTS OF CANCER ON THE NERVOUS SYSTEM

Remote effects of cancer on the nervous system produce varying clinical syndromes of neuromyopathy. Symptoms may develop after those of the neoplasm, or they may precede other evidence of the neoplastic disease. Muscular weakness and wasting occur, with weakness most frequently in the limb girdle and proximal muscles rather than distally. Pain may accompany the weakness. There is a striking tendency for symptoms to remit for long or short periods without regard to the course of the tumor or its treatment.[1] Therefore, it is important to teach the patient exercises as preventive measures against the effects of disuse and lack of range of motion.

Coupled with the remote effects of cancer on the nervous system and the muscles are similar effects of medications such as the steroids and the cancer chemotherapeutic agents. Supervoltage radiation therapy also may create malfunction in the nervous system. Peripheral nerve dysfunction tends to be late in appearance and may not occur until a year or longer after the radiation therapy has been administered, whereas effects on the central nervous system are earlier and are directly related to the degree of change within the cell of the central nervous system itself.

PSYCHOSOCIAL AND VOCATIONAL PROBLEMS

The patient with cancer may develop great psychologic problems. Anxiety results from the fear of prolonged suffering, mutilation, and death. The individual patient is likely to relate his problem and his future to what he has learned from others with cancer diagnosis. He may fear being unwanted or considered different by friends or family, and may be so distressed as to attempt complete withdrawal.

An effort must be made to gain the support of the family to promote the patient's participation in social activities and work. Help can be obtained from religious advisors and mutual assistance groups, such as the "ostomy clubs" and laryngectomy and mastectomy group. As previously described, social service personnel can direct in assisting and understanding these needs.

The cost of hospital, surgical, and rehabilitation care overburdens most budgets. Community resources should be fully utilized by families and offered by those concerned with the care of the patient before shortage of funds has created financial hardship and delay in complete care and rehabilitation.[9, 10]

A vocational outlook is important to the patient who may otherwise fear that because of his illness his family will have to make sacrifices of such things as his children's education, recreation, and needed clothing. The vocational counselor will aid immensely here to help in motivation of the patient toward rehabilitation and work. Not to be overlooked by vocational rehabilitation services is the patient with a limited work expectancy who has hope for at least temporary employment. Common sense will dictate plans commensurate with the prognosis rather than elimination of the patient as ineligible.[25] The majority of patients referred in the rehabilitation program at Memorial Hospital have been in the employable age group (Fig. 30-2).

EVALUATION FOR TREATMENT AND RESPONSE

On initial examination each patient should be classified according to expected treatment goals. Factors to be considered are cancer diagnosis and stage of disease, presence or absence of known metastases, age, occupation, health history, other disease diagnoses, x-ray and laboratory determinations, and the findings on physical examination. It has been determined that 75% of patients can be given goals of restorative or supportive result.[18] Patient's classified for a "restorative" goal are those considered to have excellent potential for cancer control and for return to independence for either appreciable or indefinite periods. The group classified for a "supportive" goal can expect the progress of disease and the effect of disability being kept under at least temporary control. Twenty-five percent of patients can only expect "palliation" as the result of care.

Evaluation of response to rehabilitation care should be made with consideration of disease status, age, total handicap, and work or activity outlook as background for the physical response. Grades can be given, arbitrarily using numerical values from 0 to 4, as listed in Table 30-1. The response of the individual case and the

Table 30-1. Grade of response and distribution of evaluation of unselected patients with cancer-related disability (1019 patients)

Grade	Number	Legend
0	211	No change or improvement
1	110	Slight improvement, marked disability
2	320	Moderate improvement and appropriate response to reha-bilitation care, residual disability
3	270	Marked improvement, residual disability
4	108	Fully independent, no residual disability
	1019	

total evaluation according to groups and classifications of goals have indicated appreciable benefit derived by the majority of unselected patients. Eight out of 10 respond with improvement, 7 with full expected response according to goal, and 3 with either marked improvement or full unhandicapped independence. Only 2 in 10 allow no change or improvement (Table 30-1).

In a hospital for acute diseases, length of time available for provision of a rehabilitation medicine program can range from one day to several weeks. Only 1 in 5 patients is hospitalized long enough to receive treatment for over 30 days, and half are given rehabilitation treatment for up to 2 weeks. A small number of patients cannot be accepted for treatment because of individual findings or complications or because of their refusal of care.

Persons needing further rehabilitation following discharge should be treated as outpatients or by a physical therapist or visiting nurse at home. Occasional patients are candidates for admission to a rehabilitation center for intensive care. The evaluation of the responses in this group of patients gives results similar to the previously discussed inpatient population.

CONCLUSION

It is essential to establish early medical staff recognition of patient disability and of rehabilitation potential and to stimulate prompt referral for rehabilitation care. Rehabilitation therapy started early in the period of disability can eliminate development of periods of hopelessness, frustration, and despair. Preoperative training and counsel for the patient about to undergo surgery will add both psychologic and physiologic benefits.

Disability from cancer or its treatment can be considered by the same criteria as are used for disabilities not connected with cancer. Goals for restorative and supportive care can be set, and there is need for rehabilitation care in the form of palliative assistance for the patient with incurable and even terminal cancer.

No accurate judgment of life expectancy or the length of time for engaging in useful activities can be estimated for a patient with cancer despite the general tables available. Therefore it is unrealistic and contrary to good rehabilitation practice to defer the provision of rehabilitation services for a waiting period to determine the status of the disease or the question of its spread.

It would be a great service for the cancer patient if we would change our current basic appraisal or concept of the results of cancer treatment from that of "cure" to that of "control." Analogous to tuberculosis, recurrent disease at the original or another site is always a possibility, but after disease "control" the patient may be returned to a relatively normal life. Such consideration would eliminate the unfortunate problems created by an arbitrary time limit of 2 to 5 years in consideration of eligibility of a cancer patient for a positive rehabilitation program or for vocational rehabilitation.

Social service can remove otherwise insurmountable obstacles, and vocational rehabilitation should be employed even in the case of limited life expectancy if by so doing the patient may again become temporarily employed or return to maintenance of a home situation, such as that of a mother with children. Each patient should be considered on an individual basis, with evaluation of the medical findings, the prognosis, and maximum eventual gain to the patient, his family, and the community. Comprehensive medical care will then have been provided for that patient to effect improvement in his quality of survival.

REFERENCES

1. Brain, W. R. B., and Norris, F. H., editors: The remote effects of cancer on the nervous system, contemporary neurology symposia, vol. 1, New York, 1965, Grune & Stratton.
2. Burgess, E. M., and Romano, R. L.: Immediate postsurgical fitting of children and adolescence following lower-extremity amputations, Committee on Prosthetics Research and Development, Inter-Clin. Inform. Bull. 7:1, 1967.
3. Dietz, J. H., Jr.: Rehabilitation of the patient with disability resulting from cancer or its treatment, World Committee on Research in Rehabilitation, Tenth World Congress of the International Society, Wiesbaden, West Germany, Sept., 1966.
4. Dohan, F. C., Moss, N. H., et al.: Surgical convalescence, Ann. N. Y., Acad. Sci. 73:381, 1964.
5. Goldsmith, H. S., Santos, R. de los, and Beattie, E. J., Jr.: Omental transposition in the control of chronic lymphedema, J.A.M.A. 203:1119, 1968.
6. Haas, A., Gayrard, P., and Dietz, J. H., Jr.: The role of the physiatrist in post-thoracic surgery. Read before the Georgia Tuberculosis Society and Georgia Medical Association, Columbus, Ga., May, 1966.
7. Hirschberg, G. G., Lewis, L., and Thomas, D.: Rehabilitation, Philadelphia, 1964, J. B. Lippincott Co.
8. Klieger, P. A.: Cancer, vol. 2, Washington, D. C., 1965, Vocational Rehabilitation Administration.
9. Krusen, F. H., Kottke, F. J., and Ellwood, P. M., Jr.: Handbook of physical medicine and rehabilitation, Philadelphia, 1965, W. B. Saunders Co.
10. Mozden, P. J.: Neoplasms. In Myers, J. S., editor: An orientation to chronic disease and disability, New York, 1965, The Macmillan Co.
11. Healey, J. E., Jr., editor: Ecology of the cancer patient, The Interdisciplinary Communications Program, The Smithsonian Institution, Washington, D. C., 1970, The Interdisciplinary Communications Associates.
12. Proceedings of the Conference on Research Needs in Rehabilitation of Persons With Disabilities Resulting From Cancer, Nov., 1965, United States Department of Health, Education, and Welfare, Vocational Rehabilitation Administration and the Institute of Physical Medicine and Rehabilitation, New York University Medical Center.
    a. Converse, J. M.: Rehabilitation of the patient with facial disfigurement resulting from cancer.
    b. Farrow, J. H.: Rehabilitation following radical breast surgery.
    c. Higinbotham, N. L.: Rehabilitation following amputation.

d. Klieger, P. A.: Background material on cancer research and rehabilitation.
e. Levin, N. M.: Rehabilitation after total laryngectomy.
f. Postel, A. H., Grier, W., Robson, N., and Localio, S. A.: Rehabilitation of the colostomy patient.
g. Urban, J.: Rehabilitation of the mastectomized patient.

13. Prosthetic fitting of children amputated for malignancy, New York, 1965, Research Division, School of Engineering and Science, New York University.
14. Rusk, H. A.: Rehabilitation medicine, ed. 2, St. Louis, 1964, The C. V. Mosby Co.
15. Russek, A. S.: Investigation of immediate prosthetic fitting and early ambulation following amputation in the lower extremity, New York, 1969, Publications Unit, Institute of Rehabilitation Medicine, New York University Medical Center.
16. Shimkin, M. B.: Science and cancer, Public Health Service Pub. no. 1162, Washington, D. C., 1964, United States Department of Health, Education, and Welfare.
17. Friedmann, L. W.: Rehabilitation of amputees. In Licht, S., editor: Rehabilitation and medicine, New Haven, Conn., 1968, Elizabeth Licht, Publisher.
18. Dietz, J. H., Jr.: Rehabilitation of the cancer patient, Med. Clin. N. Amer. 53:607, 1969.
19. Burgess, E. M., Romano, R. L., and Zettl, J. H.: The management of lower-extremity amputations, TR 10-6, Washington, D. C., 1969, United States Government Printing Office.
20. Turnbull, R. B., Jr.: Instructions to the colostomy patient, Cleveland Clin. Quart. 28:134, 1961.
21. Katona, E. A.: Learning colostomy control, Amer. J. Nurs. 67:534, 1967.
22. Postel, A. H., Grier, W. R. N., and Localio, S. A: A simplified method of irrigation of the colonic stoma, Surg. Gynec. Obstet. 121:595, 1965.
23. Zegarelli, E. V., and others: Maintaining the oral and general health of the oral cancer patient (part 1), CA 19:168, 1969.
24. Brompton Hospital. Physiotherapy for medical and surgical thoracic conditions, 3rd rev., 1967. Available at Brompton Hospital Physiotherapy Department, London, S. W. 3.
25. Dietz, J. H., Jr.: Cancer rehabilitation in the work force, Trans. Bull. (Industrial Hygiene Foundation of America) no. 43, p. 61, 1969.
26. Lauder, E.: Self-help for the laryngectomee, ed. 2. Available from author at 6334 Dove Hill Drive, San Antonio, Texas 78238.
27. Diedrich, W. M., and Youngstrom, K. A.: A laryngeal speech, Springfield, Ill., 1966, Charles C Thomas, Publisher.
28. Turnbull, R. B.: Construction and care of the ileostomy, Hosp. Med. 2:38, April, 1966.
29. Druss, R. G., O'Connor, J. F., and Stern, L. O.: Psychologic response to colectomy, Arch. Gen. Psychiat. 20:419, 1969

CHAPTER 31

GERIATRIC REHABILITATION

The philosophy and technics used in the rehabilitation of older people are essentially the same as those in general rehabilitation, modified to comply with the physiology of the aging patient. Modern medicine is far removed from the pessimistic early Roman concept that maintained that "old age is a disease itself" *(senectus ipsa est morbus)*. It is increasingly being realized that aging is not synonymous with disease nor are there any illnesses exclusively characteristic of old age. Nonwithstanding this, clinical and statistical experience demonstrates that people in the older age group have more than a proportionate share of disabling diseases in relation to their younger contemporaries. This preponderance is further aggravated by the fact that older disabled patients frequently suffer from one or more chronic illnesses in addition to their disability. Such complications have a major bearing on their rehabilitation procedures and prognoses. Whereas their cardiorespiratory and other functional limitations must always be carefully observed, these patients should not, a priori, be excluded from the benefits of restorative medicine. Experience indicates that although such patients often start and progress slowly, the final results are most encouraging. Without rigid insistence on vocational goals and with emphasis on the highest possible degree of functional restoration that will enable the patient to take care of his own daily needs, the physician will find the physical restoration of the elderly patient a rewarding addition to everyday practice.

AGING AND PROLONGED IMMOBILIZATION

One must be aware of the physiologic changes that occur with senescence in evaluating any elderly individual. Impairment of many functions begins at about 30 years of age. Although functional decrements at the cellular level may not exceed 15%, total organic performance may be depreciated by 40% to 60%. Graphic

659

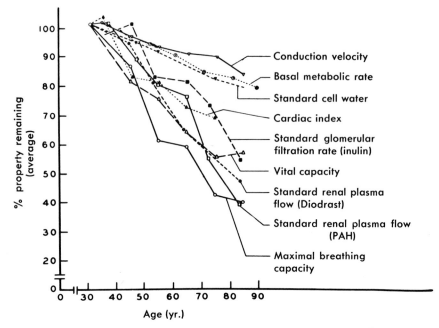

Fig. 31-1. Composite diagrams of the percent of various (human) functional capacities or properties remaining at various ages. (Based on data from Shock[7]; from Strehler, B. L.: Time, cells, and aging, New York, 1962, Academic Press, Inc.)

representation of physiologic decline in older patients is shown in Fig. 31-1, based on the work of Shock and collaborators.[1]

Studies of pulmonary function in the aged consistently reveal an increase in residual volume and a decrease in the maximal voluntary ventilatory capacity.[2] Resting cardiac output diminishes by 30%.[3] Changes in the velocity of nerve conduction have been observed, with that of the ulnar nerve reduced by 15%.[4] Muscle strength is decreased,[5] reaction time is prolonged, and the speed of motion is significantly reduced.[6]

Since the reserve capacity of the older individual is diminished, his response to stress likewise is reduced. Many physiologic characteristics that show little or no change at rest are significantly altered by stressful situations and return only slowly to resting values.

Shock[7] very succinctly states:

> In my opinion, the impairments in performance in humans associated with advancing age are due to (a) dropping out of functional units in key organ systems, (b) some impairments in the functional capacities of cells remaining in the body of the aged, and (c) breakdown of neural and endocrine integrative functions in the individual. Although advancing age is accompanied by biological impairments that offer fertile ground for the development of disease and pathology, there are compensatory devices which can maintain effective behavior in the human into advanced old age. Investigation of these as yet unmeasured and little

understood inner resources over the entire life span of the individual is
the goal of research in gerontology.*

The clinical implications are clear. In addition to pathologic changes, the natural
decline in physiologic functions in senescence must be considered in the total assessment of the aged patient.

Despite its danger, bed rest still remains the major nonspecific therapeutic tool
in the management of disease. The often quoted work of Deitrick and associates[8] is
the first experimentally supported study that establishes the detrimental effects of
inactivity. Among the elderly disabled the hazards of prolonged periods in bed have
further serious clinical implications. They include the increased incidence of venous
thrombosis, pulmonary embolism, hypostatic pneumonia, urinary retention, abdominal distention, contractures, decubiti, weakness, and constipation.

One of the clinician's primary responsibilities is to shorten the time of total
immobilization and to introduce early measures to prevent the development of
secondary complications. The adverse effect of prolonged immobility is more easily
prevented than reversed.[9]

SPECIAL PROBLEMS

While it is true that all types of disabling diseases can be found among older
people, there are a few that deserve special mention because of their high incidence.[10]

Hemiplegia. Since pathologic changes of the brain resulting in hemiplegia occur
predominantly among elderly persons, the rehabilitation procedures concerning
this disability, which are discussed in Chapter 29, have been presented with their
needs specifically in mind. In 1963, of the 201,166 deaths from stroke, 162,755
(80%) were people 65 years of age and over; and of these, 73,388 occurred in the
75 to 84 age group.[11] The rate of survival among people who suffer strokes is 80%,
and a major part of the resulting individual and social burden could be obviated
with the timely application of preventive and rehabilitative measures.[12]

In general the rehabilitation of geriatric patients is necessarily geared to the social
and economic realities of old age in the modern world. The impact of permanent
disability on the patient and his family has widespread implications. Economic
resources are usually limited, and the patient's need for physical care and psychologic support constitutes a constant demand on the family and the community.

The treatment of geriatric patients with hemiplegia must be carried out in the
context of their physiologic conditions and social circumstances. Specific concomitant
conditions as well as varying degrees of physiologic deterioration can be expected to
complicate the effects of brain damage resulting from the stroke. Diabetes mellitus,
arthritic conditions, and hypertensive cardiac abnormalities are common in the
elderly stroke patients. It is self-evident that although rehabilitation procedures
should be dynamically utilized, they must at the same time be clinically realistic and
closely geared to social planning. Discharge to the patient's family is usually but not
always preferable or even possible. Frequently there is no surviving spouse or receptive children, and in some situations the patient's condition dictates continued

---

*From Shock, N. W.: Aging—some social and biological aspects, Pub. no. 65, Washington,
D. C., 1960, American Association for the Advancement of Science, p. 255.

hospitalization. Wherever the patient lives, maintenance rehabilitation should be continued.

Hip fractures. Fracture of the proximal end of the femur (hip fracture) is one of the most prevalent traumatic conditions among the elderly, particularly among women.[13-16] A survey of 10,250 patients of both sexes in the municipal hospitals in New York City revealed that 5% were found to have recent or old hip fractures, and that 87% of all of the hip fractures found were in people 55 years of age and over.[17] Advancement in surgical technics has greatly reduced mortality in patients with hip fractures. However, unless early surgical care and appropriate restorative measures are instituted patients may suffer permanent loss of ambulation.

The period of postsurgical bed rest should be as short as possible to prevent cardiopulmonary complications. Without regard to the surgical procedure used, in the absence of significant contraindications passive range-of-motion exercises should be started at the earliest possible time following the operative procedure. A patient may stand in the parallel bars with nonweight bearing after 48 hours of postoperative rest, provided there is no surgical complication. Caution should be used at all times to prevent accidental falls.

The duration of nonweight-bearing ambulation depends on the surgical procedures used and radiographic findings. Checking by radiographic methods is desirable once every 2 months. The nature of the fracture, proximation of fragments, and location of the nail, pins, and plate must be taken into consideration when determining weight bearing. The absence of callus formation does not contraindicate weight bearing. If after 6 or 8 months no radiologic evidences of callus formation are detectable, the clinician is faced with a delicate decision. Obviously, the absence of radiologic signs of this formation should not commit the patient to immobility for the rest of his life. If the fragments show a good or fair approximation, despite the absence of a good callus, it is our practice to proceed with careful partial weight bearing and, wherever possible, full weight-bearing ambulation. The recent development of the adaptable ischial weight-bearing brace with Patton foot has permitted geriatric patients to ambulate with crutches as early as several days after the operative procedure. In general it is common to specify "partial weight bearing" for hip fracture patients. However, since the tendency of geriatric patients is to place more weight on the affected limb than is desirable, the entire problem of retraining them in ambulation must be carried out under the most careful supervision. It must always be remembered that geriatric hip fracture patients present special problems by virtue of limitations of balance and endurance as well as sensory deficits.

When a patient complains of pain during weight bearing, an investigation is always warranted. One must be aware that aseptic necrosis of the femur head, infection, or slippage of the nail or screws may occur. If any of these complications do occur, proper surgical correction should be instituted if possible. Occasionally an ischial weight-bearing brace is indicated for these cases instead of surgical reintervention.

Since for many reasons elderly people are predisposed to hip fracture, one of the most important goals in the care of the geriatric population is the prevention of this condition. The presence of osteoporosis in the elderly population is a frequent underlying causative factor.[18] Dizziness, syncope, poor vision, and shuffling gait are common characteristics of elderly people that may result in falls. Environmental conditions such as small rugs, waxed floors and other hazardous conditions, poor lighting, crowded living quarters, and dilapidated housing are also conducive to causing hip fracture and should be avoided. Falls among the elderly can be minimized by training them to be cautious and by improving the environment in which they live.

Amputations. The special problems that arise in the rehabilitation of the amputee have been reviewed in the earlier chapter concerning the rehabilitation of the patient with musculoskeletal problems. The prolongation of life expectancy has as its corollary an increase in the amputation of lower extremities among the elderly resulting from the higher incidence of vascular insufficiency related to arteriosclerosis and diabetes.

Geriatric amputees are more likely to present special problems that must be dealt with in prosthetic rehabilitation. Decrease in cardiopulmonary capacity, poor neuromuscular coordination, visual defects, weakened musculature, or limitations in range of joint movement impose restrictions in the intensity of preprosthetic and prosthetic training programs. The goals of prosthetic rehabilitation for geriatric amputees should be realistic. Their most common needs are retraining in the Activities of Daily Living in terms of their individual needs and circumstances, and recreation. Vocational aspects are generally minimal.

Within the limitations listed, most elderly below-knee amputees can be helped, even if the amputation is bilateral. Likewise, those with unilateral above-knee amputations can usually be trained, although a degree of disturbance in ambulation can be expected. The prognosis for unilateral above-knee amputees is poor because of the danger of amputation of the remaining leg. Experience with geriatric bilateral above-knee amputees is that while functional return is attainable insofar as the Activities of Daily Living are concerned, ambulation is usually precluded because of the presence of proprioceptive instabilities and increased energy cost.

The trend is toward immediate fitting for younger lower extremity amputees, although some investigators have successfully used immediate fitting for selected older patients. In prescribing a prosthesis, certain components such as the single-axis knee joint, SACH foot, or suction suspension may not be the best choice for geriatric amputees. Instead, the Otto-Bock knee or manual knee lock, single-axis ankle with wooden foot and toe-break, and pelvic band suspension may be better choices.

During the period of preprosthetic training it is essential that cardiac capacity, motivation, and the ability to use a prosthesis be determined. Work-capacity evaluation and dynamic electrocardiographic study are useful tools in estimating cardiac capacity. Pylons should be used as soon as possible despite the possibility that intermittent claudication might occur. In general, ambulation by any means is considered to be acceptable for older patients if it is safe and functional.

There is a high incidence of cardiac complications among elderly amputees. Therefore, to make them as functional and as active as possible one should not hesitate to consider providing a wheelchair along with a prosthesis when either or both are indicated. Upon discharge from the hospital it is desirable to follow them in the community by utilizing some type of home care program to meet prosthetic and general health needs.

When there has been a single amputation, preventive foot care for the remaining limb is of vital importance, and both the patient and his family should be trained in all aspects of this activity. The patient's remaining foot should be inspected by a professional person at frequent and regular intervals whether he is at home or in the hospital.

Gait patterns. Often older people develop peculiar gait patterns that at times can become incapacitating. The existence of these gait anomalies have long been known and repeatedly studied.[19, 20] These pathologic gaits create difficulties in walking and in climbing stairs and can result in serious accidents. This condition, which is characterized by a decrease in motor power and unsteadiness, is often erroneously referred to as senile paraplegia. The term is inaccurate and misleading, since these patients do not have any of the clinical characteristics of paraplegia.

Critchley,[21] who studied this problem quite extensively, classified the gait disorders according to the site of the central nervous or skeletal system involvement as follows: (1) cortical, (2) subcortical, (3) spinal, and (4) muscular. Another classification had been recommended earlier by Lhermitte,[22] a French investigator. The etiology of gait disorders among geriatric patients could be further defined in terms of the wide variety of decline of function and conditions that mark the physiology of the older person.

Senile gait disorders are likely to escape the examiner's attention during the course of routine clinical examination. On examination in the customary sitting or reclining position, certain older patients will show little, if any, diminution of muscle strength. However, when the patient's gait is tested, obvious changes can be observed. When these people begin walking, they look frantically for support and take short shuffling steps; frequently after a few such steps their knees buckle. This gait pattern was referred to by nineteenth century French clinicians as *l'astasie trepidante,* sometimes called Petrén's gait.[23]

If this condition remains uncorrected, as it frequently does, the patient is confined to his bed, with all of the ill effects consequent to this. Bracing is ineffective and often aggravates the condition. Treatment consists of the strengthening of muscles impaired by disuse atrophy, the liberation of joints restricted by fibrous ankylosis, and as much amelioration of the poor gait pattern as is attainable. Individual elements of gait, standing stance, and swing phase must be practiced separately and then fused into the best rhythmic motion possible for the patient being treated. Personal supervision and constant encouragement by the therapist are essential prerequisites.

• • •

Other disabling conditions commonly found among older people require essentially the same treatment as in younger patients. The appropriate chapters of this book should be followed as a guide, always keeping in mind that, although the disability may be identical, the altered physiology of older persons and complicating pathologic factors may require many changes in the accepted regimen.

GENERAL CONSIDERATIONS

The widely prevailing practice of medical care for the disabled older person is paradoxical in that it combines the finest technics of medical science for the preservation of life with negative and antiquated concepts about functional living. Beyond the acute phase of illness, too much reliance is placed in *vis medicatrix naturae*. This approach has become increasingly invalid in the face of current developments in the rehabilitation of older persons.

> The fundamental study of long-term illness by the Commission on Chronic Illness once again reaffirms the fact that hemiplegic patients, for example, are ordinarily discharged from the general hospital as soon as the acute medical crisis is passed without being given any retraining. The report continued that the patient "may deteriorate quite rapidly to the helpless, speechless, bedridden invalid with incontinence, contractures, and bedsores, in spite of the fact that we know that with proper rehabilitative care many of these patients can learn again to talk and walk, to control their body functions, and lead a fairly normal and productive life.*

Rehabilitation practice with older patients has amply demonstrated the enormous needs that exist, as well as the possibilities for meeting them. Work in this area has generally been of a pioneering nature, usually handicapped by inadequate funds and insufficient personnel.

Patients who fall within the purview of geriatric rehabilitation have been classified by Dacso[25] into three principal groups: (1) the obviously handicapped patient (those with hemiplegia, etc., arthritides, fractures, amputations, and neuromuscular diseases), (2) the chronically ill patient without signs of a manifest disability (those with chronic cardiac disease, chronic pulmonary diseases, etc.), and (3) the elderly person who is not obviously ill but whose physical fitness is impaired. Although there has been some movement in the direction of including patients in group 2, at the present time the use of the comprehensive rehabilitation approach is largely restricted to the patients who fall in group 1. Logically there is no reason why all three groups should not ultimately receive the benefits of comprehensive rehabilitation. Indeed, the importance of early rehabilitation can hardly be overemphasized, since the condition of patients in each of these groups tends to deteriorate rapidly in the absence of dynamic care.

Restoration of maximum function is the ultimate goal for people of all ages. In the case of older people, maximum function is more likely to mean an adjustment to the demands of daily living and psychologic adjustment rather than a return to gainful employment. Some older persons will continue to work; where retirement is contemplated, it should be planned and gradual if possible. For most disabled older people, however, self-sufficiency in ambulation, washing, eating, dressing, and toilet activities is the primary goal. When the patient has attained his maximum performance in these areas, as well as having made the best psychosocial adjustment possible for himself, he may be said to be rehabilitated. To maintain self-sufficiency as long as possible is the challenge.

Recognition of the relative nature of rehabilitation progress in older patients is of major importance. Since return to work is not usual among them, this cannot be

*From Commission on Chronic Illness: Care of the long-term patient, vol. 2, Cambridge, Mass., 1956, Harvard University Press, p. 15.

used as a dependable index of successful rehabilitation. Adequate functioning in daily living becomes the principal measure of success. Improvement in personal functioning is related to age and physical condition. Likewise, psychologic condition and social resources are vital factors in determining the degree to which the patient can return to normal living. The older patient who learns to be self-sufficient in various aspects of self-care may frequently have made a greater relative gain than a younger person who has returned to work. This is true whether measurement is in terms of the patient's personal adjustment or of the adjustment of those who would otherwise have to care for him.

Goals must be frequently modified in the rehabilitation of older patients. This applies to the time allowed for specific gains as well as to the final objectives. In the rehabilitation of older people, retraining in the capacity to carry on the Activities of Daily Living can be successfully accomplished, provided there are appropriate modifications in both tempo and goals. This is of particular importance, since transfer from a wheelchair to the bed or toilet, eating, shaving, dressing, and washing are significant contributions to the older person's comfort and efficiency as well as to those who would otherwise have to carry out or assist him with such functions.

The prevention of disability is always preferable to its alleviation or cure. In the older part of the population, prevention reaches back into the middle years of life and continues to its termination. Proper diet, adequate safety rules, freedom from excessive physical and emotional strain, adequate income, and satisfying recreation and social relationships are all essential for ensuring health in later years. Of extreme importance is the early detection and, if possible, elimination of any form of disease or maladjustment.

The teaching of health habits and health information as it affects older people should begin early and should be clear and practical. Proper nutrition and dental care[26] have their roots in the early years, and special care should be taken that they are maintained later on. Malnutrition is frequently related to dental as well as economic causes.

A certain degree of biologic regression is an inevitable consort of advancing years. This being so, efforts directed at rejuvenation have always failed and should not be encouraged. What elderly people need is a mode of life that will allow them to prevent, or at least to postpone, the occurrence of disabling chronic diseases.

In chronic illness, efforts at prevention are unfortunately not so successful as in many infectious diseases. The main tool in forestalling severe complications of chronic disease is their early discovery. This fact has long been realized by authorities in preventive medicine and public health, but unfortunately it is still not sufficiently recognized by all physicians. The sources of many disabling chronic diseases can be traced back to simple origins, at which point appropriate countermeasures could have prevented their full development.

Widespread misconceptions notwithstanding, there is no dietary regimen specifically designed for the aging person. The diet prescribed for the elderly patient should be geared to individual physiologic or pathologic requirements. It is not age but rather the disease that will determine the prescription. For example, this means that the prescription will call for a diabetic, ulcer, or gout diet rather than one specifically designed for a person of any particular age. A review of elderly patients' eating habits may reveal that, as a result of self-imposed dietary regimens or—what is even worse—fads, the intake of essential food elements is severely curtailed. This insidious practice must be corrected, lest it result in complete or partial malnutrition with all its far-reaching consequences. It is true that in some instances the aging gastrointestinal system will not properly absorb and utilize all essential food elements. If such a situation is detected by the physician, proper diet, often augmented by parenteral administration of some essential food elements and/or vitamins, will become necessary. It is important, however, that such measures should be taken only if the need for them has been definitely established. To follow dietary fads or to use nutritional additives unnecessarily, as older people sometimes do, may lead to an unbalanced and insufficient diet. Diet prescription is as much a medical function as drug prescription, and it must be followed by continued supervision by a physician or some other specially trained professional person.

Not only is total physical inactivity contraindicated in most older people, but the beneficient effects that are derived from appropriate activity are very real. The growth of recreation centers and clubs for older people is useful in this connection. Perhaps more important than organized groups is the maintenance by each individual of an active interest and participation in his own life and that of those around him.

Work with older people in the psychosocial area yields excellent results. This is noteworthy in the light of the social and psychologic isolation that so often exists. Until comparatively recently, interest in working with older patients has been incidental in the fields of psychiatry, psychology, and social work. This situation is rapidly changing as it is being increasingly demonstrated that older patients can be helped with their social and emotional problems.

The twentieth century has brought a rapidly increasing proportion and number of older people in the general population. As a result a multiplicity of problems that run the gamut of individual adjustment have come to the fore. Widespread concern on every level has resulted in intensive efforts to solve the medical, economic, social, and psychologic difficulties that older people face.

A problem of particular importance in working with older disabled patients today is the paucity of resources in the community. In a setting in which youth, energy, and productivity are most prized, age and weakness are unconsciously and often openly depreciated. A predominantly rural society that has been urbanized and industrialized has set off a chain of reactions that has weakened the position of the older person. Sudden and compulsory retirement, limited skills, small apartments, quick population turnover, and the general weakening of family ties are a few of the factors that have increased the problems of older persons and their families. When illness and disability are added to normal senescence, the stress and strain on family relationships are intensified.

As a result, constructive discharge from the hospital after rehabilitation is completed is frequently a serious problem. What should be striven for is the nearest approach to normal living that is feasible for the patient and obtainable in the community. Placement should be of such a nature as to preserve the gains that have been achieved and to prevent or deter further deterioration. This is too frequently neglected, and rehabilitated patients are faced with the loss of their gains after discharge. The development of improved resources in the community is badly needed, and all efforts in this direction must be intensified.

Discharge should be to the patient's own home or to that of a friend or relative when possible. Beyond that, there are boarding, nursing, and old-age homes, as well as a number of developments in family and congregate living that are now going on in this country and abroad. For example, the development of Medicare and Medicaid as well as the proliferation of extended care facilities, meals-on-wheels, and similar programs serve to facilitate the adjustment of geriatric patients.

When rehabilitation services for older people are offered, vigorous steps must be taken to interpret the service to the professional public. This is essential at the present time, since a great deal of misapprehension exists about the possibilities for rehabilitating older people. The realistic potential of older disabled patients for making significant functional gains is widely underestimated. On the other hand, there is a tendency to refer patients for active rehabilitation whose real need is for custodial care with a maintenance rehabilitation program.

It is particularly important that personnel should be temperamentally suited to and interested in working with older people. Rehabilitation workers should always be carefully chosen because work with older disabled people places a special demand on the patience and maturity of personnel dealing with them. In the absence of suitable personnel, the care and treatment of older people is likely to degenerate into apathy and neglect.

The desirability of having separate geriatric rehabilitation services finds its justification in the needs of research and the more subtle differences to be found in working with older people. The weight of prevailing opinion seems to be that geriatrics should not be developed as a medical specialty. This opinion is based on the view that older patients do not suffer from unique diseases, nor do they require a fundamentally different kind of medical care. On the other hand, it is argued that an analogy can be made between a need for geriatrics and the need for pediatrics as separate specialties. Children have common developmental patterns and characteristic diseases. In addition, factors of temperament and interest enter into the selection of any specialty by individual practitioners. It can be similarly argued that the declining years of life have common factors that have not yet been fully conceptualized. When they are, it may be found that although there are no unique diseases in the aging group, there are enough special aspects of the same diseases to require special skills and knowledge. Factors of suitable temperament and interest in concentrating on the treatment of older people are likewise important.

Whatever the final outcome of the question as to whether geriatrics should become a medical specialty, there is much to be learned about the older patient without regard to the often too rigid organization of medical practice. Continued research is needed in many areas, including the physiologic, pharmacologic, clinical, rehabilitative, etc. Going beyond a narrow definition of medical practice, the problems presented by the aging population must be studied on many fronts. The entire spectrum of living is involved, and many types of research and services are needed. These must include the psychosocial and economic, as well as the medical.

REFERENCES

1. Strehler, B. L.: Time, cells, and aging, New York, 1962, Academic Press, Inc.
2. Gilson, J. C.: Medicine in old age. In Agate, J. N., editor: The aging lung, Proceedings of a conference held at the Royal College of Physicians of London, 18th and 19th June, 1965.
3. Brandfonbrener, M., Landowne, M., and Shock, N. W.: Changes in cardiac output with age, Circulation **12:**557, 1955.
4. Norris, A. H., Shock, N. W., and Wagman, I. H.: Age changes in the maximum conduction velocity of motor fibers of human ulnar nerves, J. Appl. Physiol. **5:**589, 1953.
5. Fisher, M. B., and Birren, J. E.: Age and strength, J. Appl. Psychol. **31:**490, 1947.
6. Pierson, W. R., and Montoye, H. J.: Movement time, reaction time and age, J. Geront. **13:**418, 1958.
7. Shock, N. W.: Aging—some social and biological aspects, Pub. no. 65, Washington, D. C., 1960, American Association for the Advancement of Science.
8. Deitrick, J. E., Whedon, G. D., and Shorr, E.: Effects of immobilization upon various metabolic and physiologic functions of normal men, Amer. J. Med. **4:**3, 1948.
9. Kottke, F. J.: Deterioration of the bedfast patient; causes and effects, Public Health Rep. **80:**437, 1956.
10. Holmes, E. F., Lee, M., and Fleming, W. L.: Long-term illness and the effect of the aging process on health. In Leavell, H. R., and Clark, E. G., editors: Preventive medicine for the doctor in his community, ed. 3, New York, 1965, McGraw-Hill Book Co.
11. The President's Commission on Heart Disease, Cancer, and Stroke, vol. 1, Washington, D. C., Dec., 1964, United States Government Printing Office.
12. Itoh, M., and Lee, M. H. M.: The future role of rehabilitation medicine in community health, Med. Clin. N. Amer. **53:**719, 1969.
13. Freyberg, R. H., and Levy, M. D., Jr.: Medical management of the patient with fracture of the hip, J.A.M.A. **137:**1190, 1948.
14. Howorth, M. B.: Textbook of orthopedics, Philadelphia, 1952, W. B. Saunders Co.
15. Itoh, M., and Dacso, M. M.: Rehabilitation of patients with hip fracture, Postgrad. Med. **28:**138, 1960.
16. Dacso, M. M.: Restorative medicine in geriatrics, Springfield, Ill., 1963, Charles C Thomas, Publisher.
17. Rusk, H. A., Silson, J. E., Novey, J., and Dacso, M. M.: Hospital patient survey, New York, 1956, Goldwater Memorial Hospital.
18. Moldawer, M.: Senile osteoporosis: the physiologic basis for treatment, Arch. Intern. Med. **96:**202, 1955.
19. Ross, A. T.: Some neurologic causes for difficulty in walking, Postgrad. Med. **15:**40, 1954.
20. Critchley, M.: Neurologic changes in the aged, J. Chronic Dis. **3:**459, 1956.
21. Critchley, M.: On senile disorders of gait, including the so-called "senile paraplegia," Geriatrics **3:**364, Nov.-Dec., 1948.
22. Lhermitte, J.: Etude sur la paraplegies des vieillard, Paris, 1907, Maretheux; La myosclérose rétractile des vieillards et ses syndromes acinéto-hypertoniques, Encephale **23:**89, 1928.
23. Petrén, K.: Ueber den Zusammenhang zwischen anatomischbedingten und functionelen Gangstörung im Greisenalter, Arch. Psychol. **33:**818, 1900; **34:**444, 1900.
24. Commission on Chronic Illness: Care of the long-term patient, vol. 2, Cambridge, Mass., 1956, Harvard University Press.
25. Dacso, M. M.: Clinical problems in geriatric rehabilitation, Geriatrics **8:**179, April, 1953.
26. Douglas, B. L.: Dental care for the aged, New York State Dent. J. **29:**53, 151, 1963.

Additional references

Agate, J.: The practice of geriatrics, Springfield, Ill., 1963, Charles C Thomas, Publisher.
Anderson, W. F.: Practical management of the elderly, Philadelphia, 1967, F. A. Davis Co.

Birren, J. E., editor: Handbook of aging and the individual; psychological and biological aspects, Chicago, 1959, University of Chicago Press.

Blumenthal, H. T., editor: Medical and clinical aspects of aging, New York, 1962, Columbia University Press.

Cowdry, E. V., and Steinberg, F. U., editors: The care of the geriatric patient, ed. 4, St. Louis, 1971, The C. V. Mosby Co.

Field, M.: Aging with honor and dignity, Springfield, Ill., 1968, Charles C Thomas, Publisher.

Freeman, J. T., editor: Clinical features of the older patient, Springfield, Ill., 1965, Charles C Thomas, Publisher.

Hazell, K.: Social and medical problems of the elderly, ed. 2, Springfield, Ill., 1966, Charles C Thomas, Publisher.

Isaacs, B.: An introduction to geriatrics, Baltimore, 1965, The Williams & Wilkins Co.

Johnson, W. M., editor: The older patient, New York, 1960, Paul B. Hoeber, Inc.

Kaplan, J., and Aldridge, G. J., editors: Social welfare of the aging, New York, 1962, Columbia University Press.

Koller, M. R.: Social gerontology, New York, 1968, Random House, Inc.

Rosow, I.: Social integration of the aged, London, 1967, The Free Press.

Rudd, J. L., and Margolin, R. J., editors: Maintenance therapy for the geriatric patient, Springfield, Ill., 1968, Charles C Thomas, Publisher.

Shock, N. W., editor: Biological aspects of aging, New York, 1962, Columbia University Press.

Stotsky, B. A.: The elderly patient, New York, 1968, Grune & Stratton, Inc.

Tibbitts, C., editor: Handbook of social gerontology; societal aspects of aging, Chicago, 1960, University of Chicago Press.

Tibbitts, C., and Donahue, W., editors: Social and psychological aspects of aging, New York, 1962, Columbia University Press.

Vedder, C. B., and Lefkowitz, A. S.: Problems of the aged, Springfield, Ill., 1965, Charles C Thomas, Publisher.

Williams, R. H., Tibbitts, C., and Donahue, W., editors: Processes of aging, New York, 1963, Atherton Press.

INDEX

Kyphoscoliosis
in patient with congestive heart failure, effect of oxygenation of, 545
respiratory insufficiency in, 548
Kyphosis in rheumatoid spondylitis, 354, 355, 356

L

Lability, assessing in hemiplegic patients, 30
Lactate dehydrogenase (LDH) and muscular dystrophy, 430
Lambert's cosine law, 66
Laminectomy, early, in paraplegics, 325-326
Laryngectomy, 649-650
Larynx, cancer of
laryngectomy for, 649-650
radical neck dissection for, 650-651
LDH; *see* Lactate dehydrogenase
Lead poisoning, 389-390
Learning disabilities in cerebral palsy, 486-487
Leukemia, 654
Levers, muscle, 78-79
Limb-girdle dystrophy, 423, 428
Linguistics, systems comprising, 239
Lisfranc's amputation, prosthesis for, 213
Locomotor system; *see* Neuromuscular and locomotor systems
Lofstrand crutches, 97, 144, 145
Looser's zones, 394
Lovett's system of grading muscle function, 7, 10
Lower extremity, alignment of, 200
Lung
bronchopulmonary segments of, 516
cancer, 652-653
disease, chronic obstructive, 501-503
volumes in patients with neuromuscular skeletal disease, 549, 550
Lymphatic flow, effect of massage on, 72
Lymphedema associated with mastectomy, 643
Lymphomas, 654

M

Massage, 71-75
application, principles of, 73-74
clinical conditions beneficial in, 74
compression, 73
of connective tissue, 75
contraindications, 75
friction, 73
indications, 74
kneading, 73
manual, Mennell's classification of, 73
mechanical and reflex effects of
on circulation, 72
on connective tissue, 72
on edema, 72

Massage—cont'd
mechanical and reflex effects of—cont'd
on lymphatic flow, 72
on muscle, 72
on pain, 72
percussion, 73
stroking, 73
syncardial, 605
in treatment of peripheral vascular disease, 605
vibration, 73
writing prescription for, 309
Mastectomy, 641-645
Mat exercise, 92, 93
Mechanical overload and respiratory insufficiency, 549-554
Mechanotherapy in treatment of peripheral vascular disease
intermittent venous occlusion, 604
massage, 605
oscillating bed, 605
pressure-suction boot, 604
syncardial massage, 605
vasopneumatic compression, 605
MED; *see* Minimal erythematous dose
Medical history, structure of, 4-5
Memory and concentration tests in hemiplegic patients, 29
Meningitis, 456
Mennell's classification of manual massage, 73
Mental retardation in cerebral palsy, 484-485
Mercury lamps
cold quartz, 65
hot quartz, 65
Metabolic diseases; *see* Arthritis; Bone disease, metabolic; Diabetes mellitus; Hematologic diseases; Poisoning
Metabolic disturbances, electromyographic patterns in, 47
Microwave therapy, 62
biophysical considerations, 62
clinical applications, 63
Micturition, disturbances of, surgical correction in paraplegic patients, 338-340
Milkman's syndrome, 394
Minimal erythematous dose (MED), 307-308
Mitral stenosis, cardiac work in, 590
MOE; *see* Myocardial oxygen extraction rate
Motion(s)
analysis of; *see* specific motions
needed in ADL performance, 156-161
requirements, basic principles of, 161-162
test, functional, 48, 55, 162-163
charts, 51-52, 53-54
Motivation, lack of, 272-275
Motor capacity and behavior, variability of, in cerebral palsy, 479-480